PACIFIC SYMPOSIUM ON BIOCOMPUTING 2021

PACIFIC SYMPOSIUM ON BIOCOMPUTING 2021

3 – 7 January 2021

Edited by

Russ B. Altman
Stanford University, USA

A. Keith Dunker
Indiana University, USA

Lawrence Hunter
University of Colorado Health Sciences Center, USA

Marylyn D. Ritchie
University of Pennsylvania, USA

Tiffany Murray
Stanford University, USA

Teri E. Klein
Stanford University, USA

World Scientific

NEW JERSEY • LONDON • SINGAPORE • BEIJING • SHANGHAI • HONG KONG • TAIPEI • CHENNAI • TOKYO

Published by

World Scientific Publishing Co. Pte. Ltd.
5 Toh Tuck Link, Singapore 596224
USA office: 27 Warren Street, Suite 401-402, Hackensack, NJ 07601
UK office: 57 Shelton Street, Covent Garden, London WC2H 9HE

Online ISSN: 2335-6936
Print ISSN: 2335-6928

British Library Cataloguing-in-Publication Data
A catalogue record for this book is available from the British Library.

BIOCOMPUTING 2021
Proceedings of the Pacific Symposium

ISBN 978-981-123-270-1 (ebook)
ISBN 978-981-123-269-5 (print)

COMPUTATIONAL CHALLENGES AND ARTIFICIAL INTELLIGENCE IN PRECISION MEDICINE

PATTERN RECOGNITION IN BIOMEDICAL DATA FOR DISCOVERY

WHAT ABOUT THE ENVIRONMENT? LEVERAGING MULTI-OMIC DATASETS TO CHARACTERIZE THE ENVIRONMENT'S ROLE IN HUMAN HEALTH

WORKSHOPS

PACIFIC SYMPOSIUM ON BIOCOMPUTING 2021

2021 marks the 26th Pacific Symposium on Biocomputing (PSB). Unfortunately, circumstances surrounding the Covid-19 pandemic prevent us from gathering together on the Big Island to engage in the traditional scientific sessions, invited lectures, workshops and collegiality-by-the-beach (and pool) that has become the hallmark of PSB. Fortunately, we still received very high-quality submissions and these will be presented virtually and allow continued (socially distanced) interaction and scientific exchange. We are also arranging virtual workshops and a virtual poster session. We have shortened the total length of the meeting, to help participants avoid "zoom fatigue" but many of the presentations will be recorded and available, so the reach and impact of the meeting may in some ways be extended.

Covid-19 has changed life on earth in 2020 in ways that few of us could have imagined. Many of the most basic life and work practices and assumptions have been challenged. The biomedical informatics, bioinformatics and computational biology communities have played a special role in contributing to the fight against the virus. While many of our experimental colleagues had to shut down their labs for weeks to months during the pandemic, many computational labs were able to continue their work with access to their computing facilities and a move to virtual meetings. There are many examples of early contributions to understanding the biology of the SARS-Cov-2 virus based on computational analysis of its genome, epitope targets for vaccine development, and its proteome and connections to the host. In addition, the tsunami of publications (particularly preprints) has created challenges in automatic understanding and integration (and triage) of scientific findings, in order to help scientists track the rapidly evolving landscape of our understanding of the virus. Finally, there have been opportunities in tracking population health and the delivery of clinical care—in many cases, informatics technologies have helped decision makers understand the details of the pandemic, and how best to deploy resources. At the same time, the pandemic has exposed weaknesses in our health information infrastructure—for both communicating health data, storing it, integrating it and analyzing it. For example, computational researchers who may have never sent or received a fax (!) are now trying to help figure out how to automate their analysis (many Covid-19 case reports are still faxed to local health departments) and ensure that they are soon replaced as first-line communication methods. There are also great challenges in modeling the pandemic and its impacts on health and the economy. Never before have analytic and computational capabilities in biology and medicine been so important. So, we persevere under difficult circumstances to continue the mission of PSB to bring scientists together as they approach some of the most challenging and important problems.

In addition to being published by World Scientific and indexed in PubMED, the proceedings from all PSB meetings are available online at http://psb.stanford.edu/psb-online/. PSB has 1227 papers listed in PubMED (as of today). These papers are routinely cited in archival journal articles and often represent important early contributions in new subfields—many times before there is an established literature in more traditional journals; for this reason, many papers have garnered hundreds of citations.

The Twitter handle PSB 2021 is @PacSymBiocomp and the hashtag this year will be #psb21.

The efforts of a dedicated group of session organizers have produced an outstanding program. The sessions of PSB 2021 and their hard-working organizers are as follows:

Advanced Methods for Big Data Analytics in Women's Health
Organizers: Graciela Gonzalez-Hernandez, Karin Verspoor, Maricel G. Kann, Su Golder, Lisa Levine, Mary Regina Boland, Natalia Villanueva-Rosales, Karen O'Connor

Achieving Trustworthy Biomedical Data
Organizers: Dennis Wall, Nicholas Tatonetti, Jan Liphardt, Bethany Percha, Serena Yeung, Peter Washington

Pattern Recognition in Biomedical Data for Discovery
Organizers: Brett Beaulieu-Jones, Christian Darabos, Dokyoon Kim, Shilpa Kobren, Anurag Verma

What about the environment? Leveraging multi-omic datasets to characterize the environment's role in human health
Organizers: Kristin Passero, Shefali Setia Verma, Kimberly McAllister, Arjun Manrai, Chirag Patel, Molly A. Hall

Biocomputing and AI for infectious disease modelling and therapeutics
Organizers: Gil Alterovitz, Wei-Lun Alterovitz, Gail H. Cassell, Lixin Zhang, A. Keith Dunker

Computational Challenges and Artificial Intelligence in Precision Medicine
Organizers: Olga Afanasiev, Joanne Berghout, Steven E. Brenner, Martha L. Bulyk, Dana Crawford, Jonathan H. Chen, Roxana Daneshjou, Łukasz Kidziński

We are also pleased to present four workshops in which investigators with a common interest come together to exchange results and new ideas in a format that is more informal than the peer-reviewed sessions. For this year, the workshops and their organizers are:

Bioinformatics of Corals
Organizers: Lenore J. Cowen, Judith Klein-Seetharaman, Hollie Putnam

Translational Bioinformatics
Organizers: Jason Moore, Ju Han Kim, Dokyoon Kim

Making Tools that People Will Use: User-Centered Design in Computational Biology Research
Organizers: Mary Goldman, Nils Gehlenborg

Raising the Stakeholders: Improving Patient Outcomes Through Interprofessional Collaborations in AI for Healthcare
Organizers: Carly A. Bobak, Kristine A. Giffin, Marek Svoboda, Jason Moore, Dennis P. Wall

Establishing the Reliability of Algorithms in Biomedical Informatics
Organizers: Lara Mangravite, Sean Mooney, Iddo Freidberg, Justin Guinney

Tiffany Murray has managed the peer review process and assembly of the proceedings since 2001 and plays a key role in many aspects of the meeting. We are grateful for the support of the National Institutes of Health[1]. The Research Parasite Awards benefit from support from GigaScience, Jeff Stibel, Mr. and Mrs. Stephen Canon, and Drs. Casey and Anna Greene. The Research Symbiont Awards benefit from support from the Wellcome Trust, Springer-Nature and the DragonMaster Foundation.

We are particularly grateful to the PSB staff Cynthia Paulazzo and Ryan Whaley for their assistance. We also acknowledge the many busy researchers who reviewed the submitted manuscripts on a very tight schedule. The partial list following this preface does not include many who wished to remain anonymous, and of course we apologize to any who may have been left out by mistake.

We look forward to a great virtual meeting and very much hope to see you all again on the Big Island in 2022. Aloha!

Pacific Symposium on Biocomputing Co-Chairs,
October 8, 2020

Russ B. Altman
Departments of Bioengineering, Genetics, Medicine & Biomedical Data Science, Stanford University

A. Keith Dunker
Department of Biochemistry and Molecular Biology, Indiana University School of Medicine

Lawrence Hunter
Department of Pharmacology, University of Colorado Health Sciences Center

Marylyn D. Ritchie
Department of Genetics and Institute for Biomedical Informatics, University of Pennsylvania

Teri E. Klein
Departments of Biomedical Data Science & Medicine, Stanford University

[1]Funding for this conference was made possible (in part) by R13LM006766 from the National Library of Medicine. The views expressed in written conference materials or publications and by speakers and moderators do not necessarily reflect the official policies of the Department of Health and Human Services; nor does mention by trade names, commercial practices, or organizations imply endorsement by the U.S. Government."

Thanks to the reviewers...

Finally, we wish to thank the scores of reviewers. PSB aims for every paper in this volume to be reviewed by three independent referees. Since there is a large volume of submitted papers, paper reviews require a great deal of work from many people. We are grateful to all of you listed below and to anyone whose name we may have accidentally omitted or who wished to remain anonymous.

Abubakar Abid
Rocky Aikens
Tatsuya Akutsu
Wei-Lun Alterovitz
Raquel Aoki
Brittany Baur
Amber Beitelshees
Mary Regina Boland
Nicolas Buchler
Sophie Burkhardt
Laura Cao
Nyasha Chambwe
Ed Chan
Tianchi Chen
Xiangyin Chen
Yu Wai Chen
Abu Chowdhury
Evan Cofer
Jessica Cooke Bailey
Dana Crawford
Sara Cromer
Gargi Datta
Devendra Singh Dhami
Alexander D. Diehl
Jessilyn Dunn
Jason Ernst
Nicole Ferraro
Noah R. Flynn
Juan Fuxman Bass
Ethan Garner
Mario Giacobini
Benjamin Glicksberg
Graciela Gonzalez-Hernandez
Molly Hall
Jie Hao
Bryan He
Ting Hu
Tai-huang Huang
Hyungsoon Im
Alokkumar Jha
Mingon Kang
Lukasz Kidzinski
Benjamin Kompa
Dhireesha Kudithipudi
Chirag Lakhani
Avantika Lal
Jerry Lee
Tom Lenaerts
Lisa Levine
Joshua Levy
Li Li
Pinghua Liu
David Magnus
Bill Majoros
Dmitry Maslov
Magdalena Matusiak
Jason Miller
Riccardo Miotto
Elias Chaibub Neto
Karen O'Connor
Chris Oldfield
Andrew Patterson
Vivek Philip
Vasiliki Nataly Rahimzadeh
Maya Ramchandran
David Rhoiney
Pedro Romero
Daniel Rotroff
Max Salfinger
Indra Neil Sarka
Jacob Schreiber
Alejandro Schuler
Lulu Shang
Li Shen
Masaki Shimomura
Kyung-Ah Sohn
Yosuke Tanigawa
Ling Teng
Elizabeth Torrres
Anna Tyler
Martie VanderWalt
Karin Verspoor
Kailas Vodrahalli
Jennifer Wagner
Haohan Wang
Yixin Wang
Jennifer Williams
Kevin Wu
Jeff Xia
Eva Zhang
Lixin Zhang
Ren Zhiyun
Xinxin (Katie) Zhu
Marinka Zitnik

Achieving Trustworthy Biomedical Data Solutions

Peter Washington
Department of Bioengineering, Stanford University
Stanford, CA, 94305, USA
Email: peterwashington@stanford.edu

Serena Yeung
Department of Biomedical Data Science, Stanford University
Stanford, CA, 94305, USA
Email: syyeung@stanford.edu

Bethany Percha
Department of Medicine, Icahn School of Medicine at Mount Sinai
New York, NY 10029, USA
Email: bethany.percha@mssm.edu

Nicholas Tatonetti
Department of Biomedical Informatics, Columbia University
New York, NY 10032, USA
Email: nick.tatonetti@columbia.edu

Jan Liphardt
Department of Bioengineering, Stanford University
Stanford, CA, 94305, USA
Email: jan.liphardt@stanford.edu

Dennis P. Wall
Departments of Pediatrics (Systems Medicine), Psychiatry and Behavioral Sciences, and Biomedical Data Science, Stanford University
Stanford, CA, 94305, USA
Email: dpwall@stanford.edu

Privacy and trust of biomedical solutions that capture and share data is an issue rising to the center of public attention and discourse. While large-scale academic, medical, and industrial research initiatives must collect increasing amounts of personal biomedical data from patient stakeholders, central to ensuring precision health becomes a reality, methods for providing sufficient privacy in biomedical databases and conveying a sense of trust to the user is equally crucial for the field of biocomputing to advance with the grace of those stakeholders. If the intended audience does not trust new precision health innovations, funding and support for these efforts will inevitably be limited. It is therefore crucial for the field to address these issues in a timely manner. Here we describe current research directions towards achieving trustworthy biomedical informatics solutions.

Keywords: privacy; trust; data security; biomedical systems; bioinformatics; artificial intelligence (AI); trustworthy AI

1. Introduction

The importance of trust in biomedical and healthcare technologies, especially consumer-facing artificial-intelligence (AI) software, cannot be overstated. Issues of privacy and trust with regard to large-scale data capture and analysis, particularly passive data capture by mobile devices and social media, have recently come to the forefront of public and academic discourse across multiple domains [1-4]. Such issues are especially important for healthcare, where solutions must prioritize patient privacy. At a minimum, biomedical tools in the United States must satisfy the Health Insurance Portability and Accountability Act of 1996 (HIPAA), which mandates a set of regulations regarding the privacy of patient health data [5]. While satisfying legal constraints is necessary, the true metric of achieving satisfactory patient trust will come from the patients themselves, who may request more stringent solutions.

In recent years, the biomedical research community has produced a wide array of research findings relating to trustworthy biomedical data, spanning multiple fields and subdomains. Work in these areas has included genomic data storage [6], privacy and sharing of protected health information (PHI) [7-9], cryptography solutions to sharing genetic data that allow public querying while protecting patient privacy [10], ethical considerations of new technologies and paradigms [11], and privacy-preserving machine learning methods [12-13]. However, the increasing prevalence of large-scale biomedical data collection capabilities and efforts (such as the continued decrease in sequencing costs), coupled with the explosion of applied machine learning systems and products, continually creates demand for innovations in trustworthy methods which can handle growing technological capabilities.

Here, we focus on four active themes in biomedical data science where the importance of trust in data has taken center stage: (1) preserving privacy and explaining the decisions of artificial intelligence algorithms, (2) sharing genomic and health records, (3) deploying digital health solutons, and (4) crowdsourcing healthcare. For each research theme, we describe several core methodological approaches (Figure 1) for building trustworthy biomedical data solutions which apply across the data science pipeline: (1) data transformation (e.g., dimension reduction and image modification), (2) access control (e.g., federated learning and cryptography), (3) data aggregation (e.g., aggregate queries and differential privacy), and (4) transparency (e.g., explainable AI). We discuss how these trust-enabling methodologies can and should be invoked and describe prior efforts. We conclude with a brief discussion of the bioethics literature.

2. Preserving Privacy and Explaining Decisions of Artificial Intelligence

AI in healthcare is increasingly rising in importance for solving challenges in the medical workflow including clinicial decision support, preventing errors, and scaling redundant tasks. Privacy preservation and explainability are crucial when machine learning algorithms are deployed in these settings. We describe three common machine learning paradigms for attaining and preserving patient privacy when biomedical data are used to train algorithms: (1) transformation of the data, (2) federated learning, and (3) differential privacy. We also discuss efforts to attain explainable AI.

If the data can be transformed in such a way that the downstream model still yields high predictive performance, simply altering the data to obfuscate the identity of the subject may be the most desired option. For example, when using computer vision for use in activity recognition in hospital bedside settings [14-15], Yeung et al. leverage thermal [16] and depth [17] sensors to create privacy-preserved video streams. Washington et al. simply place a face box over the patients' faces and pitch shift the audio when generating behavioral phenotypes of children with autism using machine learning and crowdsourcing [18], only minimally degrading performance compared to when using unaltered videos. Machine learning models should be trained and tested on the maximally private alteration of the data while maintaining acceptable performance.

Federated learning as a privacy enhancing technique has garnered widespread attention for achieving privacy in distributed mobile devices that may collect multimedia data streams. In federated machine learning, several distributed machines train models based on local data and share only model weights, which do not contain any protected information, on either the other distributed devices or a centralized server [19]. Federated learning has been applied to analyze data from electronic health records [20-22], recognize activity patterns based on data from wearable devices [23], and improve the interpretation of medical images [24].

A third commonly used privacy preserving technique is differential privacy. Differential privacy involves injecting random noise into the training dataset such that the identifiability of each individual record is destroyed while the aggregate properties of the dataset are preserved [25]. Examples of applying differential privacy to protect patient privacy in the biomedical domain include injecting noise into data from wearable sensors [26], genome wide association studies [27], and healthcare social networks [28]. This session includes a paper by Shi et al. that explores the tradeoffs between the performance of commonly used machine learning models and the level of privacy attained using differential privacy.

Another crucial property of trustworthy machine learning is explainability, including but not limited to interpretability. Some machine learning algorithms are inherently explainable. In classification with logistic regression, for example, the exact prediction can be calculated from the input values by plugging them into an equation. Making the coefficients associated with each variable transparent to the patient in a user-friendly manner would increase trust. However, with a large dataset of high complexity, explainable algorithms may not be sufficient, requiring more powerful yet less interpretable algorithms like neural networks. While components of certain neural networks can be interpreted, such as by visualizing the weights and activations of feature maps in the intermediate layers of a convolutional neural network, making neural networks explainable is an emerging active area of research [29]. Creating explainable AI has enabled increased reasoning about the decision making process behind stroke prediction algorithms [30], further understanding of changes in the skin microbiome [31], and elucidation of the reasoning of algorithms trained on electronic health record [32]. In some cases, explainable AI can lead to scientific discovery, for example by elucidating complex disease pathways in autism [33]. As explainable AI is becoming a popular research direction across computing research fields, we expect more translatable innovations in the coming years that safely embed AI in a variety of sectors of the healthcare ecosystem.

3. Sharing Genomic and Health Records

The genome is a core foundation of precision healthcare, and shared human DNA records are essential to advancements in human health. Millions of human genomes have been sequenced, either through direct-to-consumer DNA platforms (e.g., 23andme and Ancestry) or through a healthcare provider, with the number likely to exponentially increase as genomic sequencing becomes progressively more affordable and more speedy, improving at a rate faster than Moore's Law [34]. Genomic data are exceptionally sensitive, and increasingly so as advancements in bioinformatics methods can uncover a patient's identity in a dataset with a small number of queries [35-39] through approaches like membership inference attack [40]. Addressing secure storage and sharing of genomic data to solve such issues is a key research challenge required to advance genomics-based precision health and medicine pipelines to the clinic [41]. Several methods for preserving genetic privacy have been published, including differential privacy-based approaches [42-44], perturbing the data with Bayesian statistics and Markov Chain Monte Carlo techniques [45], applying cryptographic protocols and frequency-based clinical genetics [10], and encrypting the data before offloading it to the cloud [46].

While the genome is a key data modality for precision health, it must be tightly tied to the phenotype, perhaps best embodied in electronic medical record (EMR) data. EMR can be mined to make data driven predictions about important biomedical issues such as the risk for diseases at the heart of immediate public health crises (i.e., COVID-19) [47-49], understudied and unknown adverse drug interactions [50-51], and psychiatric and behavioral conditions with a small number of behavioral biomarkers [52-56], including in underserved countries with differing laws and expectations about data sharing [57]. EMR are susceptible to attack, for example by inferring disease heritability from exposed pedigree information [58]. Previously explored solutions to addressing the sensitive nature of such records include only performing inference on common medical events while keeping the remainder private [59], reducing the dimensionality of the dataset [60-61], transforming the dataset with the use of generative adversarial networks [62], giving the patient control over who has access to the electronic health records [63], only allowing aggregate queries without revealing the underlying dataset [64], and deploying cryptography schemes such as symmetric key or asymmetric key encryption [65].

	Data Transformation	Access Control	Data Aggregation	Transparency
Preserving Privacy and Explaining Decisions of Artificial Intelligence	✓	✓	✓	✓
Sharing Genomic and Health Records	✓	✓	✓	✓
Deploying Digital Health Solutions	✓	✓	x	✓
Crowdsourcing Healthcare	x	✓	x	x

Figure 1. An opportunity space for innovation in methods for achieving trustworthy biomedical data solutions. We list the 4 most active areas where security and trust in the exchange of data is highest: private and explainable artificial intelligence; sharing and integration of genomic and medical records; construction and use of digital health tools; and crowdsourcing of healtcare management. In all 4, methodologies of data transformation, access control, data aggregation, and transparency can and should be deployed.

4. Deploying Digital Health Solutions

While EMR are traditionally generated in the clinic, digital health solutions are increasingly deployed to home settings [66-68]. As digital devices continue to receive FDA approval for medical use [69-70], it is inevitable, and exciting, that large portions of EMR data will be acquired through consumer devices such as smartphones and embedded hardware. Digital devices can longitudinally quantify patient symptoms when away from the clinic for conditions such as brain-mediated neurological and psychiatric disorders [71-72], cardiovascular disease [73-74], and infectious disease [75], among others. Examples of digital health solutions used in sensitive settings include therapeutic devices administered by clinicians [76], therapeutic tools administered in home settings [77-79], monitoring systems in hospital settings [80-81], dual-purpose interventions which explicitly collect patient health information to train machine learning models [82-84], pediatric healthcare interventions disguised to the child as a game [85-86], and wearable devices [87]. Many of these therapeutic and diagnostic devices collect potentially sensitive audio, image, and video streams for clinical use [88-91], and these data streams are often shared with clinicians or even crowdsourced with the consent of the patient. Furthermore, several digital therapies are used in home settings, and such rich data streams are filled with protected health information accompanied by potentially sensitive identifiable information such as the patient's face, images and video of the patient's home, and audio recordings of the patient or their family while using the device. It is therefore crucial to ensure patient privacy when these data leave the patient's device and are introduced into clinical workflows. Best practices discussed by Martínez-Pérez et al. include creating role-based access to data, making the privacy policy precise and clear to the user, transferring data with TLS using 256-bit encryption, erasing the data after it has been used for its intended purpose, and creating a data breach notification system [92].

Because consumer health technologies do not have direct oversight by clinicians, biased and deliberately inaccurate reporting by the target audience can be a risk. Therefore, it is particularly important to assess the quality of incoming data to garner the trust of healthcare providers and scientists, using those data for healthcare management and innovation. Algorithms that perform quality control to safeguard against biased or inaccurate reporting must go hand-in-hand with digital innovations. It is crucial for researchers to easily identify invalid or unintended data. For both consumers and scientists to gain confidence in the generalized applicability of digital tools, the data must be representative of the target population, making it pertinent to collect data that are balanced across race, ethnicity, geography, gender, and other relevant demographics.

5. Crowdsourcing Healthcare

Crowdsourcing is another approach used increasingly in clinical workflows [93-97]. Digital health and telemedical solutions that can scale through crowdsourcing approaches will become a norm for healthcare. The use of crowdsourcing in healthcare can be broadly partitioned into three categories: (1) crowdsourcing to achieve consensus on the presence or absence of medical conditions; (2) crowdsourced capture (whether active or passive, or a combination) of longitudinal data streams from from a large target cohort; (3) crowdsourcing the construction of training libraries of robustly labeled health data (e.g., radiological images), that enable progressive improvement of predictive models that can augment or replace decision points in the healthcare process.

Crowdsourcing appears in diverse healthcare settings and has been used for measurement of autism symptoms for diagnostic decision support [98-101], ranking adverse drug reactions [102], and COVID-19 contact tracing and surveillance [103-105]. Despite the strong clinical utility of crowdsourcing approaches, studies of trust and privacy for text, audio, image, and video streams rated on crowdsourcing platforms (e.g., Amazon Mechanical Turk [106-107] and Microworkers.com [108]) are lacking in the literature, especially with respect to biomedical research. As with digital consumer technologies, labeled data from crowdsourcing pipelines have the potential to suffer from low quality [109], requiring methods to filter crowd workers and into a trusted workforce of repeatedly high quality workers. This session includes a paper by Washington et al. which introduces quantitative metrics for evaluating crowd workers for their trustworthiness and reliability and provides behavioral metrics for identifying a valuable subset of crowd workers for inclusion in private clinical workflows. We hope that this study will inspire further work toward ensuring trustworthy crowd-powered telemedicine. Figure 1 highlights that research into trustworthy biomedical crowdsourcing is relatively light. In particular, privacy-preserved crowdsourced annotation of transformed data and on aggregate data is a currently unexplored yet fruitful research direction.

6. Considering the Bioethics

It is important to keep sight of the ethical considerations and formal bioethical perspectives with respect to biomedical innovations using trustworthy methods, or the lack thereof. Bioethical arguments are typically grounded in traditional ethical theories. Deontology is an ethical theory that considers actions as moral if they pass a series of conditions or rules [110]. A contrasting family of ethical theories, consequentialism, requires that moral actions maximize the public good and the utility of the action to all relevant stakeholders [110]. A third category, virtue ethics, states that moral actions should be a manifestation of a virtuous character trait [110]. While all ethical theories sound optimal in isolation, bioethical decisions may often satisfy one ethical theory while violating another. For example, heavy COVID-19 surveillance will maximize the good to all people (Utilitarianism, a type of consequentialism) while violating a core principle (deontological ethics) of the right to privacy. Bioethical analyses have been applied to genome sequencing for newborn screening [111-112], clinical machine learning [113-114], precision medicine [115-116], wearables and mobile health [117-118], and crowdsourcing [119-120].

This session includes a paper by Greenbaum et al. discussing the implications of expanded access programs with respect to COVID-19, a particularly timely topic. We hope that informaticians and scientists will interact more often with bioethicists to understand the societal implications of their work.

7. Anticipating the Future

Trustworthy biomedical data solutions will be crucial for realizing wide adoption of emerging technologies and methodologies for precision health. This session includes promising directions of exploration for the biomedical informatics research community. We have summarized some of the methods for building trust in key parts of the data analysis pipeline: data analysis(for artificial intelligence), data sharing (of genomic and health records), data capture (through digital devices), and data labeling (through crowdsourcing).

The study of trustworthy biomedical data science is in its infancy and ripe for innovations. We hope that this session will inspire further work in this important area, complementing the public's broader discussion of privacy and security considerations related to large-scale data collection and analysis. We anticipate that research that aims to improve the trustworthiness of biocomputing methods will become a major part of the PSB and a major focus for biomcomputing research in the coming years.

References

1. Bradshaw, Samantha, and Philip N. Howard. "Challenging truth and trust: A global inventory of organized social media manipulation." *The Computational Propaganda Project* 1 (2018).
2. Milne, Richard, Katherine I. Morley, Heidi Howard, Emilia Niemiec, Dianne Nicol, Christine Critchley, Barbara Prainsack et al. "Trust in genomic data sharing among members of the general public in the UK, USA, Canada and Australia." *Human genetics* 138, no. 11 (2019): 1237-1246.
3. Ovide, Shira. "Will More Data Make Us Healthier?" *The New York Times*. August 28, 2020.
4. Yan, Wei Qi. *Introduction to intelligent surveillance: Surveillance data capture, transmission, and analytics*. Springer, 2019.
5. Annas, George J. "HIPAA regulations—a new era of medical-record privacy?." (2003): 1486-1490.
6. Lin, Chi, Zihao Song, Houbing Song, Yanhong Zhou, Yi Wang, and Guowei Wu. "Differential privacy preserving in big data analytics for connected health." *Journal of medical systems* 40, no. 4 (2016): 97.
7. Lu, Rongxing, Xiaodong Lin, and Xuemin Shen. "SPOC: A secure and privacy-preserving opportunistic computing framework for mobile-healthcare emergency." *IEEE transactions on parallel and distributed systems* 24, no. 3 (2012): 614-624.
8. Drazen, Jeffrey M. "Sharing individual patient data from clinical trials." *New England Journal of Medicine* 372, no. 3 (2015): 201-202.
9. El Emam, Khaled, Sam Rodgers, and Bradley Malin. "Anonymising and sharing individual patient data." *bmj* 350 (2015): h1139.
10. Jagadeesh, Karthik A., David J. Wu, Johannes A. Birgmeier, Dan Boneh, and Gill Bejerano. "Deriving genomic diagnoses without revealing patient genomes." *Science* 357, no. 6352 (2017): 692-695.
11. Navarro, Robert. "An ethical framework for sharing patient data without consent." *Journal of Innovation in Health Informatics* 16, no. 4 (2008): 257-262.
12. Yeom, Samuel, Irene Giacomelli, Matt Fredrikson, and Somesh Jha. "Privacy risk in machine learning: Analyzing the connection to overfitting." In *2018 IEEE 31st Computer Security Foundations Symposium (CSF)*, pp. 268-282. IEEE, 2018.
13. Beaulieu-Jones, Brett K., Zhiwei Steven Wu, Chris Williams, Ran Lee, Sanjeev P. Bhavnani, James Brian Byrd, and Casey S. Greene. "Privacy-preserving generative deep neural networks support clinical data sharing." *Circulation: Cardiovascular Quality and Outcomes* 12, no. 7 (2019): e005122.
14. Singh, Amit, Albert Haque, Alexandre Alahi, Serena Yeung, Michelle Guo, Jill R. Glassman, William Beninati, Terry Platchek, Li Fei-Fei, and Arnold Milstein. "Automatic detection of hand hygiene using computer vision technology." *Journal of the American Medical Informatics Association* 27, no. 8 (2020): 1316-1320.
15. Yeung, Serena. "Visual Understanding of Human Activity: Towards Ambient Intelligence in AI-assisted Hospitals." PhD diss., Stanford University, 2018.
16. Yeung, Serena, N. Lance Downing, Li Fei-Fei, and Arnold Milstein. "Bedside computer vision-moving artificial intelligence from driver assistance to patient safety." *N Engl J Med* 378, no. 14 (2018): 1271-3.
17. Yeung, Serena, Francesca Rinaldo, Jeffrey Jopling, Bingbin Liu, Rishab Mehra, N. Lance Downing, Michelle Guo et al. "A computer vision system for deep learning-based detection of patient mobilization activities in the ICU." *NPJ digital medicine* 2, no. 1 (2019): 1-5.
18. Washington, Peter, Qandeel Tariq, Emilie Leblanc, Brianna Chrisman, Kaitlyn Dunlap, Aaron Kline et al. "Crowdsourced feature tagging for scalable autism diagnoses." *Under review*. 2020.
19. Yang, Qiang, Yang Liu, Tianjian Chen, and Yongxin Tong. "Federated machine learning: Concept and applications." *ACM Transactions on Intelligent Systems and Technology (TIST)* 10, no. 2 (2019): 1-19.

20. Brisimi, Theodora S., Ruidi Chen, Theofanie Mela, Alex Olshevsky, Ioannis Ch Paschalidis, and Wei Shi. "Federated learning of predictive models from federated electronic health records." *International journal of medical informatics* 112 (2018): 59-67.
21. Huang, Li, Yifeng Yin, Zeng Fu, Shifa Zhang, Hao Deng, and Dianbo Liu. "Loadaboost: Loss-based adaboost federated machine learning on medical data." *arXiv preprint arXiv:1811.12629* (2018).
22. Liu, Dianbo, Timothy Miller, Raheel Sayeed, and Kenneth D. Mandl. "Fadl: Federated-autonomous deep learning for distributed electronic health record." *arXiv preprint arXiv:1811.11400* (2018).
23. Chen, Yiqiang, Xin Qin, Jindong Wang, Chaohui Yu, and Wen Gao. "Fedhealth: A federated transfer learning framework for wearable healthcare." *IEEE Intelligent Systems* (2020).
24. Kaissis, Georgios A., Marcus R. Makowski, Daniel Rückert, and Rickmer F. Braren. "Secure, privacy-preserving and federated machine learning in medical imaging." *Nature Machine Intelligence* (2020): 1-7.
25. Dwork, Cynthia. "Differential privacy: A survey of results." In *International conference on theory and applications of models of computation*, pp. 1-19. Springer, Berlin, Heidelberg, 2008.
26. Lin, Zhen, Art B. Owen, and Russ B. Altman. "Genomic research and human subject privacy." (2004): 183-183.
27. Tramèr, Florian, Zhicong Huang, Jean-Pierre Hubaux, and Erman Ayday. "Differential privacy with bounded priors: reconciling utility and privacy in genome-wide association studies." In *Proceedings of the 22nd ACM SIGSAC Conference on Computer and Communications Security*, pp. 1286-1297. 2015.
28. Phan, NhatHai, Yue Wang, Xintao Wu, and Dejing Dou. "Differential privacy preservation for deep auto-encoders: an application of human behavior prediction." In *Aaai*, vol. 16, pp. 1309-1316. 2016.
29. Gunning, David. "Explainable artificial intelligence (xai)." *Defense Advanced Research Projects Agency (DARPA), nd Web* 2 (2017): 2.
30. Prentzas, Nicoletta, Andrew Nicolaides, Efthyvoulos Kyriacou, Antonis Kakas, and Constantinos Pattichis. "Integrating Machine Learning with Symbolic Reasoning to Build an Explainable AI Model for Stroke Prediction." In *2019 IEEE 19th International Conference on Bioinformatics and Bioengineering (BIBE)*, pp. 817-821. IEEE, 2019.
31. Carrieri, Anna Paola, Niina Haiminen, Sean Maudsley-Barton, Laura-Jayne Gardiner, Barry Murphy, Andrew Mayes, Sarah Paterson et al. "Explainable AI reveals key changes in skin microbiome associated with menopause, smoking, aging and skin hydration." *bioRxiv* (2020).
32. Lauritsen, Simon Meyer, Mads Kristensen, Mathias Vassard Olsen, Morten Skaarup Larsen, Katrine Meyer Lauritsen, Marianne Johansson Jørgensen, Jeppe Lange, and Bo Thiesson. "Explainable artificial intelligence model to predict acute critical illness from electronic health records." *Nature communications* 11, no. 1 (2020): 1-11.
33. Spencer, Matt, Saad Khan, Zohreh Talebizadeh, and Chi-Ren Shyu. "Explainable AI: Mining of Genotype Data Identifies Complex Disease Pathways—Autism Case Studies." *Application Of Omics, Ai And Blockchain In Bioinformatics Research* 21 (2019): 11.
34. Muir, Paul, Shantao Li, Shaoke Lou, Daifeng Wang, Daniel J. Spakowicz, Leonidas Salichos, Jing Zhang et al. "The real cost of sequencing: scaling computation to keep pace with data generation." *Genome biology* 17, no. 1 (2016): 1-9.
35. Al Aziz, Md Momin, Reza Ghasemi, Md Waliullah, and Noman Mohammed. "Aftermath of bustamante attack on genomic beacon service." *BMC medical genomics* 10, no. 2 (2017): 43.
36. Backes, Michael, Pascal Berrang, Mathias Humbert, and Praveen Manoharan. "Membership privacy in MicroRNA-based studies." In *Proceedings of the 2016 ACM SIGSAC Conference on Computer and Communications Security*, pp. 319-330. 2016.
37. Deznabi, Iman, Mohammad Mobayen, Nazanin Jafari, Oznur Tastan, and Erman Ayday. "An inference attack on genomic data using kinship, complex correlations, and phenotype information." *IEEE/ACM transactions on computational biology and bioinformatics* 15, no. 4 (2017): 1333-1343.
38. Humbert, Mathias, Erman Ayday, Jean-Pierre Hubaux, and Amalio Telenti. "Addressing the concerns of the lacks family: quantification of kin genomic privacy." In *Proceedings of the 2013 ACM SIGSAC conference on Computer & communications security*, pp. 1141-1152. 2013.
39. Shringarpure, Suyash S., and Carlos D. Bustamante. "Privacy risks from genomic data-sharing beacons." *The American Journal of Human Genetics* 97, no. 5 (2015): 631-646.

40. Shokri, Reza, Marco Stronati, Congzheng Song, and Vitaly Shmatikov. "Membership inference attacks against machine learning models." In *2017 IEEE Symposium on Security and Privacy (SP)*, pp. 3-18. IEEE, 2017.
41. Altman, Russ B., Snehit Prabhu, Arend Sidow, Justin M. Zook, Rachel Goldfeder, David Litwack, Euan Ashley et al. "A research roadmap for next-generation sequencing informatics." *Science translational medicine* 8, no. 335 (2016): 335ps10-335ps10.
42. Almadhoun, Nour, Erman Ayday, and Özgür Ulusoy. "Differential privacy under dependent tuples—the case of genomic privacy." *Bioinformatics* 36, no. 6 (2020): 1696-1703.
43. He, Zaobo, Yingshu Li, and Jinbao Wang. "Differential privacy preserving genomic data releasing via factor graph." In *International Symposium on Bioinformatics Research and Applications*, pp. 350-355. Springer, Cham, 2017.
44. Raisaro, Jean Louis, Gwangbae Choi, Sylvain Pradervand, Raphael Colsenet, Nathalie Jacquemont, Nicolas Rosat, Vincent Mooser, and Jean-Pierre Hubaux. "Protecting privacy and security of genomic data in I2B2 with homomorphic encryption and differential privacy." *IEEE/ACM transactions on computational biology and bioinformatics* 15, no. 5 (2018): 1413-1426.
45. Simmons, Sean, Bonnie Berger, and Cenk S. Sahinalp. "Protecting Genomic Data Privacy with Probabilistic Modeling." In *PSB*, pp. 403-414. 2019.
46. Wang, Bing, Wei Song, Wenjing Lou, and Y. Thomas Hou. "Privacy-preserving pattern matching over encrypted genetic data in cloud computing." In *IEEE INFOCOM 2017-IEEE Conference on Computer Communications*, pp. 1-9. IEEE, 2017.
47. Vaid, Akhil, Sulaiman Somani, Adam J. Russak, Jessica K. De Freitas, Fayzan F. Chaudhry, Ishan Paranjpe, Kipp W. Johnson et al. "Machine Learning to Predict Mortality and Critical Events in COVID-19 Positive New York City Patients." *medRxiv* (2020).
48. Yoo, Edwin, Bethany Percha, Max Tomlinson, Victor Razuk, Stephanie Pan, Madeleine Basist, Pranai Tandon et al. "Development and calibration of a simple mortality risk score for hospitalized COVID-19 adults." *medRxiv* (2020).
49. Zietz, Michael, and Nicholas P. Tatonetti. "Testing the association between blood type and COVID-19 infection, intubation, and death." *MedRxiv* (2020).
50. Basile, Anna O., Alexandre Yahi, and Nicholas P. Tatonetti. "Artificial intelligence for drug toxicity and safety." *Trends in pharmacological sciences* 40, no. 9 (2019): 624-635.
51. Percha, Bethany, and Russ B. Altman. "Informatics confronts drug–drug interactions." *Trends in pharmacological sciences* 34, no. 3 (2013): 178-184.
52. Duda, M., N. Haber, J. Daniels, and D. P. Wall. "Crowdsourced validation of a machine-learning classification system for autism and ADHD." *Translational psychiatry* 7, no. 5 (2017): e1133-e1133.
53. Duda, M., J. A. Kosmicki, and D. P. Wall. "Testing the accuracy of an observation-based classifier for rapid detection of autism risk." *Translational psychiatry* 4, no. 8 (2014): e424-e424.
54. Lyalina, Svetlana, Bethany Percha, Paea LePendu, Srinivasan V. Iyer, Russ B. Altman, and Nigam H. Shah. "Identifying phenotypic signatures of neuropsychiatric disorders from electronic medical records." *Journal of the American Medical Informatics Association* 20, no. e2 (2013): e297-e305.
55. Tariq, Qandeel, Jena Daniels, Jessey Nicole Schwartz, Peter Washington, Haik Kalantarian, and Dennis Paul Wall. "Mobile detection of autism through machine learning on home video: A development and prospective validation study." *PLoS medicine* 15, no. 11 (2018): e1002705.
56. Wall, Dennis Paul, J. Kosmicki, T. F. Deluca, E. Harstad, and Vincent Alfred Fusaro. "Use of machine learning to shorten observation-based screening and diagnosis of autism." *Translational psychiatry* 2, no. 4 (2012): e100-e100.
57. Tariq, Qandeel, Scott Lanyon Fleming, Jessey Nicole Schwartz, Kaitlyn Dunlap, Conor Corbin, Peter Washington, Haik Kalantarian, Naila Z. Khan, Gary L. Darmstadt, and Dennis Paul Wall. "Detecting developmental delay and autism through machine learning models using home videos of Bangladeshi children: Development and validation study." *Journal of medical Internet research* 21, no. 4 (2019): e13822.
58. Polubriaginof, Fernanda CG, Rami Vanguri, Kayla Quinnies, Gillian M. Belbin, Alexandre Yahi, Hojjat Salmasian, Tal Lorberbaum et al. "Disease heritability inferred from familial relationships reported in medical records." *Cell* 173, no. 7 (2018): 1692-1704.
59. Tatonetti, Nicholas, Russ B. Altman, and Guy Haskin Fernald. "Signal detection algorithms to identify drug effects and drug interactions." U.S. Patent 9,305,267, issued April 5, 2016.

60. Johnson, Kipp W., Jessica K. De Freitas, Benjamin S. Glicksberg, Jason R. Bobe, and Joel T. Dudley. "Evaluation of patient re-identification using laboratory test orders and mitigation via latent space variables." In *PSB*, pp. 415-426. 2019.
61. Washington, Peter, Kelley Marie Paskov, Haik Kalantarian, Nathaniel Stockham, Catalin Voss, Aaron Kline, Ritik Patnaik et al. "Feature selection and dimension reduction of social autism data." In *Pac Symp Biocomput*, vol. 25, pp. 707-718. 2020.
62. Bae, Ho, Dahuin Jung, and Sungroh Yoon. "AnomiGAN: Generative adversarial networks for anonymizing private medical data." *arXiv preprint arXiv:1901.11313* (2019).
63. Demuynck, Liesje, and Bart De Decker. "Privacy-preserving electronic health records." In *IFIP International Conference on Communications and Multimedia Security*, pp. 150-159. Springer, Berlin, Heidelberg, 2005.
64. Luthria, Gaurav, and Qingbo Wang. "Implementing a Cloud Based Method for Protected Clinical Trial Data Sharing." In *Pacific Symposium on Biocomputing. Pacific Symposium on Biocomputing*, vol. 25, pp. 647-658. NIH Public Access, 2020.
65. Fernández-Alemán, José Luis, Inmaculada Carrión Señor, Pedro Ángel Oliver Lozoya, and Ambrosio Toval. "Security and privacy in electronic health records: A systematic literature review." *Journal of biomedical informatics* 46, no. 3 (2013): 541-562.
66. Istepanian, Robert, Swamy Laxminarayan, and Constantinos S. Pattichis, eds. *M-health: Emerging mobile health systems*. Springer Science & Business Media, 2007.
67. Lupton, Deborah. *Digital health: critical and cross-disciplinary perspectives*. Routledge, 2017.
68. Murray, Elizabeth, Eric B. Hekler, Gerhard Andersson, Linda M. Collins, Aiden Doherty, Chris Hollis, Daniel E. Rivera, Robert West, and Jeremy C. Wyatt. "Evaluating digital health interventions: key questions and approaches." (2016): 843-851.
69. Maisel, William H. "Medical device regulation: an introduction for the practicing physician." *Annals of internal medicine* 140, no. 4 (2004): 296-302.
70. Zuckerman, Diana M., Paul Brown, and Steven E. Nissen. "Medical device recalls and the FDA approval process." *Archives of internal medicine* 171, no. 11 (2011): 1006-1011.
71. Stark, David E., Rajiv B. Kumar, Christopher A. Longhurst, and Dennis P. Wall. "The quantified brain: a framework for mobile device-based assessment of behavior and neurological function." *Applied clinical informatics* 7, no. 2 (2016): 290.
72. Torous, John, and Laura Weiss Roberts. "Needed innovation in digital health and smartphone applications for mental health: transparency and trust." *JAMA psychiatry* 74, no. 5 (2017): 437-438.
73. McConnell, Michael V., Mintu P. Turakhia, Robert A. Harrington, Abby C. King, and Euan A. Ashley. "Mobile health advances in physical activity, fitness, and atrial fibrillation: moving hearts." *Journal of the American College of Cardiology* 71, no. 23 (2018): 2691-2701.
74. McConnell, Michael V., Anna Shcherbina, Aleksandra Pavlovic, Julian R. Homburger, Rachel L. Goldfeder, Daryl Waggot, Mildred K. Cho et al. "Feasibility of obtaining measures of lifestyle from a smartphone app: the MyHeart Counts Cardiovascular Health Study." *JAMA cardiology* 2, no. 1 (2017): 67-76.
75. Ngwatu, Brian Kermu, Ntwali Placide Nsengiyumva, Olivia Oxlade, Benjamin Mappin-Kasirer, Nhat Linh Nguyen, Ernesto Jaramillo, Dennis Falzon, and Kevin Schwartzman. "The impact of digital health technologies on tuberculosis treatment: a systematic review." *European Respiratory Journal* 51, no. 1 (2018).
76. Washington, Peter, Catalin Voss, Nick Haber, Serena Tanaka, Jena Daniels, Carl Feinstein, Terry Winograd, and Dennis Wall. "A wearable social interaction aid for children with autism." In *Proceedings of the 2016 CHI Conference Extended Abstracts on Human Factors in Computing Systems*, pp. 2348-2354. 2016.
77. Kalantarian, Haik, Khaled Jedoui, Peter Washington, and Dennis P. Wall. "A mobile game for automatic emotion-labeling of images." *IEEE Transactions on Games* (2018).
78. Kline, Aaron, Catalin Voss, Peter Washington, Nick Haber, Hessey Schwartz, Qandeel Tariq, Terry Winograd, Carl Feinstein, and Dennis P. Wall. "Superpower glass." *GetMobile: Mobile Computing and Communications* 23, no. 2 (2019): 35-38.
79. Washington, Peter, Catalin Voss, Aaron Kline, Nick Haber, Jena Daniels, Azar Fazel, Titas De, Carl Feinstein, Terry Winograd, and Dennis Wall. "SuperpowerGlass: a wearable aid for the at-home therapy of children with autism." *Proceedings of the ACM on interactive, mobile, wearable and ubiquitous technologies* 1, no. 3 (2017): 1-22.

80. Daniels, Jena, Nick Haber, Catalin Voss, Jessey Schwartz, Serena Tamura, Azar Fazel, Aaron Kline et al. "Feasibility testing of a wearable behavioral aid for social learning in children with autism." *Applied clinical informatics* 9, no. 1 (2018): 129.
81. Waran, Vicknes, Nor Faizal Ahmad Bahuri, Vairavan Narayanan, Dharmendra Ganesan, and Khairul Azmi Abdul Kadir. "Video clip transfer of radiological images using a mobile telephone in emergency neurosurgical consultations (3G Multi-Media Messaging Service)." *British journal of neurosurgery* 26, no. 2 (2012): 199-201.
82. Kalantarian, Haik, Khaled Jedoui, Peter Washington, Qandeel Tariq, Kaiti Dunlap, Jessey Schwartz, and Dennis P. Wall. "Labeling images with facial emotion and the potential for pediatric healthcare." *Artificial intelligence in medicine* 98 (2019): 77-86.
83. Voss, Catalin, Peter Washington, Nick Haber, Aaron Kline, Jena Daniels, Azar Fazel, Titas De et al. "Superpower glass: delivering unobtrusive real-time social cues in wearable systems." In *Proceedings of the 2016 ACM International Joint Conference on Pervasive and Ubiquitous Computing: Adjunct*, pp. 1218-1226. 2016.
84. Kalantarian, Haik, Khaled Jedoui, Kaitlyn Dunlap, Jessey Schwartz, Peter Washington, Arman Husic, Qandeel Tariq, Michael Ning, Aaron Kline, and Dennis Paul Wall. "The Performance of Emotion Classifiers for Children With Parent-Reported Autism: Quantitative Feasibility Study." *JMIR Mental Health* 7, no. 4 (2020): e13174.
85. Daniels, Jena, Jessey N. Schwartz, Catalin Voss, Nick Haber, Azar Fazel, Aaron Kline, Peter Washington, Carl Feinstein, Terry Winograd, and Dennis P. Wall. "Exploratory study examining the at-home feasibility of a wearable tool for social-affective learning in children with autism." *NPJ digital medicine* 1, no. 1 (2018): 1-10.
86. Kalantarian, Haik, Peter Washington, Jessey Schwartz, Jena Daniels, Nick Haber, and Dennis Wall. "A gamified mobile system for crowdsourcing video for autism research." In *2018 IEEE international conference on healthcare informatics (ICHI)*, pp. 350-352. IEEE, 2018.
87. Voss, Catalin, Jessey Schwartz, Jena Daniels, Aaron Kline, Nick Haber, Peter Washington, Qandeel Tariq et al. "Effect of wearable digital intervention for improving socialization in children with autism spectrum disorder: a randomized clinical trial." *JAMA pediatrics* 173, no. 5 (2019): 446-454.
88. Brown, Stephen James. "Interactive video based remote health monitoring system." U.S. Patent 7,979,284, issued July 12, 2011.
89. Kalantarian, Haik, Peter Washington, Jessey Schwartz, Jena Daniels, Nick Haber, and Dennis P. Wall. "Guess what?." *Journal of Healthcare Informatics Research* 3, no. 1 (2019): 43-66.
90. Ramanujam, Bhargavi, Deepa Dash, and Manjari Tripathi. "Can home videos made on smartphones complement video-EEG in diagnosing psychogenic nonepileptic seizures?." *Seizure* 62 (2018): 95-98.
91. Rao, Sira P., Nikil S. Jayant, Max E. Stachura, Elena Astapova, and Anthony Pearson-Shaver. "Delivering diagnostic quality video over mobile wireless networks for telemedicine." *International Journal of Telemedicine and Applications* 2009 (2009).
92. Martínez-Pérez, Borja, Isabel De La Torre-Díez, and Miguel López-Coronado. "Privacy and security in mobile health apps: a review and recommendations." *Journal of medical systems* 39, no. 1 (2015): 181.
93. Celi, Leo Anthony, Andrea Ippolito, Robert A. Montgomery, Christopher Moses, and David J. Stone. "Crowdsourcing knowledge discovery and innovations in medicine." *Journal of medical Internet research* 16, no. 9 (2014): e216.
94. Créquit, Perrine, Ghizlène Mansouri, Mehdi Benchoufi, Alexandre Vivot, and Philippe Ravaud. "Mapping of crowdsourcing in health: systematic review." *Journal of medical Internet research* 20, no. 5 (2018): e187.
95. Ranard, Benjamin L., Yoonhee P. Ha, Zachary F. Meisel, David A. Asch, Shawndra S. Hill, Lance B. Becker, Anne K. Seymour, and Raina M. Merchant. "Crowdsourcing—harnessing the masses to advance health and medicine, a systematic review." *Journal of general internal medicine* 29, no. 1 (2014): 187-203.
96. Swan, Melanie. "Health 2050: The realization of personalized medicine through crowdsourcing, the quantified self, and the participatory biocitizen." *Journal of personalized medicine* 2, no. 3 (2012): 93-118.
97. Wazny, Kerri. "Applications of crowdsourcing in health: an overview." *Journal of global health* 8, no. 1 (2018).
98. David, Maude M., Brooke A. Babineau, and Dennis P. Wall. "Can we accelerate autism discoveries through crowdsourcing?." *Research in Autism Spectrum Disorders* 32 (2016): 80-83.

99. Washington, Peter, Haik Kalantarian, Qandeel Tariq, Jessey Schwartz, Kaitlyn Dunlap, Brianna Chrisman, Maya Varma et al. "Validity of online screening for autism: crowdsourcing study comparing paid and unpaid diagnostic tasks." *Journal of medical Internet research* 21, no. 5 (2019): e13668.
100. Washington, Peter, Emilie Leblanc, Kaitlyn Dunlap, Yordan Penev, Aaron Kline, Kelley Paskov, Min Woo Sun et al. "Precision Telemedicine through Crowdsourced Machine Learning: Testing Variability of Crowd Workers for Video-Based Autism Feature Recognition." *Journal of personalized medicine* 10, no. 3 (2020): 86.
101. Washington, Peter, Natalie Park, Parishkrita Srivastava, Catalin Voss, Aaron Kline, Maya Varma, Qandeel Tariq et al. "Data-driven diagnostics and the potential of mobile artificial intelligence for digital therapeutic phenotyping in computational psychiatry." *Biological Psychiatry: Cognitive Neuroscience and Neuroimaging* (2019).
102. Gottlieb, Assaf, Robert Hoehndorf, Michel Dumontier, and Russ B. Altman. "Ranking adverse drug reactions with crowdsourcing." *Journal of medical Internet research* 17, no. 3 (2015): e80.
103. Budd, Jobie, Benjamin S. Miller, Erin M. Manning, Vasileios Lampos, Mengdie Zhuang, Michael Edelstein, Geraint Rees et al. "Digital technologies in the public-health response to COVID-19." *Nature medicine* (2020): 1-10.
104. Hegde, Ajay, Ramesh Masthi, and Darshan Krishnappa. "Hyperlocal Postcode Based Crowdsourced Surveillance Systems in the COVID-19 Pandemic Response." *Frontiers in Public Health* 8 (2020): 286.
105. Sun, Kaiyuan, Jenny Chen, and Cécile Viboud. "Early epidemiological analysis of the coronavirus disease 2019 outbreak based on crowdsourced data: a population-level observational study." *The Lancet Digital Health* (2020).
106. Kittur, Aniket, Ed H. Chi, and Bongwon Suh. "Crowdsourcing user studies with Mechanical Turk." In *Proceedings of the SIGCHI conference on human factors in computing systems*, pp. 453-456. 2008.
107. Paolacci, Gabriele, Jesse Chandler, and Panagiotis G. Ipeirotis. "Running experiments on amazon mechanical turk." *Judgment and Decision making* 5, no. 5 (2010): 411-419.
108. Hirth, Matthias, Tobias Hoßfeld, and Phuoc Tran-Gia. "Anatomy of a crowdsourcing platform-using the example of microworkers. com." In *2011 Fifth international conference on innovative mobile and internet services in ubiquitous computing*, pp. 322-329. IEEE, 2011.
109. Raykar, Vikas C., and Shipeng Yu. "Eliminating spammers and ranking annotators for crowdsourced labeling tasks." *The Journal of Machine Learning Research* 13, no. 1 (2012): 491-518.
110. Beauchamp, Tom L., and James F. Childress. *Principles of biomedical ethics*. Oxford University Press, USA, 2001.
111. Botkin, Jeffrey R., and Erin Rothwell. "Whole genome sequencing and newborn screening." *Current genetic medicine reports* 4, no. 1 (2016): 1-6.
112. Howard, Heidi Carmen, Bartha Maria Knoppers, Martina C. Cornel, Ellen Wright Clayton, Karine Sénécal, and Pascal Borry. "Whole-genome sequencing in newborn screening? A statement on the continued importance of targeted approaches in newborn screening programmes." *European Journal of Human Genetics* 23, no. 12 (2015): 1593-1600.
113. Char, Danton S., Nigam H. Shah, and David Magnus. "Implementing machine learning in health care—addressing ethical challenges." *The New England journal of medicine* 378, no. 11 (2018): 981.
114. Cohen, I. Glenn, Ruben Amarasingham, Anand Shah, Bin Xie, and Bernard Lo. "The legal and ethical concerns that arise from using complex predictive analytics in health care." *Health affairs* 33, no. 7 (2014): 1139-1147.
115. Korngiebel, Diane M., Kenneth E. Thummel, and Wylie Burke. "Implementing precision medicine: the ethical challenges." *Trends in pharmacological sciences* 38, no. 1 (2017): 8-14.
116. Minari, Jusaku, Kyle B. Brothers, and Michael Morrison. "Tensions in ethics and policy created by National Precision Medicine Programs." *Human genomics* 12, no. 1 (2018): 1-10.
117. Kreitmair, Karola V., Mildred K. Cho, and David C. Magnus. "Consent and engagement, security, and authentic living using wearable and mobile health technology." *Nature biotechnology* 35, no. 7 (2017): 617-620.
118. Torous, John, and Laura Weiss Roberts. "The ethical use of mobile health technology in clinical psychiatry." *The Journal of nervous and mental disease* 205, no. 1 (2017): 4-8.
119. Fort, Karën, Gilles Adda, and K. Bretonnel Cohen. "Amazon mechanical turk: Gold mine or coal mine?." *Computational Linguistics* 37, no. 2 (2011): 413-420.

120. Kreitmair, Karola V., and David C. Magnus. "Citizen science and gamification." *Hastings center report* 49, no. 2 (2019): 40-46.

Selection of trustworthy crowd workers for telemedical diagnosis of pediatric autism spectrum disorder

Peter Washington[1], Emilie Leblanc[2,3], Kaitlyn Dunlap[2,3], Yordan Penev[2,3], Maya Varma[4], Jae-Yoon Jung[2,3], Brianna Chrisman[1], Min Woo Sun[3], Nathaniel Stockham[5], Kelley Marie Paskov[3], Haik Kalantarian[2,3], Catalin Voss[4], Nick Haber[6], Dennis P. Wall[2,3]

Department of Bioengineering[1], Stanford University, Palo Alto, CA, 94305, USA
Department of Pediatrics (Systems Medicine)[2], Stanford University, Palo Alto, CA, 94305, USA
Department of Biomedical Data Science[3], Stanford University, Palo Alto, CA, 94305, USA
Department of Computer Science[4], Stanford University, Palo Alto, CA, 94305, USA
Department of Neuroscience[5], Stanford University, Palo Alto, CA, 94305, USA
School of Education[6], Stanford University, Palo Alto, CA, 94305, USA
Email: dpwall@stanford.edu

Crowd-powered telemedicine has the potential to revolutionize healthcare, especially during times that require remote access to care. However, sharing private health data with strangers from around the world is not compatible with data privacy standards, requiring a stringent filtration process to recruit reliable and trustworthy workers who can go through the proper training and security steps. The key challenge, then, is to identify capable, trustworthy, and reliable workers through high-fidelity evaluation tasks without exposing any sensitive patient data during the evaluation process. We contribute a set of experimentally validated metrics for assessing the trustworthiness and reliability of crowd workers tasked with providing behavioral feature tags to unstructured videos of children with autism and matched neurotypical controls. The workers are blinded to diagnosis and blinded to the goal of using the features to diagnose autism. These behavioral labels are fed as input to a previously validated binary logistic regression classifier for detecting autism cases using categorical feature vectors. While the metrics do not incorporate any ground truth labels of child diagnosis, linear regression using the 3 correlative metrics as input can predict the mean probability of the correct class of each worker with a mean average error of 7.51% for performance on the same set of videos and 10.93% for performance on a distinct balanced video set with different children. These results indicate that crowd workers can be recruited for performance based largely on behavioral metrics on a crowdsourced task, enabling an affordable way to filter crowd workforces into a trustworthy and reliable diagnostic workforce.

Keywords: Crowdsourcing; Machine Learning; Diagnostics; Trust; Privacy; Autism

1. Introduction

Autism spectrum disorder (ASD, or autism) is a pediatric developmental condition affecting 1 in 40 children in the United States [1], with prevalence continuing to rise [2]. While access to care relies on a formal diagnosis from a clinician, an uneven distribution of diagnostic resources across the United States contributes to increasingly long waitlists. Some evidence suggests that 80% of counties lack sufficient diagnostic resources [3], with underserved communities disproportionately affected by this shortage [4]. Telemedicine has the potential to minimize this gap by capitalizing on the increasing pervasiveness and affordability of digital devices. Such diagnostic solutions are especially pertinent during times of pandemic, most notably the coronavirus, which further hinders access to diagnosis and care.

Mobile digital autism interventions administered on smartphones [5-12] and on ubiquitous devices [13-27] passively collect structured home videos of children with neuropsychiatric conditions for use in subsequent diagnostic data analysis [27-28]. In order for the video data collected from digital therapies to become widely used, trustworthy data sharing methodologies must be incorporated into the diagnostic pipeline [29]. One possible approach, which we realize in the present study, is to carefully recruit a trustworthy set of workers to transform the video streams into a secure, quantitative, and structured format. While modern computer vision algorithms could handle this task in several domains, extracting complex behavioral features from video is currently beyond the scope of state-of-the-art machine learning methods and therefore requires human labor. However, the collected videos naturally contain highly sensitive data, requiring careful selection of trustworthy and reliable labelers who are allowed access to protected health information (PHI) after completion of Health Insurance Portability and Accountability Act (HIPAA) training, Collaborative Institutional Training Initiative (CITI) human subjects training, and whole disk encryption.

In the present study, we examine strategies for quantitatively determining the credibility and reliability of crowd workers whose labels can be trusted by researchers. It is important that the metrics for evaluating workers are speedy and simple, as formally credentialing recruited crowd workers through institutional channels is laborious and slow. We crowdsource the task of providing categorical feature labels to videos of children with autism and matched controls. For each crowdsourced worker, we evaluate correlations of their mean classifier probability of the correct class (PCC) using their answers as input with (1) the mean L1 distance between their responses to the same video spaced one month apart, (2) the mean L1 distance between their answer vector to each video and all other videos they rated, (3) the mean time spent rating videos, and (4) the mean time and L1 distance of answers when the worker is explicitly warned about not spending enough time rating a video and provided with a chance to revise their response. We then feed the metrics which are correlated with PCC into a linear regression model predicting the PCC.

2. Methods

2.1. *Clinically representative videos*

We used a set of 24 publicly available videos from YouTube of children with autism and matched neurotypical controls (6 females with autism, 6 neurotypical females, 6 males with autism, and 6 neurotypical males). Criteria for video selection and inclusion were that (1) the child's hand and face must be visible, (2) opportunities for social engagement must be present, and (3) an opportunity for using an object such as a toy or utensil must be present. Child diagnosis was determined through the video title and description. The videos were short, with a mean duration of 47.75 seconds (SD = 30.71 seconds). The mean age of children in the video was 3.65 years (SD = 1.82 years).

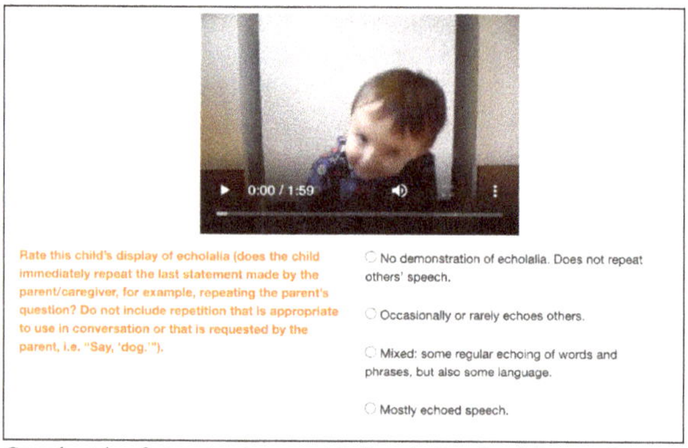

Fig. 1. Crowd worker feature tagging user interface deployed on Microworkers.com. Each worker answered a series of multiple-choice questions corresponding to each input feature of a gold standard classifier.

2.2. *Crowdsourcing task for Microworkers*

Prior work has validated the capability of subsets of the crowd recruited from the Amazon Mechanical Turk crowdsourcing platform [30] to provide feature tags of children with autism comparable to clinical coordinators working with children with autism on a daily basis [31-32]. We instead recruited workers from Microworkers.com, as Microworkers consists of a diverse representation of worker nationalities [33] compared to Mechanical Turk, which contains workers mostly from the United States and India [34]. Furthermore, Microworkers provides built in functionality for allowing workers to revise their answers if a requester is unsatisfied but believes the worker can redeem their response. This functionality was crucial for our trustworthiness metric.

The task consisted of a series of 13 multiple choice questions identified, in prior work which employed feature selection algorithms on electronic health records [35-44], as salient categorical ordinal features for autism prediction. Workers were asked to watch a short video and answer the multiple-choice questions using the interface depicted in Fig. 1. Microworkers automatically records the time spent on each task.

Through a pilot study of internal lab raters providing 9,374 video ratings for which we logged labeling times, we observed that the mean time per video was 557.7 seconds (9 minutes 18 seconds), with a standard deviation of 929.7 seconds (15 minutes 30 seconds). The pilot task consisted of answering 31 multiple choice questions, while the Microworkers task only contained 13 questions; the proportional mean time is 233.9 seconds (3 minutes 54 seconds). We therefore required workers to spend at least 2 minutes per video, a time threshold significantly below the 233.9 second mean proportional time. If any crowd worker spent less than 2 minutes rating a video, we leveraged the built-in functionality on Microworkers to prompt these users to revise their answers and sent them a warning message disclosing that we know the "*Impossibly short time spent on task.*" We measured the additional time spent by the worker, if any, as well as the changes in the answer vector (L1 distance) after receiving this message.

We posted all tasks for all 24 videos exactly 30 days after the original task, allowing workers who completed the first task to complete the task again while minimizing the chance that they could use the memory of their prior responses to bias the test. Previous studies which evaluate test-retest reliability consider 2 weeks to be sufficient time to prevent memorization of prior administrations of the questionnaire [45-48], and we increased this time frame to 30 days to minimize the likelihood that any memory of the workers' previous answers remained. The same video of the child was provided for both administrations of the task. Workers were not provided with their original answers for reference. The difference between the worker's original answers and their revised answers on the same video served as quantitative information about the *reliability* of the worker.

2.3. *Classifier to evaluate performance*

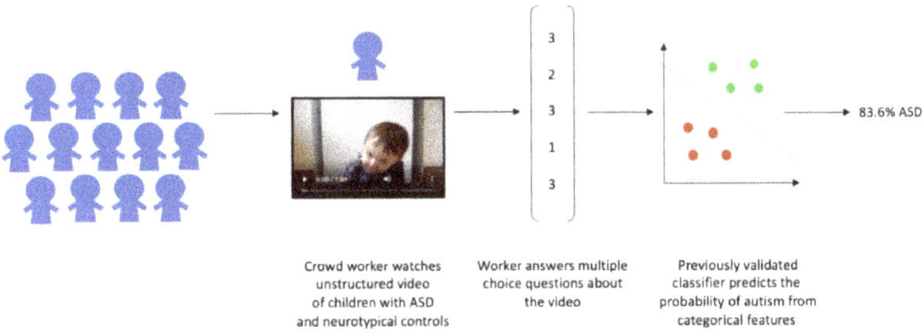

Fig. 2. Process for collecting the data needed to evaluate trust and reliability metrics for crowd workers. Each crowd worker watches unstructured videos of children with autism and neurotypical controls, answering multiple choice questions about each video. These multiple-choice answers serve as categorical ordinal feature vectors for a previously validated logistic regression classifier, trained on clinician-filled electronic health records, that predicts the probability that a child has autism.

For a gold standard, we use a previously published and validated [49-54] logistic regression classifier (Fig. 2), trained on electronic health record databases of autism diagnostic scoresheets filled out by expert clinicians, which emits a probability score of autism using the crowd workers' multiple-choice responses as categorical ordinal feature vectors. Because logistic regression classifiers produce a probability, we treat the probability as a confidence score of the crowdsourced workers' responses. We analyze the probability of the correct class (referred to as PCC), which is p when the true class is autism and $1-p$ when the true class is neurotypical. When assessing classifier predictions, we use a threshold of 0.5. We use a worker's average PCC for videos the worker has rated as a metric of the worker's video tagging capability, with a higher mean PCC corresponding to greater mean performance by the worker.

2.4. *Metrics evaluated*

We strive to develop metrics which only take input parameters that do not depend on *a priori* knowledge about the correct classification score of the videos. We test the following metrics for correlation with the PCC, where N is the number of videos rated by a worker, M is the number of questions per video rating task (inputs to the diagnostic classifier), and $A_{i,j,k}$ is the answer for video i and question j for the kth time.

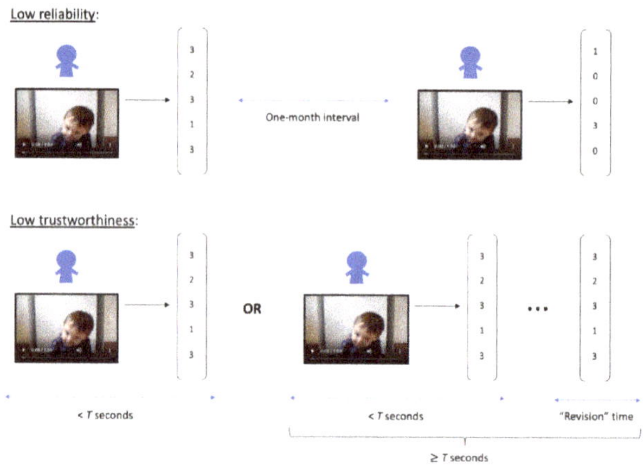

Fig. 3. Process for calculating trust and reliability metrics for crowd workers. The reliability of workers is determined by how different their answers are when rating the same video one month apart. The trustworthiness of workers is determined by whether they spend the minimal amount of time needed to properly answer the questions, whether they spend sufficient time when receiving a warning, and whether their original answers change after receiving the warning.

Mean same-child L1 distance ($MSCL_1$): We asked crowd workers to rate the same child at least one month apart. Workers did not have access to their originally recorded answers and were unaware that they would be asked to rate the same video a second time when providing the first set of ratings. We observe the mean deviation for all videos between a worker's original ratings for the video and their subsequent ratings one month later. We call this metric the *mean same-child L1 distance ($MSCL_1$)*, which we consider as a metric of the worker's *test-retest reliability*. Higher values for the $MSCL_1$ correspond to greater variation in worker responses when re-rating the same video one month apart. Formally, $MSCL_1$ is calculated as:

$$MSCL_1 = \frac{\sum_{i=1}^{N}\sum_{j=1}^{M}|A_{i,j,2} - A_{i,j,1}|}{N}$$

Mean pairwise internal L1 distance ($MPIL_1$): To analyze the reliability of the worker's answers across videos, we look at the mean L1 distance between a worker's answer to each video and all other videos they rated. We call this metric the *mean pairwise internal L1 distance ($MPIL_1$)*. $MPIL_1$ is high when workers provide a wide variety of answer patterns across videos. If the worker answers all questions the same way per video, the $MPIL_1$ will be 0. Formally, $MPIL_1$ is calculated as:

$$MPIL_1 = \frac{\sum_{i1=1}^{N} \sum_{i2=1}^{N} \sum_{j=1}^{M} |A_{i2,j} - A_{i1,j}|}{0.5\, N\, (N-1)}, i1 < i2$$

Penalized time (PT): We aimed to build a metric that prioritizes rewarding workers who spent sufficient time rating the first time while rewarding, to a lesser extent, workers who spend sufficient time rating after receiving a warning. We also aimed to penalize workers who either do not spend more time rating after receiving a warning or who do not sufficiently update their answers. We create a metric of worker *trustworthiness* taking both of these factors into account which we call the *penalized time (PT)*. If workers spend longer than a time threshold T rating, then they are not asked to revise their answers and receive a baseline score M. If they do not spend a sufficient time (T) rating, then they are asked to spend more time and to revise their answers. In this case, the metric consists of two terms, balanced by a weighting constant c. The first term is the "revision" mean same-child L1 distance ($RMSCL_1$) between initial and revised answers only for videos that the worker was explicitly asked to revise. The second term is the mean of the total time spent rating, which is the time spent initially (t_1) and the time spent revising the answers (t_2). Formally, PT is calculated as:

$$PT = \begin{cases} M, & t_1 \geq T \\ \frac{t_1 + t_2}{N} + c\, RMSCL_1, & t_1 < T \end{cases}$$

Time spent: Finally, we record the mean amount of time spent rating per video, in seconds. We hypothesized that workers who spend more time on the rating task will tend towards achieving higher performance.

We hypothesized that all four metrics are correlated with PCC. We only calculate metrics for workers who rated at least 10 videos. Because 13 questions were asked, an $MSCL_1$ or $MPIL_1$ of 13 means that, on average, the worker's answer differed by 1 categorical ordinal answer choice per question (e.g., the difference between "*Mixed: some regular echoing of words and phrases, but also some language*" and "*Mostly echoed speech*" in Fig. 1).

2.5. *Prediction of crowd worker performance from metrics*

We train and test a linear regression model to predict the mean PCC of the workers using 5-fold cross validation. We evaluate all non-empty subsets of the correlative metrics described in section 2.4 as inputs to the model. Since not all workers reopened the task after receiving a warning and not all workers conducted the second task in the series, we evaluated our model both using all available workers with complete data for all metrics as well as using the subset of 55 workers with data for all metrics.

3. Results

3.1. *Correlation between metrics and probability of the correct class*

Correlations of each of the worker metrics with their mean PCC are displayed in Fig. 4. Mean values per worker are only plotted and analyzed if at least 5 data points are available for the worker. MSCL_1, MPIL_1, and mean time spent were all significantly correlated with PCC ($r=0.31$, $p=0.0212$ for MSCL_1; $r=0.57$, $p<0.0001$ for MPIL_1; $r=0.16$, $p=0.0284$ for time), supporting the predictive power of these metrics. Intuitively, this means that higher variability in worker answers for the same video and across videos correlates with increased worker performance. We note that only MPIL_1 passes Bonferroni correction. Penalized time was not significantly correlated with PCC ($r=0.17$, $r=0.1413$ for penalized time).

Interestingly, Fig. 4 reveals that the presence of enough data to calculate certain metrics is in itself predictive of worker performance. Fig. 4C shows that there are several workers who had a mean PCC below 50%. However, none of these workers appear in the plot for MSCL_1 (Fig. 4A), MPIL_1 (Fig. 4B), or penalized time (Fig. 4D), indicating that workers with low average performance did not rate videos again after one month and did not revise their answers when prompted.

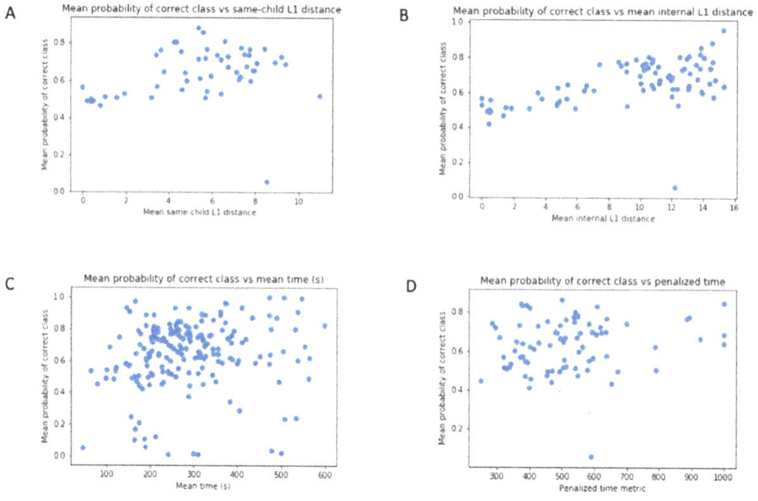

Fig. 4. Correlations between metrics and probability of the correct class (PCC). (A) Correlation between mean same-child L1 distance and PCC. (B) Correlation between mean pairwise internal L1 distance and PCC. (C) Correlation between time spent (s) and PCC. (D) Lack of correlation between penalized time and PCC.

We evaluate all values of the weighting constant *c* for the penalized time metric in the interval [0.05, 10.0] using a step size of 0.05. No value resulted in a metric that positively correlates with

PCC. To investigate, we review the correlation between both terms of penalized time: (1) the mean total time spent rating post-warning and (2) the mean L1 distance between the answer vector before and after the warning (Fig. 5). Neither of these metrics are correlated with PCC ($r=-0.10$, $p=0.3414$ for revision L1 distance; $r=0.11$, $p=0.2908$ for total time), explaining the inability of the penalized time metric to predict PCC regardless of the parameters chosen.

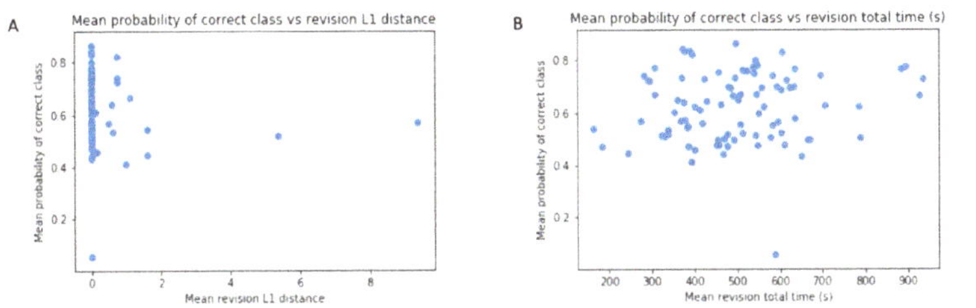

Fig. 5. Lack of correlation between PCC and (A) the total time spent rating post-warning and (B) the L1 distance between the answer before and after the warning.

3.2. *Regression prediction of the mean probability of the correct class*

Table 1 contains the mean average error (MAE) of a linear regression model predicting the probability of the correct class for each worker using metrics on the same set of videos. There were 55 workers with data for all 3 metrics used in the regression model. For these workers, all metrics predicted the PCC with less than 10% MAE.

The MAE when using all 3 features performs nearly identically, to two decimal places, compared to using only $MSCL_l$ and $MPIL_l$. Mean time does not contribute much predictive power given the other metrics. Interestingly, the most predictive input configuration when using the same 55 workers is $MPIL_l$ together with mean time (6.97% MAE), followed by $MPIL_l$ alone as a close second (6.98% MAE). This is a testament to the success of the $MPIL_l$ metric.

Input Features	**5-fold MAE (%) All data points**	**5-fold MAE (%) 55 workers with all metric data**	**N**
$MSCL_l$, $MPIL_l$, mean time	7.51	7.51	55
$MSCL_l$, mean time	8.89	8.89	55
$MPIL_l$, mean time	7.43	6.97	81
$MSCL_l$, $MPIL_l$	7.51	7.51	55
$MSCL_l$	9.24	9.24	55
$MPIL_l$	7.39	6.98	81
Mean time	15.56	9.83	193

Table 1. 5-fold cross validated mean average error (MAE) of a linear regression model predicting the probability of the correct class for each worker using metrics on the same set of videos.

Table 2 contains the mean average error of a linear regression model predicting the probability of the correct class for each worker using metrics from one set of children and mean probability of the correct class calculations for a distinct set of children. The most predictive input feature configuration ($MSCL_I$ and $MPIL_I$) results in a MAE of 10.41%, only 3.44% higher than the best MAE when training and testing on the same set of videos and workers using cross-validation (Table 1). $MPIL_I$ is involved in all of the top-4 input metric configurations resulting in the lowest MAE, again verifying the success of the $MPIL_I$ metric.

Input Features	**MAE (%) All data points**
$MSCL_I$, $MPIL_I$, mean time	10.93
$MSCL_I$, mean time	13.03
$MPIL_I$, mean time	11.50
$MSCL_I$, $MPIL_I$	10.41
$MSCL_I$	11.87
$MPIL_I$	10.91
Mean time*	12.10

Table 2. Mean average error (MAE) of the linear regression model predicting the probability of the correct class for each worker using the same metric data and resulting classifier weights *for the workers and videos used in Table 1* and mean probability of the correct class calculations for a *distinct set of videos* for a *distinct set of workers*. *Mean time as the only feature is the only configuration of input features that requires a different set of data points: N=102 instead of a subset of size N=62 for all other configurations.

4. Discussion and Future Work

We identify three metrics which are individually highly correlated with the mean probability of the worker's categorical behavioral feature tags predicting the correct class. In particular, one of our two reliability metrics - the mean pairwise internal L1 distance, which is the mean L1 distance between a worker's answer to each video and all other videos they rated - stood out as the most predictive metric. Mean pairwise internal L1 distance alone can predict a worker's PCC within 7% MAE when trained on the same set of workers as in the test set but with different videos, and it can predict PCC within 11% MAE when trained on one group of workers and tested on an entirely district set of workers and videos. This metric alone therefore provides a powerful behavioral predictor of worker performance and is therefore likely to be useful for rapidly filtering workers. The positive correlation shown in Fig. 4B suggests that unreliable workers will provide the same or similar patterns of answer sequences for each task. We see that an increasing diversity of answers between tasks results in a higher PCC for the entire spectrum of possible L1 distances. Intuitively, this may be a result of the diverse set of features exhibited by the heterogeneous behavioral characteristics of the children in our dataset.

Interestingly, the raw time metric is not particularly correlative with PCC, indicating that analyzing the answer domain is more informative than the time domain. For workers who received a warning for low time spent, neither the time spent revising post-warning nor the L1 distance between the original and revised set of answers was predictive of the workers' final performance. It is possible that once workers are aware that their time is tracked, they idly keep the rating interface open, accumulating time without accumulating thoughtful work. This hypothesis is speculative, and more fine-grained timing information must be recorded to evaluate such hypotheses.

Future work should evaluate workers on a larger scale, which will validate the preliminary findings of the present study. It is possible that predictive time-based trustworthiness metrics exist. Evaluation on a larger scale in conjunction with more fine-tuned worker metrics will lead to more precise predictions.

5. Conclusion

We demonstrate that behavioral metrics about crowd workers can predict, with a high degree of accuracy, the performance of crowd workers on behavioral feature extraction tasks for the binary diagnosis of autism. Metrics like these can be used for quickly and efficiently identifying crowd workers who are trustworthy and reliable enough for exposure to highly sensitive PHI based on a quantification of their reliability.

6. Acknowledgments

This work was supported by awards to DPW by the National Institutes of Health (R01EB025025, R01LM013083, and R21HD091500). Additionally, we acknowledge the support of grants to DPW from The Hartwell Foundation, the David and Lucile Packard Foundation Special Projects Grant, Beckman Center for Molecular and Genetic Medicine, Coulter Endowment Translational Research Grant, Berry Fellowship, Spectrum Pilot Program, Stanford's Precision Health and Integrated Diagnostics Center (PHIND), Wu Tsai Neurosciences Institute Neuroscience: Translate Program, and Stanford's Institute of Human Centered Artificial Intelligence as well as philanthropic support from Mr. Peter Sullivan. PW would like to acknowledge support from the Schroeder Family Goldman Sachs Stanford Interdisciplinary Graduate Fellowship (SIGF).

References

1. Kogan, Michael D., Catherine J. Vladutiu, Laura A. Schieve, Reem M. Ghandour, Stephen J. Blumberg, Benjamin Zablotsky, James M. Perrin et al. "The prevalence of parent-reported autism spectrum disorder among US children." *Pediatrics* 142, no. 6 (2018).
2. Fombonne, Eric. "The rising prevalence of autism." *Journal of Child Psychology and Psychiatry* 59, no. 7 (2018): 717-720.
3. Ning, Michael, Jena Daniels, Jessey Schwartz, Kaitlyn Dunlap, Peter Washington, Haik Kalantarian, Michael Du, and Dennis P. Wall. "Identification and quantification of gaps in access to autism resources in the United States: an infodemiological study." *Journal of Medical Internet Research* 21, no. 7 (2019): e13094.
4. Howlin, Patricia, and Anna Moore. "Diagnosis in autism: A survey of over 1200 patients in the UK." *autism* 1, no. 2 (1997): 135-162.
5. Escobedo, Lizbeth, David H. Nguyen, LouAnne Boyd, Sen Hirano, Alejandro Rangel, Daniel Garcia-Rosas, Monica Tentori, and Gillian Hayes. "MOSOCO: a mobile assistive tool to support children with autism practicing social skills in real-life situations." In *Proceedings of the SIGCHI Conference on Human Factors in Computing Systems*, pp. 2589-2598. 2012.
6. Hashemi, Jordan, Kathleen Campbell, Kimberly Carpenter, Adrianne Harris, Qiang Qiu, Mariano Tepper, Steven Espinosa et al. "A scalable app for measuring autism risk behaviors in young children: a technical validity and feasibility study." In *Proceedings of the 5th EAI International Conference on Wireless Mobile Communication and Healthcare*, pp. 23-27. 2015.
7. Kalantarian, Haik, Khaled Jedoui, Kaitlyn Dunlap, Jessey Schwartz, Peter Washington, Arman Husic, Qandeel Tariq, Michael Ning, Aaron Kline, and Dennis Paul Wall. "The Performance of Emotion Classifiers for Children With Parent-Reported Autism: Quantitative Feasibility Study." *JMIR Mental Health* 7, no. 4 (2020): e13174.
8. Kalantarian, Haik, Khaled Jedoui, Peter Washington, and Dennis P. Wall. "A mobile game for automatic emotion-labeling of images." *IEEE Transactions on Games* (2018).

9. Kalantarian, Haik, Khaled Jedoui, Peter Washington, Qandeel Tariq, Kaiti Dunlap, Jessey Schwartz, and Dennis P. Wall. "Labeling images with facial emotion and the potential for pediatric healthcare." *Artificial intelligence in medicine* 98 (2019): 77-86.
10. Kalantarian, Haik, Peter Washington, Jessey Schwartz, Jena Daniels, Nick Haber, and Dennis P. Wall. "Guess what?." *Journal of Healthcare Informatics Research* 3, no. 1 (2019): 43-66.
11. Kalantarian, Haik, Peter Washington, Jessey Schwartz, Jena Daniels, Nick Haber, and Dennis Wall. "A gamified mobile system for crowdsourcing video for autism research." In *2018 IEEE international conference on healthcare informatics (ICHI)*, pp. 350-352. IEEE, 2018.
12. Li, Wei, Farnaz Abtahi, Christina Tsangouri, and Zhigang Zhu. "Towards an "in-the-wild" emotion dataset using a game-based framework." In *2016 IEEE Conference on Computer Vision and Pattern Recognition Workshops (CVPRW)*, pp. 1526-1534. IEEE, 2016.
13. Boyd, LouAnne E., Alejandro Rangel, Helen Tomimbang, Andrea Conejo-Toledo, Kanika Patel, Monica Tentori, and Gillian R. Hayes. "SayWAT: Augmenting face-to-face conversations for adults with autism." In *Proceedings of the 2016 CHI Conference on Human Factors in Computing Systems*, pp. 4872-4883. 2016.
14. Daniels, Jena, Nick Haber, Catalin Voss, Jessey Schwartz, Serena Tamura, Azar Fazel, Aaron Kline et al. "Feasibility testing of a wearable behavioral aid for social learning in children with autism." *Applied clinical informatics* 9, no. 1 (2018): 129.
15. Daniels, Jena, Jessey N. Schwartz, Catalin Voss, Nick Haber, Azar Fazel, Aaron Kline, Peter Washington, Carl Feinstein, Terry Winograd, and Dennis P. Wall. "Exploratory study examining the at-home feasibility of a wearable tool for social-affective learning in children with autism." *NPJ digital medicine* 1, no. 1 (2018): 1-10.
16. El Kaliouby, Rana, and Peter Robinson. "The emotional hearing aid: an assistive tool for children with Asperger syndrome." *Universal Access in the Information Society* 4, no. 2 (2005): 121-134.
17. Haber, Nick, Catalin Voss, and Dennis Wall. "Making emotions transparent: Google Glass helps autistic kids understand facial expressions through augmented-reaiity therapy." *IEEE Spectrum* 57, no. 4 (2020): 46-52.
18. Kline, Aaron, Catalin Voss, Peter Washington, Nick Haber, Hessey Schwartz, Qandeel Tariq, Terry Winograd, Carl Feinstein, and Dennis P. Wall. "Superpower glass." *GetMobile: Mobile Computing and Communications* 23, no. 2 (2019): 35-38.
19. Madsen, Miriam, Rana El Kaliouby, Matthew Goodwin, and Rosalind Picard. "Technology for just-in-time in-situ learning of facial affect for persons diagnosed with an autism spectrum disorder." In *Proceedings of the 10th international ACM SIGACCESS conference on Computers and accessibility*, pp. 19-26. 2008.
20. Nyström, Pär, Emilia Thorup, Sven Bölte, and Terje Falck-Ytter. "Joint attention in infancy and the emergence of autism." *Biological psychiatry* 86, no. 8 (2019): 631-638.
21. Ravindran, Vijay, Monica Osgood, Vibha Sazawal, Rita Solorzano, and Sinan Turnacioglu. "Virtual reality support for joint attention using the Floreo Joint Attention Module: Usability and feasibility pilot study." *JMIR pediatrics and parenting* 2, no. 2 (2019): e14429.
22. Strobl, Maximilian AR, Florian Lipsmeier, Liliana R. Demenescu, Christian Gossens, Michael Lindemann, and Maarten De Vos. "Look me in the eye: evaluating the accuracy of smartphone-based eye tracking for potential application in autism spectrum disorder research." *Biomedical engineering online* 18, no. 1 (2019): 1-12.
23. Voss, Catalin, Nick Haber, and Dennis P. Wall. "The Potential for Machine Learning–Based Wearables to Improve Socialization in Teenagers and Adults With Autism Spectrum Disorder—Reply." *Jama Pediatrics* 173, no. 11 (2019): 1106-1106.
24. Voss, Catalin, Jessey Schwartz, Jena Daniels, Aaron Kline, Nick Haber, Peter Washington, Qandeel Tariq et al. "Effect of wearable digital intervention for improving socialization in children with autism spectrum disorder: a randomized clinical trial." *JAMA pediatrics* 173, no. 5 (2019): 446-454.
25. Voss, Catalin, Peter Washington, Nick Haber, Aaron Kline, Jena Daniels, Azar Fazel, Titas De et al. "Superpower glass: delivering unobtrusive real-time social cues in wearable systems." In *Proceedings of the 2016 ACM International Joint Conference on Pervasive and Ubiquitous Computing: Adjunct*, pp. 1218-1226. 2016.
26. Washington, Peter, Catalin Voss, Nick Haber, Serena Tanaka, Jena Daniels, Carl Feinstein, Terry Winograd, and Dennis Wall. "A wearable social interaction aid for children with autism." In *Proceedings of the 2016 CHI Conference Extended Abstracts on Human Factors in Computing Systems*, pp. 2348-2354. 2016.
27. Washington, Peter, Catalin Voss, Aaron Kline, Nick Haber, Jena Daniels, Azar Fazel, Titas De, Carl Feinstein, Terry Winograd, and Dennis Wall. "SuperpowerGlass: a wearable aid for the at-home therapy of children with autism." *Proceedings of the ACM on interactive, mobile, wearable and ubiquitous technologies* 1, no. 3 (2017): 1-22.
28. Nag, Anish, Nick Haber, Catalin Voss, Serena Tamura, Jena Daniels, Jeffrey Ma, Bryan Chiang et al. "Toward Continuous Social Phenotyping: Analyzing Gaze Patterns in an Emotion Recognition Task for Children With Autism Through Wearable Smart Glasses." *Journal of Medical Internet Research* 22, no. 4 (2020): e13810.
29. Washington, Peter, Natalie Park, Parishkrita Srivastava, Catalin Voss, Aaron Kline, Maya Varma, Qandeel Tariq et al. "Data-driven diagnostics and the potential of mobile artificial intelligence for digital therapeutic phenotyping in computational psychiatry." *Biological Psychiatry: Cognitive Neuroscience and Neuroimaging* (2019).
30. Paolacci, Gabriele, Jesse Chandler, and Panagiotis G. Ipeirotis. "Running experiments on amazon mechanical turk." *Judgment and Decision making* 5, no. 5 (2010): 411-419.

31. Washington, Peter, Emilie Leblanc, Kaitlyn Dunlap, Yordan Penev, Aaron Kline, Kelley Paskov, Min Woo Sun et al. "Precision Telemedicine through Crowdsourced Machine Learning: Testing Variability of Crowd Workers for Video-Based Autism Feature Recognition." *Journal of personalized medicine* 10, no. 3 (2020): 86.
32. Washington, Peter, Haik Kalantarian, Qandeel Tariq, Jessey Schwartz, Kaitlyn Dunlap, Brianna Chrisman, Maya Varma et al. "Validity of online screening for autism: crowdsourcing study comparing paid and unpaid diagnostic tasks." *Journal of medical Internet research* 21, no. 5 (2019): e13668.
33. Hirth, Matthias, Tobias Hoßfeld, and Phuoc Tran-Gia. "Anatomy of a crowdsourcing platform-using the example of microworkers. com." In *2011 Fifth international conference on innovative mobile and internet services in ubiquitous computing*, pp. 322-329. IEEE, 2011.
34. Ipeirotis, Panagiotis G. "Analyzing the amazon mechanical turk marketplace." *XRDS: Crossroads, The ACM Magazine for Students* 17, no. 2 (2010): 16-21.
35. Abbas, Halim, Ford Garberson, Stuart Liu-Mayo, Eric Glover, and Dennis P. Wall. "Multi-modular Ai Approach to Streamline Autism Diagnosis in Young children." *Scientific reports* 10, no. 1 (2020): 1-8.
36. Abbas, Halim, Ford Garberson, Eric Glover, and Dennis P. Wall. "Machine learning approach for early detection of autism by combining questionnaire and home video screening." *Journal of the American Medical Informatics Association* 25, no. 8 (2018): 1000-1007.
37. Duda, Marlena, Jena Daniels, and Dennis P. Wall. "Clinical evaluation of a novel and mobile autism risk assessment." *Journal of autism and developmental disorders* 46, no. 6 (2016): 1953-19
38. Duda, M., N. Haber, J. Daniels, and D. P. Wall. "Crowdsourced validation of a machine-learning classification system for autism and ADHD." *Translational psychiatry* 7, no. 5 (2017): e1133-e1133.
39. Duda, M., J. A. Kosmicki, and D. P. Wall. "Testing the accuracy of an observation-based classifier for rapid detection of autism risk." *Translational psychiatry* 4, no. 8 (2014): e424-e424.
40. Duda, M., R. Ma, N. Haber, and D. P. Wall. "Use of machine learning for behavioral distinction of autism and ADHD." *Translational psychiatry* 6, no. 2 (2016): e732-e732.
41. Paskov, Kelley M., and Dennis P. Wall. "A low rank model for phenotype imputation in autism spectrum disorder." *AMIA Summits on Translational Science Proceedings* 2018 (2018): 178.
42. Wall, Dennis P., Rebecca Dally, Rhiannon Luyster, Jae-Yoon Jung, and Todd F. DeLuca. "Use of artificial intelligence to shorten the behavioral diagnosis of autism." *PloS one* 7, no. 8 (2012): e43855.
43. Wall, Dennis Paul, J. Kosmicki, T. F. Deluca, E. Harstad, and Vincent Alfred Fusaro. "Use of machine learning to shorten observation-based screening and diagnosis of autism." *Translational psychiatry* 2, no. 4 (2012): e100-e100.
44. Washington, Peter, Kelley Marie Paskov, Haik Kalantarian, Nathaniel Stockham, Catalin Voss, Aaron Kline, Ritik Patnaik et al. "Feature selection and dimension reduction of social autism data." In *Pac Symp Biocomput*, vol. 25, pp. 707-718. 2020.
45. Deyo, Richard A., Paula Diehr, and Donald L. Patrick. "Reproducibility and responsiveness of health status measures statistics and strategies for evaluation." *Controlled clinical trials* 12, no. 4 (1991): S142-S158.
46. Paiva, Carlos Eduardo, Eliane Marçon Barroso, Estela Cristina Carneseca, Cristiano de Pádua Souza, Felipe Thomé Dos Santos, Rossana Verónica Mendoza López, and Sakamoto Bianca Ribeiro Paiva. "A critical analysis of test-retest reliability in instrument validation studies of cancer patients under palliative care: a systematic review." *BMC medical research methodology* 14, no. 1 (2014): 8.
47. Polit, Denise F. "Getting serious about test–retest reliability: a critique of retest research and some recommendations." *Quality of Life Research* 23, no. 6 (2014): 1713-1720.
48. Vilagut, Gemma. "Test-retest reliability." *Encyclopedia of quality of life and well-being research* (2014): 6622-6625.
49. Bone, Daniel, Matthew S. Goodwin, Matthew P. Black, Chi-Chun Lee, Kartik Audhkhasi, and Shrikanth Narayanan. "Applying machine learning to facilitate autism diagnostics: pitfalls and promises." *Journal of autism and developmental disorders* 45, no. 5 (2015): 1121-1136.
50. Fusaro, Vincent A., Jena Daniels, Marlena Duda, Todd F. DeLuca, Olivia D'Angelo, Jenna Tamburello, James Maniscalco, and Dennis P. Wall. "The potential of accelerating early detection of autism through content analysis of YouTube videos." *PLOS one* 9, no. 4 (2014): e93533.
51. Kosmicki, J. A., V. Sochat, M. Duda, and D. P. Wall. "Searching for a minimal set of behaviors for autism detection through feature selection-based machine learning." *Translational psychiatry* 5, no. 2 (2015): e514-e514.
52. Levy, Sebastien, Marlena Duda, Nick Haber, and Dennis P. Wall. "Sparsifying machine learning models identify stable subsets of predictive features for behavioral detection of autism." *Molecular autism* 8, no. 1 (2017): 65.
53. Tariq, Qandeel, Jena Daniels, Jessey Nicole Schwartz, Peter Washington, Haik Kalantarian, and Dennis Paul Wall. "Mobile detection of autism through machine learning on home video: A development and prospective validation study." *PLoS medicine* 15, no. 11 (2018): e1002705.
54. Tariq, Qandeel, Scott Lanyon Fleming, Jessey Nicole Schwartz, Kaitlyn Dunlap, Conor Corbin, Peter Washington, Haik Kalantarian, Naila Z. Khan, Gary L. Darmstadt, and Dennis Paul Wall. "Detecting developmental delay and autism through machine learning models using home videos of Bangladeshi children: Development and validation study." *Journal of medical Internet research* 21, no. 4 (2019): e13822.

Differential Privacy Protection Against Membership Inference Attack on Machine Learning for Genomic Data

Junjie Chen[1], Wendy Hui Wang[2] and Xinghua Shi[1]*
[1]*Department of Computer and Informatics Sciences, Temple University, Philadelphia, PA 19122, USA.*
[2]*Department of Computer Science, Stevens Institute of Technology, Hoboken, NJ 07030, USA.*
* *To whom correspondence should be addressed. E-mail: mindyshi@temple.edu*

Machine learning is powerful to model massive genomic data while genome privacy is a growing concern. Studies have shown that not only the raw data but also the trained model can potentially infringe genome privacy. An example is the membership inference attack (MIA), by which the adversary can determine whether a specific record was included in the training dataset of the target model. Differential privacy (DP) has been used to defend against MIA with rigorous privacy guarantee by perturbing model weights. In this paper, we investigate the vulnerability of machine learning against MIA on genomic data, and evaluate the effectiveness of using DP as a defense mechanism. We consider two widely-used machine learning models, namely Lasso and convolutional neural network (CNN), as the target models. We study the trade-off between the defense power against MIA and the prediction accuracy of the target model under various privacy settings of DP. Our results show that the relationship between the privacy budget and target model accuracy can be modeled as a log-like curve, thus a smaller privacy budget provides stronger privacy guarantee with the cost of losing more model accuracy. We also investigate the effect of model sparsity on model vulnerability against MIA. Our results demonstrate that in addition to prevent overfitting, model sparsity can work together with DP to significantly mitigate the risk of MIA.

Keywords: Differential privacy; Membership inference attack; Machine learning; Genomics.

1. Introduction

Genomics has emerged into a frontier of data analytics empowered by machine learning and deep learning, thanks to the rapid growth of genomic data that contains individual-level sequences or genotypes at large scale. To build powerful and robust machine learning models for genomics analysis, it is critical to collect, aggregate, and deposit sufficiently large assembly of genomic data. However, genetic privacy is a growing and legitimate concern that prevents wide sharing and aggregation of genomic data. Since genomic data is naturally sensitive and private, the sharing of such data can potentially disclose an individual's sensitive information such as identity, disease susceptibility or family history.[1,2] The current strategies of protecting genomic privacy is centered around relevant regulations and guidelines (i.e. HIPAA[3]), together with the controlled access of individual-level genomic data (e.g. dbGaP[4]). However, we are in great need of new techniques for protecting genetic privacy toward an overarching goal of achieving trustworthy biomedical data sharing and analysis. Specifically, it is imperative to develop computational strategies to mitigate leakage of genetic privacy including the following two types of privacy leakage:

- *Privacy leakage via sharing data*: an individual's genomic data record may be leaked by sharing raw genomic data or summary statistics data; and
- *Privacy leakage via sharing models*: the information that an individual's genomic data is included in the training dataset of a particular machine learning model, may be leaked by sharing the model.[5]

While most of the prior works focus on the former type of privacy leakage resulted from sharing data,[6–8] in this study, we mainly focus on the latter type of privacy leakage from sharing machine learning models. Several studies have recently showed that trained models might memorize training data and thus disclose privacy of data records.[9,10] Although there exists a wide spectrum of attacks on machine learning models, the *membership inference attack* (MIA)[11] has recently attracted research efforts that induces privacy leakage when sharing machine learning models. More specifically, MIA refers to an attack to infer if the target record was included in the target model's training dataset. MIA has been demonstrated as an effective attack on images and relational data.[5,11,12] However, it remains unclear if MIA is effective on genomic data that significantly differ from conventional data.

Although less explored in genomics study, membership privacy leakage does pose an emerging risk given the increasing application and sharing of machine learning models in genomic data analysis. One particular scenario is that a publicly accessible model trained on valuable patient data may leak the privacy of patient.[13] For example, suppose a cancer treatment center builds a machine model to predict therapeutic responses based on patients' genomic and other biomedical data. The cancer center then releases the trained model to the public (e.g. for publications or depositing the model into a public model repository) or deploys the model as a machine-learning-as-a-service platform (e.g. Amazon Web service, Microsoft Azure, Google Cloud). An adversary may use the model's output to infer if a person, whose genomic data the adversary has access to, is a cancer patient or cancer survivor, and such information may provide the adversary some additional information that can be exploited. Hence, in this study, we will investigate the efficiency of MIA on machine learning models for phenotype prediction based on genomic data, a widely assessed prediction task carried out in agriculture, animal breeding, and biomedical science.

To defend against various attacks including MIA, a few techniques have been developed to mitigate privacy leakage such as homomorphic encryption,[14] federated learning,[15] and differential privacy (DP).[16] While homomorphic encryption and federated learning are mainly used to provide privacy protection for data sharing,[17,18] DP provides a popular solution for publicly sharing information not only about the data[19] but also the models.[20] The idea behind DP is that the query results cannot be used to infer information about any single individual, if the effect of perturbing in the database is small enough.[16] Recently, multiple defense mechanisms against MIA[21–23] have been explored, with DP[16] standing out as an efficient strategy that provides a rigorous privacy guarantee against MIA.[11] Previous studies on imaging data[24,25] have shown that DP is an effective solution for granting wider access to machine learning models and results, while keeping them private. Therefore, we will mainly consider DP as a defense mechanism against MIA, given its theoretical privacy guarantee and its applicability for data and models. In this study, we investigate the effectiveness of using DP as a defense mechanism against MIA for phenotype prediction on genomic data to prevent the risk of sharing two widely-used machine learning methods including Lasso[26]) and convolutional neural network (CNN[27]). The **main contributions** of our study lie in two folds:

First, we investigate the vulnerability of machine learning against MIA on genomic data, and evaluate the effectiveness of using DP as a defense mechanism. Particularly, we evaluate the trade-off between the defense power against MIA and the prediction accuracy of the target model under various privacy settings of DP. Our results show that the relationship between the privacy budget and target model accuracy can be modeled as a log-like curve, and hence there exists a trade-off between privacy and accuracy near the turning point.

Second, we evaluate the effect of model sparsity on privacy vulnerability to effectively defend against MIA. Genomic data is primarily high dimensional, where the feature size is significantly larger than sample size. Hence, adding sparsity (e.g. the regularization terms in Lasso models) to machine learning models is a critical and effective strategy to alleviate the curse of dimensionality and avoid overfitting high-dimensional genomic data. Our results show that model sparsity together with DP can significantly mitigate the risk of MIA, in addition to providing robust and effective models for genomic data analysis.

2. Related Work

Membership inference attack (MIA). MIA is a privacy-leakage attack that predicts whether a given record was used in training a target model based on the output of the target model for the given record.[11] Shokri *et al.*[11] is the first work that defines MIA and inspires a few follow-up studies. For example, Truex *et al.*[28] characterize the attack vulnerability with respect to the types of learning models, data distribution, and transferability. Salem *et al.*[5] design new variants of MIA by relaxing the assumptions of model types and data. Long *et al.*[12] generalize MIA by identifying vulnerable records and indirect inference. While most existing works focus on MIA against discriminative models, relatively fewer works have considered MIA against generative models.[29,30] Liu *et al.*,[31] Song *et al.*[32] and Hayes *et al.*[33] propose new MIA variants against deep learning models including variational autoencoders (VAEs) and generative adversarial networks (GANs). These MIA attacks require only black-box access to a trained model. In practice, many studies usually release their models with white-box access.[17] Such white-box access provides many additional properties of the training models, which make an MIA attack even easier.

Differential privacy (DP). DP[16] has become the most widely-used approach that measures the disclosure of privacy pertaining to individuals. The guarantee of a DP algorithm lies in that anything the algorithm might output on a database containing some individual's information, is almost as likely to have come from a database without that individual's information. DP strategies have been applied to preserve genome privacy in genome-wide association studies

(GWAS).[8] For example, Johnson *et. al*[34] developed privacy-preserving algorithms for computing the number and location of single nucleotide polymorphisms (SNPs) that are significantly associated with certain diseases. Uhlerop *et. al*[7] proposed a method that allows for the release of aggregate GWAS data without compromising an individual's privacy. Various DP mechanisms also have been developed[35] to preserve model privacy, including a logistic regression with DP[36] and a random forest algorithm with DP.[37] Going beyond classic machine learning models, Shokri *et al.*[38] adapted DP to deep neural networks. Abadi *et al.*[25] developed a differentially private stochastic gradient descent (SGD) algorithm for the TensorFlow framework.

3. Methods

In this section, we introduce the methods used in our study, including differential privacy and membership inference attack. The supplementary materials and source code are available at `https://github.com/shilab/DP-MIA.git`.

3.1. *Membership inference attack (MIA).*

As illustrated in **Fig. 1**, MIA assumes that a target machine learning model is trained on a set of labeled samples from a certain population. The adversary utilizes the output of the target model of a given sample to infer the membership of the sample (i.e., the given sample was included in the training dataset of the target model). Formally, let $f_{target}()$ be the target model trained on a private dataset D^{train}_{target} which contains labeled samples $(\mathbf{x}, \mathbf{y})$. The output of the target model is a probability vector $\mathbf{y} = f_{target}(\mathbf{x})$ whose size is the number of classes. Let $f_{shadow}()$ be the shadow model trained on a dataset D^{train}_{shadow}, that is generated by the attacker to mimic the target model $f_{target}()$ (i.e. take similar input and output of the target model). We use the same assumption as in the pioneering work,[11] that the shadow dataset is disjoint from the private target dataset used to train the target model (i.e., $D^{train}_{shadow} \cap D^{train}_{target} = \emptyset$). Let $f_{attack}()$ be the attack model. Its input $\mathbf{x}_{attack}$ is composed of a predicted probability vector and a true label, where the distribution of predicted probability vectors heavily depends on the true label. Since the goal of the attack is membership inference, the attack model is a binary classifier, in which the output 1 indicates that the target record is in the training dataset, and 0 otherwise.

To construct the MIA model, a shadow training technique is often applied to generate the ground truth of membership inference. One or multiple shadow models are built to imitate the target model. In this study, we consider the white-box setting, where the adversary has the full knowledge of the target model including its hyperparameters and network structure. This white-box threat setting reflects the observations that researchers often share their full models and accidentally white-box representations of models may fall into the hands of an adversary via means such as a security breach.

3.2. *Differential privacy (DP)*

DP describes the statistics of groups while withholding individuals' information within the dataset.[16] Informally, DP ensures that the outcome of any data analysis on two databases

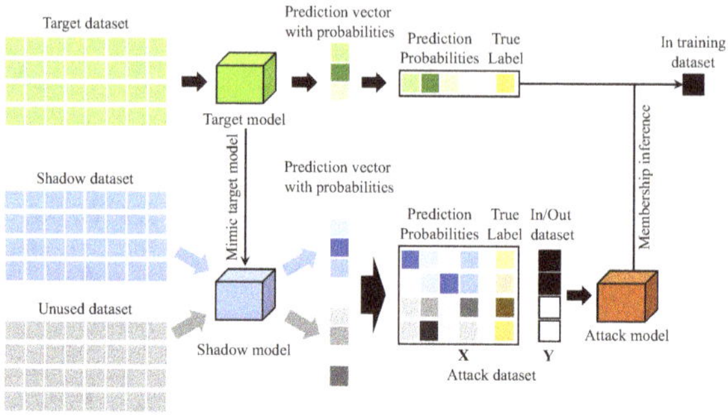

Fig. 1. **An illustration of the membership inference attack.** A record in the target dataset is fed into the target model and outputs a predicted probability vector. The shadow dataset and unused dataset are either simulated or selected from publicly available datasets that have the same distribution as the target dataset. A shadow model is built on the shadow and unused datasets to mimic the target model. The attack dataset is composed of the probability vectors and true labels. The attack model performs a binary classification (in/out) to determine whether a data record is included in the training dataset (in) or not (out).

differing in a single record does not vary much. Formally, a randomized algorithm $\mathcal{M}: \mathcal{D} \to \mathcal{R}$ with domain $\mathcal{D}$ and range $\mathcal{R}$ is (ε, δ)-differentially private if for all subsets of $\mathcal{S} \subseteq \mathcal{R}$ and for all database inputs $d, d' \in \mathcal{D}$ such that $\|d - d'\|_1 \leq 1$ satisfied with $\Pr[\mathcal{M}(d) \in \mathcal{S}] \leq e^{\varepsilon}\Pr[\mathcal{M}(d') \in \mathcal{S}] + \delta$. Here, $\|d - d'\|_1$ requires that the number of records that differ between d and d' is at most 1. The parameter ε is called the *privacy budget* and a lower ε indicates stronger privacy protection. The parameter δ controls the probability that ε-differential privacy is violated. A lower δ value signifies greater confidence of differential privacy. If $\delta = 0$, we say $\mathcal{M}$ is ε-differentially private, and simplify $(\varepsilon, 0)$-differential privacy as ε-differential privacy. A rule of thumb for setting δ is that it is smaller than the inverse of the training data size (i.e. $1/\|d\|$).[25]

4. Experimental Setup

4.1. *Dataset*

We evaluate the effectiveness of DP against MIA on a widely-used yeast genomic dataset.[39] We choose this yeast dataset because it provides an ideal scenario for evaluating the power and privacy of phenotype prediction with well-controlled genetic background and phenotype quantifications, without worries about complex genetic background and the hard-to-defined phenotypes in humans. We extract and filter missing values of the original genotypes[39] and organize them into a matrix that contains genotypes of 28,820 genetic variants or features (with values of 1 and 2 representing the allele comes from a laboratory strain or a vineyard strain respectively) from 4,390 individuals. Similar to any typical human genomic data, the yeast data is high dimensional where the feature size (28,820) is much larger than the sample size (4,390). We also obtain phenotypes or labels of these 4,390 individuals for 20 traits,[39] where we pick the trait of copper sulfate as our target phenotype in this study. This trait represents the growth of yeast by measuring the normalized colony radius at a 48-hour endpoint in agar plates with different concentrations of copper sulfate.[39] Since MIA is mainly launched on classification models, we binarize the quantitative phenotype values as 1 if they are larger than the mean value and 0 otherwise.

4.2. *Implementation of target models*

For the target models of MIA, we implement a Lasso model[26,40] as an example of sparse learning models, and a CNN model[27,41,42] as an example of deep learning model, that are widely-used in analyzing high-dimensional genomics data.

Lasso is a regression analysis method that performs variable selection with a regularization term using ℓ_1 norm.[26] Lasso minimizes the residual sum of squares subject to the sum of the absolute value of the coefficients being less than a constant. The general objective of Lasso is $\min_{\beta} \frac{1}{2}\|y - \mathbf{X}\beta\|_2^2 + \lambda\|\beta\|_1$, where $\mathbf{X}$ is the feature matrix, β is the coefficient vector, and y is the label vector. λ is the coefficient of ℓ_1 norm which controls the model sparsity. Lasso uses an ℓ_1 norm regularization to shrink the parameters of the majority of features to zero which are trivial, and those variants corresponding to non-zero terms are selected as the identified important features. We set λ to be 0 (without model sparsity) and 0.001352 (with model sparsity selected using the glmnet package in R[43]).

CNN has shown its capability to capture local patterns in genomic data.[27] For demonstration, the CNN model in this study includes one CNN layer, followed by a dense layer as an output layer. To improve model robustness, the ℓ_1 norm is applied to all layers to shrink small weights to zero. We utilize a grid search with 5-fold cross validation to find the optimized hyperparameters. In particular, we use two different learning rates (0.01 and 0.001) and two micro batch sizes (50% and 100% of batch size). Regarding ℓ_2 norm clipping which determines the maximum amounts of ℓ_2 norm clipped to cumulative gradient across all network parameters from each microbatch, we use four unique ℓ_2 norm clipping values (0.6, 1.0, 1.4, and 1.8 respectively). For CNN models, we use two different kernel sizes (5 and 9), and two different numbers of kernels (8 and 16). Furthermore, we set the values of λ as 0 (without model sparsity) and 0.001352 (with model sparsity chosen using glmnet[43]).

4.3. *Implementation of DP*

We implement DP on both Lasso and CNN models with and without ℓ_1 norm respectively, using a Python library called TensorFlow-privacy.[44] DP is implemented in these models by adding a standard Gaussian noise on each gradient of the SGD optimizer. The major process for training a model with parameters θ by minimizing the empirical loss function $L(\theta)$ with differentially private SGD, is summarized as the following: at each step of computing the SGD: 1) compute the gradient $\nabla_\theta L(\theta, x_i)$ for a random subset of examples; 2) clip the ℓ_2 norm of each gradient; 3) compute the average of gradients; 4) add some noise in order to protect privacy; 5) take a step in the opposite direction of this average noisy gradient; 6) in addition to outputting the model, compute the privacy loss of the mechanism based on the information maintained by the privacy accountant.

In the DP implementation, the privacy budget is determined by a function that takes multiple hyperparameters as the input. These hyperparameters include the number of epochs, batch size and noise multiplier. The noise multiplier controls the amount of noises added in each training batch. In general, adding more noise leads to better privacy and lower utility. The hyperparameters used in this study are: two epoch sizes (50 and 100), two batch sizes (8 and 16) and five noise multipliers (0.4, 0.6, 0.8, 1.0, 1.2). We set the value of the parameter δ as the inverse of training dataset size (i.e. $\delta = 0.00066489$).[25]

4.4. *Implementation of MIA*

To train differentially private machine learning models and perform MIA, we split the whole dataset into two disjoint subsets, one as the private target dataset and the other one as the public shadow dataset.[11] We randomly split the public shadow dataset, with 80% used for model training and 20% used to generate the ground truth of the attack model. We focus on a white-box model attack, where the target model's architecture and weights are accessible, to evalute how much privacy will be leaked in the worst case. Hence, the shadow model has the same architecture and hyperparameters as the target model. We use an open-source library of MIA[45] to conduct MIA attacks on the Lasso and CNN models. We build one shadow model on the shadow dataset to mimic the target model, and generate the ground truth to train the attack model. The attack dataset is constructed by concatenating the probability vector output from the shadow model and true labels. If a sample is used to train the shadow model, the corresponding concatenated input for the attack dataset is labeled 'in', and 'out' otherwise. For the attack model, we build a random forest with 10 estimators and a max depth of 2. Each MIA attack is randomly repeated 5 times.

4.5. *Evaluation metrics*

Our evaluation metrics include: (1) the mean accuracy of 5-fold cross validation of the target model on the private target dataset, and (2) the mean of MIA accuracy of 5 MIA attacks. The accuracy of the target model on the training (testing, resp.) data is measured as the precision (i.e., the fraction of classification results that are correct) of the prediction results on the training (testing, resp.) data. We follow the pioneering work[11] and use the *attack accuracy* to measure MIA performance. All samples in the target dataset are fed into the attack model.

5. Results

5.1. *Vulnerability of target model against MIA without DP protection*

We investigate the vulnerability of Lasso and CNN models against MIA for predicting the target phenotype without any DP protection. **Table 1** shows the accuracy of the two target models without DP and attack accuracy of MIA on these models. When the models are not sparse ($\lambda = 0$), Lasso and CNN achieves a similar accuracy on the target dataset (0.7910 vs. 0.7894). The attack accuracy of MIA on Lasso and CNN with no sparsity is 0.5728 and 0.5726 respectively, which is better than random guess (0.5) and on a par with MIA accuracy reported in other areas.[11] The high dimensionality of genomic data makes MIA on genomic data much harder than other types of datasets, since shadow models hardly mimic the target model on a high dimensional dataset. Nonetheless, with such a MIA accuracy, the adversary still has a chance to infer the membership in a genomic dataset. After introducing model sparsity by adding an ℓ_1 norm ($\lambda = 0.001352$) to coefficients (in Lasso) or weights (in CNN), the target accuracy of both models is slightly improved and their attack accuracy is reduced.

Table 1. **Model performance against MIA (without DP).**

Methods	Target model Accuracy	Std.	Attack model Accuracy	Std.
Lasso ($\lambda = 0$)	0.7910	0.0123	0.5728	0.0071
Lasso ($\lambda = 0.001352$)	0.7963	0.0157	0.5631	0.0042
CNN ($\lambda = 0$)	0.7894	0.0199	0.5726	0.0059
CNN ($\lambda = 0.001352$)	0.7936	0.0225	0.5628	0.0050

5.2. *Impact of privacy budget on the target model accuracy*

In order to evaluate the impact of DP on the accuracy of the target model, we conduct a grid search to find different privacy budgets and quantitatively investigate the impact of privacy budget. As summarized in **Fig. 2(a)**, we observe that the fitting curve between the privacy budget and the target accuracy can be represented as a log-like curve. The performance of all target models rapidly deteriorates as the privacy budget becomes smaller. When the privacy budget is large, both non-sparse Lasso ($\lambda = 0$) and non-sparse CNN ($\lambda = 0$) models achieve similar target accuracy. Compared with non-sparse models, the target accuracy of sparse Lasso ($\lambda = 0.001352$) and sparse CNN ($\lambda = 0.001352$) models, is downgraded by DP to a more extent even when the privacy budget is large. This is because sparse models only keep coefficients or weights which are higher than λ, and shrink those coefficients or weights that are smaller than λ to 0. Therefore, adding a noise to those large weights will have a more significant impact on the accuracy of the target model.

5.3. *Effectiveness of DP against MIA*

To assess the effectiveness of DP against MIA, we conduct MIA on the target models with different DP budgets. Our results (**Fig. 2(b)**) show that, for Lasso models, the fitting curve between the privacy budget and the target accuracy can be represented as a log-like curve. For CNN, we notice that the curve of attack accuracy is different from that of Lasso, since the attack accuracy becomes unstable when the epsilon is smaller than 10. However, CNN with DP still can provide strong privacy protection. In both Lasso and CNN models, we observe that DP can defend against MIA effectively by perturbing the prediction vector output from the target model, so that the adversary cannot easily infer the membership from such noisy predictions.

According to results in **Fig. 2**, we choose the turning point with a maximum curvature in the log curve as a trade-off between privacy budget and model accuracy. As the privacy budget becomes tight, the target accuracy is rapidly dropped after this turning point, while the target model with DP can still provide sufficient protection against MIA. Based on this observation, we choose the privacy budget of 10 that best addresses the trade-off between privacy and target accuracy in this study.

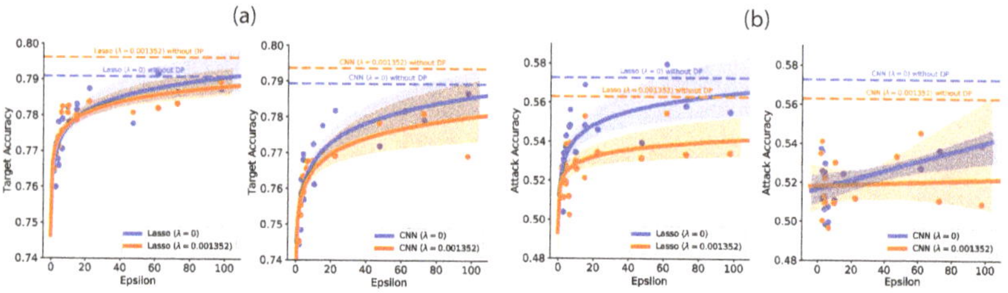

Fig. 2. **Accuracy values of the (a) target model and (b) attack model respectively under various privacy budgets (5-fold cross validation).** Curves indicate the fitted regression lines; shadow areas represent the 95% confidence intervals for corresponding regressions. Horizontal dotted lines represent model performances without DP.

5.4. *Effect of model sparsity*

We investigate the effect of model sparsity by adding an ℓ_1 norm to model coefficients or weights. Due to the large hyperparameter searching space, we only use the value of $\lambda = 0.001352$ for both Lasso and CNN, chosen using the glmnet package.[43] Our results (**Table 1**) show that adding sparsity to a model can improve the accuracy of the target model and reduce the attack accuracy of MIA when DP is not deployed. This is because that on the high-dimensional dataset, a Lasso or CNN model with no sparsity (i.e. $\lambda = 0$) can overfit the training data. However, by introducing model sparsity, the overfitting of the model is reduced, leading to better accuracy of the target model.

We further explore the impact of model sparsity on the accuracy of the target model when DP is deployed. We observe that sparse models with DP have slightly worse model accuracy compared with those non-spare models with DP (**Fig. 2(a)**). This is because each weight in a sparse model is important to prediction results; and any perturbation to these weights can significantly impact model accuracy. We also find that when the privacy budget is smaller than the trade-off (e.g. $\varepsilon < 10$ in our results), the accuracy of the target model is relatively insensitive to model sparsity compared with larger privacy budgets (i.e., $\varepsilon > 10$). Next, we evaluate the impact of model sparsity on the defense power of DP against MIA. As shown in **Fig. 2(b)**, sparse models provide better privacy protection compared with those models without sparsity, given the same DP budget ε.

6. Conclusion

We investigate the vulnerability of trained machine learning models for phenotype prediction on genomic data against a new type of privacy attack named membership inference attack (MIA), and evaluate the effectiveness of using differential privacy (DP) as a defense mechanism against MIA. We find the MIA can successfully infer if a particular individual is included in the training dataset for both Lasso and CNN models, and DP can defend against MIA on genomic data effectively with a cost of reducing accuracy of the target model. We also evaluate the trade-off between privacy protection against MIA and the prediction accuracy of the target model. Moreover, we observe that introducing sparsity into the target model can further defend against MIA in addition to implementing the DP strategy.

Using yeast genomic data as a demonstration, our study provides a novel computational framework that allows for investigating not only the privacy leakage induced from MIA attacks on machine learning models, but also the efficiency of classical defending mechanisms like DP against these new attacks. Nonetheless, there are several limitations of our current study. We are limited to white-box setting where hyperparameters and model architectures are accessible to an adversary in this study. In the future, we will also evaluate black-box access where the adversary simply uses the target model as a black-box for query without any inside information of the model. We will comprehensively explore the relationship between privacy budget and model accuracy, under various combinations of model hyperparameters space and phenotypes. We will apply the framework to analyze large-scale human genomic data where privacy is of a realistic concern. We will investigate whether DP gives unequal privacy benefits to genomes from minority groups compared with those from majority groups. We will investigate other factors (e.g., the number of classes) and conventional genomic analysis (e.g. associations studies, risk prediction) to assess the attack power of MIA and the effectiveness of appropriate defense mechanisms.

References

1. J. C. Lee, D. Biasci, R. Roberts, R. B. Gearry, J. C. Mansfield, T. Ahmad, N. J. Prescott, J. Satsangi, D. C. Wilson, L. Jostins *et al.*, Genome-wide association study identifies distinct genetic contributions to prognosis and susceptibility in crohn's disease, *Nature genetics* **49**, p. 262 (2017).
2. S. Sanchez-Roige, P. Fontanillas, S. L. Elson, A. Pandit, E. M. Schmidt, J. R. Foerster, G. R. Abecasis, J. C. Gray, H. de Wit, L. K. Davis *et al.*, Genome-wide association study of delay discounting in 23,217 adult research participants of european ancestry, *Nature neuroscience* **21**, 16 (2018).
3. R. B. Ness, J. P. Committee *et al.*, Influence of the hipaa privacy rule on health research, *Jama* **298**, 2164 (2007).
4. M. D. Mailman, M. Feolo, Y. Jin, M. Kimura, K. Tryka, R. Bagoutdinov, L. Hao, A. Kiang, J. Paschall, L. Phan *et al.*, The ncbi dbgap database of genotypes and phenotypes, *Nature genetics* **39**, p. 1181 (2007).
5. A. Salem, Y. Zhang, M. Humbert, P. Berrang, M. Fritz and M. Backes, Ml-leaks: Model and data independent membership inference attacks and defenses on machine learning models, *arXiv preprint arXiv:1806.01246* (2018).
6. N. Homer, S. Szelinger, M. Redman, D. Duggan, W. Tembe, J. Muehling, J. V. Pearson, D. A.

Stephan, S. F. Nelson and D. W. Craig, Resolving individuals contributing trace amounts of dna to highly complex mixtures using high-density snp genotyping microarrays, *PLoS genetics* **4**, p. e1000167 (2008).
7. C. Uhlerop, A. Slavković and S. E. Fienberg, Privacy-preserving data sharing for genome-wide association studies, *The Journal of privacy and confidentiality* **5**, p. 137 (2013).
8. X. Shi and X. Wu, An overview of human genetic privacy, *Annals of the New York Academy of Sciences* **1387**, 61 (2017).
9. D. Arpit, S. Jastrzebski, N. Ballas, D. Krueger, E. Bengio, M. S. Kanwal, T. Maharaj, A. Fischer, A. C. Courville, Y. Bengio and S. Lacoste-Julien, A closer look at memorization in deep networks, *ArXiv* **abs/1706.05394** (2017).
10. C. Song, T. Ristenpart and V. Shmatikov, Machine learning models that remember too much, *Proceedings of the 2017 ACM SIGSAC Conference on Computer and Communications Security* (2017).
11. R. Shokri, M. Stronati, C. Song and V. Shmatikov, Membership inference attacks against machine learning models, in *2017 IEEE Symposium on Security and Privacy (SP)*, 2017.
12. Y. Long, V. Bindschaedler, L. Wang, D. Bu, X. Wang, H. Tang, C. A. Gunter and K. Chen, Understanding membership inferences on well-generalized learning models, *arXiv preprint arXiv:1802.04889* (2018).
13. S. Truex, L. Liu, M. Gursoy, L. Yu and W. Wei, Demystifying membership inference attacks in machine learning as a service, *IEEE Transactions on Services Computing* , 1 (2019).
14. J. H. Cheon, A. Kim, M. Kim and Y. Song, Homomorphic encryption for arithmetic of approximate numbers, in *International Conference on the Theory and Application of Cryptology and Information Security*, 2017.
15. J. Xu and F. Wang, Federated learning for healthcare informatics, *arXiv preprint arXiv:1911.06270* (2019).
16. C. Dwork, A. Roth *et al.*, The algorithmic foundations of differential privacy, *Foundations and Trends® in Theoretical Computer Science* **9**, 211 (2014).
17. M. Nasr, R. Shokri and A. Houmansadr, Comprehensive privacy analysis of deep learning: Passive and active white-box inference attacks against centralized and federated learning, in *2019 IEEE Symposium on Security and Privacy (SP)*, 2019.
18. L. Melis, C. Song, E. De Cristofaro and V. Shmatikov, Exploiting unintended feature leakage in collaborative learning, in *2019 IEEE Symposium on Security and Privacy (SP)*, 2019.
19. Y. Wang, J. Wen, X. Wu and X. Shi, Infringement of individual privacy via mining differentially private gwas statistics, in *International Conference on Big Data Computing and Communications*, 2016.
20. Y. Wang, C. Si and X. Wu, Regression model fitting under differential privacy and model inversion attack, in *Twenty-Fourth International Joint Conference on Artificial Intelligence*, 2015.
21. M. Nasr, R. Shokri and A. Houmansadr, Machine learning with membership privacy using adversarial regularization, in *Proceedings of the 2018 ACM SIGSAC Conference on Computer and Communications Security*, 2018.
22. A. Salem, Y. Zhang, M. Humbert, P. Berrang, M. Fritz and M. Backes, Ml-leaks: Model and data independent membership inference attacks and defenses on machine learning models, in *In Proceedings of the 2019 Network and Distributed System Security Symposium (NDSS)*, 2019.
23. J. Jia, A. Salem, M. Backes, Y. Zhang and N. Z. Gong, Memguard: Defending against black-box membership inference attacks via adversarial examples, *arXiv preprint arXiv:1909.10594* (2019).
24. N. Phan, Y. Wang, X. Wu and D. Dou, Differential privacy preservation for deep auto-encoders: an application of human behavior prediction., in *AAAI*, 2016.
25. M. Abadi, A. Chu, I. Goodfellow, H. B. McMahan, I. Mironov, K. Talwar and L. Zhang, Deep learning with differential privacy, in *Proceedings of the 2016 ACM SIGSAC Conference on Com-*

puter and Communications Security, 2016.
26. R. F. Barber, M. Reimherr, T. Schill *et al.*, The function-on-scalar lasso with applications to longitudinal gwas, *Electronic Journal of Statistics* **11**, 1351 (2017).
27. Z. J and T. OG., Predicting effects of noncoding variants with deep learning-based sequence model., *Nat Methods.* **12**, 931 (2015).
28. S. Truex, L. Liu, M. E. Gursoy, L. Yu and W. Wei, Demystifying membership inference attacks in machine learning as a service, *IEEE Transactions on Services Computing* (2019).
29. B. Hilprecht, M. Härterich and D. Bernau, Monte carlo and reconstruction membership inference attacks against generative models, *Proceedings on Privacy Enhancing Technologies* **2019**, 232 (2019).
30. D. Chen, N. Yu, Y. Zhang and M. Fritz, Gan-leaks: A taxonomy of membership inference attacks against gans, *arXiv preprint arXiv:1909.03935* (2019).
31. K. S. Liu, B. Li and J. Gao, Performing co-membership attacks against deep generative models, *arXiv preprint arXiv:1805.09898* (2018).
32. L. Song, R. Shokri and P. Mittal, Membership inference attacks against adversarially robust deep learning models, in *2019 IEEE Security and Privacy Workshops (SPW)*, 2019.
33. J. Hayes, L. Melis, G. Danezis and E. De Cristofaro, Logan: Membership inference attacks against generative models, *Proceedings on Privacy Enhancing Technologies* **2019**, 133 (2019).
34. A. Johnson and V. Shmatikov, Privacy-preserving data exploration in genome-wide association studies, in *Proceedings of the 19th ACM SIGKDD international conference on Knowledge discovery and data mining*, 2013.
35. F. McSherry and K. Talwar, Mechanism design via differential privacy, in *Foundations of Computer Science, 2007. FOCS'07. 48th Annual IEEE Symposium on*, 2007.
36. K. Chaudhuri and C. Monteleoni, Privacy-preserving logistic regression, in *Advances in Neural Information Processing Systems*, 2009.
37. A. Patil and S. Singh, Differential private random forest, in *2014 International Conference on Advances in Computing, Communications and Informatics (ICACCI)*, 2014.
38. R. Shokri and V. Shmatikov, Privacy-preserving deep learning, in *Proceedings of the 22nd ACM SIGSAC conference on computer and communications security*, 2015.
39. J. S. Bloom, I. Kotenko, M. J. Sadhu, S. Treusch, F. W. Albert and L. Kruglyak, Genetic interactions contribute less than additive effects to quantitative trait variation in yeast, *Nature Communications* **6**, p. 8712 (2015).
40. J. Chen and C. Nodzak, Statistical and machine learning methods for eqtl analysis, in *eQTL Analysis*, (Springer, 2020) pp. 87–104.
41. J. Chen and X. Shi, A sparse convolutional predictor with denoising autoencoders for phenotype prediction, in *Proceedings of the 10th ACM International Conference on Bioinformatics, Computational Biology and Health Informatics*, 2019.
42. J. Chen and X. Shi, Sparse convolutional denoising autoencoders for genotype imputation, *Genes* **10**, p. 652 (2019).
43. G.-X. Yuan, C.-H. Ho and C.-J. Lin, An improved glmnet for l1-regularized logistic regression, *The Journal of Machine Learning Research* **13**, 1999 (2012).
44. A. Galen, C. Steve and P. Nicolas, Tensorflow privacy: Library for training machine learning models with privacy for training data `https://github.com/tensorflow/privacy`, (2019), Accessed: 2020-01-30.
45. K. Bogdan and Y. Mohammad, Mia: A library for running membership inference attacks against ml models `https://github.com/spring-epfl/mia`, (2019), Accessed: 2020-01-30.

Making Compassionate Use More Useful: Using real-world data, real-world evidence and digital twins to supplement or supplant randomized controlled trials

Dov Greenbaum JD PhD
Zvi Meitar Institute for Legal Implications of Emerging Technologies, IDC Herzliya
Herzliya, Israel
Harry Radzyner Law School, IDC Herzliya
Herzliya, Israel
Molecular Biophysics and Biochemistry, Yale University
New Haven, CT, USA
Email: dov.greenbaum@yale.edu

The coronavirus pandemic has placed renewed focus on expanded access (EA) programs to provide compassionate use exceptions to the waves of patients seeking medical care in treating the novel disease. While commendable, justifiable, and compassionate, EA programs are not designed to collect the necessary vital clinical data that can be later used in the New Drug Application process before the U.S. Food and Drug Administration (FDA). In particular, they lack the necessary rigor of properly crafted and controlled randomized controlled trials (RCT) which ensure that each patient closely monitored for side effects and other potential dangers associated with the drug, that the data is documented, stable and are traceable and that the patient population is well defined with the defined target condition. Overall, while RCTs is deemed to be of the most reliable methodologies within evidence-based medicine, morally, however, they are problematic in EA programs. Nevertheless, actionable data ought to be collected from EA patients. To this end, we look to the growing incorporation of real-world data real-world evidence as increasingly useful substitutes for data collected via RCTs, including the ethical, legal and social implications thereof. Finally, we suggest the use of digital twins as an additional method to derive causal inferences from real-world trials involving expanded access patients.

Keywords: Real-World Data, Real-World Evidence, Randomized Clinical Trials, Randomized Controlled Trials, FDA, Bioethics, Digital Twin, Machine Learning, GAN

1. Introduction

1.1 Compassionate use

Compassionate use is a catchall lay term[1] for various legal shortcuts in providing access to experimental and limited-access medications.[2] Many jurisdictions worldwide provide for different levels of compassionate use of in-clinical-trial or unapproved pharmaceuticals under varied legal oversight by their respective regulatory bodies.[3] Broadly, these regulatory programs provide limited access exceptions —alternative legal means to access that missed opportunity— particularly for desperate patients who can't otherwise legally obtain a medical product. Most commonly when a patient is unable to join an ongoing clinical trial.

These loopholes have legal limitations. In the United States, for example, there is no constitutional right to compel access to said pharmaceuticals, even for terminally ill patients.[4] It remains up to the various stakeholders in the process, such as the doctors, pharmaceutical companies, institutional review boards and regulators to decide whether to help the patient.[5] In some cases, courts have allowed pharmaceutical companies to terminate access even while patients are still using the drug, arguably effectively.[6]

In the US there are several compassionate use programs including a federal Expanded Access (EA) program that is ultimately administered by the Food and Drug Administration (FDA) for medical products under Investigational New Drug Applications.[7] The EA program is distinct from the similarly sounding and acronymed, but rarely implemented[8] Emergency Use Authorization (EUA) which allows the FDA to facilitate broad access to an unapproved or differently labeled drug during a declared state of emergency, such as a pandemic;[9] in contrast to the 'effectiveness' standard for FDA approval under conventional conditions, EUAs require a much lower 'may be effective' standard to be approved.[10] EUA access to medication circumvents much of the minimal infrastructure of EA access, and so is not part of this analysis.

The FDA's EA program, enacted in 1987,[11] sought to codify a long-standing ad hoc system.[12] Per 21 CFR 312.300 et seq, the FDA was tasked with facilitating "the availability of such drugs to patients with serious diseases or conditions when there is no comparable or satisfactory alternative therapy ..." If preconditions are met, the FDA allows for distribution of the drug even prior to market approval of the drug.[13] These access programs are not necessarily small or limited in size or scope: in some cases, thousands of patients were provided investigational drugs prior to their final FDA approval.[14]

The approval process for EAs can be relatively onerous. It requires a physician to sign on to the project, the acquiescence of the drug company to provide the drug, and the eventual approval of an institutional ethics review board (IRB) and the FDA. The patient must have exhausted all their options before an EA opportunity is even considered. Federal and state laws also provide for even less onerous paths to access investigational drugs through various Right to Try (RTT) regulations.[15] However, in contrast to EAs with their at least tenuous ties to FDA oversight, RTT wholly abandons the FDA's gatekeeper role, requiring no IRB (as per the federal statute, although state statutes vary) or any FDA approval for the requested access, just the approval of the treating physician and the drug manufacturer.

In EAs, given the possible negative outcomes, both the doctor and the drug company are disincentivized to approve a Hail Mary use of an unproven drug for an individual that otherwise did not qualify to be part of a clinical trial. Some manufactures fear both the repercussions to their subsequent new drug application (NDA) as well as bad PR given the probability of poor patient outcomes. (Historically, the former fear has been unfounded; there have been less than a handful of cases where an EA program had a negative effect on the drug labeling.[16]) As such, only a percentage of requests ever end up at the final step of seeking FDA approval. The FDA approves more than 98% of all requests that reach the threshold, depending on the year, of the around a thousand per year expanded access requests[17]. The FDA has even recently mandated additional efforts to further facilitate access to the EA programs. [18] A website was recently developed to facilitate the process for patients and their advocates.[19]

But even if the EA programs do not have a proximate effect on the NDA or the pharmaceutical company's bottom line, they ultimately take away limited resources and possibly even consume limited drug supply that could have gone to additional patients within a structured clinical trial.[20] Moreover, EA efforts don't produce much useful data for the final drug approval. Thus, while such programs may be immediately helpful for a small number of desperate patients, they are often unhelpful for the much larger group of patients that will benefit from the 75% of EA drugs that are eventually approved by the FDA.[21]

1.2 Compassionate use during the pandemic

The coronavirus pandemic has placed renewed focus on the FDA's EA programs. While commendable, justifiable, and compassionate, from a utilitarian point of view —which is an underlying philosophy for other FDA regulations as well[22]— EAs are arguably wasted opportunities and wasted resources. EA programs are rarely able to collect the necessary vital clinical data that can be later used in the New Drug Application (NDA) process before the U.S. Food and Drug Administration (FDA). In particular, they lack the necessary rigor of properly crafted and controlled randomized controlled trials (RCT) which ensure that each patient is closely monitored for side effects and other potential dangers associated with the drug, that the data is documented, stable and are traceable and that the patient population is well defined with the defined target condition.

1.3 What is an RCT?

The FDA considers the RCT to be the best research program for use in generating data for an NDA for a number of reasons, including: (i) RCTs are optimally largely separate from routine clinical practice without its concomitant confusing data. Further, (ii) through the rigorous nature of its development, the RCT is specifically designed to control variability and to maximize data quality. And, in contrast to EA programs, (iii) RCTs have restrictive eligibility to limit participants to certain characteristics and homogeneity such that the detection of an effect of a drug, if any, is more concretely determinable.[23] Although this can also become a problem when an NDA is approved and an untested portion of the population reacts unexpectedly. Also, (iv) there is division between the research and the clinical through particular procedures and protocols, data collection systems and even the use of non-clinical personnel.

It is because of these central characteristics that RCTs —thought to have been around at least since 18th century when James Lind conducted controlled experiments relating to scurvy— are the universal gold standard in establishing efficacy and safety data for an entire population, and trusted to answer the important NDA questions: does the drug actually work;

do the benefits of the drug outweigh the risk; and, what is the optimally safe dosage and regimen. Importantly, for FDA labelling, the evidence must be fully supportive of the conclusions. The RCT has long provided that necessary support. Nevertheless, RCTs are evolving. Versions now include hybrid designs that collect less standardized data, as well as pragmatic-styled trials that may more closely reflect clinical rather than research standards.

1.3 EA data and NDAs

EA data has historically not been seen as particularly relevant to the NDA application.[24] In contrast to RCTs, EA data is currently not seen to be fully supportive of the necessary conclusions: they are not well controlled, patients are less well defined, neither the participants not the researchers are blinded making it harder to support casual inferences, adherence to regimens are far from assured, and they lack the organizational support of standard RCTs with their monitoring and evaluations.[25] And while there have already been some efforts to include EA data within the regulatory review of therapeutics, the data is often weak.[26]

That ought to change. With an increase in the frequency of requests for EAs, the FDA ought to consider both practical changes in the way data is, if at all, systematically collected from EA programs, and regulatory changes that would allow this new data collection to be better included in an NDA.

One possibility is the development of EA programs designed to effectively collect relevant and even actionable real-world data (RWD) —which can originate from non-standard data sources. Data collected from EA participants, both prospectively and retrospectively could potentially be used as real-world evidence (RWE) that would support efficacy and safety determinations applicable to the FDA drug approval process.

2. Real-World Information

2.1 Real-world data in trials

The immediacy and urgency of the COVID-19 pandemic has resulted in a growing appreciation for the need to collect more data faster. Even without the need/opportunity to collect data from EA programs, RCTs are inherently tedious to design and implement. The use of RWD to create RWE would provide an additional source of usable data to help push the regulatory decision-making process forward as well as providing valuable post-market data. The growing push to include RWE[27] has been proposed for pharmaceuticals, vaccines and medical devices.[28]

Pharmaceutical RWD falls across a broad spectrum of confidence. The data is rarely robust enough to allow for casual inferences, especially difficult when the treatment effects from the drug are not large in general. But RWD, even in EA situations, do not have to be poor versions of data collected via RCTs. EA-based trials can be designed to create RWD with sufficient confidence levels that they can be included in the new drug application, if not even replace RCTs. Non-RCT trials have already been proposed and used in the NDA process: these include non-interventional clinical observational-type studies, historical retrospective analyses, and pragmatic trials that are more clinical than research in nature.

The incorporation of RWD and RWE is mandated. The 21st Century Cures Act (Cures Act) requires the FDA to set "standards and methodologies for collection and analysis of real-world evidence" while providing broad leeway to find other applicability for this type of data. Section 505F of the Cures Act specifically defines this RWE broadly any "data regarding the usage, or the potential benefits or risks, of a drug derived from sources other than randomized clinical trials." As per the Cures Act, the FDA is obligated to seek alternatives to the expensive, narrow and rigid RCT paradigm and, among other efforts, incorporate RWE into its approval process. The FDA and other third parties have already developed numerous guidance documents,[29] initiatives and frameworks including apps, to this end.[30]

However, without reassessing how RWD is collected and extracted, and without designing robust EA programs that focus on extracting actionable, reliable and transparent RWE from RWD we are far off from achieving the Act's goals. Currently there is no unified system that allows evaluation and quality comparison across various RWD sets; we are far from replacing the RCT through RWD extracting trials, although there are efforts.[31]

2.2 Real-world data and real-world evidence

The FDA, in its guidance documents draws a distinction between RWE and the RWD that support it. In particular, RWE must be evaluated in light of the reliability and relevance of the underlying data. To this end, the FDA defines RWD as "data relating to patient health status and/or the delivery of health care routinely collected from a variety of sources," and RWE as "the clinical evidence about the usage and potential benefits or risks of a medical product derived from analysis of RWD."[32]

RWD can be extracted from numerous sources, including: electronic health records (EHR), although EHRs will typically only collect major events, and not daily relevant data outside of hospitalization,[33] various data associated with the administrative provision of health care, including billing, claims, and insurance data, self-identifying information provided by individuals to patient registries, groups, social media pages and the like, and information collected by professional and recreational internet of things (Iot) devices ranging from insulin pumps to Apple Watches and Fitbits.

RWD fits in well with continually expanding universe of Big Data: Broadly speaking, big data is defined by at least 4 V's: velocity, variety, volume, and (lack of) veracity. RWD is similarly defined, it can be collected in real, or near-real time, from a host of diverse sources, providing innumerous data points, but due to its sources and structure, lacks the veracity of standard clinical data. Notably, the FDA does not yet see RWE as sufficient to stand on its own, but rather as simply further support and additional data collected through "randomized trials (e.g., large simple trials, pragmatic clinical trials) and observational studies (prospective or retrospective)."

Moreover, the use or RWD may not, as per the FDA, always result in RWE that can be used directly toward a clinical drug trial. Still, the FDA already sees value in RWD in optimizing the trial process itself, even if it cannot be used towards an outcome. For example, in generating new hypothesis to test via RCTs, identifying relevant biomarkers and prognostic indicators, or assessing various inclusion/exclusion criteria.

But these limitations can be overcome. The use of RWD within clinical drug trials themselves, while still limited, is expanding in many different jurisdictions and have even been incorporated into a handful of NDA submissions. To its credit, the UK, under the Early Access to Medicines Scheme, was the first to allow RWD from a compassionate use program to be officially considered as part of regulatory submission.

The FDA has published numerous recent papers relating to RWD and RWE, signaling their intent to promote their use.[34] The FDA has shown additional interest in this area in the recently published a Funding Opportunity Announcement that seeks to examine a number of potential applications of RWD in the drug regulatory process.[35] Even more recently, the FDA further announced its participation in the COVID-19 Diagnostics Evidence Accelerator organized by the Reagan-Udall Foundation.

2.2 Real-world limitations

The RWE and RWD FDA frameworks are intended, as per the Cures Act to be developed in consultation with stakeholders. Outside of their incorporation into EA-based clinical trials, RWD and RWE are thought to also be useful expanding clinical trials into more rare diseases[36] and cancers[37-39] where RCTs are harder to develop, developing tools to estimate the effectiveness of treatments, increasing the diversity of the clinical trials in general, expanding the usability of the results into new interventions and comparisons of alternative interventions, creating actionable data where RCT opportunities are limited, adding evidence from broader studies, learning more about safety concerns in the broader population, doing more comprehensive risk/benefit analyses beyond simply efficacy, and appreciating how the drug actually acts under less constrained conditions than the ones provided in an RCT.

To accomplish this, major limitations associated with RWE and RWD need to be attended to, especially lack of structure and standardization, biases, and confounding factors, and concerns with clarity and the relevant clinical granularity from the sources. This is non-trivial: Biases and confounding factors are typically dealt with through randomization, eligibility criteria and follow-up audits, which are much harder to accomplish with RWD.

These biases include, information biases stemming from errors in data capture, lack of standardization, or incomplete data retrieval due to holes in the data, limited access to the relevant data; attrition biases resulting from patients that wholly drop out of any structured surveillance, compliance or performance bias, given the unstructured nature of RWD collection, patients might not be as incentivized to adhere as effectively to treatment; confounding biases as a result of the heterogeneity of the patients, including but not limited to patient demographics, their environment, their health environment including their provider and clinical settings, the use of alternative therapies, and the patients' comorbidities; immortal time biases when we cannot ascertain when patients began being tracked, and selection bias of the patients and the choice of therapies made by their physicians, among many other potential biases. RWD also creates, at least in the outset, additional costs especially relating to sourcing, capturing, standardizing, cleaning, integrating, analyzing via data science tools, bioinformatics, natural language processing and machine learning the data, and even encrypting the data such that patient privacy is protected.

Patient privacy is another non-trivial concern. RWE requires the collection of a wide variety of hard to anonymize datasets such as insurance records, social media information and electronic health records. Anonymization of this data can raise costs and hinder its utility. And re-identification from correlating data with other pubic databases is always a possibility creating additional regulatory hurdles: The European General Data Protection Regulation (GDPR) is particularly onerous here.

3.0 Making RWD Work

But for all of its limitations and complications RWD may become an invaluable source of data for pre-clinical drug trials and NDAs, especially RWD culled from EA related trials. Desperate times allow for the development and implementation of methods and technologies that heretofore have not found mainstream approval. A number of these technological and regulatory wallflowers have recently found greater traction, including distance learning, telemedicine, universal basic income,[40] and potentially now RWE and RWD.

RWD is a technology that has been waiting for an opportunity to spread its metaphorical wings. More than just providing more of the same, RWD is potentially less demographically homogenous than standard RCTs which often underrepresent minorities and often do not represent a spectrum of clinical presentations, and can miss additional useful datapoints that might not be collected within the rigid structure of the RCT. RCTs are expensive, unwieldy and often difficult to implement especially in low-incidence diseases.

Now is the opportunity for the FDA to set standards for data collection related to EA programs with a focus on reducing design flaws and biases. Under EA programs, the FDA can incorporate requirements on both the patient, the managing physician and the pharmaceutical company to follow guidelines to limit the number of incomplete data sets and variabilities in data collection.

Additionally, data should be collected and curated such that it is meaningful and actionable. Standards such as ICD-10[41] and HCPCS[42] should be employed when applicable, and data should be expedited such that it can be collected and shared in real-time, and transparent in that it can be reproduced and replicated, in addition to being verifiable by auditors.[43] This last is especially important as outside of the controlled environment of clinical studies, there is a concern that physicians and/or patients will consciously or subconsciously cherry-pick data to report.

One of the greatest limitations of RWE relates to the difficulty in making causal inferences from the data, especially when there is no placebo data to counterbalance the collected data. The incorporation of placebos are especially ethically problematic when the RWE comes from compassionate use programs.

3.1 Digital twins

One way to circumvent this particular limitation is through the use of digital twins, i.e., the development of in silico representations of real-world objects. In silico digital twins began being developed in earnest around the turn of the century, originally conceived for NASA space vehicle product lifecycle management[44] they have been heretofore used primarily in engineering fields wherein devices can be stress-tested without building a second device.[45]

The engineering concept was designed such that the virtual and physical systems would be linked throughout the entirety of the product lifecycle, from creation, through production and operation and eventually disposal.[46] Nevertheless, digital twins don't have to necessarily be linked and mirror the physical device exactly. They can also be predictive wherein a range of potential future states can be created and tested independently on the digital twin.

This ability to run computer simulations on virtual objects allows devices to be tested even before they are fully built and/or deployed. Digital twins range from small devices to even cities, with varying degrees of complexity. The virtual system is designed to mirror, as closely as possible all the complexities of the original system.

Digital twins are non-trivial to design and they require complicated modeling, advanced computing power and huge amounts of data to accurately reflect the physical object. The complexities are further exacerbated when creating a digital twin of a living organism that also exists and operates in a complex dynamic living environment. Nevertheless, published patent applications[47] and some early papers[48] suggest that there are significant efforts in the early development of patient digital twins.

One promising example are simulator engines being developed by Unlearn.AI with the goal of simulating patient populations, disease progressions, and/or predicted responses to various medical treatments.[49] The company uses an unsupervised machine learning model called a Conditional Restricted Boltzmann Machine (CRBM) to simulate detailed patient trajectories.[50] These digital twins are intended specifically for the treatment of neurodegenerative diseases, including, Alzheimer's Disease and Multiple Sclerosis. Termed digital subjects by Unlearn.AI, they provide "... a computationally generated clinical trajectory with the same statistical properties as clinical trajectories from actual patients. ... they present no risk of revealing private health information and make it possible to quickly simulate patient cohorts of any size and characteristic."[51] Others have developed synthetic individuals from real-life data to predict aging and mortality trajectories via tracking predicted health deficits that accumulate through damage.[52]

Succinctly, in the medical context, demographic data, family history and other unstructured health data, electronic health records, laboratory results, physiologic measurements and insurance data, imaging and signal data as well as substantial 'omic data (genome, microbiome, transcriptome, proteome, metabalome and others[53] (e.g., all the stuff that can also be termed RWD) can be collected and used to develop as close as representation to the original patient as possible, for example, via machine learning technologies.[54]

In a number of examples, generative adversarial networks (originally developed by Goodfellow et al[55] and heretofore used primarily in imaging processing to generate synthetic content) have been proposed.[56] In one early study, a GAN was employed to predict clinical outcomes via the determination of the trajectory of laboratory tests, based on data culled from thousands of patients.[57] In another, a GAN was used to create synthetic electronic medical records that closely fit real data.[58]

However, with all their promise, digital twins and similar predictive efforts are still limited by the heterogeneity of data and inability to generalize across widely different datasets ostensibly collecting the same data. Further, the data privacy concerns discussed above also relate to digital twin development, especially as data needs to be collected that not only relates to the physical patient and their in silico twin, but also the development of the technology itself will require the collection of data on various non-trial related individuals in optimizing the algorithms.

4.0 Conclusions

The ability to extract casual inferences is fundamental to clinical trials. The validity of these inferences are bolstered by many of the attributes of clinical trials that have become de facto in the industry, including: double blind randomized controls via placebos, homogenous sample populations, transparency, standardization and oversight. The recent pandemic has highlighted the limitations of these types of trials, especially when regulatory bodies provide broad expanded access to promising therapies in the second and third stages of their clinical trials.

While our humanity demands that we do our utmost to help patients in need, the provision of unproven trial drugs to patients that cannot be included in the data-creating trials creates numerous practical, legal and ethical problems. Practically, EA programs take potentially scarce resources, such as trial drugs that do not yet have dedicated manufacturing platforms, as well as clinical personnel, away from a clinical trial. This redistribution of scarce resources, often politically motivated, or influenced by the potential for both positive[59] and negative public relations, is aggravated by the current practical and legal inability to parlay the potential information collectable by those receiving compassionate use of the drugs into the dataset of the clinical trial and toward the NDA.

COVID-19 has created increasing demand for EAs and the resulting data applicability to NDAs is unclear at this early stage.[60] In one example, Gilead Sciences was flooded with expanded access requests for an unapproved investigational drug Veklury (remdesivir), which it initially had to halt, due to the overwhelming nature of the demand and its effect on the concurrent clinical trials. Subsequently, with the declaration of a public health emergency, the FDA issued an EUA for remdesivir for hospitalized patients with COVID-19.[61] While the FDA does not have the authority to force a pharmaceutical company to provide the drug,[62] its likely more difficult from a PR standpoint to refuse an EUA.

While the risk benefit calculus for the individual patients often favors granting EUAs[63] as well as EAs. From a utilitarian viewpoint of maximizing social benefit, they can be seen as problematic as they place the immediate and statistically unproven need of the few receiving expanded access medications ahead of the need of a general population that may benefit from the final approved drug. (Notably, COVID-19 has resulted in other wasted resources due to the fast and furious rush to publish in the field.)[64,65] If the EA programs and their associated inability to generate actionable data push off the NDA, then there is the possibility for substantial harm to the broader population, especially when the drug is a promising opportunity to minimize the effects of the pandemic.

But EUA and EA patients need not be deprioritized. There is a regulatory solution: the decision by regulatory agencies to develop usable methods to extract RWD from EA programs that have been heretofore axiomatically unable to approach the rigor of an RCT to create RWE that can be used toward an NDA. There is already a regulatory drive to find opportunities to include RWD and RWE into the NDA process, for example in vaccine development.[66] Clearly, pandemics provide the opportunity for regulatory agencies like the FDA to set up standards and best practices as to how to extract the most useful and actionable data from the heterogenous often mortally-ill patients that access pre-clinical drugs through the various expanded access programs.

Moreover, with the moral problems with providing placebos to expanded access patients, the FDA and other regulatory agencies should pursue the emerging area of biomedical digital twins. Already a maturing technology in areas such as aerospace, digital twins and other in silico biomedical data predicting and data synthesizing technologies can provide opportunities to test placebos on virtual rather than real patients to minimize confounding factors typically associated with RWD and RWE and finally enabling the enhanced derivation of casual inferences from real-world non-clinical data. Such efforts might even promote enhanced internal and external validity,[67] something that we cannot as yet accomplish from RCTs alone.

References

[1] *US—Securities and Exchange Commission v. Ferrone*, No. 11 C 5223 (N.D. Ill. Oct. 10, 2014).
[2] *Abigail Alliance for Better Access v. Eschenbach*, 469 F.3d 129 (D.C. Cir. 2006).
[3] Mussa Rahbari & Nuh N Rahbari, Compassionate use of medicinal products in Europe: current status and perspectives *Bulletin of the World Health Organization* 2011;89:163-163.
[4] *Abigail Alliance v. von Eschenbach* (US 2008).
[5] *Cacchillo v. Insmed Inc.*, Civil Action No. 1: 10-CV-01199 (N.D.N.Y. Feb. 19, 2013).
[6] *Abney v. Amgen, Inc.*, 443 F.3d 540 (6th Cir. 2006).
[7] FDA, Expanded Access Program Report May 2018.
[8] Rome, B.N. and Avorn, J., 2020. Drug evaluation during the Covid-19 pandemic. *NEJM*
[9] §§ 564, 564A, 564B of the Federal Food, Drug, and Cosmetic Act
[10] Office of the Commissioner, et al., Guidance Document Emergency Use Authorization of Medical Products and Related Authorities, January 2017.
[11] 21 CFR 312 section & 21 CFR part 812
[12] Zoffer W. Recent Legislation that Secured a 'Right to Try' Unapproved Drugs: Why the 'Fuss' Over a 'Fix' of What 'Ain't Broke'?. Wake Forest J. of L. & Pol., 2019
[13] 21 CFR 312.320
[14] Young, F.E., et al., 1988. The FDA's new procedures for the use of investigational drugs in treatment. *JAMA*, *259*(15), pp.2267-2270
[15] Trickett Wendler, et al., Right to Try Act of 2017 (S. 204, Pub.L. 115–176)
[16] Jarow, J.P. and Moscicki, R., 2017. Impact of expanded access on FDA regulatory action and product labeling. *Therapeutic innovation & regulatory science*, *51*(6), pp.787-789.
[17] McKee, A.E., et al. P., 2017. How often are drugs made available under the Food and Drug Administration's expanded access process approved?. *J. of Clinical Pharm.*, *57*, pp.S136-S142.
[18] Press Release, FDA announces Project Facilitate to assist physicians seeking access to unapproved therapies for patients with cancer, 2019

[19] Reagan-Udall Foundation for the FDA, Expanded Access Navigator
[20] Darrow, J.J., et al., 2015. Practical, legal, and ethical issues in expanded access to investigational drugs. *NEJM 372*(3), p.279.
[21] Miller, J.E., et al., 2017. Characterizing expanded access and compassionate use programs for experimental drugs. *BMC research notes*, *10*(1), p.350
[22] Warchol, J.M., 2019. Should a Law Governing the Pharmaceutical Market Be Ethically Examined Based on Its Intent or Its Practical Applications?. *AMA J. of Ethics*, *21*(8),661-667.
[23] FDA, Framework for FDA's Real-World Evidence Program 2018.
[24] Polak, T.B., van Rosmalen, J. and Uyl–de Groot, C.A., 2020. Expanded Access as a source of real-world data: An overview of FDA and EMA approvals. *British J. of Clin. Pharm.*
[25] Chapman, C.R., et al. 2019. What compassionate use means for gene therapies. *Nat. Biotech.*, *37*(4), pp.352-355.
[26] Kurd, R., et al., 2020. Compassionate Use of Opaganib For Patients with Severe COVID-19. *medRxiv*.
[27] NAS, 2018. "Examining the Impact of Real-World Evidence on Medical Product Development: II. Practical Approaches: Proceedings of a Workshop—in Brief."
[28] Sharon Hensley Alford et al., 2020 Medical Device Innovation Consortium Real-World Clinical Evidence Generation: Advancing 5 Regulatory Science and Patient Access for In Vitro Diagnostics (IVDs)
[29] U.S. Dept. HHS et al., 2019. Submitting Documents Using Real-World Data and Real-World Evidence to FDA for Drugs and Biologics,
[30] Baumfeld Andre, et al., 2019. Trial designs using real-world data: The changing landscape of the regulatory approval process. *Pharmacoepidemiology and Drug Safety*.
[31] Gliklich, R.E. and Leavy, M.B., 2020. Assessing Real-World Data Quality: The Application of Patient Registry Quality Criteria to Real-World Data and Real-World Evidence. *Therapeutic Innovation & Regulatory Science*, pp.1-5.
[32] FDA, Framework for FDA's Real-World Evidence Program 2018
[33] CDER, et al., 2017. Use of Electronic Records and Electronic Signatures in Clinical Investigations Under 21 CFR Part 11 – Questions and Answers, June
[34] Klonoff DC. 2020 The new FDA real-world evidence program to support development of drugs and biologics. *Journal of diabetes science and technology*. 14(2):345-9.
[35] FDA, 2020. Exploring the use of Real-World Data to Generate Real-World Evidence in Regulatory Decision-Making (U01) Clinical Trials.
[36] EMA. Public Assessment Report: Kolbam. 2015 (Exceptional Circumstances)
[37] Ison G, Beaver JA, McGuinn WD, et al. FDA approval: uridine triacetate for the treatment of patients following fluorouracil or Capecitabine overdose or exhibiting early-onset severe toxicities following Administration of these Drugs. *Clin Cancer Res*. 2016;22(18):
[38] Nowakowski, G.S., et al, 2020. RE-MIND study: A propensity score-based 1: 1 matched comparison of tafasitamab+ lenalidomide (L-MIND) versus lenalidomide monotherapy (real-world data) in transplant-ineligible patients with relapsed/refractory (R/R) diffuse large B-cell lymphoma (DLBCL).
[39] Gyawali, B., et al., 2017. Real-world evidence and randomized studies in the precision oncology era: the right balance. *JCO Precision Oncology*, *1*, pp.1-5.
[40] Dov Greenbaum, After decades of being unappreciated, the pandemic could give distance learning a new lease on life, Calcalist CTECH June 9, 2020
[41] WHO, 2004. *International statistical classification of diseases and related health problems* (Vol. 1).
[42] Centers for Medicare & Medicaid Services, 2003. *Healthcare Common Procedure Coding System (HCPCS)*. Centers for Medicare & Medicaid Services.
[43] Schneeweiss S, Glynn RJ. 2018 Real-world data analytics fit for regulatory decision-making. *American journal of law & medicine*. May;44(2-3):197-217.

[44] Glaessgen E, Stargel D. The digital twin paradigm for future NASA and US Air Force vehicles. In53rd AIAA/ASME/ASCE/AHS/ASC structures, structural dynamics and materials conference 20th AIAA/ASME/AHS adaptive structures conference 14th AIAA 2012 Apr 16 (p. 1818).
[45] Grieves MW. 2005 Product lifecycle management: the new paradigm for enterprises. *International Journal of Product Development*.2(1-2):71-84.
[46] Grieves M, Vickers J. Digital twin: Mitigating unpredictable, undesirable emergent behavior in complex systems. In Transdisciplinary perspectives on complex systems 2017 (pp. 85-113). Springer, Cham.
[47]EP3646331A1, Zimmerman, J., Methods and systems for generating a patient digital twin General Electric Company
[48] Bruynseels, K., Santoni de Sio, F. and van den Hoven, J., 2018. Digital twins in health care: ethical implications of an emerging engineering paradigm. *Front. in genetics*, *9*, p.31.
[49] US Patent application 20190220733A1 Systems and Methods for Modeling Probability Distributions
[50] Fisher CK, Smith AM, Walsh JR. 2019. Machine learning for comprehensive forecasting of Alzheimer's Disease progression. *Scientific reports*.9(1):1-4.
[51] Walsh JR, et al. Generating Digital Twins with Multiple Sclerosis Using Probabilistic Neural Networks. *arXiv preprint arXiv*:2002.02779. 2020 Feb 4.
[52]Farrell S, et al, 2020 Generating individual aging trajectories with a network model using cross-sectional data. *bioRxiv*.
[53] Greenbaum D, et al. 2001 Interrelating different types of genomic data, from proteome to secretome:'oming in on function. Genome research. 11(9):1463-8.
[54] Rajkomar A, et al., 2018, Scalable and accurate deep learning with electronic health records. *NPJ Digital Medicine*. 1(1):18.
[55] Ian Goodfellow, et al., 2014 Generative adversarial nets. In *Advances in neural information processing systems*, pp. 2672–2680.
[56] Georges-Filteau J, Cirillo E. 2020. Generative Adversarial Networks Applied to Observational Health Data. *CoRR*. May 11.
[57] Yahi A, et al. 2017 Generative adversarial networks for electronic health records: A framework for exploring and evaluating methods for predicting drug-induced laboratory test trajectories. *arXiv preprint arXiv:*1712.00164..
[58] Guan J, Li R, Yu S, Zhang X. 2019. A Method for Generating Synthetic Electronic Medical Record Text. IEEE/ACM Transactions on Computational Biology and Bioinformatics.
[59] Pluristem Expands its Compassionate Use Program: Treated First COVID-19 Patient in U.S. Under FDA Single Patient Expanded Access Program BioSpace April 13, 2020 /
[60] Shah, N.N., et al 2020. Abstract PO-092: Expanded access trial of tocilizumab in COVID19+ hospitalized cancer patients.
[61] Denise Hinton, Emergency Use Authorization Letter, August 28, 2020
[62] Office of the Commissioner, et al. Guidance Document Emergency Use Authorization of Medical Products and Related Authorities, January 20173
[63] Webb, J., Shah, L.D. and Lynch, H.F., 2020. Ethically allocating COVID-19 drugs via pre-approval access and emergency use authorization. *The American Journal of Bioethics*, *20*(9), pp.4-17.
[64] London, A.J. and Kimmelman, J., 2020. Against pandemic research exceptionalism. Science, 368(6490), pp.476-477.
[65] Rozenberg, O. and Greenbaum, D., 2020. Making It Count: Extracting Real World Data from Compassionate Use and Expanded Access Programs. *AJOB 20*(7), pp.89-92.
[66] Food and Drug Administration, et al, Considerations for the Use of Real-World Evidence to Assess the Effectiveness of Preventive Vaccines, September 17 - 18, 2020
[67] Steckler, A. and McLeroy, K.R., 2008. The Importance of External Validity. *American Journal of Public Health*, *98*(1), p.9.

Advanced Methods for Big Data Analytics in Women's Health

Mary Regina Boland†
Perelman School of Medicine, University of Pennsylvania
Philadelphia, PA, USA
Email: bolandm@pennmedicine.upenn.edu

Karin Verspoor
University of Melbourne
Melbourne, Australia

Maricel G Kann
University of Maryland
Baltimore, MD, USA

Su Golder
University of York
York, UK

Lisa Levine, Karen O'Conner
Perelman School of Medicine, University of Pennsylvania
Philadelphia, PA, USA

Natalia Villanueva-Rosales
University of Texas
El Paso, TX, USA

Graciela Gonzalez-Hernandez
Perelman School of Medicine, University of Pennsylvania
Philadelphia, PA, USA
Email: gragon@pennmedicine.upenn.edu

Women's health is an often-overlooked aspect of medicine. The National Institutes of Health has emphasized the importance of investigating 'sex as a biological variable' in all new research grants. This has placed emphasis once again on the need for more nuanced studies that explore the role of sex as a biological variable on study outcomes. This session sought to elicit participation from researchers with strong backgrounds in women's health and informatics to develop methods that harness big datasets and 'big data techniques' including machine learning and artificial intelligence and apply those tools to women's health questions. Some important questions discussed in this section include Intimate Partner Violence (IPV) and the importance of early identification along with C-section deliveries and the importance of emergency vs. elective procedures.

† Work partially supported by the University of Pennsylvania (MRB)

Keywords: Text mining; machine learning; electronic health records; women's health; biomedicine; text mining; natural language processing; precision medicine.

1. Introduction

Recent advances in data science and digital epidemiology have unlocked an unprecedented amount of data for analysis, and uncovered previously unseen sex-specific patterns that point at marked differences in disease symptoms, progression and care that affect women of all ages. In 2016, the NIH published a guidance document[1] and changed its policy for reviewing proposals whereby accounting for "sex as a biological variable" became a required and scorable aspect of the research strategy, highlighting that "an over-reliance on male animals and cells may obscure understanding of key sex influences on health processes and outcomes". Dr. Kathryn Rexrode, chief of the Division of Women's Health at Brigham and Women's Hospital, is quoted[2] as succinctly stating the enormity of the problem: "without the inclusion of women, all the way through from basic research to clinical research, we can't be sure we really have the right answers for 51 percent of the population."

Aside from x-linked inheritable diseases, where women generally are carriers rather than express the disease[3], there are various aspects of women's health that challenge current methods. Recent research shows that variations in physiology may alter the pharmacokinetics or pharmacodynamics that determines drug dosing and effect for women, both in general and particularly during pregnancy[4], as hormonal and other biological differences may influence the impact of drugs, their effectiveness and their side effects. Over two-thirds of women receive prescription drugs while pregnant, with treatment and dosing strategies based on data from healthy male volunteers and non-pregnant women[5]. A paucity of research exists for optimizing prescription usage during pregnancy and more methods are needed that utilize artificial intelligence and machine learning [6-8]. In addition, health processes unique to women, such as pregnancy and pregnancy loss, menstruation and menopause require differential approaches to data representation and analysis. Disorders related to pregnancy and menstruation (such as miscarriage and heavy bleeding, which have a significant impact on women's health) have been recently found to be related to specific genetic mutations and are just being explored[9,10]. Furthermore, it has become clear through numerous recent studies that many diseases (cardiovascular disease, asthma, eating disorders, lung cancer, and autoimmune disorders, among others) impact women differently than men. Advanced data science methods specifically designed for exploring the influence that sex hormones and a women's physiology can have on the pathophysiology of these processes diseases and on their treatment are essential to advance our understanding of key processes in women's health, and, at the same time, the contrast could also shed light on the specific mechanisms that affect men.

This session highlights original research in the form of presentations and papers on the subject of big data and women's health. These include the use of machine learning methods to predict intimate partner violence (IPV) over 1 year before that violence occurs, along with pattern mining to determine patterns of IPV and co-occurrence with other subgroups of IPV, including sexual violence. Another method explores the role of emergency vs. elective C-section deliveries on the study of C-sections as an adverse outcome of delivery. These studies together enable further understanding of processes and diseases that are specific to women or differentially impact women. In harmony with the focus of PSB, the session emphasizes methodological advances and applications in data science, emphasizing reproducibility and validation.

2. Session Summary

The session includes three full-length papers competitively selected for inclusion that are focused on exploring problems associated with the complex problem of intimate partner violence, including patterns and injury prediction (2 distinct papers) and another study focused on deconstructing Cesarean sections into emergency versus elective to better understand this complex health outcome. We selected these important contributions that are applicable to utilize big datasets on studying women's health outcomes.

2.1. *Full-length papers*

In *Co-occurrence Patterns of Intimate Partner Violence,* the authors present a method that learns patterns of survivors of intimate partner violence (IPV) [11]. The main data-source for their study is the National Intimate Partner and Sexual Violence Survey (NISVS). The algorithm then clusters IPV into 5 different subgroups, and the authors compare these algorithm-chosen subgroups to traditional categories of IPV including physical violence, psychological aggression, sexual violence and micro-aggression. An important finding of their pattern analysis and co-occurrence pattern mining is that physical violence often co-occurs with psychological aggression and co-occurs less often with micro-aggression. In addition, the authors found that sexual violence tended to be a mutually exclusive form of IPV. Furthermore, this exclusive nature of sexual violence was so strong that it formed a single connected component in their subsequent network analysis. Overall, the findings from this study underscore the importance of breaking down IPV into type of IPV (e.g., physical violence, psychological aggression, sexual violence and micro-aggression) as these different types of IPV have different co-occurrence patterns and could be important for subsequent studies that link IPV to other health outcomes. The authors results suggest that their method effectively clusters types of IPV patients into subgroups that pertain to the type of IPV experienced by the patient and underscore the importance of co-occurrence patterns in IPV.

In *Intimate Partner Violence and Injury Prediction from Radiology Reports,* **Chen et al.** present an algorithm to predict which patients will experience injury as a result of IPV[12]. Because there are different types of IPV and not all IPV results in an injury to the partner, this method would be useful in determining *a priori* what patients will be likely to experience injuries as a result of IPV. This study differs from the previous study in that **Chen et al.'s** algorithm utilizes data from a large academic hospital's violence prevention support program from Jan 2013 - Jun 2018. For information on the subsequent injuries, the authors also had access to the patients' radiology reports. The authors develop a machine learning model assess IPV patients for risk of injury. Their method was successfully able to predict IPV 1.34 years before entrance into a violence prevention program with 95% sensitivity and 71% specificity. There are future plans to deploy their model as a clinical risk model for early detection of IPV.

In *Not All C-sections are the same: Investigating Emergency vs. Elective C-section Deliveries*, **Canelón et al.** present a method that utilizes Electronic Health Records (EHR) data to breakdown Cesarean sections (C-sections) into emergency vs. elective C-sections [13]. This breakdown is important because C-sections are often deemed an 'adverse outcome' across the board. However, there can be situations where it is the best outcome for a particular patient. Therefore, detailing out the important difference between a patient with an elective or planned C-section (e.g., in the case of a patient with complex comorbidities) versus an emergency C-section (e.g., as the result of an amniotic fluid embolism) is important when determining if the C-section is an adverse delivery outcome or not. In this study, the authors confirm that they adequately capture the differences between emergency and elective C-section by comparing the rates on weekday versus weekend, observing the expected drop in elective C-sections on the weekends. In addition, they modeled emergency deliveries in general as an adverse outcome and found that the following patient characteristics increased the risk of an emergency delivery: preterm birth, being younger than 25, identifying as Black/African American, Asian, or Other/Mixed, after adjusting for pregnancy number and C-section number for each patient. Interestingly, later pregnancies and repeat cesareans decreased the risk of an emergency delivery, and identifying as White, Hispanic, and Native Hawaiian/Pacific Islander patients appeared to lower the risk of an emergency delivery. The same risk factors and trends were found also for Cesarean deliveries (when looking at emergencies as the outcome) except that Asian patients did not have an increased risk of an emergency delivery in the C-section population, and Native Hawaiian/Pacific Islander patients did not have a reduced risk in this group. Overall, modeling the relationship between emergency vs. elective deliveries is important to understanding the relationship between other comorbidities and risk factors for C-sections. In addition, it is important for breaking down C-sections into those that are likely adverse events (e.g., emergencies) versus those that are due to comorbidities or other patient health issues (e.g., elective or planned).

3. Discussion

Informatics and 'Big Data Analytics' algorithms as applied and developed specifically for women's health questions such as those presented in this session enable novel approaches of existing data from diverse sources including EHR and survey data sources. These methods can be used for early prediction of IPV (over 1-year before violence occurs) and these methods have potential to be implemented in clinics for early identification of at-risk patients. Before these methods can be implemented, care must be taken that these machine learning algorithms have not 'learned' any features or other signals that may be indicative of patterns of care that may be biased against women or other minority or otherwise disadvantaged groups. However, the work presented in this session does represent important first steps towards early risk prediction for a complex issue such as IPV.

Overall, the research presented in this session focuses on different clinical questions that pertain to women and women's health, including IPV and C-section as an adverse outcome following delivery or birth. The studies presented explore the complexity and the need to take these larger groups (either IPV or C-sections) and further break them down into meaningful subclusters, in the case of IPV that would be breaking it down into physical violence vs. sexual violence and so forth. In the case of C-sections, it requires breaking it down into emergency vs. elective C-sections. This highlights the complexity of these outcomes and the importance of developing novel informatics algorithms to study these important women's health outcomes. The overarching goal will be to use these findings and algorithms to improve clinical care in the form of enhanced understanding of risk factors or to predict patients at risk for IPV for early identification at the point of care.

References

1. NIH. Consideration of Sex as a Biological Variable in NIH-funded Research [Internet]. . *Available from:* https://orwh.od.nih.gov/sites/orwh/files/docs/NOT-OD-15-102_Guidance.pdf. 2015.
2. Esposito L. 7 Major Gaps in Women's Health Research [Internet] Available from: https://health.usnews.com/health-care/patient-advice/slideshows/7-major-gaps-in-womens-health-research. 2017.
3. X-Linked Recessive Disorders [Internet] Available from: https://www.sciencedirect.com/topics/neuroscience/x-linked-recessive-disorders. 2020.
4. Louis GMB, Yeung E, Kannan K, et al. Patterns and Variability of Endocrine-disrupting Chemicals During Pregnancy: Implications for Understanding the Exposome of Normal Pregnancy. *Epidemiology.* 2019;30:S65-S75.
5. Feghali M, Venkataramanan R, Caritis S. Pharmacokinetics of drugs in pregnancy. Paper presented at: Seminars in perinatology2015.
6. Davidson L, Boland MR. Enabling pregnant women and their physicians to make informed medication decisions using artificial intelligence. *Journal of Pharmacokinetics and Pharmacodynamics.* 2020:1-14.
7. Boland MR, Polubriaginof F, Tatonetti NP. Development of a machine learning algorithm to classify drugs of unknown fetal effect. *Scientific reports.* 2017;7(1):1-15.
8. Duan R, Boland MR, Moore JH, Chen Y. ODAL: A one-shot distributed algorithm to perform logistic regressions on electronic health records data from multiple clinical sites. Paper presented at: PSB2019.
9. Maybin JA, Boswell L, Young VJ, Duncan WC, Critchley HO. Reduced transforming growth factor-β activity in the endometrium of women with heavy menstrual bleeding. *The Journal of Clinical Endocrinology & Metabolism.* 2017;102(4):1299-1308.
10. Husseini-Akram F, Haroun S, Altmäe S, et al. Hyaluronan–binding protein 2 (HABP2) gene variation in women with recurrent miscarriage. *BMC women's health.* 2018;18(1):143.
11. Hacaliefendioglu A, Ylmaz S, Koyuturk M, Karakurt G. Co-occurrence Patterns of Intimate Partner Violence. Paper presented at: PSB2021.
12. Chen IY, Alsentzer E, Park H, et al. Intimate Partner Violence and Injury Prediction From Radiology Reports. *PSB.* 2021.
13. Canelon S, Boland MR. Not All C-sections Are the Same: Investigating Emergency vs. Elective C-section deliveries as an Adverse Pregnancy Outcome. *PSB.* 2021.

Intimate Partner Violence and Injury Prediction From Radiology Reports

Irene Y. Chen[1*], Emily Alsentzer[2], Hyesun Park[3], Richard Thomas[4], Babina Gosangi[5], Rahul Gujrathi[3], and Bharti Khurana[3,6]

[1]*Electrical Engineering and Computer Science, Massachusetts Institute of Technology, Cambridge, MA 02139, USA*

[2]*Health Sciences and Technology, Massachusetts Institute of Technology, Cambridge, MA 02139, USA*

[3]*Department of Radiology, Brigham and Women's Hospital, Boston, MA 02115, USA*

[4]*Department of Radiology, Lahey Health Medical Center, Burlington, MA 01805, USA*

[5]*Department of Radiology, Yale University, New Haven, CT 06510, USA*

[6]*Department of Radiology, Harvard Medical School, Boston, MA 02115, USA*

Intimate partner violence (IPV) is an urgent, prevalent, and under-detected public health issue. We present machine learning models to assess patients for IPV and injury. We train the predictive algorithms on radiology reports with 1) IPV labels based on entry to a violence prevention program and 2) injury labels provided by emergency radiology fellowship-trained physicians. Our dataset includes 34,642 radiology reports and 1479 patients of IPV victims and control patients. Our best model predicts IPV a median of 3.08 years before violence prevention program entry with a sensitivity of 64% and a specificity of 95%. We conduct error analysis to determine for which patients our model has especially high or low performance and discuss next steps for a deployed clinical risk model.

Keywords: intimate partner violence; radiology; risk stratification; natural language processing; contextual word embeddings.

1. Introduction

Intimate partner violence (IPV) is defined as physical, sexual, psychological, or economic violence that occurs between former or current intimate partners. While men can also be affected, IPV is a gendered phenomenon largely perpetrated against women by male partners.[1] The Centers for Disease Control report that more than 1 in 3 women, and 1 in 10 men in the U.S. will experience physical violence, sexual violence, psychological violence, and/or stalking by an intimate partner during their lifetime.[2] IPV victims have a greater risk of health problems including higher rates of mental health illnesses, chronic pain, reproductive difficulties, and generally poorer health.[3–5] According to the United Nations, half of the women

*Corresponding author email: iychen@mit.edu.

who are intentionally killed globally are killed by their intimate partners or family members.[6] It is essential to detect IPV victims early to provide timely intervention.

Healthcare providers have the opportunity to screen patients for IPV, but several barriers at both patient and provider levels limit the effectiveness. IPV victims often seek treatment within healthcare settings;[7] however, despite its high prevalence, IPV is substantially underdiagnosed due to underreporting of violence by the victim to health care providers. Because IPV victims generally do not present with obvious trauma, even in emergency departments,[8] they do not readily receive IPV-specific resources.

Imaging studies provide an objective measurement of patient status, especially for vulnerable individuals who are not forthcoming.[9] In a prior observational study, researchers identified IPV-related injury patterns including soft-tissue and musculoskeletal injuries from imaging studies of victims who visited the emergency department. They also found that IPV victims receive more radiology studies than a comparable control cohort.[10]

In this work, we present algorithms to predict IPV and injury from radiology reports. We predict IPV from a dataset of 24,131 radiology reports from 262 IPV victims who enrolled into a violence prevention support program and 794 controls from the same hospital who were age and sex-matched based on a subset of the IPV victims. We demonstrate strong quantitative results with our best model achieves a mean area under the received operator curve (AUC) of 0.852. With a sensitivity of 64% and a specificity of 95%, we are able to predict IPV a median of 3.08 years in advance of entry into the violence prevent support program. To better detect severe forms of IPV, we predict injury from a dataset of radiology reports from only IPV victims with labels from four emergency radiology fellowship-trained radiologists. Our best model achieves a mean AUC of 0.887.

We analyze our models for validity and usability. Because IPV can manifest differently across race,[11] gender,[12] age,[13] and marital status,[14] we present error analysis comparing accuracy, sensitivity, and specificity across these groups using demographic information extracted from the clinical record. As IPV continues to affect vulnerable individuals—especially in times of great crisis[15,16]—we demonstrate how automated predictive algorithms can be used to identify patients at high risk of IPV and injury.

2. Related Work

2.1. *Intimate partner violence*

Early detection in IPV is critical to facilitate early intervention in the cycle of abuse, thereby preventing worsening health conditions,[3–5] life threatening injuries, and potentially homicides.[17] The main obstacle to early intervention is underreporting by the patient due to variety of factors including shame, economic dependency, or lack of trust in healthcare providers.[18] Automated screening can help physicians identify high risk individuals—potentially from radiology studies,[19] substance abuse disorders,[20] or other clinical data—and intervene quickly. Prior work has focused on analyzing associative patterns among IPV victims.[10] To our knowledge, we present the first work to present an algorithm for IPV and injury prediction.

2.2. *Clinical prediction*

Machine learning methods can assess patients and other individuals for different levels of risk to allocate resources and improve clinical workflows.[21,22] The strength of machine learning lies in its ability to learn latent patterns from observational data and make robust predictions on new and previously unseen patients. Researchers have shown promising results about the use of machine learning on chronic diseases like diabetes,[23] diagnosis from radiology reports,[24] rare conditions like preterm infant illnesses,[25] and public health concerns like child welfare.[26,27] In particular, supervised learning models excel in structured settings with large datasets and clearly defined labels, e.g. radiology report text and whether the patient ultimate enters a violence prevention program.

2.3. *Natural language processing*

Natural language processing (NLP) techniques can extract information from unstructured text.[28] In healthcare settings, researchers have leveraged NLP on clinical text such as nursing notes, discharge summaries, and radiology and pathology reports for disease surveillance,[24,29] cohort creation,[30,31] prediction of adverse events,[32–34] and diagnosis.[35,36]

A promising new area of natural language processing research is the use of contextual word embeddings. Whereas traditional approaches represent text as a non-sequential bag of words or a sequence of static word embeddings, more recent approaches construct unique representations for each word (or sub-word) depending on its surrounding context. For instance, the abbreviation "MS" may refer to mitral stenosis or multiple sclerosis depending on the surrounding context. BERT,[37] RoBERTa,[38] AlBERT,[39] and numerous other recent models are pretrained on large amounts of text using language modelling objectives and then fine-tuned on a smaller task-specific dataset. Among other examples, large open-source clinical datasets[40] have enabled researchers to release clinical contextual word embedding models. ClinicalBERT is a publicly available BERT model initialized from BioBERT[41] and further trained on intensive care unit notes.[42]

3. Dataset

We predict IPV using a dataset of IPV victims and age-matched control patients. We predict injury using a dataset of only IPV victims, with labels from emergency radiologists.

3.1. *IPV patient selection*

The study cohort consisted of victims who were referred to a large academic hospital's violence prevention support program between January 2013 and June 2018. For the early detection of IPV through IPV prediction, we randomly selected 265 women reporting physical abuse. We excluded all victims without any radiological studies from both groups because our algorithm seeks to predict IPV from radiology reports. The final IPV dataset consists of 262 patients.

For injury prediction, we examine a wider set of patients from two groups of victims referred to a large academic hospital's violence prevention support program between January 2013 and June 2018. For the first group, we randomly selected 940 victims out of 2948 reporting any type of IPV-physical, psychosocial, or sexual. The second group comprised of all 308 IPV victims (including 265 women) reporting physical abuse. We excluded all victims without any radiological studies from both groups. The final IPV dataset consists of 530 patients.

3.2. *Control group selection*

We age-matched against 265 women with physical abuse and filtered for patients with at least one radiology study that was not canceled. We selected the first 795 of the resulting 1006 patients to build our control cohort. Note that the control cohort was matched against the 265 female IPV victims and does not contain any men.

3.3. *Injury labels*

The full set of radiological studies and reports of the injury prediction patient cohort were analyzed for the presence of injury for each study. Any radiological findings unrelated to potential physical injury such as pancreatitis, malignancy, subarachnoid hemorrhage due to aneurysm rupture, etc. were not recorded as "injury". All images were reviewed by four emergency radiology fellowship-trained radiologists who were aware of history of IPV but were blinded to the date of identification of IPV and clinical notes. The readers had full access to the radiology reports. The radiologists also recorded any injuries such as soft tissue swelling, rib fracture, etc. which might be overlooked or not mentioned in the original radiology reports. Each report was reviewed separately and labeled with an injury or not. Of the 15,639 radiology reports reviewed, 2.57% of them were found to have an injury.

3.4. *Data cleaning*

For each radiology report, we remove extraneous information to improve clarity for the predictive models. We remove all header and footer information, punctuation, and line breaks. We change the text to lowercase and create tokens from each word through bag of words or clinicalBERT.[42] Radiology reports that lack meaningful information after this cleaning are removed from the dataset. Patients who do not have any radiology reports after this step are removed from the dataset completely.

3.5. *Demographic data*

We extract demographic data from IPV victims and controls including age, gender, race, and marital status. To structure free-form responses for some fields, we consolidate each field into several categories. For age, we discretize the field into < 30, 30-50, 51-65, and 66+. The average age of patients in dataset is 43.8 ± 18.5, with IPV victims average age at 40.9 ± 13.3 and control population average age at 46.3 ± 4.7. For race, we consider white, Black, Hispanic, and "other" categories with patients allowed to belong to more than one group. For marital status, we categorize single, married, and other. Note that because our control population was sex and age-matched against a cohort of female IPV victims, our control population contains no men. We do not use demographic information for predictions and use only radiology reports. For summary statistics about the dataset, see Table 1.

Table 1. Summary statistics for dataset, with percentages of radiology reports.

		IPV Prediction			Injury Prediction		
		Total	IPV	Control	Total	Injury	No Injury
# Patients		1,056	262	794	530	135	395
# Radiology Reports		24,131	5,127	19,004	10,009	172	9,837
Age	< 30	6.8%	14.4%	4.8%	9.8%	10.5%	7.6%
	30-50	32.0%	47.4%	27.8%	53.6%	58.7%	58.6%
	51-65	37.2%	32.1%	38.5%	29.3%	21.5%	25.7%
	66+	23.9%	6.1%	28.7%	7.3%	9.3%	8.1%
Gender	Female	100.0%	100.0%	100.0%	93.4%	90.7%	94.8%
	Male	0.0%	0.0%	0.0%	6.6%	9.3%	5.2%
Race	Black	50.8%	34.6%	55.2%	28.3%	29.1%	23.6%
	Hispanic	12.0%	24.7%	8.6%	22.2%	19.2%	21.9%
	White	10.0%	29.4%	4.8%	38.7%	40.7%	43.6%
	Other	27.6%	11.6%	32.0%	11.6%	11.6%	12.0%
Marital Status	Single	45.0%	56.6%	41.8%	50.8%	57.0%	47.9%
	Married	36.1%	19.4%	40.7%	29.7%	27.3%	36.2%
	Other	18.9%	24.0%	17.5%	19.5%	15.7%	15.9%

4. Methodology

4.1. *Experiment setup*

We train our models on 60% of the patients, validate and select hyperparameters based on 20% of the patients, and report test performance on 20% of the patients. To avoid data leakage, we split our data based on patient rather than radiology study. Once a patient is assigned to train, validation, or test dataset, we assign all radiology reports and labels for that patient to the corresponding dataset. We perform analysis on five trials with shuffled splits of the data. All models are compared against the same five dataset splits.

4.2. *Models*

We compare two tasks and five models. We predict IPV and injury based on collected labels. We consider data from extracted demographic data, radiology reports, and a combination of the two. We use logistic regression, random forest, gradient boosted trees, neural network with bag of words representation, and neural network with clinicalBERT[42] representation.

For logistic regression, we search over hyperparameters including regularization constant $C = [0.001, 0.01, 0.1, 1., 2., 5.]$ and regularization type of L1 or L2. For random forest, we search over maximum depth of trees of 10, 50, 100, 500, or no maximum depth. For gradient boosted trees, we search over hyperparameters learning rate of 0.01, 0.1, 0.5, or 1 and maximum depth of 2, 3, and 4. We use the sklearn-learn Python package[43] with otherwise default settings.

We train two neural network models using the AllenNLP library.[44] Both models contain an embedding layer followed by two feed forward layers with rectified linear unit function and linear activations. The first model represents each note as a vector of word frequencies ("Bag of Words") projected down to a lower dimensional vector while the second model leverages clinicalBERT's contextual word embeddings to represent each note.

To facilitate more rapid training on CPUs, we freeze the clinicalBERT embeddings and only train the feed forward layers. The first model was trained for 40 epochs with an early stopping patient of 5 epochs, and the second model was trained for 10 epochs due to computational constraints. Gradient norms were rescaled to a max of 5.0, and training examples were batched by note length to minimize excess padding. Hyperparameters were selected according to validation set performance, resulting in a learning rate of 0.001, weight decay of 0.0001 and batch size of 32 for both models.

4.3. *Evaluation*

4.3.1. *Prediction and predictive features*

We report the predictive performance as the area under the receiver operator curve (AUC) on the same train, validation, and test datasets for all models compared. We compute AUC means and standard deviations for the test datasets of the five shuffled splits of the data.

We present predictive features by finding words with high feature importance. Because many compared models are non-linear, it is difficult to use interpretability methods to find predictive words. As logistic regression performance is comparable to that of other other non-linear methods (see Table 2), we present linear coefficients of the logistic regression across five test sets of the shuffled splits of the data.

4.3.2. *Error analysis*

As clinical models face high stakes decisions, it is important that machine learning reduce health disparities[45] rather than amplify existing biases.[46] We audit our best prediction model for IPV and injury by comparing accuracy, sensitivity, and specificity for different subgroups including age, race, gender, and marital status.[47–49] We compute means and standard deviations of performance metrics for each subgroup with model sensitivity set to 0.95 because the clinical healthcare system can accommodate many false positives—e.g. offering a conversation with a social worker—whereas false negatives can be more dire—e.g. not providing an IPV victim with additional resources for help. Predicted probabilities are computed for test datasets and compared to the true labels for the five shuffled splits of the data.

4.3.3. *Report-program date gap*

One practical measure of IPV prediction is how much earlier does our model predict IPV compared to the date of patient entry into a violence prevention program. Although some patients may be reluctant to seek clinical assistance,[18] earlier detection and appropriate triage of IPV victims can help empower clinicians to intervene and provide better care for victims.[19]

Table 2. Model AUC means and standard deviations over five data splits for IPV and injury prediction using radiology reports. Bold rows indicate best performance for task.

Model	IPV	Injury
Logistic Regression	0.841 ± 0.033	0.866 ± 0.016
Random Forest	$\mathbf{0.852 \pm 0.022}$	$\mathbf{0.887 \pm 0.019}$
Gradient Boosted Trees	0.842 ± 0.027	0.858 ± 0.030
Neural Network (Bag of Words)	0.849 ± 0.026	0.879 ± 0.010
Neural Network (clinicalBERT[42])	0.843 ± 0.022	0.852 ± 0.021

For each radiology report, we compare the radiology report date with the entry date into the program. We call this difference in dates the *report-program date gap*, or simply the *date gap*. Negative date gaps denote reports that occur before program entry. A radiology report with a large magnitude date gap is one that occurs long before program entry whereas a low magnitude date gap occurs shortly before program entry. A model that can make predictions with a large magnitude date gap per IPV victim would allow us to allocate resources and support to high risk individuals more efficiently. For each IPV victim, we compute the largest date gap for which the model predicts IPV above the chosen threshold.

We select the prediction threshold to satisfy specificity constraints. A trivial way to maximize the early IPV detection would be to predict IPV for every patient in the dataset. This simplification would yield redundant results and a high sensitivity (true positive rate). Accordingly, we fix our specificity level (true negative rate) to be at least 95% and compute the corresponding model threshold. We report the median earliest date gap for all IPV victims for whom the model predicts correctly.

5. Results

5.1. *IPV and injury prediction and predictive features*

We are able to predict IPV (best mean AUC of 0.852, random forest classifier) and injury (best mean AUC of 0.887, random forest classifier). For more results, see Table 2. We find that words that are most predictive for IPV and injury match clinical literature in IPV injury patterns from radiology reports. In Table 3, we show words with highest feature importance from logistic regression for both tasks. Findings include soft-tissue abnormalities such as swelling and hematomas and musculoskeletal injuries such as fractures. These findings reflect prior research on IPV injury patterns.[10]

5.2. *Error analysis*

We find differences in performance in subgroups of age, gender, race in our error analysis (see Table 4). We focus on sensitivity because in cases of IPV and injury, it is much more important to detect all true positives. In particular, older patients (51-65, 66+) have lower sensitivity for both IPV and injury prediction. Other groups have low sensitivity for either

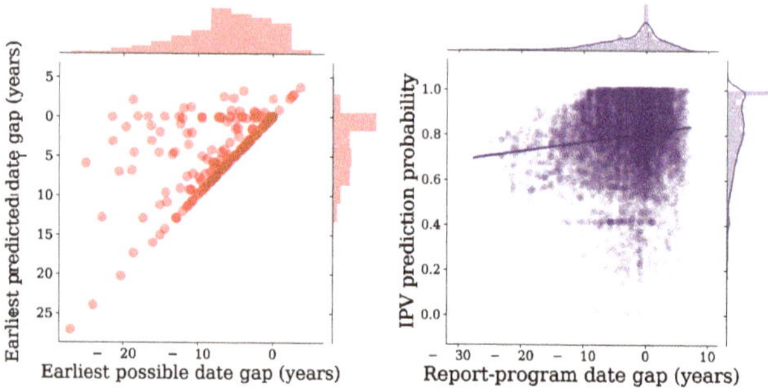

Fig. 1. Scatterplots and marginal histograms for random forest classifier for IPV prediction. **Left:** Earliest possible report-program date gap per patient (x-axis) compared to earliest predicted date gap (y-axis) with sensitivity of 64% and specificity of 95%. **Right:** Report-program date gap (x-axis) and IPV prediction probability (y-axis) for all radiology reports of IPV victims.

Table 3. Predictive words for IPV and injury averaged across five trials based on linear coefficients of logistic regression. Underline indicates words consistent with clinical literature.[10]

Task	Predictive words
IPV	ordering, final, trauma, hematoma, technique, swelling, cell, fracture, type, fractures, lymphoma, electronically, male, pancreatitis, reason, gms, implants, unresponsive, assault, none, cancer, pregnancy, mca
Injury	hematoma, fracture, fractures, swelling, trauma, subchorionic, foreign, ankle, third, hand, nondisplaced, fall, stab, phalanx, finger, deformity, skullbase, fifth, wound, laceration, sob, digit, measuring

IPV or injury prediction, but not both. For example, Black patients have lower sensitivity for injury prediction. White patients have low sensitivity for IPV prediction. It appears that patients who are not single or married (e.g. widowed, separated) have lower sensitivity for injury whereas married patients have lower sensitivity for IPV prediction.

5.3. *Report-program date gap*

We can detect IPV from radiology reports much earlier than a patient's entry into a violence prevention program. We compute the report-program date gap with specificity threshold of the random forest classifier set to 95% and find a median date gap of 3.08 years, compared to the median earliest possible date gap of 5.83 years. For visual representations of the predicted date gap compared to the earliest possible date gap and the prediction scores, see Figure 1.

Table 4. Error analysis for IPV and injury predictions from random forest classifier. Means and standard deviations of accuracy, sensitivity (TPR), and specificity (TNR) computed over 5 data splits with overall model sensitivity set to 0.95. Bold indicates subgroups with particularly low metrics.

		IPV Prediction			Injury Prediction		
		Accuracy	TPR	TNR	Accuracy	TPR	TNR
Age	< 30	83.6 ± 4%	97.7 ± 1%	53.6 ± 11%	62.5 ± 11%	93.9 ± 9%	61.2 ± 11%
	30-50	87.4 ± 1%	96.3 ± 1%	49.6 ± 5%	54.1 ± 12%	94.8 ± 1%	52.9 ± 13%
	51-65	**71.8 ± 5%**	92.5 ± 2%	49.1 ± 2%	41.4 ± 18%	**89.5 ± 3%**	40.4 ± 19%
	66+	**60.9 ± 5%**	**84.4 ± 2%**	45.2 ± 9%	**33.5 ± 16%**	98.0 ± 4%	31.1 ± 17%
Gender	Female	77.2 ± 1%	94.6 ± 1%	48.4 ± 4%	50.0 ± 15%	93.4 ± 1%	48.9 ± 15%
	Male	—	—	—	**31.7 ± 21%**	96.2 ± 4%	28.8 ± 21%
Race	Black	**72.3 ± 2%**	95.6 ± 0%	41.4 ± 7%	47.8 ± 14%	**88.3 ± 3%**	46.5 ± 14%
	Hispanic	91.1 ± 2%	97.9 ± 0%	51.9 ± 11%	58.0 ± 13%	96.9 ± 3%	57.4 ± 13%
	White	84.6 ± 1%	**90.5 ± 2%**	43.0 ± 5%	41.6 ± 18%	95.1 ± 3%	39.8 ± 18%
	Other	**68.7 ± 3%**	98.0 ± 0%	55.1 ± 5%	58.3 ± 13%	95.0 ± 6%	57.4 ± 13%
Marital Status	Single	81.5 ± 2%	95.4 ± 0%	45.5 ± 7%	49.6 ± 13%	95.3 ± 1%	48.1 ± 14%
	Married	**70.6 ± 1%**	**92.2 ± 2%**	49.8 ± 3%	49.2 ± 18%	92.6 ± 7%	48.5 ± 19%
	Other	83.4 ± 2%	95.4 ± 2%	49.8 ± 9%	46.5 ± 16%	**88.7 ± 3%**	45.4 ± 17%

6. Discussion and conclusion

We present a range of findings on the use of prediction algorithms to address IPV in the clinical setting through the analysis of radiology reports. Our results demonstrate several main takeaways. First, with a dataset of 34,642 reports and 1,479 patients, we are able accurately predict IPV and injury with AUCs of 0.852 and 0.887, respectively. Second, while our algorithm demonstrates some bias in the form of differences in accuracy, sensitivity, and specificity with respect to age, gender, race, and marital status, we are able to predict a median report-program date gap of over 3.08 years with sensitivity of 64% and specificity of 95%.

Our work leads naturally to many directions for future research. One limitation of our current work is that we consider one radiology report at a time for IPV and injury prediction and exclude clinical history. Because IPV victims seek greater medical care from clinical settings like the emergency department,[7,8] patient data including previous visits, clinical notes, and diagnoses could yield more accurate predictions and therefore earlier detection.[19] Additionally, predictive algorithms can help identify the best intervention for an IPV victim. Currently screening programs for IPV vary in execution and effect,[50] and once screened, IPV victims face many obstacles before leaving an abusive relationship.[51] Deeper understanding of targeted interventions could provide a crucial contribution to patient advocacy.

Deployment of a predictive model for IPV and injury detection faces several practical challenges. As with many machine learning algorithms in clinical settings, question of generalization across hospitals[22] and across subgroups[47] raise concerns about robustness and fairness.

Moreover, better understanding of physician reliance on, distrust of, and confusion towards predictive models in clinical settings is an active area of research.[52] We have shown in our analysis that automated detection through machine learning can predict IPV and injury from radiology reports. We look forward to future work towards the deployment of an IPV early detection model in a clinical setting.

References

1. E. Fulu, R. Jewkes, T. Roselli, C. Garcia-Moreno *et al.*, Prevalence of and factors associated with male perpetration of intimate partner violence: findings from the un multi-country cross-sectional study on men and violence in asia and the pacific, *The lancet global health* **1**, e187 (2013).
2. M. Black, K. Basile, M. Breiding, S. Smith, M. Walters, M. Merrick, J. Chen and M. Stevens, National intimate partner and sexual violence survey: 2010 summary report (2011).
3. J. C. Campbell, Health consequences of intimate partner violence, *The lancet* **359**, 1331 (2002).
4. K. Tollestrup, D. Sklar, F. J. Frost, L. Olson, J. Weybright, J. Sandvig and M. Larson, Health indicators and intimate partner violence among women who are members of a managed care organization, *Preventive medicine* **29**, 431 (1999).
5. M. Ellsberg, H. A. Jansen, L. Heise, C. H. Watts, C. Garcia-Moreno *et al.*, Intimate partner violence and women's physical and mental health in the who multi-country study on women's health and domestic violence: an observational study, *The lancet* **371**, 1165 (2008).
6. U. N. O. on Drugs and Crime, *Global Study on Homicide: Gender-related Killing of Women and Girls* (UNODC, United Nations Office on Drugs and Crime, 2018).
7. C. Wisner, T. Gilmer, L. Saltzman and T. Zink, Intimate partner violence against women, *Journal of family practice* **48**, 439 (1999).
8. S. R. Dearwater, J. H. Coben, J. C. Campbell, G. Nah, N. Glass, E. McLoughlin and B. Bekemeier, Prevalence of intimate partner abuse in women treated at community hospital emergency departments, *Jama* **280**, 433 (1998).
9. A. Russo, A. Reginelli, M. Pignatiello, F. Cioce, G. Mazzei, O. Fabozzi, V. Parlato, S. Cappabianca and S. Giovine, Imaging of violence against the elderly and the women, in *Seminars in Ultrasound, CT and MRI*, (1)2019.
10. E. George, C. H. Phillips, N. Shah, A. Lewis-O'Connor, B. Rosner, H. M. Stoklosa and B. Khurana, Radiologic findings in intimate partner violence, *Radiology* **291**, 62 (2019).
11. S. Lipsky, R. Caetano, C. A. Field and G. L. Larkin, The role of intimate partner violence, race, and ethnicity in help-seeking behaviors, *Ethnicity and Health* **11**, 81 (2006).
12. P. Tjaden and N. Thoennes, Prevalence and consequences of male-to-female and female-to-male intimate partner violence as measured by the national violence against women survey, *Violence against women* **6**, 142 (2000).
13. C. M. Rennison, *Intimate partner violence and age of victim, 1993-99* (US Department of Justice, Office of Justice Programs, Bureau of Justice ..., 2001).
14. A. Salomon, S. S. Bassuk and N. Huntington, The relationship between intimate partner violence and the use of addictive substances in poor and homeless single mothers, *Violence Against Women* **8**, 785 (2002).
15. N. Van Gelder, A. Peterman, A. Potts, M. O'Donnell, K. Thompson, N. Shah and S. Oertelt-Prigione, Covid-19: Reducing the risk of infection might increase the risk of intimate partner violence, *EClinicalMedicine* **21** (2020).
16. B. Gosangi, H. Park, R. Thomas, R. Gujrathi, C. P. Bay, A. S. Raja, S. E. Seltzer, M. C. Balcom, M. L. McDonald, D. P. Orgill *et al.*, Exacerbation of physical intimate partner violence during covid-19 lockdown, *Radiology* , p. 202866 (2020).

17. H. Stöckl, K. Devries, A. Rotstein, N. Abrahams, J. Campbell, C. Watts and C. G. Moreno, The global prevalence of intimate partner homicide: a systematic review, *The Lancet* **382**, 859 (2013).
18. D. Hien and L. Ruglass, Interpersonal partner violence and women in the united states: An overview of prevalence rates, psychiatric correlates and consequences and barriers to help seeking, *International journal of law and psychiatry* **32**, p. 48 (2009).
19. B. Khurana, S. E. Seltzer, I. S. Kohane and G. W. Boland, Making the 'invisible'visible: transforming the detection of intimate partner violence, *BMJ quality & safety* **29**, 241 (2020).
20. F. L. Kraanen, E. Vedel, A. Scholing and P. M. Emmelkamp, Prediction of intimate partner violence by type of substance use disorder, *Journal of substance abuse treatment* **46**, 532 (2014).
21. M. Ghassemi, T. Naumann, P. Schulam, A. L. Beam, I. Y. Chen and R. Ranganath, A review of challenges and opportunities in machine learning for health, *AMIA Summits on Translational Science Proceedings* **2020**, p. 191 (2020).
22. M. Ghassemi, T. Naumann, P. Schulam, A. L. Beam, I. Y. Chen and R. Ranganath, Practical guidance on artificial intelligence for health-care data, *The Lancet Digital Health* **1**, e157 (2019).
23. N. Razavian, S. Blecker, A. M. Schmidt, A. Smith-McLallen, S. Nigam and D. Sontag, Population-level prediction of type 2 diabetes from claims data and analysis of risk factors, *Big Data* **3**, 277 (2015).
24. K. L. Kehl, H. Elmarakeby, M. Nishino, E. M. Van Allen, E. M. Lepisto, M. J. Hassett, B. E. Johnson and D. Schrag, Assessment of deep natural language processing in ascertaining oncologic outcomes from radiology reports, *JAMA oncology* **5**, 1421 (2019).
25. S. Saria, A. K. Rajani, J. Gould, D. Koller and A. A. Penn, Integration of early physiological responses predicts later illness severity in preterm infants, *Science translational medicine* **2**, 48ra65 (2010).
26. A. Chouldechova, D. Benavides-Prado, O. Fialko and R. Vaithianathan, A case study of algorithm-assisted decision making in child maltreatment hotline screening decisions, in *Conference on Fairness, Accountability and Transparency*, 2018.
27. A. Brown, A. Chouldechova, E. Putnam-Hornstein, A. Tobin and R. Vaithianathan, Toward algorithmic accountability in public services: A qualitative study of affected community perspectives on algorithmic decision-making in child welfare services, in *Proceedings of the 2019 CHI Conference on Human Factors in Computing Systems*, 2019.
28. S. Wu, K. Roberts, S. Datta, J. Du, Z. Ji, Y. Si, S. Soni, Q. Wang, Q. Wei, Y. Xiang *et al.*, Deep learning in clinical natural language processing: a methodical review, *Journal of the American Medical Informatics Association* **27**, 457 (2020).
29. S. Gao, M. T. Young, J. X. Qiu, H.-J. Yoon, J. B. Christian, P. A. Fearn, G. D. Tourassi and A. Ramanthan, Hierarchical attention networks for information extraction from cancer pathology reports, *Journal of the American Medical Informatics Association* **25**, 321 (2018).
30. L. Chen, Y. Gu, X. Ji, C. Lou, Z. Sun, H. Li, Y. Gao and Y. Huang, Clinical trial cohort selection based on multi-level rule-based natural language processing system, *Journal of the American Medical Informatics Association* **26**, 1218 (2019).
31. N. Afzal, V. P. Mallipeddi, S. Sohn, H. Liu, R. Chaudhry, C. G. Scott, I. J. Kullo and A. M. Arruda-Olson, Natural language processing of clinical notes for identification of critical limb ischemia, *International journal of medical informatics* **111**, 83 (2018).
32. C. Poulin, B. Shiner, P. Thompson, L. Vepstas, Y. Young-Xu, B. Goertzel, B. Watts, L. Flashman and T. McAllister, Predicting the risk of suicide by analyzing the text of clinical notes, *PloS one* **9**, p. e85733 (2014).
33. K. Huang, J. Altosaar and R. Ranganath, Clinicalbert: Modeling clinical notes and predicting hospital readmission, *arXiv preprint arXiv:1904.05342* (2019).
34. A. Rumshisky, M. Ghassemi, T. Naumann, P. Szolovits, V. Castro, T. McCoy and R. Perlis,

Predicting early psychiatric readmission with natural language processing of narrative discharge summaries, *Translational psychiatry* **6**, e921 (2016).
35. A.-D. Pham, A. Névéol, T. Lavergne, D. Yasunaga, O. Clément, G. Meyer, R. Morello and A. Burgun, Natural language processing of radiology reports for the detection of thromboembolic diseases and clinically relevant incidental findings, *BMC bioinformatics* **15**, 1 (2014).
36. W. Boag, D. Doss, T. Naumann and P. Szolovits, What's in a note? unpacking predictive value in clinical note representations, *AMIA Summits on Translational Science Proceedings* **2018**, p. 26 (2018).
37. J. Devlin, M.-W. Chang, K. Lee and K. Toutanova, Bert: Pre-training of deep bidirectional transformers for language understanding, *arXiv preprint arXiv:1810.04805* (2018).
38. Y. Liu, M. Ott, N. Goyal, J. Du, M. Joshi, D. Chen, O. Levy, M. Lewis, L. Zettlemoyer and V. Stoyanov, Roberta: A robustly optimized bert pretraining approach, *arXiv preprint arXiv:1907.11692* (2019).
39. Z. Lan, M. Chen, S. Goodman, K. Gimpel, P. Sharma and R. Soricut, Albert: A lite bert for self-supervised learning of language representations, *arXiv preprint arXiv:1909.11942* (2019).
40. A. E. Johnson, T. J. Pollard, L. Shen, H. L. Li-Wei, M. Feng, M. Ghassemi, B. Moody, P. Szolovits, L. A. Celi and R. G. Mark, Mimic-iii, a freely accessible critical care database, *Scientific data* **3**, 1 (2016).
41. J. Lee, W. Yoon, S. Kim, D. Kim, S. Kim, C. H. So and J. Kang, Biobert: a pre-trained biomedical language representation model for biomedical text mining, *Bioinformatics* **36**, 1234 (2020).
42. E. Alsentzer, J. R. Murphy, W. Boag, W.-H. Weng, D. Jin, T. Naumann and M. McDermott, Publicly available clinical bert embeddings, *arXiv preprint arXiv:1904.03323* **2019** (2019).
43. F. Pedregosa, G. Varoquaux, A. Gramfort, V. Michel, B. Thirion, O. Grisel, M. Blondel, P. Prettenhofer, R. Weiss, V. Dubourg, J. Vanderplas, A. Passos, D. Cournapeau, M. Brucher, M. Perrot and E. Duchesnay, Scikit-learn: Machine learning in Python, *Journal of Machine Learning Research* **12**, 2825 (2011).
44. M. Gardner, J. Grus, M. Neumann, O. Tafjord, P. Dasigi, N. F. Liu, M. Peters, M. Schmitz and L. S. Zettlemoyer, Allennlp: A deep semantic natural language processing platform2017.
45. I. Y. Chen, S. Joshi and M. Ghassemi, Treating health disparities with artificial intelligence, *Nature Medicine* **26**, 16 (2020).
46. A. Rajkomar, M. Hardt, M. D. Howell, G. Corrado and M. H. Chin, Ensuring fairness in machine learning to advance health equity, *Annals of internal medicine* **169**, 866 (2018).
47. I. Y. Chen, P. Szolovits and M. Ghassemi, Can ai help reduce disparities in general medical and mental health care?, *AMA journal of ethics* **21**, 167 (2019).
48. I. Y. Chen, M. Agrawal, S. Horng and D. Sontag, Robustly extracting medical knowledge from ehrs: A case study of learning a health knowledge graph, in *Pac Symp Biocomput*, 2020.
49. M. Hardt, E. Price and N. Srebro, Equality of Opportunity in Supervised Learning, in *Proceedings of the 30th International Conference on Neural Information Processing Systems*, NIPS'16 (Curran Associates Inc., USA, June 2016). Barcelona, Spain.
50. L. O'Doherty, K. Hegarty, J. Ramsay, L. L. Davidson, G. Feder and A. Taft, Screening women for intimate partner violence in healthcare settings, *Cochrane database of systematic reviews* (2015).
51. J. Kim and K. A. Gray, Leave or stay? battered women's decision after intimate partner violence, *Journal of Interpersonal Violence* **23**, 1465 (2008).
52. P. Tschandl, C. Rinner, Z. Apalla, G. Argenziano, N. Codella, A. Halpern, M. Janda, A. Lallas, C. Longo, J. Malvehy *et al.*, Human–computer collaboration for skin cancer recognition, *Nature Medicine* , 1 (2020).

Not All C-sections Are the Same: Investigating Emergency vs. Elective C-section deliveries as an Adverse Pregnancy Outcome*

Silvia P. Canelón, PhD and Mary Regina Boland, MA, MPhil, PhD, FAMIA
Department of Biostatistics, Epidemiology, and Informatics, University of Pennsylvania
423 Guardian Drive, PA, 19104, USA
Corresponding Email: bolandm@upenn.edu

Electronic Health Records (EHR) contain detailed information about a patient's medical history and can be helpful in understanding clinical outcomes among populations generally underrepresented in research, including pregnant individuals. A cesarean delivery is a clinical outcome often considered in studies as an adverse pregnancy outcome, when in reality there are circumstances in which a cesarean delivery is considered the safest or best choice given the patient's medical history, situation, and comfort. Rather than consider all cesarean deliveries to be negative outcomes, it is important to examine other risk factors that may contribute to a cesarean delivery being an adverse event. Looking at emergency admissions can be a useful way to ascertain whether or not a cesarean delivery is part of an adverse event. This study utilizes EHR data from Penn Medicine to assess patient characteristics and pregnancy-related conditions as risk factors for an emergency admission at the time of delivery. After adjusting for pregnancy number and cesarean number for each patient, preterm birth increased risk of an emergency admission, and patients younger than 25, or identifying as Black/African American, Asian, or Other/Mixed, had an increased risk. Later pregnancies and repeat cesareans decreased the risk of an emergency delivery, and White, Hispanic, and Native Hawaiian/Pacific Islander patients were at decreased risk. The same risk factors and trends were found among cesarean deliveries, except that Asian patients did not have an increased risk, and Native Hawaiian/Pacific Islander patients did not have a reduced risk in this group.

Keywords: Electronic Health Records; pregnancy; cesarean section; C-section; emergency admission; population health.

1. Background and Significance

Electronic Health Records (EHR) contain rich information on patient medical history and treatment and can be used to study effects of prenatal exposures on delivery-related outcomes. These databases chronicle a patient's medical history, and therefore information at the pregnancy-level for each of a patient's pregnancies must be extracted from patient-specific medical information. This study utilizes an algorithm designed to extract delivery episode details from the EHR [1]. Our previously developed algorithm enables multiple deliveries to be extracted per patient from the EHR and does not limit the data to one pregnancy per patient, which is an improvement over other algorithms in the field. The purpose of this study is to assess the impact of pregnancy-specific maternal morbidity and patient-specific characteristics on experiencing an emergency admission at the time of delivery and its relationship to Cesarean section (C-section) deliveries.

---

* This work is supported by the University of Pennsylvania.

The United States has one of the highest rates of maternal mortality among developed nations at 24.7% [2,3] and high rates of C-section deliveries at 31.6%[4]. The World Health Organization found that a country-level C-section rate of greater than 10% was not associated with reductions in maternal and newborn mortality rates[5] and the American College of Obstetricians and Gynecologists expressed concern for the potential that C-sections were being overused after observing the rapid increase of C-sections between 1996 and 2011 without clear evidence of concomitant decreases in maternal morbidity or mortality rates [6,7]. Some suggest financial incentives [8–10] and the resource and scheduling convenience associated with C-section procedures [11–13] may play a role.

Primary C-sections, or individuals' first C-section, have been associated with some increased risk in morbidity, and subsequent or repeat C-sections in the future pose even greater risk[14]. There also exists consensus within the medical community that a C-section procedure is sometimes the best approach, as in placenta previa or uterine rupture [7]. Understanding that not every C-section can be considered an adverse pregnancy outcome, it is important to consider other factors that may be indicative of an adverse event. In this study, we examine emergency admissions as an adverse event among the general population as well as the population of patients with C-sections while considering a variety of patient- and pregnancy-specific characteristics as risk factors. We investigate preterm birth, multiple birth, and stillbirth diagnoses as risk factors along with patient-specific characteristics (at time of birth) including age, marital status, and race/ethnicity. The decision to investigate a patient's race or ethnicity as a risk factor has no biological basis but rather is grounded in an effort to explore how systemic racism[15,16] may be reflected in the health outcomes studied. Importantly, there are no race-based or ancestry-specific genetic factors that have been implicated in increasing the risk of C-section deliveries.

2. Methods

We identified pregnant patients who delivered via a C-section using structured EHR data that included a combination of inpatient and outpatient encounters within the health system. This data was coupled with information about type of admission to the clinic (i.e. elective or emergency), patient race/ethnicity, and patient age and marital status at the time of the encounter. We also determined if each pregnancy resulted in a multiple birth, preterm birth, or stillbirth using structured billing codes. We constructed a generalized logistic model to explore the relationship between these predictors and an emergency admission as a binary outcome variable.

All code for this analysis and data visualization was implemented in R[17] (version 4.0.2) using the tidyverse collection of packages[18], and EHR data was stored on a HIPAA secure server in a MySQL database. This study was approved by the Institutional Review Board of the University of Pennsylvania.

2.1. *Dataset characteristics*

We obtained EHR data for 1,060,100 female patients with visits to inpatient or outpatient clinics within the Penn Medicine system between 2010 and 2017. Previously, we developed and validated an algorithm to extract delivery episode information and delivery dates for each patient (accuracy of 98.6% and F-1 score of 92.1%) called MADDIE [1]. This algorithm identified 50,560 female patients with 63,334 distinct deliveries. The predominant race/ethnicity descriptions of the patients with deliveries were non-Hispanic Black or African American (47.3% of deliveries) and non-Hispanic White (33.9% of deliveries). We were able to identify pregnant patients who delivered by C-section and found that 35.52% (17,951 of 50,560) of patients delivered at least once via C-section and 32.99% (20,894 of 63,334) of all deliveries were via C-section (Table 1).

Table 1. Demographics of Patients with Deliveries at Penn Medicine

	All deliveries		C-section deliveries	
	Patients (%)	Deliveries (%)	Patients (%)	Deliveries (%)
Demographics	50560 (100)	63334 (100)	17951 (100)	20894 (100)
Patient race/ethnicity[a]				
Black/African American	23777 (47.0)	29965 (47.3)	8220 (45.8)	9502 (45.5)
White	17034 (33.7)	21443 (33.9)	6413 (35.7)	7626 (36.5)
Hispanic	4031 (8.0)	4985 (7.9)	1403 (7.8)	1611 (7.7)
Asian	3305 (6.5)	4073 (6.4)	1110 (6.2)	1269 (6.1)
Other or Mixed	2426 (4.8)	2883 (4.6)	569 (3.2)	638 (3.1)
Native Hawaiian/Pacific Islander	75 (0.15)	94 (0.15)	36 (0.2)	39 (0.2)
American Indian/Alaskan Native	61 (0.12)	81 (0.13)	19 (0.1)	28 (0.1)
Unknown	865 (1.71)	971 (1.53)	270 (1.5)	291 (1.4)
Patient age	29.5 ± 6.1	N/A	30.6 ± 6.1	N/A

[a]Race/ethnicity descriptions are 'non-Hispanic' unless otherwise indicated

2.2. *Identification of delivery outcomes*

Each delivery episode comprised a window of time containing an inferred delivery date. This delivery episode window consists of a start and end date corresponding to the start and end dates of when delivery codes were assigned. We needed to use an episode window because the visit to the hospital related to a delivery often can cross over multiple days and, in some cases, can last for several days. This is especially true for preterm deliveries where an attempt is made to delay the delivery, but is often unsuccessful. We use several outcomes (defined in subsections below) in this study. To link a patient delivery to a specific outcome, we required that the outcome diagnostic code be assigned within the delivery episode window for a particular delivery. We conducted our study at the pregnancy-level rather than the patient-level. However, later analysis looks at the effects of a prior C-section or a prior-pregnancy on subsequent pregnancy outcomes (thereby incorporating patient-level information).

2.2.1. *Cesarean section deliveries*

We used the U.S.-modified *International Classification of Diseases* version 9 (ICD-9) and version 10 (ICD-10) codes to identify all records that were assigned a C-section diagnosis or procedure code, and that had a C-section code assigned within the delivery episode window or time frame. In the event that a C-section code was assigned on more than one date within a delivery episode, the date closest to the patient delivery date was selected as the C-section date.

2.2.2. *Preterm birth, stillbirth, and multiple birth deliveries*

In the absence of gestational weeks in the structured data, we used ICD-9 and ICD-10 codes to identify records that were assigned a preterm birth diagnosis code within the delivery episode, and created a binary variable accordingly. The same process was used to identify a stillbirth or multiple birth within the delivery episode. These three variables were included as predictors in the regression models.

2.3. *Integration of data from encounter records*

All delivery records were matched with admission type details in the encounter data to determine if patients had "emergency" or "elective" admissions to the hospital. Delivery admissions of type "emergency" were categorized as **emergency** deliveries while those recorded as "elective", "routine/elective", or "routine/elective admission" were categorized as **elective** deliveries. Categorization as an emergency admission was modeled as a binary response variable in the logistic regression models.

Each encounter date was mapping to the day of the week information (i.e. Monday, Thursday, Saturday, etc.) using R. Additional details within the encounter records were used to extract the patient's race/ethnicity as well as their age and marital status at the time of the delivery encounter. Patient age was included in the regression model as a categorical predictor variable with categories "<25 years", "25-34 years", and ">35 years", with "25-34 years" serving as the reference variable. This age breakdown was chosen to assess whether patients younger or older than the majority of pregnant patients in our cohort[1] were at a different risk of emergency admission. Marital status was considered only so far as whether the patient was 'Single' at the time of the encounter, and included in the model as a binary predictor variable.

2.4. *Generalized regression models*

We constructed a binomial multivariate logistic regression model to explore the relationship between a variety of predictor variables and emergency admission as the binary **response**, within the delivery population. *Age*, *race/ethnicity*, marital status *single*, *preterm birth*, *multiple birth*, and *stillbirth* diagnoses were all modeled as **predictors** of an emergency admission.

A similar model was constructed to explore the risk of an emergency admission specifically among patients with C-sections. *Age*, *race/ethnicity*, marital status *single*, *preterm birth*, *multiple birth*, and *stillbirth* diagnoses were all modeled as predictors of an emergency admission. To account for any prior deliveries and/or C-sections, we also created adjusted models that included the *delivery number* and *C-section number* as **predictors**. All predictors were binary with the exception of *age* which was categorical, and *delivery number* (ranging from 0-7 deliveries) and *C-section number* (ranging from 0-5 C-sections) which were both continuous.

Patients' first deliveries were also modeled as a separate group to consider the possibility that a patient's first experience giving birth could relate differently to the risk of an emergency admission.

The odds ratio for each predictor in all models was estimated by exponentiating the coefficients produced by the regression models.

3. Results

3.1. *Utilization of cesarean section codes*

We found that 10 unique ICD-9 codes and 6 unique ICD-10 codes were utilized to record a C-section diagnosis or procedure within the EHR. Among ICD-9 codes, the most common diagnosis code was 649.81 "Spontaneous labor with planned C-section-delivered", and the most common procedure code was 74.1 "Low cervical C-section" (Figure 1A). Among ICD-10 C-section codes, which were utilized starting in 2015, the most common diagnosis code was O82 "Encounter for Cesarean delivery without indication," and the most common procedure code was 10D00Z1 "Extraction of products of conception, low, open approach" (Figure 1B). Overall, the most common codes were procedure codes ICD-9 74.1 and ICD-10 10D00Z1 (Figure 1C).

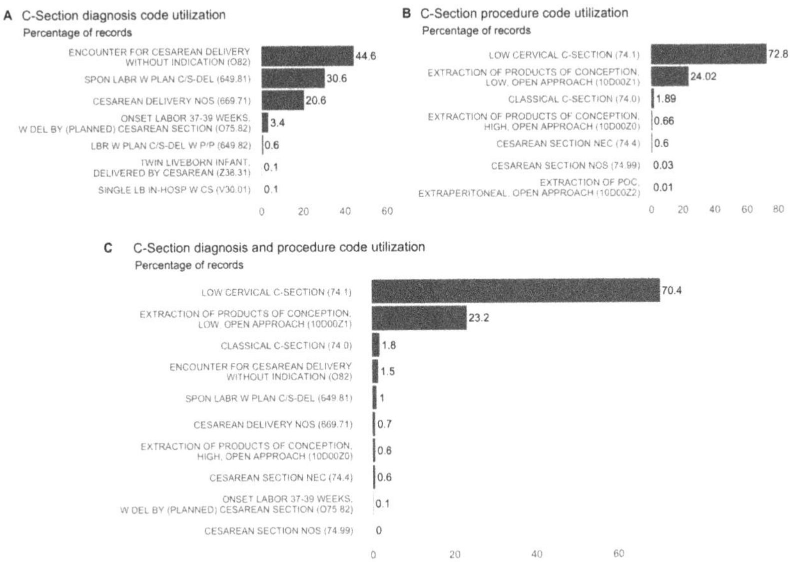

Fig 1. Distribution of ICD-9/10 codes most commonly utilized to code for a C-section delivery.

3.2. *Admission types recorded in encounter records*

The encounter records revealed 62 distinct admission types (excluding the empty field) among all delivery records and 47 among C-sections. The most common admission types recorded in the EHR at the time of the encounter for both groups included "emergency", "elective", and "routine elective admission". Among all deliveries, "emergency" made up 25.3% of records, "elective" made up 4.8%, and "routine elective admission" made up 0.9%. The most common admission types and a similar pattern were seen among C-section deliveries, with "emergency" making up 22.1%, "elective" making up more encounters compared to all deliveries at 10.1%, and "routine elective admission" making up 1.3% of records (Table 2). We grouped all admission types that were not explicitly emergency and not explicitly elective into an 'Other' admission type for the purposes of our study.

Table 2. Ten Most Common Admission Types Recorded in the Encounter Records

Admission type	Encounters	Patients	Deliveries
All deliveries	N = 78505	N = 50560	N = 63334
PREGNANCY	37699 (48%)	30688 (60.7%)	35856 (56.6%)
EMERGENCY	19873 (25.3%)	17250 (34.1%)	19766 (31.2%)
(empty field)	6930 (8.8%)	6477 (12.8%)	6645 (10.5%)
OTHER	3912 (5%)	3879 (7.7%)	3894 (6.1%)
ELECTIVE	3806 (4.8%)	3541 (7%)	3614 (5.7%)
RETURN OB	2295 (2.9%)	2237 (4.4%)	2269 (3.6%)
NON STRESS TEST	1610 (2.1%)	1594 (3.2%)	1606 (2.5%)
ROUTINE ELECTIVE ADMISSION	688 (0.9%)	655 (1.3%)	657 (1%)
INDUCTION	436 (0.6%)	430 (0.9%)	430 (0.7%)
US LIMITED	295 (0.4%)	292 (0.6%)	293 (0.5%)
C-section deliveries	N = 27034	N = 17951	N = 20895
PREGNANCY	11905 (44%)	10213 (56.9%)	11216 (53.7%)
EMERGENCY	5971 (22.1%)	5447 (30.3%)	5883 (28.2%)
(empty field)	2960 (10.9%)	2760 (15.4%)	2798 (13.4%)
ELECTIVE	2717 (10.1%)	2461 (13.7%)	2526 (12.1%)
OTHER	1137 (4.2%)	1126 (6.3%)	1128 (5.4%)
NON STRESS TEST	700 (2.6%)	692 (3.9%)	696 (3.3%)
RETURN OB	670 (2.5%)	639 (3.6%)	644 (3.1%)
ROUTINE ELECTIVE ADMISSION	364 (1.3%)	334 (1.9%)	335 (1.6%)
US LIMITED	131 (0.5%)	129 (0.7%)	129 (0.6%)
INDUCTION	113 (0.4%)	107 (0.6%)	107 (0.5%)

3.3. *Age distribution by delivery admit type*

Among all deliveries, the average age at the time of delivery was 27.9 ± 6.3 years for emergency deliveries, 31.6 ± 5.9 years for elective deliveries, and 30.1 ± 5.8 years for "Other" admission types. Within C-sections, the average age was higher for all admission categories with an average age of 29.2 ± 6.5 years for emergency deliveries, 32.1 ± 5.5 years for elective deliveries, and 30.9 ± 5.9 years for other admissions (Figure 2).

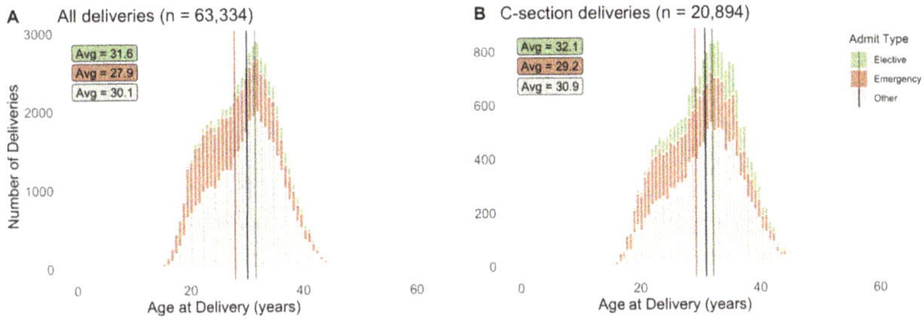

Fig 2. Distribution of patient age at time of delivery by admit type for (A) all deliveries and (B) C-sections.

3.4. *Number of deliveries by weekday and admit type*

Overall, most deliveries occurred during the work week from Monday to Friday with a noticeable decline on Saturday and Sunday, a trend further emphasized within C-sections (Figure 3). The decrease in elective admissions between weekdays and the weekend was 2.25x greater among C-section deliveries (12.4% vs. 5.5% for all deliveries). This difference between C-section deliveries and all deliveries was similar for the modest increase in emergency admissions on the weekend (1.6% vs. 0.7% for all deliveries). This transition between weekday and weekend with regards to emergency vs. elective C-section deliveries was expected given that C-sections are not scheduled for the weekend except in the case of an emergency. Most deliveries were associated neither with an elective nor an emergency admission but one of the "Other" admission types (Table 3).

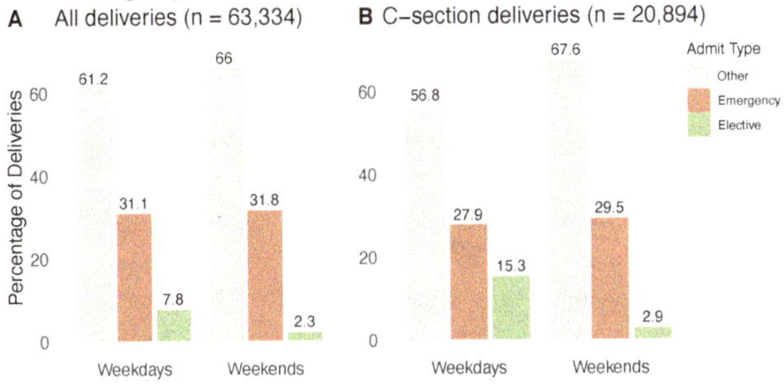

Fig 3. Deliveries on weekdays compared to weekends by admit type for (A) all deliveries and (B) C-sections.

Table 3. Proportion of Deliveries by Weekday and Admit Type

Weekday	Elective	Emergency	Other
All deliveries			
Avg. Weekday	777 (7.8%)	3107.2 (31.1%)	6118.4 (61.2%)
Avg. Weekend	150.5 (2.3%)	2115 (31.8%)	4395 (66.0%)
C-section deliveries			
Avg. Weekday	544.8 (15.4%)	993.6 (27.9%)	2020.6 (56.8%)
Avg. Weekend	45.5 (2.9%)	457.5 (29.5%)	1047.5 (67.6%)

4. Generalized regression model

Figure 4 presents odds ratio estimates for risk of an emergency delivery from the logistic regression models constructed for three groups of deliveries: first deliveries, all deliveries, and C-section deliveries. Among first deliveries for all patients, preterm birth and age <25 years increased the risk, and patients Black/African American, Other or Mixed, or Asian were at increased risk. Patients >35 years of age, single, White, Hispanic, or Native Hawaiian/Pacific Islander were at a decreased risk.

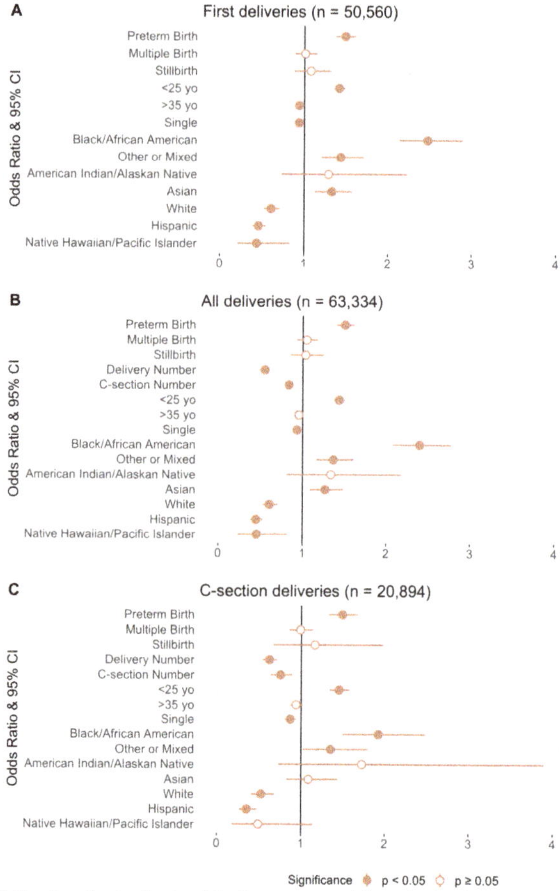

Fig 4. Odds ratio estimates showing risk of an emergency delivery for first deliveries (A), all deliveries (B), and C-section deliveries (C).

These trends persisted when considering all deliveries together and also after adjusting for the delivery number and C-section number, the only difference being that patients >35 years were no longer at decreased risk of an emergency admission.

In the C-section subgroup, all the same significant risk factors were identified, with the exceptions that Asian patients were no longer at increased risk and Native Hawaiian/Pacific Islander patients were no longer found to be at decreased risk of an emergency admission (Table 4).

Across all three groups, preterm birth, age, and single marital status were found to be significant risk factors for an emergency admission, as well as identifying as Black/African American, Other, or Mixed, White, or Hispanic. All deliveries and the C-section subgroup also shared in common the number of delivery and number of C-section as significant risk factors. Notably, each model reflects that Black/African American patients were at a higher risk of having an emergency delivery than any other racial/ethnic group. Hispanic patients were the least likely to experience an emergency delivery, followed closely by White patients.

Table 4. Logistic Regression Model Results

	Original Model		Adjusted Model	
Predictor	OR (95% CI)	P-value	OR (95% CI)	P-value
All deliveries				
Preterm Birth	1.52 (1.42-1.64)	<0.001	1.51 (1.41-1.62)	<0.001
Multiple Birth	0.98 (0.87-1.10)	0.709	1.05 (0.93-1.18)	0.437
Stillbirth	1.08 (0.90-1.30)	0.409	1.04 (0.86-1.25)	0.716
Age <25 years	1.52 (1.45-1.58)	<0.001	1.44 (1.38-1.51)	<0.001
Age >35 years	0.93 (0.88-0.97)	0.003	0.96 (0.91-1.01)	0.091
Marital Status Single	0.94 (0.90-0.98)	0.009	0.93 (0.89-0.98)	<0.01
Black/African American	2.16 (1.88-2.50)	<0.001	2.40 (2.08-2.78)	<0.001
Other or Mixed	1.30 (1.11-1.53)	0.001	1.37 (1.17-1.61)	<0.001
American Indian/Alaskan Native	1.19 (0.72-1.92)	0.491	1.34 (0.80-2.18)	0.245
Asian	1.21 (1.04-1.42)	0.015	1.27 (1.09-1.49)	0.002
White	0.58 (0.50-0.67)	<0.001	0.61 (0.53-0.58)	<0.001
Hispanic	0.42 (0.36-0.50)	<0.001	0.45 (0.38-0.53)	<0.001
Native Hawaiian/Pacific Islander	0.43 (0.22-0.77)	0.008	0.46 (0.23-0.82)	0.014
Delivery Episode	N/A	N/A	0.55 (0.53-0.58)	<0.001
C-section Episode	N/A	N/A	0.84 (0.81-0.87)	<0.001
C-section deliveries				
Preterm Birth	1.55 (1.38-1.74)	<0.001	1.49 (1.33-1.68)	<0.001
Multiple Birth	0.99 (0.86-1.15)	0.935	0.99 (0.86-1.15)	0.922
Stillbirth	1.15 (0.66-1.94)	0.690	1.17 (0.67-1.98)	0.577
Age <25 years	1.50 (1.38-1.62)	<0.001	1.46 (1.34-1.58)	<0.001
Age >35 years	0.94 (0.86-1.02)	0.128	0.94 (0.87-1.02)	0.156
Marital Status Single	0.89 (0.82-0.96)	0.004	0.87 (0.80-0.95)	<0.001
Black/African American	1.77 (1.38-2.29)	<0.001	1.93 (1.50-2.49)	<0.001
Other or Mixed	1.33 (1.00-1.76)	0.050	1.36 (1.02-1.80)	0.035
American Indian/Alaskan Native	1.35 (0.58-2.99)	0.467	1.73 (0.73-3.90)	0.194
Asian	1.06 (0.80-1.40)	0.690	1.09 (0.83-1.44)	0.538
White	0.50 (0.39-0.65)	<0.001	0.53 (0.41-0.68)	<0.001
Hispanic	0.34 (0.25-0.46)	<0.001	0.36 (0.27-0.48)	<0.001
Native Hawaiian/Pacific Islander	0.49 (0.18-1.12)	0.117	0.49 (0.18-1.14)	0.127
Delivery Episode	N/A	N/A	0.62 (0.54-0.72)	<0.001
C-section Episode	N/A	N/A	0.76 (0.64-0.90)	<0.001

4.1. *Surgical Incision Type for C-section and Effect on Emergency Admission*

Not all C-section procedures are the same with regards to the surgical incisions, so we explored whether the type of C-section incision was indicative of an elective vs. emergency delivery. Low C-section procedures have become the default procedure compared to the classical/high approach[19]. Figure 1B showed the two most common categories of C-section procedures corresponded to low C-section procedures and classical/high C-section procedures which were much less common. Including both ICD-9 and ICD-10 codes, low C-section procedures made up nearly 97% of all C-section records. In contrast, classical/high C-section procedures only made up roughly 2.5% of records. After categorizing these two types of procedures by admission type, we did not find that surgical incision type varied much by admission type (elective vs. emergency delivery): 10.7% vs. 13.6% of classical vs. low C-sections were elective deliveries and 28.4% vs. 28.0% of classical vs. low C-sections were emergency deliveries (Table 5). From this, we conclude that the emergency vs. elective admission type confers different information then surgical incision type.

Table 5. Proportion of C-section Patients and Deliveries by Procedure Type and Admit Type

Procedure type	Elective	Emergency	Other
Patients			
Low C-section	2669 (15.3%)	5261 (30.2%)	10668 (61.1%)
Classical (high) C-section	54 (11.0%)	142 (28.8%)	301 (61.1%)
Other C-section	192 (24.4%)	143 (18.2%)	457 (58.0%)
Deliveries			
Low C-section	2745 (13.6%)	5665 (28.0%)	11810 (58.4%)
Classical (high) C-section	54 (10.7%)	143 (28.4%)	307 (60.9%)
Other C-section	192 (24.2%)	143 (18.0%)	458 (57.8%)

5. Discussion

The extraction of diagnosis and procedure records, encounter records, and delivery date information from the EHR facilitates the study of adverse pregnancy-related outcomes with patient-specific as well as pregnancy-specific information. This information serves to provide rich context for patient's healthcare experience and makes it possible to investigate outcomes with a broader perspective. A C-section procedure as the mode of delivery is an example of a health outcome that requires richer context. There may be multiple reasons for a patient and their healthcare provider to consider a C-section delivery over a vaginal delivery, which may include a medical indication or patient preference. Therefore, automatically categorizing all C-sections, as adverse pregnancy outcomes would not be appropriate because not all C-sections are the same. It is important when studying pregnancy-related outcomes to explore additional factors that may contribute to an adverse experience. The approach taken by this study considers **emergency** deliveries to be the adverse event rather than C-sections more generally speaking, and evaluates a number of patient-specific and pregnancy-specific details as risk factors for an emergency admission at the time of delivery.

In addition to investigating C-sections as a subset of all deliveries, we also studied a subset containing only the first delivery from each patient in the dataset. Because our dataset includes EHR data for patients at Penn Medicine, the first delivery of each patient in our cohort is the first delivery that Penn Medicine has on record for that patient. This is a limitation because it means our dataset does not include deliveries that may have occurred prior to that first record or outside of Penn Medicine. The first deliveries subset provides a baseline perspective and accounts for the possibility that a patient's first delivery experience at Penn Medicine may itself relate to an emergency delivery. A limitation to note here is potential selection bias with our cohort if patients had an extremely negative delivery experience at Penn Medicine and chose not to return for future pregnancy care.

Our logistic regression models found that patients with a *preterm birth* diagnosis, younger than *25 years*, and identifying as *Black/African American or Other/Mixed*, were at an increased risk of an emergency delivery among first deliveries, all deliveries, and C-section deliveries. A greater risk among patients with a preterm birth diagnosis is expected, as public health efforts to prevent preterm birth have suggested as an intervention the elimination of early elective deliveries [20]. For related reasons, multiple birth and stillbirth diagnoses were also included in the analysis though neither were found to increase risk of emergency C-sections. A greater risk at a younger age may be due in part to a lack of familiarity with the birth process and anxiety in anticipation of birth [21]. This may cause them to choose to be admitted through the emergency department when entering labor and have an elective delivery (C-section or otherwise) that is ultimately captured as an emergency admission. This theory is supported by the decreased risk for patients with more deliveries or repeat C-sections among all deliveries and C-sections. This was unexpected as repeat C-sections have been associated with other adverse outcomes[14], suggesting other risk factors are more strongly correlated with emergency deliveries. The health disparities evident in the results of this study align with patterns identified in pregnancy care[22] and more broadly throughout healthcare[16,23].

Patients who have experienced more births (multiparous) may have a lower risk of an emergency delivery because they are more informed about what to expect and perhaps more confident in advocating for themselves and/or finding support in their delivery experience. Patients with more births may also have had prior positive experiences at Penn Medicine and/or suffer less disease overall and be able to sustain more pregnancies as a result. Relative to all deliveries combined, patients with C-sections were on average older regardless of admission type, and there was a clearer distinction between elective and emergency deliveries. When considering deliveries throughout the week, we confirmed that most deliveries occurred on weekdays (Monday–Friday), including C-sections. For both groups, the proportion of elective deliveries dropped substantially from weekdays to the weekend. Among C-sections the drop was more pronounced showing it is less likely for a patient to have an elective C-section scheduled on the weekend, but instead during the conventional work week. These last findings support the hypothesis that resource and scheduling conveniences of C-section procedures contribute to overall C-section rates[11–13].

This study elucidated the importance of considering a variety of risk factors contributing to a patient's adverse experience during delivery, and the benefit of considering admission type as a way to distinguish between elective and emergency C-sections. It also generated opportunities to further explore, including: the decreased risk of an emergency delivery with later pregnancies and C-sections, further understanding of "other" admission types (i.e., not emergency or elective), and the relationship between repeat C-sections and emergency deliveries. We believe leveraging pregnancy-specific details extracted from the EHR is critical in understanding pregnancy-related outcomes at the patient level, and a useful approach to exploring deliveries with a greater level of granularity. In conclusion, our methodological approach enabled the findings presented in this study that support the importance of examining emergency vs. elective C-sections and assessing emergency C-sections as an adverse outcome rather than assuming that all C-sections are adverse events.

References

[1] S.P. Canelón, H.H. Burris, L.D. Levine, et al., Development and Evaluation of MADDIE: Method to Acquire Delivery Date Information from Electronic Health Records, MedRxiv. (2020). https://doi.org/10.1101/2020.07.30.20165381.

[2] N.J. Kassebaum, R.M. Barber, Z.A. Bhutta, et al., Global, regional, and national levels of maternal mortality, 1990-2015: a systematic analysis for the Global Burden of Disease Study 2015, Lancet. 388 (2016) 1775–1812. https://doi.org/10.1016/S0140-6736(16)31470-2.

[3] NPR, ProPublica, The Last Person You'd Expect To Die In Childbirth, (2017).

[4] M.P. Hehir, C. V. Ananth, Z. Siddiq, et al., Cesarean delivery in the United States 2005 through 2014: a population-based analysis using the Robson 10-Group Classification System, Am J Obstet Gynecol. 219 (2018) 105.e1-105.e11. https://doi.org/10.1016/j.ajog.2018.04.012.

[5] World Health Organization, WHO Statement on Caesarean Section Rates, 2015.

[6] K.D. Gregory, S. Jackson, L. Korst, et al., Cesarean versus vaginal delivery: Whose risks? whose benefits?, Am J Perinatol. 29 (2012) 7–18. https://doi.org/10.1055/s-0031-1285829.

[7] American College of Obstetricians and Gynecologists, Safe Prevention of the Primary Cesarean Delivery, 2014.

[8] E.M. Johnson, M.M. Rehavi, Physicians Treating Physicians: Information and Incentives in Childbirth, Am Econ J Econ Policy. 8 (2016) 115–141. https://doi.org/10.1257/pol.20140160.

[9] P.K. Foo, R.S. Lee, K. Fong, Physician prices, hospital prices, and treatment choice in labor and delivery, Am J Heal Econ. 3 (2017) 422–453. https://doi.org/10.1162/ajhe_a_00083.

[10] E. Oster, W.S. McClelland, Why the C-Section Rate Is So High - The Atlantic, Atl. (2019).

[11] A. Arrieta, A. García Prado, Non-elective C-sections in public hospitals: capacity constraints and doctor incentives, Appl Econ. 48 (2016) 4719–4731. https://doi.org/10.1080/00036846.2016.1164820.

[12] US News, High C-Section Rates at Birth Raise Questions About Hospitals, Health, Heal Communities. (2019).

[13] US News, The Rise of C-Sections and What It Means, Heal Communities. (2019).

[14] R.M. Silver, M.B. Landon, D.J. Rouse, et al., Maternal Morbidity Associated With Multiple Repeat Cesarean Deliveries, Obstet Gynecol. 107 (2006) 1226–1232. https://doi.org/10.1097/01.AOG.0000219750.79480.84.

[15] J. Feagin, Z. Bennefield, Systemic racism and U.S. health care, Soc Sci Med. 103 (2014) 7–14. https://doi.org/10.1016/j.socscimed.2013.09.006.

[16] Z.D. Bailey, N. Krieger, M. Agénor, et al., Structural racism and health inequities in the USA: evidence and interventions, Lancet. 389 (2017) 1453–1463. https://doi.org/10.1016/S0140-6736(17)30569-X.

[17] R Core Team, R: A language and environment for statistical coding, R Found Stat Comput. (2019).

[18] H. Wickham, M. Averick, J. Bryan, et al., Welcome to the Tidyverse, J Open Source Softw. 4 (2019) 1686. https://doi.org/10.21105/joss.01686.

[19] D. Peleg, Y.Z. Burke, I. Solt, et al., The history of the low transverse Cesarean section: The pivotal role of Munro Kerr, Isr Med Assoc J. 20 (2018) 316–319.

[20] B. Jacobsson, J.L. Simpson, Preterm birth: A clinical enigma and a worldwide public health concern, Int J Gynecol Obstet. 150 (2020) 1–2. https://doi.org/10.1002/ijgo.13194.

[21] M. Laursen, C. Johansen, M. Hedegaard, Fear of childbirth and risk for birth complications in nulliparous women in the Danish National Birth Cohort, BJOG An Int J Obstet Gynaecol. 116 (2009) 1350–1355. https://doi.org/10.1111/j.1471-0528.2009.02250.x.

[22] Centers for Disease Control and Prevention, Racial and Ethnic Disparities Continue in Pregnancy-Related Deaths, CDC Online Newsroom. (2019).

[23] Center for American Progress, Health Disparities by Race and Ethnicity, 2020.

Co-occurrence Patterns of Intimate Partner Violence

Ahmet Hacıaliefendioğlu[1], Serhan Yılmaz[1], Mehmet Koyutürk[1,2], and Günnur Karakurt[†,3,4]

(1) Department of Computer and Data Sciences, (2) Center for Proteomics and Bioinformatics, (3) Department of Psychiatry, Case Western Reserve University, Cleveland, OH, USA
(4) University Hospitals, Cleveland, OH, USA
[†]*E-mail:* gunnur.karakurt@case.edu

Intimate partner violence (IPV) is an important social and public health problem, affecting millions of women worldwide. Violence in a relationship can occur in multiple ways, including physical violence, psychological aggression, and sexual violence. In this study, utilizing data from the National Intimate Partner and Sexual Violence Survey (NISVS), we comprehensively investigate the interplay between physical, psychological, and sexual violence, in terms of their co-occurrence patterns, their relation to trauma symptoms and overall health of victims. For this purpose, we perform network analysis and develop a visualization technique that enables in-depth navigation of the three-dimensional (physical, psychological, sexual) space of violence. Our findings show that physical violence tends to significantly co-occur with psychological abuse, and violence intensifies when both are present. We also find that sexual violence tends to overlap less with other types of violence, particularly with physical violence. Milder forms of psychological abuse are prominent in the population and seem to represent a separate type of abuse (micro-aggression) in terms of its occurrence patterns. Finally, we observe that trauma symptoms and health problems tend to be reported more by survivors at the presence of intense psychological aggression. Our findings can be useful in developing treatments that target different patterns of IPV.

Keywords: Intimate partner violence, psychological aggression, physical violence, sexual violence, micro-aggression, co-occurrence, network analysis, clustering, data visualization

1. Introduction

Intimate partner violence (IPV), also commonly referred to as domestic violence, is a significant public health issue that adversely affects the well-being of millions of women across the world. IPV is often defined as physical, sexual, and psychological aggression by a current or former intimate partner. According to CDC data, during their lifetime, one in every four women experience severe forms of physical violence.[1] Breiding *et al.*[1] define physical violence as using physical force with the intent to harm, inflict injury or cause death. Physical violence encompasses behaviors such as pushing, punching, kicking and using weapons.[2] Psychological aggression is defined as using communication, both verbal and non-verbal, with intent to mentally and emotionally harm another person. They also include exerting control into their definition.[1] Psychological aggression encompasses explosive anger, coercive control, degradation and isolation.[3,4] The definition of sexual violence includes any sexual acts either committed or attempted without the informed consent of the victim and/or despite their refusal.[1] Sexual violence encompasses but is not limited to intentional unwanted sexual touching, pressuring for sex, and forced penetration.[1,2] The intensity of IPV cases range in severity from executing threats to committing homicide.[5]

Harmful effects of IPV on the physical health of women are often linked to acute injuries including bruises, lacerations, fractions, as well as chronic conditions including chronic pain syndrome, hypertension, and fibromyalgia.[5] IPV is also detrimental to sexual health and is frequently linked with sexually transmitted infections and urinary tract infections.[6] Furthermore, IPV's harmful effects on mental health are often associated with depression, anxiety, post-traumatic stress disorders, excessive stress, and suicidality[47].[8]

The co-occurrence of multiple types of violence is also common.[9] Different types of violence can co-occur with varying ranges of intensity in a relationship[9].[10] Indeed, past research indicated high positive correlation with psychological and physical abuse.[11] Identifying patterns of different types of IPV that are simultaneously occurring in the relationships can help immensely with treatment efforts.[12] However, elucidation of these complex co-occurrence patterns require comprehensive computational analyses on large scale data. In this paper, capitalizing on the availability of data from the National Intimate Partner and Sexual Violence Survey (NISVS), we aim to comprehensively characterize the co-occurrence patterns of IPV.

NISVS surveyed thousands of women in the United States to collect comprehensive data on the manifestation of different types of violence. While these large-scale data have been useful in assessing the prevalence and intensity of different types of violence, little is known on the interplay between these different types. Here, we develop a comprehensive computational framework to systematically characterize the interplay between different types of violence. Our computational framework and contributions include the following components:

(1) Using contingency analysis, we comprehensively quantify the overlap between four different types of violence (also including micro-aggression (MA) in addition to the other three types that are explicitly measured, as our analysis suggests that MA comprises an individual type of violence in terms of its prevalence and occurrence patterns).
(2) Using network analysis, we investigate the co-occurrence of individual violence items and identify the items that are central to each violence type and characteristic of the interplay between different violence types.
(3) We develop a radial visualization technique that quantifies the intensity and the type of IPV (reported by a survivor), which allows elaborate visualization of the interplay between different violence types and subgroups, as well as the projection of other variables (trauma symptoms, health problems) to the space defined by violence type and intensity.
(4) Using clustering, we identify subgroups of survivors who are similar in terms of their reported violence and assess how the resulting subgroups align with violence types.

2. Materials and Methods

2.1. *Description of Data and Pre-Processing*

Data from the National Intimate Partner and Sexual Violence Survey (NISVS) is utilized in this study.[13] This data was obtained through phone surveys of households across the United States. Randomly selected households were sent letters indicating they would be contacted for an interview. Overall, 16507 participants completed the interviews through the end.

NISVS is specifically designed to measure various characteristics related to relationship demographics, IPV and their adverse effects on health. The 39 items measuring the type, frequency and intensity of IPV are listed in Table 1. These items ask how many times the perpetrator did a specific action in the past year and answers are rated by the survivor in a scale of 0 (never), 1 (ten time), 2 (two to ten times), 3 (eleven to fifty times), and 4 (more than fifty times). We use these reported numbers directly in our analyses as an approximation to log-transformed frequencies of occurrence.

The items in the violence questionnaire are grouped into three violence types: (i) Physical violence (PV, 12 items), (ii) Sexual violence (SV, 22 items), and (iii) Psychological aggression (PA, 5 items). As we discuss in Section 3, we move some items between violence types. Based

Table 1: The questionnaire items used in our study.

- How many times did [perpetrator] ... ?
MA1: called you names like ugly, fat, crazy, or stupid
PA1: acted very angry towards you in a way that seemed dangerous
PA2: told you that you were a loser, a failure, or not good enough
PA4: insulted, humiliated, or made fun of you in front of others
PA5: told you that NO one else would want you
PA6: made threats to physically harm you
- How many times did [perpetrator] ... ?
PV2: slapped you
PV3: pushed or shoved you
PV4: hit you with a fist or something hard
PV5: kicked you
PV6: hurt you by pulling your hair
PV7: slammed you against something
PV9: tried to hurt you by choking or suffocating
PV10: beaten you
PV11: burned you on purpose
PV12: used a knife or gun on you
- How many times did [perpetrator] ... you didn't want it to happen?
SV1: exposed their sexual body parts to you, flashed you, or masturbated in front of you
SV2: made you show your sexual body parts to them
SV3: made you look at or participate in sexual photos or movies
SV4: harassed you while you were in a public place in a way that made you feel unsafe
SV5: kissed you in a sexual way?
- How many times did [perpetrator] ... you didn't want it to happen?
SV6: fondled or grabbed your sexual body parts
How many times did [perpetrator] ... when you were drunk, high, drugged, or passed out and unable to consent?
SV7: had vaginal sex with you
SV9: made you receive anal sex
SV10: made you perform oral sex
SV11: made you receive oral sex
- How many times did [perpetrator] used physical force or threats to physically harm you to make you ... ?
SV12: have vaginal sex
SV15: perform oral sex
SV16: receive oral sex
SV18a: (if male) try to make you have vaginal sex with them, but sex did not happen
SV18b: try to have (if female, vaginal) oral, or anal sex with you, but sex did not happen
- How many people have you had vaginal, oral, or anal sex with after they pressured you by ... ?
SV19: doing things like telling you lies, making promises about the future they knew were untrue, threatening to end your relationship, or threatening to spread rumors about you
SV20: wearing you down by repeatedly asking for sex, or showing they were unhappy
SV21: using their influence or authority over you, for example, your boss or your teacher
SV22: forced you to engage in sexual activity

on occurrence patterns, we also separate one item in the PA group as a separate violence type. Namely, we observe that the item "called you names like ugly, fat, crazy, or stupid" appears too frequently and lies as an outlier in the principal component space (Figure 2(b)). To facilitate thorough analysis of this frequent item with its own occurrence pattern, we separate this item as fourth violence type termed micro-aggression (MA). This results in the following number of items per violence type: 10 items for PV, 23 items for SV, 5 items for PA, and 1 item for MA. In addition, the dataset includes 16 items measuring health problems (including asthma, diabetes, irritable bowel syndrome, high blood pressure, frequent headaches, chronic pain, difficulty sleeping, stress, perceived physical and mental health) as well as items measuring trauma symptoms (including concern for safety, fear, having nightmares, and desire to avoid remembering).

Filtering the survivors. Since the study is performed on randomly selected households, most of the participants did not report any IPV. We also exclude instances where the perpetrator is not an intimate partner. Among the 16,507 participants who completed the survey, 873 of them reported at least one incidence of IPV in the past year (i.e., responded 1-4 to at least one of the 39 items in the survey). We focus on these 873 survivors in this study.

Data matrix and the computation of scores for violence types. Filtering results in a

873×39 data matrix R of survivors vs. items, where $R(i, j) \in \{0, 1, 2, 3, 4\}$ represents the response of survivor i to item j. We systematically analyze this data matrix from the perspective of survivors, as well as items. For this purpose, we call each row of this matrix a *survivor profile* and each column of this matrix an *item profile*. To assess the intensity of violence for each type, for each survivor, we compute an aggregate score averaging the responses of all items in the respective subscale. These scores, denoted $s_{PV}(i)$, $s_{PA}(i)$, $s_{SV}(i)$, and $s_{MA}(i)$ for survivor i, provide a summary statistic of the intensity of a particular violence type for the survivor.

2.2. *Co-Occurrence of Violence Types*

Here, we aim to assess whether violence types have a strong association with each other. For this purpose, for each violence type (PA, PV, SV, and MA), we identify the set of survivors who report a "high" level of violence in that category. To identify "high" levels of violence in a category, we use the population mean of s_T as a threshold. Namely, if $s_T(i) > \bar{s}_T$ for participant i, we consider that participant i reports high violence in category T, where $\bar{s}_T = \sum_{i=1}^{n} s(i)/n$. We denote the number of participants who report "high" levels of violence in type T according to this threshold as $n_H(T) = |\{i : s_T(i) > \bar{s}_T\}|$. While we report results according to population mean as the threshold, the results we obtain with different thresholds (including a threshold of zero, i.e., existence of violence or one or two standard deviation(s) above mean) are similar.

We assess the pairwise co-occurrence between two violence type T and T' as $n_{HH}(T, T') = |\{i : s_T(i) > \bar{s}_T \text{ and } s'_T(i) > \bar{s}'_T\}|$, i.e., the number of survivors who report both T and T' above population mean. To provide a baseline for expected co-occurrence, we compute the expected number of overlaps based on the assumption that the two violence types are independent, i.e., $E[N_{HH}(T, T')] = n_H(T)n_H(T')/n$, where $N_{HH}(T, T')$ denotes the random variable that represents the co-occurrence of T and T' (with observed value $n_{HH}(T, T')$).

To quantify the magnitude of the co-occurrence between T and T', we use odds ratios:[14]

$$OR(T, T') = \frac{n_{HH}(T, T')n_{LL}(T, T')}{n_{HL}(T, T')n_{LH}(T, T')}. \quad (1)$$

Here, $n_{LL}(T, T') = |\{i : s_T(i) \leq \bar{s}_T \text{ and } s'_T(i) \leq \bar{s}'_T\}|$ denotes the number of survivors who report "low" violence for both types T and T'. $n_{HL}(T, T')$ and $n_{HL}(T, T')$ are defined similarly as respectively "high T"/"low T" and "low T"/"high T". To assess the statistical significance of the odds ratios, we compute 95% confidence intervals as follows:

$$\begin{aligned} SE(T, T') &= \left(\frac{1}{n_{HH}(T, T')} + \frac{1}{n_{LL}(T, T')} + \frac{1}{n_{HL}(T, T')} + \frac{1}{n_{LH}(T, T')} \right)^{1/2} \\ OR_{\max}(T, T') &= OR(T, T')e^{SE(T,T')}, \qquad OR_{\min}(T, T') = OR(T, T')/e^{SE(T,T')}. \end{aligned} \quad (2)$$

2.3. *Co-Occurrence Network of Individual Violence Items*

To obtain the co-occurrence network between individual violence items, we first compute their Pearson correlation between the item profiles in a pairwise manner. We then construct a network by putting an edge between two items if they exhibit positive correlation greater than a given threshold (we use 0.2 in the results presented Section 3).

2.4. *Radial Visualization*

To investigate the relationship between violence types, we apply principal component analysis (PCA) to map survivors to the 2-dimensional PCA space as shown in Figure 1. In the top panel of this figure, the survivors are colored according to each violence type separately. To color the survivors according to violence type scores, we first use *rank normalization* to normalize the scores into the [0, 1] range. For this purpose, separately for each violence type T, we sort the survivors according to their s_T scores. Then, for each survivor i, we take the percentile of that survivor according to this ranking as their rank normalized score $r_T(i)$.

The brightness of the R/G/B channel for each survivor indicates is set to be proportional to this rank-normalized score for respectively PV, PA, and SV. As seen in the figure, survivors with high PV score are typically clustered on the bottom side of the PCA plane and the survivors with high PA score are typically clustered on top right side of the PCA plane. Survivors with high SV scores do not appear to be clustered. This is not surprising since SV items are given little weight by the principal components (Figure 2a) due to the relative rarity of SV.

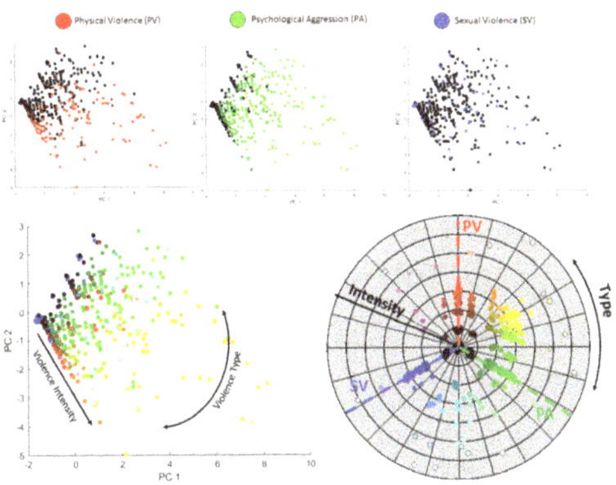

Fig. 1: Using radial projection to visualize survivors in the three-dimensional space of violence types. (Top) Distribution of physical violence (PV), psychological aggression (PA), sexual violence (SV) scores in the plane of first two principal components. Coloring indicates the intensity of the corresponding violence type (PV, PA or SV). (Bottom Left) Distribution of PV, PA and SV scores in the plane of the first two principal components. Coloring is done according to PV, PA and SV scores: Red component: PV, Green component: PA, Blue Component: SV. As it can be seen, the first two principal components reflect the violence intensity as well as the violence type. (Bottom Right) Projection of the survivors to the radial space of violence type vs. violence intensity.

When we integrate the RGB values to visualize all violence types at once, we obtain the plot in the bottom left panel of Figure 1, leading to two interesting observations: (i) The intensity of the violence (as well as the brightness of the color) typically increases as the distance from the center in the PCA plane (middle left corner) increase, and (ii) The type of the violence (as well as the hue of the color) changes depending on the angle of the surrounding arc. This means that the principal component analysis essentially captures these two inherent properties in the population.

Motivated by this observation, we develop a novel radial visualization scheme where the survivors are placed onto a two-dimensional plane with respect to their violence intensity and/or types. The objective of our approach is to present the interplay between the physical, psychological, and sexual components of violence in a visually accessible and comprehensible manner.

In order to visualize the survivors on a two-dimensional plane of violence type and intensity, we utilize a transformation scheme that is originally proposed for transforming the color space. This transformation (known as HSL) aims to represent a color on a three dimensional space having hue, saturation and luminance as axes instead of the usual red-green-blue (RGB) axes. In this space, hue indicates the color type (e.g., measures the difference between red and yellow colors), saturation indicates the color homogeneity (e.g., measures how much the color is different from gray), and the luminance indicates the brightness of the color (e.g., measures the difference between black and white).

In our case, when we consider that each of the three violence types corresponds to a different color (red, green and blue), the hue and luminance components of the HSL transformation essentially indicate the violence type and the intensity respectively. From a given set of rank normalized scores $r_{PV}(i)$, $r_{PA}(i)$, and $r_{SV}(i)$ for survivor i, we compute the HSL components as follows:[15]

$$I_{max}(i) = \max\{r_{PV}(i), r_{PA}(i), r_{SV}(i)\}, \quad I_{min}(i) = \min\{r_{PV}(i), r_{PA}(i), r_{SV}(i)\},$$
$$\text{Intensity}(i) = (I_{max}(i) + I_{min}(i))/2. \tag{3}$$

Now letting $M(i)$ denote the violence type with maximum rank-normalized score for survivor i and setting $\Delta(i) = I_{max}(i) - I_{min}(i)$, we quantify the Type of violence for survivor i as follows:

$$H'(i) = \begin{cases} \text{undefined}, & \text{if } \Delta(i) = 0, \\ ((r_{PA}(i) - r_{SV}(i))/\Delta(i)) \mod 6 & \text{if } M(i) = PV, \\ ((r_{SV}(i) - r_{PV}(i))/\Delta(i)) + 2 & \text{if } M(i) = PA, \\ ((r_{PV}(i) - r_{PA}(i))\Delta(i)) + 4 & \text{if } M(i) = SV. \end{cases} \tag{4}$$
$$\text{Type(i)} = H'(i) \times \pi/6$$

Note that Type indicates an angle, thus, it is defined in radians.

Using *Intensity* (corresponding to violence intensity) and *Type* (corresponding to violence type), we compute the location of survivor i in the two-dimensional plane as:

$$x(i) = \text{Intensity}(i) \times cos(\text{Type}(i)), \qquad y(i) = \text{Intensity}(i) \times sin(\text{Type}(i)) \tag{5}$$

The visualization of the survivors in this violence type vs. intensity space is shown in Figure 1, bottom right panel. As seen in the figure, in this space, the distance from the center ([0, 0]) indicates the intensity of violence, and the arc angle indicates the type of violence.

2.5. *Clustering of Survivors and Identification of Subgroups*

We apply clustering to identify subgroups of survivors based on their responses to the 39 items in the violence questionairre. For this purpose, we cluster the survivor profiles using K-means clustering with Euclidean distance by employing *kmeans* function in MATLAB Statistics and Machine Learning Toolbox.[16] In order to find more reliable clusters, we run K-means 100 times and select the clustering with the minimum total within-cluster distance (sums of point-to-centroid distances). We use different values of K to optimize the number of clusters using Calinski Harabasz Evaluation.[17]

2.6. *Health Problems and Trauma Symptoms*

To investigate the relationship of reported violence with health problems and trauma symptoms, we compute category scores for health problems and trauma symptoms subscale as previously described for the violence subscales. Subsequently, we bin survivors according to their location in the violence intensity vs. violence type space and compute the average scores for health problems and trauma symptoms in each bin. We use radial visualization to visualize this results. This visualization allows the investigation of health problems and trauma symptoms with respect to violence type and intensities.

3. Results

Individual item frequencies and assignment of items to violence types. We first investigate the overall reporting frequency of individual items in the IPV questionnaire. The results of this analysis are shown in Figure 2. The bar plot in Figure 2(a) shows the frequencies

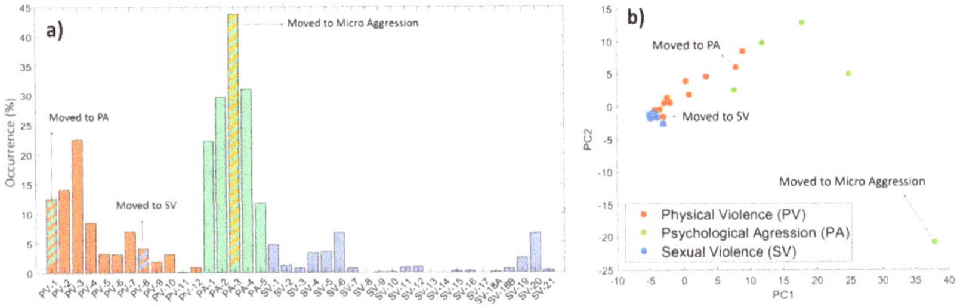

Fig. 2: Response rate and principal component analysis of the items in the questionnaire. a) Reporting frequencies of the questionnaire items among the 873 survey survivors who report an incidence of intimate partner violence. Items are grouped and colored according to their scales: Physical Violence (PV), Psychological Aggression (PA), and Sexual Violence (SV). b) The projection of items projected on the space induced by the first two principal components. Each item is colored according to their scales. The items that were moved to another scale are marked.

of all 39 items, grouped by subscales (violence types). As seen in the figure, items that belong to the Psychological Aggression (PA) subscale are most frequently reported by survivors of violence, while items in the Sexual Violence (SV) subscale are reported least frequently.

The projection of the items to the two-dimensional principal component space is shown in Figure 2(b). As seen in the figure, items that belong to the same subscale are clustered in this reduced dimensional space. If we consider these two principal components as "eigensurvivors", it is clear that these eigen-survivors tend to report similarly on all items (i.e., the items lie on a linear line with a positive slope), with the exception of PA3 ("called you names like ugly, fat, crazy, or stupid?"). This item lies as an outlier in the principal component space. Since this item is the most frequently reported item in the questionnaire and has a substantial influence on the PCA analysis, we decided to investigate it separately and labeled it as microaggression (MA). We also observe that PV1("made threats to physically harm you?") can be considered psychological aggression and lies close to psychological violence in the principal component space. Similarly, PV8 ("forced you to engage in sexual activity?") involves sexual violence and lies close to SV items in this space. For these reasons, we move these items to the respective subscales.

Co-occurrence of violence types. Once the assignment of items to violence types is finalized, we investigate the co-occurrence of violence types. The results of this analysis are shown in Figure 3. We observe significant co-occurrence of physical violence and psychological aggression, with an odds ratio of 3.62 (95% confidence interval: [2.60, 5.06]) and a linear correlation of 0.449 (P<0.001). While the co occurrence between physical violence and micro-aggression is weaker (OR=1.52, correlation=0.218, P<0.001), we observe that micro-aggression and psychological aggression occur frequently together (OR=2.97, correlation: 0.385, P<0.001). Interestingly, sexual violence tends to exhibit significantly less co-occurrence with all other types of violence, with an odds ratio below 1.0 and near zero (below 0.1) correlation for all other violence types (correlations: PV-SV=0.015, PA-SV=0.08, MA-SV=0.007). The 95% confidence intervals for the odds ratios fall completely below 1.0 for SV vs. PA and for SV vs. MA.

Co-occurrence of individual items. To assess the co-occurrence patterns of IPV at a higher resolution, we also investigate the co-occurrence at the level of individual items. For this purpose, we assess the correlation between all pairs of the 39 items, and construct a

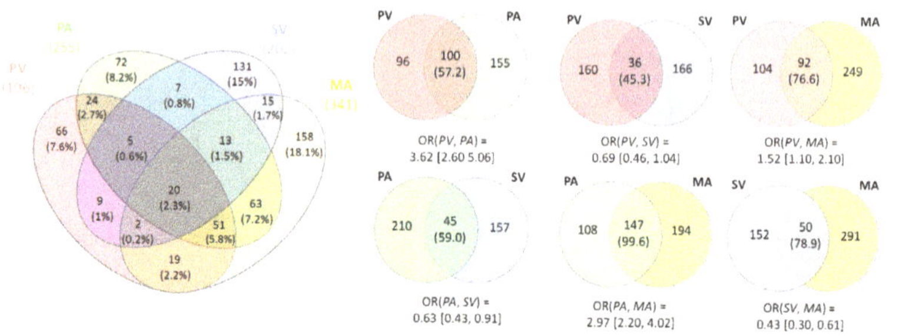

Fig. 3: Co-occurrence of different types of intimate partner violence. The sets represent Physical Violence (PV, red), Psychological Aggression (PA, green), sexual violence (SV, blue), or Micro Aggression (MA, yellow). The first number in each set shows the number of survivors who report that type of violence above population mean. For the 4-way Venn-diagram, the numbers in parentheses show the percentage of survivors (over all survivors) in the respective set. The 2-way Venn diagrams assess the significance of the overlap between pairs of violence types, where the number in parenthesis shows the expected value of the intersection given the frequencies of each type. The resulting odds ratios (and 95% confidence intervals for the ORs) are shown below the Venn diagrams.

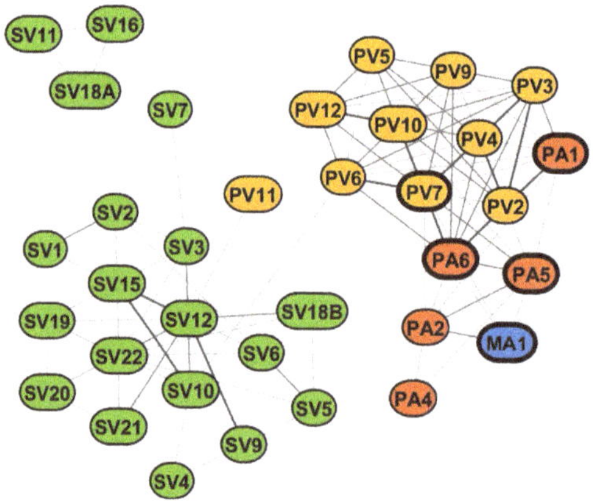

Item	Cross Degree	Total Degree	
PA6	8	11	Cross Degree
PA1	7	10	
PV7	5	13	
MA1	4	4	
PA5	4	8	
PV7	5	13	Total Degree
PA6	8	11	
PV4	3	11	
PV6	3	11	
PV10	3	11	

Fig. 4: Itemwise co-occurrence network of intimate partner violence. The nodes represent violence items and the edges indicate the existence of positive correlation (> 0.2) between survivors' responses to the corresponding pair of items. The widths of edges show the strength of correlation. The nodes (items) are colored according to their corresponding subscale: Yellow for Physical Violence (PV), red for Psychological Aggression (PA), green for Sexual Violence (SV), blue for Micro-Aggression (MA). The top five nodes with highest degree and highest cross-degree (with other violence types) are shown on the right.

network by retaining all pairs with correlation > +0.2. As seen in Figure 4, the network has two large connected components connected by a single weak edge.

One of these components represents Sexual Violence (SV), while Physical Violence (PV),

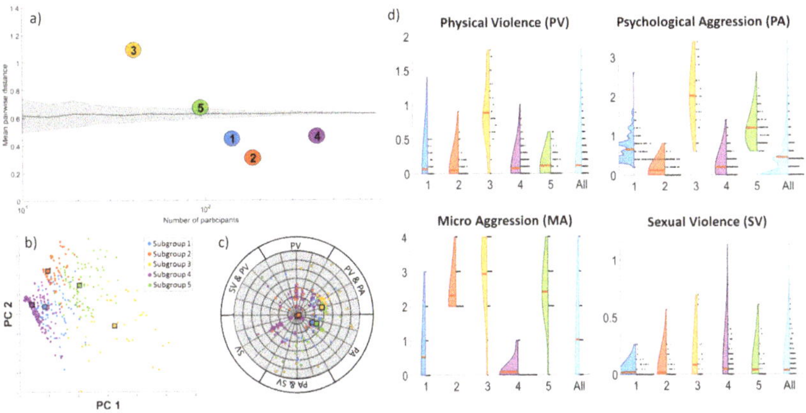

Fig. 5: Clustering of survivors and the identification of subgroups. a) Size (number of survivors, log-scaled) vs. heterogeneity (measured by mean pairwise distance between survivors) of the five clusters of survivors identified using K-means. The black line and the grey area show the mean/95% confidence interval for the heterogeneity of random groups of survivors as a function of size (100 permutations). (b, c) Visualization of clusters in the two-dimensional principal component space/radial projection. Each survivor is colored according to their cluster/subgroup. Colored squares show the centers of respective subgroups. (d) Distribution of the scores for four different violence types in the identified subgroups.

Psychological Aggression (PA), and Micro-Aggression (MA) are together represented by a single component. We observe that the correlations among items within PV are stronger, with PV7 ("slammed you against something") being the central node in the PV-PA cluster. The central item for the SV component, on the other hand, is SV12 ("used physical force or threats to physically harm you to make you have vaginal sex"). Interestingly, PV11 ("burned you on purpose ") is also connected to SV12, although it is not connected to any other PV item.

Clustering of survivors to identify violence subgroups. To understand whether the survivors induce coherent subgroups in their reporting of violence and whether these subgroups are aligned with reported violence types, we use K-means to cluster the survivors. Using Calinski-Harabasz evaluation,[17] we determine that $K = 5$ provides a reasonable balance between model fit and complexity. The resulting subgroups are shown in Figure 5.

We observe that subgroups with more survivors tend to be more homogeneous, where the smallest subgroup (#3) is significantly more heterogeneous than would be expected for a random group of survivors (Figure 5(a)). Visualization of the survivors in the subgroups in the two-dimensional principal component axes (Figure 5(b)) and radial axes (Figure 5(c)) shows that subgroups #2 and #5 are well-separated from other subgroups and this separation is reflective of the intensity of violence. In contrast, subgroups #1, #3, and #4 are separated from each other mostly based on the type of violence. Based on the distribution of the scores of violence types in each subgroup (Figure 5(d)), we annotate these subgroups as follows:

- *Subgroup #1:* Very low intensity of sexual violence, low/moderate intensity of other violence types.
- *Subgroup #2:* Very high intensity of micro-aggression, low intensity of psychological ag-

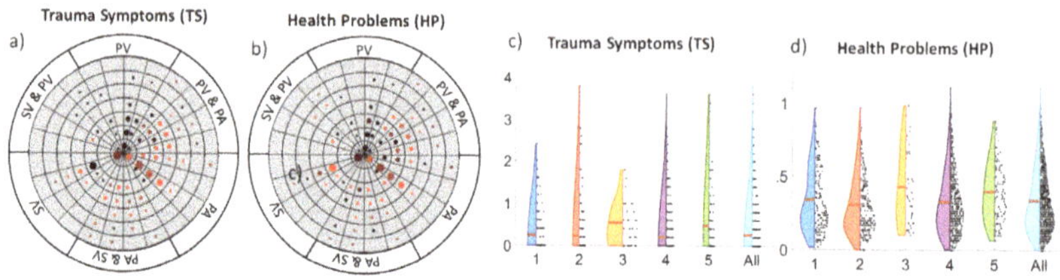

Fig. 6: Relationship between IPV types/subgroups and health problems/trauma symptoms. (a)/(b) Radial visualization of trauma symptoms/health problems reported by the survivors. survivors are binned into small groups (shown as circles) based on their violence type and intensity. Distance from the center indicates the intensity of the violence, the angle indicates the type of violence. The size of the circle indicates the number of survivors in the corresponding group. The intensity of red indicates the prevalence of trauma symptoms/health problems reported by the survivors in that group. (c)/(d) The distribution of the prevalence of trauma symptoms and health problems reported by the survivors in each subgroup identified by clustering (see Figure 5).

gression, low/moderate intensity of other violence types.

- *Subgroup #3:* Very high intensity of all violence types.
- *Subgroup #4:* Variable intensity of sexual violence, low intensity of other violence types, particularly very low intensity of micro-aggression.
- *Subgroup #5:* Very high intensity of psychological aggression and micro-aggression, low intensity of physical violence and sexual violence.

Health problems and trauma symptoms reported by survivors. In addition to the violence variables, NISVS also screens survivors for trauma symptoms and health problems. To understand how these health problems and trauma symptoms correlate with violence types and subgroups, we assess the distribution of survivors' responses to these questions in the radial axis of PV-PA-SV, and in the subgroups we identify via clustering. The results of these analyses are shown in Figure 6. As seen in Figure 6(a), trauma symptoms are most commonly reported at the presence of intense psychological aggression and this effect is more pronounced when physical violence is also present. We also observe a similar pattern for health problems in Figure 6(b); however, health problems are also amplified with the presence of sexual violence. The distributions of these two variables in the five subgroups (Figure 6(c)/(d)) also show that trauma symptoms and health problem are most frequently reported in Subgroup #3, the subgroup that is associated with most intense psychological aggression and physical violence. The other subgroup that reports trauma symptoms and health problems above the population mean is Subgroup #5, which is associated with very high intensity of psychological aggression and micro-aggression, despite having lower levels of physical violence. We also observe that Subgroups #2 and #4 have long tails for trauma symptoms, while the long tail of the entire population for health problems is carried by Subgroup #4, indicating that very high intensity of sexual violence can be associated with significant health problems.

4. Discussion

Severity and type of violence perpetrated in the relationships have been increasingly utilized to understand patterns of IPV. In this study, we aimed to identify these patterns. Our results indicated that physical violence occurs frequently with psychological agression, its co-occurrence with micro-aggression is weaker (Figure 3). We also found that sexual violence tends to overlap less with all other types of violence. We also observed that individual items for sexual violence formed a single connected component in the co-occurrence network of individual items. This is one of the important findings of this study. It is important to note that the sexual violence in our analysis only includes acts of sexual violence perpetrated by intimate partners, as we restricted our analysis to instances in which the survivor is in an intimate relationship with the perpetrator.

Our network analysis indicated that the co-occurrence of physical violence items is more common compared to other types of violence (Figure 4). Slamming the partner against something exhibited strong co-occurrence with other physical violence items, as well as psychological aggression items. Other physical violence items with "high degree" included beating and hitting with a fist or something hard. Interestingly, these items were not as frequent as the most frequent physical violence items, such as slapping, pushing, and showing (Figure 2). Thus the presence of these moderate-frequency high-co-occurrence items may be indicative of more systemic physical violence. Making threats to harm the partner was more frequent, and also exhibited strong co-occurrence with many physical violence and psychological aggression items. Acting very angry toward the partner in a way that seemed dangerous almost exclusively co-occured with physical violence. The observation that the violence items that tend to co-occur with other items are not necessarily more prevalent suggests that these co-occurrence patterns can be useful in dissecting the etiology of violence in a relationship.

The sexual violence item that most frequently co-occured with other sexual violence items was "used physical force or threats to physically harm you to make you have vaginal sex". The physical violence item "burned you on purpose" was also connected to sexual violence, although it was not connected to any other physical violence item.

Our data driven definition of micro-aggression is conceptually consistent with the widely used definition of micro-aggression. Although micro-agression is a relatively new construct and is still in the process of refinement, it draws considerable attention by researchers. Our findings can help application of this concept in relationships.

With cluster analysis, we identified five subgroups of intimate partner violence (Figure 5). These subgroups were mostly aligned with violence types, with micro-aggression claiming its own subgroup. The distribution of sexual violence in the subgroups was variable and seemed to exclude micro-aggression. An important outcome of cluster analysis was that severe psychological abuse seems to underlie two different forms of severe violence; one with intense micro-aggression and another with severe physical violence.

A longitudinal study investigating the mental health trajectories of IPV victims indicated that women who were exposed to psychological abuse were less likely to recover overtime from mental health issues such as depression, anxiety and PTSD.[18] Past research also showed higher levels of mental health deterioration when both psychological and physical violence were co-occurring.[19] Another study investigating court-involved battered women's exposure to different types of IPV and the traumatic responses such as depression, acute stress and PTSD, demonstrated that they are associated while psychological abuse explained higher variance as compared to physical abuse.[20] Intensity of the psychological, physical and sexual violence as well as the context and presence of one or more types of victimization is critical for our understanding to develop effective treatments.

In summary, it is crucial to understand the nature of the violence and develop strategies to effectively deliver treatments and support for the victims. Based on nationally representative data, we identified co-occurence patterns and subgroups of IPV. These results can be useful to develop screening tools as well as targeted and integrative treatment strategies.

5. Acknowledgments

This publication was made possible by US National Health Institutes (NIH) grant R01-LM012518 from the National Library of Medicine. Its contents are solely the responsibility of the authors and do not necessarily represent the official views of the NIH. The authors declare that they have no competing interests.

References

1. M. Breiding, K. C. Basile, S. G. Smith, M. C. Black and R. R. Mahendra, Intimate partner violence surveillance: Uniform definitions and recommended data elements. version 2.0 (2015).
2. M. Straus, S. Hamby, S. Boney-McCoy and D. Sugarman, The revised conflict tactics scales (cts2) development and preliminary psychometric data, *Journal of family issues* **17**, 283 (1996).
3. N. S. Jacobson and J. M. Gottman, *When men batter women: New insights into ending abusive relationships* (Simon and Schuster, 1998).
4. D. R. Follingstad, The impact of psychological aggression on women's mental health and behavior: The status of the field, *Trauma, Violence, & Abuse* **10**, 271 (2009).
5. J. C. Campbell, Health consequences of intimate partner violence, *The lancet* **359**, 1331 (2002).
6. J. Silverman, M. Decker, N. Kapur, J. Gupta and A. Raj, Violence against wives, sexual risk and sexually transmitted infection among bangladeshi men, *Sexually transmitted infections* **83**, 211 (2007).
7. G. Karakurt, V. Patel, K. Whiting and M. Koyutürk, Mining electronic health records data: Domestic violence and adverse health effects, *Journal of family violence* **32**, 79 (2017).
8. T. O. Afifi, H. MacMillan, B. J. Cox, G. J. Asmundson, M. B. Stein and J. Sareen, Mental health correlates of intimate partner violence in marital relationships in a nationally representative sample of males and females, *Journal of interpersonal violence* **24**, 1398 (2009).
9. C. Armour and E. Sleath, Assessing the co-occurrence of intimate partner violence domains across the life-course: Relating typologies to mental health, *European Journal of Psychotraumatology* **5**, p. 24620 (2014).
10. D. R. Follingstad, S. Coyne and L. Gambone, A representative measure of psychological aggression and its severity, *Violence and Victims* **20**, 25 (2005).
11. E. W. Gondolf, D. A. Heckert and C. M. Kimmel, Nonphysical abuse among batterer program participants, *Journal of Family Violence* **17**, 293 (2002).
12. I. T. Jolliffe, Principal component analysis, *Principal component analysis* (2002).
13. U. S. D. of Health, H. S. C. for Disease Control, P. N. C. for Injury Prevention and Control, National intimate partner and sexual violence survey (nisvs): General population survey raw data, 2010 ann arbor, mi:inter-university consortium for political and social research [distributor] https://doi.org/10.3886/icpsr34305.v1 (2016).
14. M. Szumilas, Explaining odds ratios, *Journal of the Canadian academy of child and adolescent psychiatry* **19**, p. 227 (2010).
15. M. K. Agoston and M. K. Agoston, *Computer graphics and geometric modeling* (Springer, 2005).
16. kmeans (2020), The MathWorks, Natick, MA, USA.
17. T. Caliński and J. Harabasz, A dendrite method for cluster analysis, *Communications in Statistics-theory and Methods* **3**, 1 (1974).
18. C. Blasco-Ros, S. Sánchez-Lorente and M. Martinez, Recovery from depressive symptoms, state anxiety and post-traumatic stress disorder in women exposed to physical and psychological, but not to psychological intimate partner violence alone: A longitudinal study, *BMC psychiatry* **10**, p. 98 (2010).
19. M. A. Pico-Alfonso, M. I. Garcia-Linares, N. Celda-Navarro, C. Blasco-Ros, E. Echeburúa and M. Martinez, The impact of physical, psychological, and sexual intimate male partner violence on women's mental health: depressive symptoms, posttraumatic stress disorder, state anxiety, and suicide, *Journal of women's health* **15**, 599 (2006).
20. M. A. Dutton, L. A. Goodman and L. Bennett, Court-involved battered women's responses to violence: The role of psychological, physical, and sexual abuse, *Violence and victims* **14**, 89 (1999).

AI for infectious disease modelling and therapeutics

Gil Alterovitz
Brigham and Women's Hospital / Harvard Medical School, Boston, MA, 02115, USA.
National Artificial Intelligence Institute, Department of Veterans Affairs, Washington, DC, 20005, USA.
Email: ga@alum.mit.edu

Wei-Lun Alterovitz
Center for Biologics Evaluation and Research (CBER), U.S. Food and Drug Administration, Silver Spring, MD, 20993, USA
Argentys Informatics, Gaithersburg, MD, 20877, USA

Gail H. Cassell
Department of Global Health and Social Medicine, Harvard Medical School Professor and Chair, Emeritus, Department of Microbiology, University of Alabama at Birmingham
Scientific Affairs (ret), Eli Lilly and Company

Lixin Zhang
State Key Laboratory of Bioreactor Engineering East China University of Science and Technology
Shanghai, 200237, China

A. Keith Dunker
Center for Computational Biology and Bioinformatics (Emeritus), Department of Biochemistry and Molecular Biology Indiana University School of Medicine, Indianapolis, IN, 46202, USA

AI for infectious disease modelling and therapeutics is an emerging area that leverages new computational approaches and data in this area. Genomics, proteomics, biomedical literature, social media, and other resources are proving to be critical tools to help understand and solve complicated issues ranging from understanding the process of infection, diagnosis and discovery of the precise molecular details, to developing possible interventions and safety profiling of possible treatments.

Keywords: Artificial intelligence, infectious diseases, COVID-19, 2019-nCoV, modelling, therapeutics

1. Background

Back in the 19th century, physicians and scientists used to think "bad air" was the source of infection and disease. This miasma theory was ultimately replaced by the germ theory with the advance of the microscope and the discovery of microorganisms. This switch dramatically changed our understanding of infectious disease and started the new era of public health. This year again, the outbreak of novel coronavirus 2019-nCoV has turned people's attention to the importance of surveillance, prevention, diagnosis and treatment of infectious disease.

Besides harmful viruses like the coronavirus (e.g. 2019-nCoV, SARS, MERS), HIV, Zika, Ebola virus, some bacteria (1%) cause diseases in people such as tuberculosis (*Mycobacterium tuberculosis*) and pertussis (*Bordetella pertussis*). While most antibiotic drugs were developed for bacteria-based infection, antibiotic resistance has become a growing challenge because of antibiotic misuse and poor stewardship. On the other hand, adopting new microbiome-based therapeutics is another potential risk delivering antimicrobial resistance genes to the human body and intestinal microtome via mobile genetic elements (or the other way around). For example, an important safety alert has been issued for use of a recent successful FDA-approved microbiome-based intervention, fecal microbiota transplantation (FMT) due to transmission of multi-drug resistant organisms. Disordered protein modelling is playing an important role in understanding microorganism structure and function as well.

2. *Introduction*

New computational approaches are leading to new possibilities for AI for infectious disease modelling and therapeutics, leveraging new resources such as the 2019-nCoV-released genomic sequences along with protein-protein interactions, among others. The large number of bacteria, viruses, fungi, and other microorganism genomes that are available along with clinical implications of observed mutations, make these particularly amenable to development of novel computational methods. By combining information at multiple scales, new insights have arisen via the tools, pipelines, and associated algorithm development as well. This session has a number of areas that integrated AI for infectious disease modelling and therapeutics in the proceedings:

3. *Social Media and COVID-19*

In "Characterization of Anonymous Physician Perspectives on COVID-19 Using Social Media Data" by K. J. Sullivan, et. al. [1] explored using Twitter to characterize different perspectives on COVID-19. Specifically, physician direct messages were compared to general public tweaks. The work analyzed over 513 million tweets in the process. Sentiment and n-gram analysis revealed patterns within physician vs public discourse regarding COVID-19[1].

4. *Biomedical literature and COVID-19 plus neglected tropical diseases*

Work by B. Dinakar, et. al. [2] explored the biomedical literature using an algorithm to find novel directions in disease research, with focus on COVID-19 and also neglected tropical diseases. The paper "Semantic Changepoint Detection for Finding Potentially Novel Research Publications" analyzed publications over time to find patterns where there are significant changes in direction such that a semantic changepoint can be defined by the algorithm. The software is also released via link in the paper.

5. *Genomics and HCV*

S. Sledzieski, et. al. [3] describe a new method for reconstructing transmission phylogenies that increases accuracy while maintaining scalability. The paper "TreeFix-TP: Phylogenetic Error-Correction for Infectious Disease Transmission Network Inference" applied the method to HCV outbreaks. It also released the software via link in the paper.

6. *Protein intrinsically disordered regions and SARS-CoV-2*

A. Mudide, et. al. [4] analyzed SARS-CoV-2 for certain protein regions, known intrinsically disordered regions, that have additional flexibility, that may be targets for drug candidates. The paper, "SARS-CoV-2 Drug Discovery Based On Intrinsically Disordered Regions," also leveraged different docking approaches to model this flexibility and prioritize potential drug candidates, analyzing over 290 thousand compounds.

G. Goh, et. al. [5] analyzed the shell disorder of SARS-CoV-2 and other viruses to establish shell disorder as a proxy for vaccine development feasibility. It characterized SARS-CoV-2 as having an exceptionally hard outer shell, suggesting that vaccine development for SARS-CoV-2 is likely feasible and may be easier than for several other viruses such as HIV, HSV and HCV. The work in the paper "Feasibility study of vaccine development for SARS-CoV-2 and other viruses using shell disorder analysis" also presented several ideas on how shell disorder can be leveraged to characterize virulence, immune system evasion, and potential animal hosts.

7. *Protein-protein interactions and SARS-CoV-2*

M. Kshirsagar, et. al. [6] examined protein-protein interactions between viruses and their host. By comparing SARS-CoV-2 and host interactions with other virus-host interactions a number of motifs and themes emerged. The paper, entitled, "Functional comparison of virus-host pathogen communication using sequence-feature based SARS-CoV-2 protein interaction prediction" also specifically created SARS-CoV-2-human protein-protein interaction predictor based on sequence information and validated it with an independent dataset.

References

1. Sullivan, Katherine J., et al, "Characterization of Anonymous Physician Perspectives on COVID-19 Using Social Media Data", *Pac Symp Biocomput, 2020.*
2. Dinakar, Bhavish, et al. "Semantic Changepoint Detection for Finding Potentially Novel Research Publications", *Pac Symp Biocomput, 2020.*
3. Sledzieski, Samuel, et al. "TreeFix-TP: Phylogenetic Error-Correction for Infectious Disease Transmission Network Inference", *Pac Symp Biocomput, 2020.*
4. Mudide, Anish. et al. "SARS-CoV-2 Drug Discovery Based On Intrinsically Disordered Regions", *Pac Symp Biocomput, 2020.*
5. Goh, Gerard Kian-Meng, et al. "Feasibility study of vaccine development for SARS-CoV-2 and other viruses using shell disorder analysis", *Pac Symp Biocomput, 2020.*
6. Kshirsagar, Meghana, et al. "Functional comparison of virus-host pathogen communication using sequence-feature based SARS-CoV-2 protein interaction prediction", *Pac Symp Biocomput, 2020.*

Characterization of Anonymous Physician Perspectives on COVID-19 Using Social Media Data

Katherine J. Sullivan
Data Science to Patient Value, University of Colorado School of Medicine
Aurora, CO 80045, USA
Corresponding Email: Katherine.Sullivan@CUAnschutz.edu

Marisha Burden, MD and Angela Keniston, MSPH
Division of Hospital Medicine, University of Colorado School of Medicine
Aurora, CO 80045, USA
Email: marisha.burden@cuanschutz.edu and Angela.Keniston@cuanschutz.edu

Juan M. Banda
Department of Computer Science, Georgia State University
Atlanta, Georgia, 30303, USA
Email: jbanda@gsu.edu

Lawrence E. Hunter
Computational Bioscience Program, University of Colorado School of Medicine
Aurora, CO 80045 USA
Email: Larry.Hunter@CUAnschutz.edu

Physicians' beliefs and attitudes about COVID-19 are important to ascertain because of their central role in providing care to patients during the pandemic. Identifying topics and sentiments discussed by physicians and other healthcare workers can lead to identification of gaps relating to the COVID-19 pandemic response within the healthcare system. To better understand physicians' perspectives on the COVID-19 response, we extracted Twitter data from a specific user group that allows physicians to stay anonymous while expressing their perspectives about the COVID-19 pandemic. All tweets were in English. We measured most frequent bigrams and trigrams, compared sentiment analysis methods, and compared our findings to a larger Twitter dataset containing general COVID-19 related discourse. We found significant differences between the two datasets for specific topical phrases. No statistically significant difference was found in sentiments between the two datasets, and both trended slightly more positive than negative. Upon comparison to manual sentiment analysis, it was determined that these sentiment analysis methods should be improved to accurately capture sentiments of anonymous physician data. Anonymous physician social media data is a unique source of information that provides important insights into COVID-19 perspectives.

Keywords: Social media, Covid-19, Physicians

1. Introduction

Physicians treating COVID-19 patients have unique insights into the current pandemic response, some of which may identify opportunities for immediate improvements as well as improved responses to possible future pandemics. Their insights and perspectives during this difficult time are unique in understanding the impact that COVID-19 has had on frontline healthcare workers, patients, and perhaps even the public as a whole.[1] Understanding these physicians' attitudes and beliefs can improve health outcomes and drive successful health policies, particularly as patient safety and organizations' safety culture have been shown to be deeply affected by healthcare worker beliefs.[2–4]

Social media platforms, such as Twitter, are rich resources for opinionated data with respect to a myriad of topics, and can lead to a deeper understanding of ideas, opinions, and perspectives about a specific topic of interest, including COVID-19, and do so in real-time.[5] Twitter data is publicly available and relatively accessible for download; it is an exceptional resource for evaluating public discourse and sentiments given that it is the third most popular social media platform with approximately 330 million active users per month.[6] Unfortunately, in the midst of the COVID-19 pandemic, many physicians are hesitant to publicly discuss topics such as lack of sufficient personal protective equipment, testing equipment, and other issues they are facing in their workplace for fear of being reprimanded or even fired from their jobs.[7] For this reason, social media data posted specifically by physicians about COVID-19 is less likely to be an honest representation of their beliefs and attitudes regarding the pandemic response.[8]

In light of these concerns, a user handle on Twitter was set up where administrators collected direct messages (DMs) from physicians in the United States, and posted them anonymously under that handle, giving physicians some anonymity and a platform to express their perspective as it relates to the pandemic. In this study, we analyze the topics and sentiments of anonymous COVID-19 physician tweets and compare them to a broad baseline of public comments about the disease.

2. Methods

2.1. Data Collection

We extracted all tweets from the specific Twitter user account, @Covid19Docs, wherein the administrators collected direct messages (DMs) from physicians, and posted them to the page on behalf of those physicians, giving them some anonymity and a platform to express their perspective as it relates to the COVID-19 pandemic. Our extracted anonymous physician COVID-19 tweets are in English and span from March 16-July 17, 2020, with 875 total tweets. To prepare the tweets for analysis, punctuation, special characters, URLs, and stop words were removed from tweets, and words lemmatized. A dataset containing the tweet identifiers and other relevant datasets have been deposited at http://doi.org/10.5281/zenodo.4060340.

In order to put the specific Twitter account dataset within a larger scope of Twitter COVID-19 discourse, we compared our findings against a dataset of 513 million tweets gathered from the public Twitter streaming API [9], which contains a sampled set of 1 percent of all Tweets generated in real time. The dataset which is publicly available, is the result of an international collaboration and is maintained by researchers at Georgia State University.[9] It contains the top 1000 bigrams and trigrams we used for direct comparison of frequency in our smaller dataset. Note that to extract the mentions of our terms of interest, we removed all retweets, all tweets not in the English language, and tweets from accounts that are determined to be bots. Bot accounts are identified as accounts that are very recently created, and tweet more than 1000 times per day or are described on the account as a bot.

2.2. N-gram Frequency Measures

In order to understand the subject matter of the tweets, we counted frequencies of the most common lemmatized bigrams and trigrams in the anonymous physician tweets, and compared those to the more general COVID-19 dataset. This allowed us to qualitatively assess the topics discussed among physicians versus the general public with regard to COVID-19.

In addition, an experienced hospitalist physician helped us to identify four specific topics to assess within the anonymous physician data and the general COVID-19 data. These specific topics were identified because they were recognized to be of importance within the public discourse, and particularly within the physician discourse, during the COVID-19 pandemic. These topics were personal protective equipment (PPE), unemployment, telemedicine, and racial injustice. We measured the frequency of the topics and sentiments of tweets about these topics.

2.3. Sentiment Analysis

Sentiment analysis is a tool used to analyze and understand the opinions, emotions, and sentiments of language.[8] It is often used in marketing to understand opinions of certain brands, for prediction of political candidates' likelihood to win an election, and crowd opinions about events or policies.[10] The advent of social media has created a rich data source filled with sentiments and opinions about a myriad of topics, and is increasingly being used in the healthcare arena.[11] There are different sentiment analysis approaches, which all have their own strengths and weaknesses.

We wished to evaluate the sentiments of the anonymous physician data, and garner insight about whether these approaches accurately assess the sentiments of anonymous physician tweets. To do this, we began by conducting two popular sentiment analysis methods on the anonymous physician data. The first was the National Research Council (NRC) Word-Emotion Association Lexicon, which contains 10,170 English words and their associations with eight emotions (anger, anticipation, disgust, fear, joy, sadness, surprise, and trust) and two sentiments (negative and positive), making this a lexicon-based sentiment analysis.[12] The NRC method is unique in that it measures eight emotions, taken from the psychologist, Robert Pluchik's theory that people have eight basic emotions.[13] We used the Natural Language Toolkit (NLTK) in Python 3 and the NRC-Sentiment-Emotion-Lexicons to conduct this NRC sentiment analysis.[12,14]

We then used the VADER (Valence Aware Dictionary and sEntiment Reasoner) sentiment analysis approach, which is specifically designed for analyzing social media data using a lexicon and rule-based approach to assess sentiments (negative, neutral, and positive).[15] VADER has the benefit of providing a normalized, weighted composite score of each tweet by summing valence scores of each word in its lexicon, adjusting them according to grammatical and syntactical rules, and then normalizing them to give a compound score between -1 and +1.[16] We used the Natural Language Toolkit (NLTK) for VADER (nltk.sentiment.vader) in Python 3 to carry out this analysis.[17] We then compared the sentiments from the VADER sentiment analysis approach, which is thought to be more accurate for capturing sentiments of social media data, to the general COVID-19 dataset.[15]

Manual classification of all 875 tweets in the anonymous physician tweets was done by one of the authors. Tweets for which the sentiment was unclear in the first round of annotations were then annotated by two co-authors to determine the sentiment for those tweets. Each tweet was given an overall ranking of -1 (negative), 0 (neutral), or +1 (positive). To determine whether VADER accurately captured the sentiments of these tweets compared to the manual classification of tweets, precision, recall, and F1 scores were calculated. Data collection and analyses for this project were done using Python (3.7.0), R (3.6.0), SAS (9.4), and Microsoft Excel (2016).

3. Results

3.1. Frequency of terms and n-grams

We measured the most frequent bigrams and trigrams that occurred in the anonymous physician tweets and compared them to the general COVID-19 tweets (Figure 1). The top two phrases in the anonymous physician tweets might be expected, "health care" and "covid patients;" others point to concerns that might be more specific to physicians and other frontline healthcare workers, such as "need ppe," "elective cases," and "surgical masks."

Top phrases in the more general COVID-19 dataset look a bit different. The top two phrases are perhaps obvious, "covid 19" and "coronavirus cases," with frequency of "covid 19" far exceeding the other phrases. This general Twitter dataset captures public COVID-19 chatter specifically, so the very high frequency of the phrase "covid 19" is reasonable. The more general COVID-19 tweets also have a number of phrases associated with the political sphere of COVID-19 that the physician tweets do not have, such as, "white house," "trump administration," and "dr fauci." These phrases give a qualitative understanding of what physicians are talking about compared to the general public with regard to the COVID-19 pandemic.

Taking a step further, we identified specific topics of interest that are of importance within public discourse, and particularly within the physician discourse, given the emergence of the COVID-19 pandemic (Table 1). By doing this, we can further our understanding of physicians' perspectives regarding these topics. We measured and compared the frequency of tweets containing the specific topical phrases in the anonymous physician tweets and the general COVID-19 tweets. Using a chi-square test, we found significant differences between proportions of tweets containing all of the four phrases of interest, with a larger proportion of topics discussed among the anonymous physician tweets, except for phrases surrounding "racial injustice," which had a slightly higher proportion in the general COVID-19 tweets.

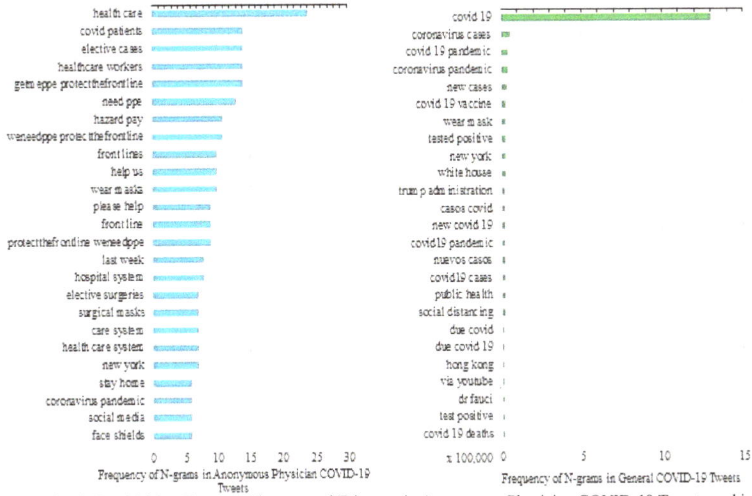

Fig. 1. Top 25 Most Frequent Bigrams and Trigrams in Anonymous Physician COVID-19 Tweets and in General COVID-19 Tweets

Table 1. Frequency of Topics in Anonymous Physician Tweets and General COVID Tweets

	Anonymous physician COVID-19 Tweets		General COVID-19 Tweets		
	Frequency (n_{total}=875)	Percent of Total Tweets	Frequency (n_{total}=73,377,056)	Percent of Total Tweets	p-value
Personal Protective Equipment	118	13.49%	22,155	0.03%	< .00001
Unemployment	15	1.60%	138,965	0.19%	< .00001
Telemedicine	12	1.37%	50,308	0.07%	< .00001
Racial Injustice	2	0.23%	198,906	0.27%	0.01

Specific terms used to capture phrases: "PPE," "personal protective equipment," "N95," "face shield"; "telemedicine," "telehealth"; "furlough," "unemployed," "pay cut"; "racial injustice," "racial discrimination," "racism," "racial inequality"

3.2. *Sentiment analysis*

We did two sentiment analyses in order to learn which sentiments these methods associated with the anonymous physician tweets, and also to determine if sentiment analysis could accurately capture sentiments of anonymous physician tweets compared to a manual assessment. We also did a sentiment analysis on the general COVID-19 Twitter data for comparison.

Figure 2 shows the number of tweets that contain each sentiment-emotion pair. Assessment of emotions is unique to the NRC method; most other methods only capture negative and positive sentiments. The most frequent emotion identified by NRC was "trust" in the anonymous physician tweets, and there were more positive words among the tweets than negative.

Fig. 2. NRC Word-Emotion Association Lexicon Sentiment Analysis of Anonymous Physician COVID-19 Tweets

Figure 3 shows negative, positive, and neutral sentiments over time for the anonymous physician data and general COVID-19 data using the VADER sentiment analysis method. We compared the VADER sentiments of the anonymous physician data to that of the general COVID- 19 tweets to discern if sentiments differed between the two datasets. We used VADER for this comparison because it is tailored toward sentiment analysis of social media data.

Over time, there was usually a slightly higher proportion of positive tweets compared to negative tweets in the general COVID-19 tweets. The anonymous physician tweet sentiments show less of a distinction among the sentiments compared to the general COVID-19 assessment.

During the week of June 1, 2020, there was a spike in positive tweets in the anonymous physician data (Figure 3). During this particular week many physicians describe the work they are doing for patients during the COVID-19 pandemic, which might explain the positive spike. Tweets from this week contain the following words and phrases: "take care of," "boost," "help save our community," "promise to do better," "beautiful children," "stand tall," "best medical care possible," "hold the hands," "equanimity and grace," "cheerfully," and "with a smile."

It is possible that with more tweets and over a longer period of time, the sentiments of the anonymous physician tweets would result in a comparable pattern over time to that of the general COVID-19 tweets. It is also possible that VADER sentiment analysis has trouble discerning sentiments from the anonymous physician tweets compared to the more general COVID-19 tweets, perhaps due to its more specialized clinical vocabulary. While this method is considered suitable for social media data, this may not be true for anonymous physician social media data specifically; we further elaborate on this point in the discussion section by comparing these results to a manual review of the tweets.

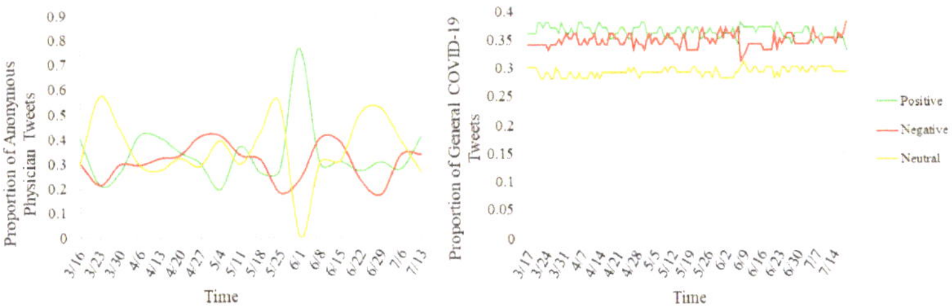

Fig. 3. Proportion of Positive, Negative, and Neutral Tweets Over Time for Anonymous Physician COVID-19 Tweets (left) and General Covid-19 Tweets (right) using VADER Sentiment Analysis

There was not a statistically significant difference in the average overall sentiments between the anonymous physician and general COVID-19 tweets, according to VADER sentiment analysis (p-value= 0.76). A majority of tweets were assessed as being positive for both datasets ($\text{Positive}_{\text{Anonymous Physician}}$ = 34.4 percent; $\text{Positive}_{\text{General COVID-19}}$ = 36.25 percent). There were 31.7 percent and 34.6 percent of tweets that were negative for the anonymous physician data and general COVID-19 data, respectively.

3.3. *Sentiments of tweets containing specific terms*

We also assessed the sentiments associated with tweets that contained specific topical phrases, which were captured using the same terms described in Table 1. The small size of the anonymous physician dataset prevented us from assessing topic-specific sentiments over time, so we measured frequencies and proportions instead (Table 2). The sentiments over time for tweets containing these phrases in the general COVID-19 tweet dataset can be seen in Figure 4.

Table 2. Frequency of Sentiments for Tweets Containing Specific Topics in Anonymous Physician Tweets According to VADER Sentiment Analysis

	Positive		Negative		Neutral	
Topic	Frequency (n)	Percent (%)	Frequency (n)	Percent (%)	Frequency (n)	Percent (%)
Personal Protective Equipment (n=118)	45	38.14%	42	35.59%	3	26.27%
Unemployment (n=14)	4	28.57%	9	64.29%	1	7.14%
Telemedicine (n=12)	5	41.67%	3	25.00%	4	33.33%
Racial Injustice(n=2)	0	0.00%	2	100.00%	0	0.00%

Percent calculated from tweets containing the specific topic of interest as denominator

There were more positive than negative tweets about telemedicine in the anonymous physician data. The general COVID-19 dataset presented a larger proportion of tweets that trended positive with regard to telemedicine too. In both datasets, the proportion of tweets about personal protective equipment were more positive, while the proportion of tweets about unemployment were more negative. There were few tweets about racial injustice captured in the anonymous physician data, but both presented as having negative sentiments. The larger COVID-19 dataset showed a much larger proportion of negative sentiments than positive sentiments over time with respect to social injustice phrases.

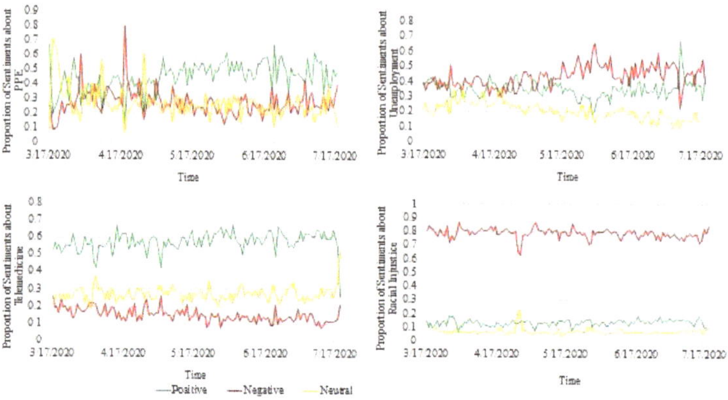

Fig. 4 Proportion of Sentiments for Tweets Containing Phrases about PPE, Unemployment, Telemedicine, and Racial Injustice for General COVID-19 Tweets

4. Discussion and Conclusion

Top phrases in the anonymous physician data, such as "help us," and "need ppe," paint a poignant picture of physician perspectives during COVID-19. This research shows how Twitter data can be used to qualitatively assess physician attitudes, beliefs, and perspectives as they relate to the COVID-19 pandemic. We showed that the discourse of anonymous physician tweets is different from the discourse of more general tweets with regard to COVID-19. The anonymous physician tweets are more clinically oriented with phrases such as "health care", "elective cases", and "covid patients."

Our analysis also showed that current lexicon and rule-based sentiment analysis methods should be improved in the future to be specifically targeted for clinically oriented social media data. As social media data is being used more often in the health arena, and healthcare professionals face stricter regulations from employers with regard to posting on social media, it would be of interest to create sentiment analysis methods that more aptly capture this specific type of data.

The uniqueness of anonymity of the physician tweets is important. Koohikamali and Gerhart (2018) found that anonymous social media data is in fact different from more general discourse of social media data, particularly during a social crisis.[8] They report that because anonymity lowers inhibitions, it often results in posting more honest opinions due to less risk of repercussion, such as job loss or unpaid suspension, which some frontline healthcare workers are currently facing.[8] This can lead to valuable insights that would otherwise not be captured in more public discourse, and paints a truer picture necessary for implementing more impactful change during the COVID- 19 pandemic, and in the future. For this reason, it will be beneficial in future work to improve sentiment analysis tools so they are capable of assessing anonymous physician social media data.

That being said, sentiment analysis is difficult because human language is complex; this is particularly apparent for social media data, where contextual understanding of the language is very meaningful to the overall sentiments.[18] For example, while VADER is tailored toward social media data, a manual assessment of the anonymous physician tweets found far more negative tweets than positive. Upon manual analysis of the anonymous physician tweets, the F1 score was 0.52 (precision = 0.63, recall = 0.56), indicating that sentiment analysis using this method might need to be improved in order to more accurately capture sentiments of anonymous physician tweets. Manual assessment of these tweets resulted in 67 percent of tweets having a negative sentiment (22 percent neutral and eleven percent positive), while VADER resulted in only 32 percent of tweets having a negative sentiment (Table 2).

The NRC method captured more positive than negative words among the tweets, but after manual review, the overall sentiments of the tweets leaned far more negatively. Additionally, the most frequent emotion NRC identified among the tweets was "trust." Upon reading the tweets, "trust" did not fit the overall emotional sentiment of these tweets. For example, "trust" was the most frequent emotion identified in the following, and "positive" was the resulting sentiment: "Top academic institution cut MD Salaries; includes frontline hospitalists and intensive care no offer of hazard pay, pay for extra shifts, and no promise of back pay." It is possible that this lexicon-based method does not capture the negations within the text. Table 4 shows some examples of how the lexicon and rule-based methods failed to capture the overall sentiment of the anonymous physician tweets compared to the manual analysis.

Table 4. Examples of Misinterpreted Paraphrased Tweets by Sentiment Analysis Compared to Manual Assessment of Tweets

Tweet	Sentiment Analysis	Manual Assessment
How is it fair for admin to silence doctors asking for help?	Positive	Negative
Physicians are afraid of losing jobs.	Positive	Negative
When faced with PPE shortages, the program director sourced 3D printers to make face shields!	Negative	Positive
Please help! No PPE!! Please help!	Positive	Negative
We are doctors and have NO PPE. Please donate if able-stop hoarding please! We need it to care for patients.	Positive	Negative

While we found some interesting and important results, there were some limitations in our study we wish to address. The first is that the tweets from the anonymous physician data were likely just a small sample of tweets representing this type of data. Anonymous physician social media data is unique, and as frontline healthcare workers in the U.S. face pressures from their administration to stay off of social media platforms, pages that allow them to share their thoughts with a more anonymous approach are an important, perhaps overlooked source of information to better understand healthcare workers' perspectives. It is certainly possible that our small sample did not represent this type of data in full; further evaluation of how well our sample represents this population is warranted.

Also, while there was some amount of anonymity to the physician tweets, administrators of the user page likely knew the user name of the physicians through the DMs. This means the tweets were unknown to only those who were not administrators of the page. Still, understanding that their name would not be attached to the posted tweet provided some amount of anonymity to the physicians. Having a sense of anonymity may also encourage physicians to post tweets that lean more negatively in sentiment, but further assessment should be done to understand if anonymity leads to a negative bias.

Another limitation is with regard to selection of the four specific topical phrases that we chose to explore. We used a small number of exact terms to capture these phrases, and they could probably be expanded to capture tweets about each of these phrases. For example, we captured very few anonymous physician tweets about the topic, "racial injustice," and it is possible that expanding the list of exact phrases would improve detection of this topic and others. There are also, of course, other topical phrases that might be of interest to assess that we did not assess in our analysis. Future studies should widen the scope of topical phrases of interest.

A final limitation is simply that social media data can be difficult to work with in many ways. It contains many informal, idiomatic phrases, special characters, emoticons, grammatical mistakes, misspellings, and abbreviations that make it challenging for text analysis methods.[18] Despite this, valuable insights and perspectives can be obtained through this rich source of data, particularly sentiments and opinions that might have impactful meaning for healthcare workers, patients, and the general public health. There are many future avenues that this body of work might take. The development of a lexicon for sentiment analysis that is specific to anonymous social media data, physician or other healthcare professional social media data, or even more specifically, anonymous physician social media data might be very useful for future sentiment analysis studies of this type of data. It would also be of interest to mine the public discourse for more general healthcare professional social media data for comparison to anonymous data, as there might be large differences in topics and sentiments discussed.

This study has identified interesting underlying topics and sentiments from anonymous physician data with regard to the COVID-19 pandemic. We found these topics and sentiments are usually different from the overall COVID-19 discourse on Twitter. It is likely that anonymous social media data from physicians and other frontline healthcare workers will become more popular as they continue to experience the effects of the COVID-19 pandemic. This is especially true as they are hesitant to post publicly their perspectives on social media about the current state of affairs in their working environment for fear of being reprimanded or fired.[7] Frontline healthcare workers have an important impact on patients' lives, and this is especially true during times of exceptional difficulty or social crisis, both of which are relevant to today's current atmosphere. Understanding frontline healthcare workers' perspectives, needs, and opinions may help improve patients' experience and health outcomes, and perhaps even guide improvements to public health strategies in the future.

5. Acknowledgments

We wish to acknowledge and thank Mayla R. Boguslav for her helpful consultations throughout the development of this project, her advice on best practices in coding, and revisions on this paper. This manuscript was supported by the Data Science to Patient Value (D2V) initiative funded by the University of Colorado School of Medicine Dean's Transformational Research Funding. Lawrence Hunter's efforts were supported by NIH grants 2R01LM008111 and 1R01LM013400.

References

1. Wu JT, McCormick JB. Why Health Professionals Should Speak Out Against False Beliefs on the Internet. *AMA Journal of Ethics*. 2018; 20 (11): E1052-E1058.
2. Wakefield JG, McLaws M-L, Whitby M, Patton L. Patient safety culture: factors that influence clinician involvement in patient safety behaviours. *Qual Saf Health Care*. 2010; 19 (6): 585-591.
3. Kirk S, Parker D, Claridge T, Esmail A, Marshall M. Patient safety culture in primary care: developing a theoretical framework for practical use. *Qual Saf Health Care*. 2007; 16 (4): 313-320.
4. Tavares APM, Moura ECC, Avelino FVSD, Lopes VCA, Nogueira LT. Patient safety culture from the perspective of the nursing team. *Rev Rene*. Published online January 15, 2018.
5. Ledford H. Computing humanity. *Nature*. 2020; 582 (June 18): 328-330.
6. Kellogg K. The 7 Biggest Social Media Sites in 2020. Published online 2020.
7. Carville BO, Larson E. Doctors and Nurses Beware: Hospitals Are Watching Your Facebook. *Bloomberg Law*. 2020: 1-6.
8. Koohikamali M, Gerhart N. Yaks versus Tweets: Sentiment Discrepancy During a Social Crisis. *Proceedings of the 51st Hawaii International Conference on System Sciences*. 2018; (March).
9. Banda JM, Tekumalla R, Wang G, et al. A large-scale COVID-19 Twitter chatter dataset for open scientific research -- an international collaboration. *ArXiv*. Published online April 7, 2020.
10. Liu B. *Sentiment Analysis and Opinion Mining*. Morgan & Claypool Publishers; 2012.
11. Garcia-Rudolph A, Laxe S, Saurí J, Guitart MB. Stroke survivors on Twitter: Sentiment and topic analysis from a gender perspective. *J Med Internet Res*. 2019; 21 (8).
12. Mohammad S, Turney P. Emotions Evoked by Common Words and Phrases: Using Mechanical Turk to Create an Emotion Lexicon. *Proceedings of the NAACL HLT 2010 Workshop on Computational Approaches to Analysis and Generation of Emotion in Text*. 2010; (California, US): 26-34.
13. Plutchik R. The nature of emotions. *Am Sci*. 2001; 89 (4): 344-350.
14. Natural Language Toolkit — NLTK 3.5 documentation.

15. Hutto CJ, Gilbert EE. VADER: A Parsimonious Rule-based Model for Sentiment Analysis of Social Media Text. Eighth International Conference on Weblogs and Social Media (ICWSM-14)." *Proceedings of the 8th International Conference on Weblogs and Social Media, ICWSM 2014*. Published online 2014.
16. Hutto CJ. vaderSentiment. Published online 2020.
17. Hutto CJ, Klein E, Pantone P, Berry G, Malavika S. Source Code for nltk.sentiment.vader. Published online 2020.
18. Burkhardt S, Siekiera J, Glodde J, Andrade-Navarro MA, Kramer S. Towards identifying drug side effects from social media using active learning and crowd sourcing. *Pac Symp Biocomput*. 2020; 25 (2020): 319-330.

Semantic Changepoint Detection for Finding Potentially Novel Research Publications

Bhavish Dinakar
Department of Chemical and Biomolecular Engineering, University of California, Berkeley
Berkeley, CA 94720
Email: bhavishdinakar@berkeley.edu

Mayla R. Boguslav
Computational Bioscience Program, University of Colorado Anschutz Medical Campus
Aurora, CO 80045
Email: mayla.boguslav@cuanschutz.edu

Carsten Görg
Department of Biostatistics and Informatics, University of Colorado Anschutz Medical Campus
Aurora, CO 80045
Email: carsten.goerg@cuanschutz.edu

Deendayal Dinakarpandian[†]
Center for Biomedical Informatics Research, Stanford University
Stanford, CA 94305
Email: dinakar@stanford.edu

How has the focus of research papers on a given disease changed over time? Identifying the papers at the cusps of change can help highlight the emergence of a new topic or a change in the direction of research. We present a generally applicable unsupervised approach to this question based on semantic changepoints within a given collection of research papers. We illustrate the approach by a range of examples based on a nascent corpus of literature on COVID-19 as well as subsets of papers from PubMed on the World Health Organization list of neglected tropical diseases. The software is freely available at: https://github.com/pdddinakar/SemanticChangepointDetection.

Keywords: Changepoint; Semantic; Novel research paper; Literature search.

1. Introduction

There are several possible motivations behind a literature search. These range from finding the answer to a highly specific question to writing a general review of a topic. One of the motivations for a literature review might be to select a topic for research, where one may choose to perform research in a well-established area, pick an emerging area or aspire to be a pioneer in uncharted territory. Another possible motivation might be for a funding agency to keep track of emerging areas

[†] Corresponding Author

of research that might merit funding in the near future. Yet another motivation might be to keep track of new insights or technologies that address an acute health need such as a pandemic or diseases that are hard to treat effectively.

What if it were possible to identify papers that strayed from the mainstream? While many of these might end up as blind alleys, a subset of these might turn out to be harbingers of innovative, influential, or impactful directions in research. A few of the potential approaches to identify outliers, first-to-report, or first-in-field papers are topic modeling[1], clustering[2], trend analysis[3,4], citation network analysis[5], and machine learning approaches for predicting high impact papers. We present a set of strategies from changepoint analysis and text embedding to address two questions. Which are the papers in a research area that are substantially different from previous work? Which papers are part of a related cluster that is substantially different from previous work? We use infectious diseases from two different time scales to illustrate the approach - COVID-19[6] over a temporal resolution of weeks, and leprosy[7], considered a neglected tropical disease by the World Health Organization. We begin with a description of the methods used, followed by results and discussion, and end with an acknowledgment of the limitations and future work to address them.

2. Methods

The overall summary of the methodology is shown in Fig. 1. Briefly, titles or abstracts from a collection are either embedded as a vector or represented in terms of word frequency distributions. Temporal changes in these representations of titles or abstracts are detected by approaches described below. Papers or terms corresponding to the temporal changes are marked as potentially novel for the corresponding time period.

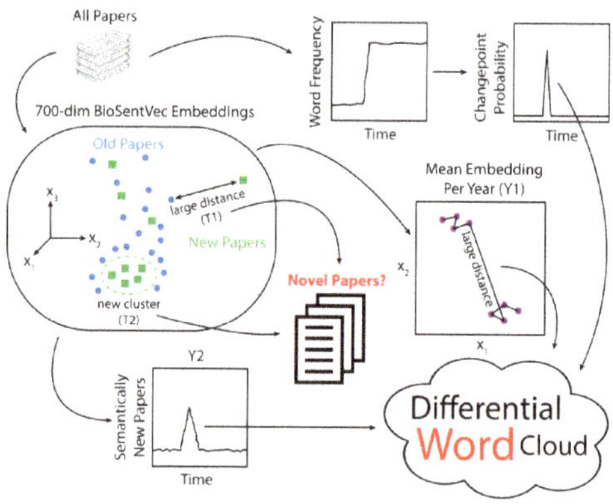

Figure 1. Overview of semantic changepoint analysis using word frequency-based and embedding space-based strategies (T1, T2, Y1, Y2).

2.1. *Data collection and general procedures*

COVID papers were downloaded from the COVID-19 SARS-CoV-2 Preprints available from bioRxiv in JSON format on 7/31/2020.[8] The title, upload date, and abstract were retrieved for each paper published in 2020, yielding a total of 7151 analyzed papers.

Leprosy papers were retrieved from the National Center for Biotechnology Information E-Utilities API. A list of UIDs for papers where the term "leprosy" appears in the title were retrieved using the ESearch method, and the title, abstract, and publication date for each UID were retrieved with the EFetch method. We only analyzed papers for which the title, abstract, and publication date were available. We also only considered papers published between 1980 and 2019 due to the low number of annual papers (less than 50 per year) published before 1980, yielding a total of 5068 analyzed papers. Plots were generated using the matplotlib package[9].

2.2. *Title and abstract entropies*

We used the scikit-learn[10] CountVectorizer tool to convert each title to a Bag-of-Words representation. We calculated the probability of each word in a year using Eq. (1).

$$p_{\text{word,year}} = \frac{\text{frequency of word in year}}{\text{total frequency of all words in year}} \tag{1}$$

The yearly entropy of word proportions in titles (or alternatively, in abstracts) was calculated by Shannon entropy, which is a popular measure of information content or variability of a distribution.

$$S_{\text{year}} = -\sum_{\text{word}} p_{\text{word,year}} \log_2(p_{\text{word,year}}) \tag{2}$$

2.3. *Bayesian changepoint analysis*

Changepoint detection aims to identify the point at which the probability distribution of a sequential variable changes. Changepoint analysis was conducting using the bayesian-changepoint-detection (bcp) Python package.[11–13] The advantage of the bcp method is that it also provides a probability of there being a changepoint at a given time point. We performed changepoint detection for each word in the Bag-of-Words model of paper titles to analyze the word frequency per title vs time (year for Leprosy, 2020 week number for COVID). Abstract changepoints were calculated in the same manner, except using abstracts as input to the Bag-of-Words model instead of titles.

2.4. *Differential word clouds*

Differential word clouds are visual depictions of changes in research paper titles between two years, denoted Year A and Year B. Two groups are selected from the paper titles: Group A contains the titles of all papers published in or before Year A, and Group B contains the titles of all papers published in Year B. The Bag-of-Words model is applied to each group to determine the frequency per title of each word, and stop-words appearing in the NLTK stop-words set are removed.[14] Word clouds were created using the Wordcloud Python package,[15] with weights for each word corresponding to the difference in frequency per title between Group B and Group A. Positive differences (increases in word frequency) are colored black, and negative differences (decreases in word frequency) are colored red.

2.5. *Title and abstract embeddings*

Titles were first pre-processed by converting all words to lowercase and removing punctuation and stop-words found in the NLTK punctuation and stop-word sets. The processed titles were then converted into 700-dimensional vectors using the BioSentVec model[16]. Abstracts were embedded in the same manner, except using the abstract as the input instead of the title. The embeddings were visualized using Principal Component Analysis as implemented by Scikit-Learn.

2.6. *Semantic novelty*

We use the following approaches to detect potentially novel papers or subtopics in one temporal window (or subcollection) with respect to another. For instance, one may compare papers in 2020 with all preceding years (novel compared to entire research legacy). Alternatively, one may compare papers published in 2020 with those published in 2019 (a change in direction of research compared to recent past). We employ 4 different strategies: T1, T2, Y1 and Y2 (see Fig. 1).

2.6.1. *Strategy T1: Novel paper detection based on semantic distance*

We first analyzed the distribution of the pair-wise distance between all titles in embedded space for the COVID corpus and examined pairs sorted by distance to determine suitable thresholds for relatedness S-rel and, conversely, S-unrel for being semantically unrelated with high probability. There is a grey area in between the two thresholds where pairs of titles at the same distance from each other are sometimes related, and sometimes not. We confirmed the consistency of the thresholds by repeating the calibration check with the leprosy dataset. To determine if a title T is novel relative to a comparator collection C (typically in a preceding time window), we first determine the minimum Euclidean distance E-min between T and all titles in C. All titles T with E-min values higher than S-rel are potentially novel (e.g., the green square in Fig. 1 labeled as T1).

2.6.2. *Strategy T2: Detection of novel papers that may constitute a trend*

This builds on strategy T1 by requiring that papers not only be distant from an 'old' neighborhood (subcollection or time window) but also be part of a 'trendy' neighborhood. In other words, a title T is considered to be part of a trend if it lies in a location that corresponds to low-density in the old neighborhood (blue dots) and high-density in the new neighborhood (green squares). This is quantified by requiring that a novel title T that is part of a trend be closer to k papers in the new neighborhood (e.g., green squares labeled T2 in Fig. 1) than k papers in the old neighborhood. The default value of k is set to 3 to correspond to an emerging trend but may be set higher if desired.

2.6.3. *Strategy Y1: Detection of a group of novel papers based on their mean vector*

Strategy Y1 tracks the location of the mean vector of papers in each time window. Long hops in embedded space may imply a substantial difference in the direction of research (e.g., see Y1 in Fig.

1). The underlying signal may be uncovered by word frequency analysis or titles in high density areas close to the mean.

2.6.4. *Strategy Y2: Proportion of novel papers*

The premise is that the novelty of a time window (or subcollection of papers) may be gauged by estimating what proportion of papers in that window are at least distance S-rel from all papers in the past. When examining successive time windows, large upward oscillations of this proportion may signal the presence of a new trend.

3. Results and Discussion

We use two different approaches to detect changes over time in the focus of research papers on a particular topic. The first approach consists of using changepoint analysis to detect changes in the frequency of words within titles or abstracts. The second approach consists of embedding titles in vector space and using the distance between titles as an approximation of the corresponding semantic difference. To illustrate these approaches, we have chosen to focus on a pair of contrasting infectious diseases, COVID-19 and leprosy. COVID-19 has a short history as a pandemic affecting millions of people, while leprosy is one of the oldest human diseases. While much progress has been made, and effective treatment is available when diagnosed early, around 200,000 new cases continue to be reported each year.[17]

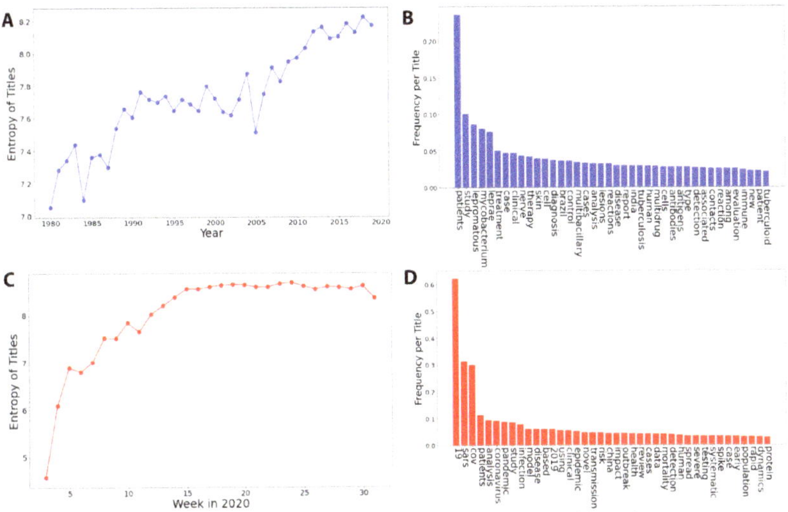

Figure 2. A) Entropy of titles on leprosy. B) Frequency of words in titles on leprosy. C) Entropy of titles on COVID-19. D) Frequency of words in titles on COVID-19.

Fig. 2 shows the increasing complexity of the information content of titles over time, corresponding to the diversity of terms shown on the right. Though the entropy of titles on leprosy has risen over a longer period of time than for COVID, it is interesting that both show a maximum value of just over 8 bits. This suggests that titles of research papers on diseases might have a similar complexity, and by extension share properties of similarity in the embedding space. The most frequent terms occurring within the titles of papers (Fig. 2B and 2D) are used for subsequent changepoint analysis.

Fig. 3 shows the temporal frequencies of some of the most frequent terms occurring within the titles of research papers. The corresponding changepoint analysis highlights points in time when there is a high probability of a significant change in the probability of the occurrence of a term in a title. The following inferences can be made from this figure. First, despite the fluctuations shown in panels A and B, changepoint analysis finds identical changepoints for "MDT" and "Therapy," indicating a surge of literature mentioning multidrug therapy for the treatment of leprosy. In fact, this corresponds to the period of excitement when (dapsone+rifampin+clofazimine) was recommended by WHO in the 1980s as curative treatment, resulting in the elimination of leprosy as a global public health problem (defined as an incidence of less than 1 in 10000) by the year 2000.[17] Fig. 4 shows the relative rise and relative fall in the frequency of words spanning a period suggested by the location of changepoints for the terms in Fig. 3. As expected, terms such as the following show an increase in frequency: (multidrug, therapy, treated, patients, paucibacillary, multibacillary) - presumably indicating the advent of successful multidrug therapy. In contrast, terms such as (lepromatous, nerve, cases, granulomatous) show a decrease, presumably corresponding to a decrease in the incidence and morbidity of the disease.

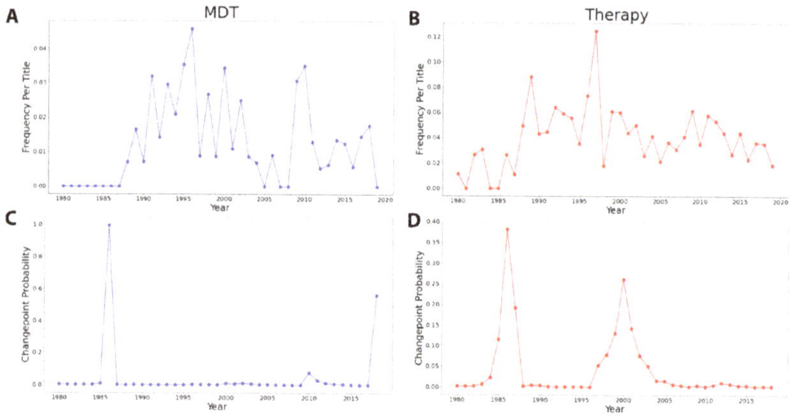

Figure 3. Examples of changepoint analysis of the frequency of the words "MDT" and "Therapy" in titles containing the word "leprosy." A) Temporal frequency of "MDT." B) Temporal frequency of "Therapy." C) Changepoint peaks mark the beginning and end of a period of relatively high frequency of "MDT." D) Changepoint peaks mark the beginning and end of a period of relatively high frequency of "Therapy."

Figure 4. Differential cloud of terms within titles containing the word "leprosy" between the years 1980 and 1990, corresponding to the first changepoint in Fig. 3. Note "multidrug" and "therapy" in lower left corner in similar font size. Black = increase in frequency, Red = decrease in frequency.

The use of term frequencies as the basis of finding temporal changes in the focus of research papers has the following limitations:

1. While considering terms independently might work well for categories such as drug names, it has limited ability to exploit the meaning of the entire title.
2. Focusing only on the higher frequency terms may miss the true harbingers of change; the long tail of low frequency terms makes it harder to find the significant ones.

Ideally, we would like to project the key focus of research papers into a shared semantic space, without having to count frequencies first. This would make it possible to highlight newly populated regions of the space as containing papers representing new directions. Several models have been published for projecting words into vector spaces[18,19,20]. More recently, it has been shown that models that directly embed a sentence are more accurate than taking the average of the constituent word vectors. Titles of papers typically encode more meaning than a set of words but often fail to form a well-formed sentence. In order to confirm if title embeddings retain inter-title similarity in the same manner as inter-sentence similarity[21], we used BioSentVec[16] to embed the titles from the bioRxiv COVID dataset and sorted pairs of titles by increasing Euclidean distance between the corresponding vectors. Representative pairs of related titles are shown in Table 1. Titles may be addressing the same objective except for minor differences in methodology or disease population. As hoped for, the embedded vectors also capture similarity based on implicit meaning. For example, associations are captured between heart injury and raised blood levels of BNP, and between kidney injury and hypertension (supplementary information). Table 2 shows a sample of titles that are unrelated, and further apart in vector space. An empirical threshold of 2.3 corresponds to an FDR of less than 1% for semantic relatedness, and distances above 4 are very unlikely to indicate relatedness (supplementary information for full table).

Table 1. Examples of related titles within the bioRxiv set of papers on COVID.

Title 1	Title 2 (Related)	L2 distance
Predicting the number of reported and unreported cases for the COVID-19 epidemics in China, South Korea, Italy, France, Germany and United Kingdom	Predicting the number of reported and unreported cases for the COVID-19 epidemic in South Korea, Italy, France and Germany	1.429
The impact of current and future control measures on the spread of COVID-19 in Germany	A first study on the impact of current and future control measures on the spread of COVID-19 in Germany	1.844
Characterization of a novel, low-cost, scalable ozone gas system for sterilization of N95 respirators and other COVID-19 use cases	Characterization of a novel, low-cost, scalable vaporized hydrogen peroxide system for sterilization of N95 respirators and other COVID-19 personal protective equipment	2.135
A 5-min RNA preparation method for COVID-19 detection with RT-qPCR	A simple RNA preparation method for SARS-CoV-2 detection by RT-qPCR,	2.229
Clinical features and outcomes of 2019 novel coronavirus-infected patients with high plasma BNP levels	Clinical features and outcomes of 2019 novel coronavirus-infected patients with cardiac injury	2.241
Clinical characteristics of Coronavirus Disease 2019 (COVID-19) patients in Kuwait	Clinical and epidemiological characteristics of Coronavirus Disease 2019 (COVID-19) patients	2.318

Table 2. Examples of unrelated titles within the bioRxiv set of papers on COVID.

Title 1	Title 2 (Unrelated)	L2 distance
Early Prediction of Disease Progression in 2019 Novel Coronavirus Pneumonia Patients Outside Wuhan with CT and Clinical Characteristics	Epidemiological and Clinical Characteristics of 17 Hospitalized Patients with 2019 Novel Coronavirus Infections Outside Wuhan, China	2.337
Preliminary epidemiological analysis on children and adolescents with novel coronavirus disease 2019 outside Hubei Province in China: an observational study utilizing crowdsourced data	Evolving epidemiology of novel coronavirus diseases 2019 and possible interruption of local transmission outside Hubei Province in China: a descriptive and modeling study	2.351
Clinical course and potential predicting factors of pneumonia of adult patients with coronavirus disease 2019 (COVID-19): A retrospective observational analysis of 193 confirmed cases in Thailand	History of Coronary Heart Disease Increases the Mortality Rate of Coronavirus Disease 2019 (COVID-19) Patients: A Nested Case-Control Study Based on Publicly Reported Confirmed Cases in Mainland China	2.419
The First Consecutive 5000 Patients with Corona virus Disease 2019 from Qatar; a Nation-wide C ohort Study	Knowledge and perceptions of coronavirus disease 2019 among the general public in the United States and the United Kingdom: A cross-sectional online s urvey	3.294
Analysis of hospitalized COVID-19 patients in t he Mount Sinai Health System using electronic medical records (EMR) reveals important progno stic factors for improved clinical outcomes	Core warming of coronavirus disease 2019 (COVID -19) patients undergoing mechanical ventilation: pr otocol for a randomized controlled pilot study	3.663

Based on the embedded space of titles, we used the following ways to identify titles that might be novel compared to the past:

T1. Find titles that are furthest away from any title in the past (Table 3).

T2. Find titles whose location corresponds to a low density area of the set of past titles but high density area in the current time period.

Strategy T1 is aimed at the identification of one of the earliest papers in a possibly new area. Each row in Table 3 lists the paper whose title was the most dissimilar to all previous titles (note the large distances in the right column from the closest title among prior papers). Most of the titles are compatible with being one of the first papers on the topic, with the embedding also highlighting subtle novelties like different types of phylogenetic research. Note that "COVID-19 spreading: a model" could be considered similar to "Spatio-temporal propagation of COVID-19 pandemics," even though there is minimal lexical overlap. Strategy T2 is aimed at the identification of a burgeoning area compared to the past since it is meant to detect several new titles that are related; it also minimizes the chance of false positives that might confound the results of strategy T1. Examples of titles yielded by strategy T2 are a surge of publications on multidrug therapy in the period 1988-1991 compared to the period until 1987 (3-neighborhood ratio of old:new of 1.35). Examples are:

Leprosy control through multidrug therapy (MDT).
Experience with WHO-recommended multidrug therapy (MDT) for multibacillary (MB) leprosy patients in the leprosy control program of the All Africa Leprosy and Rehabilitation Training Center in Ethiopia: appraisal of the recommended duration of MDT for MB patients.
Bacillaemia in leprosy and effect of multidrug therapy.

A search of PubMed confirms this trend in that the first hit for the query "leprosy AND MDT" is only in 1985.

Table 3. Most novel paper each week compared to all previous weeks, as predicted by strategy T1

Week (2020)	Title (farthest from all prior titles on COVID in bioRxiv dataset)	Distance
4	From SARS-CoV to Wuhan 2019-nCoV: Will History Repeat Itself?	6.399
5	Nucleotide Analogues as Inhibitors of Viral Polymerases	7.016
6	Phylogenomic analysis of the 2019-nCoV coronavirus	6.938
7	Transmission Dynamics of 2019-nCoV in Malaysia	7.127
8	Fractal kinetics of COVID-19 pandemic	8.479
9	Application and optimization of RT-PCR in diagnosis of SARS-CoV-2 infection	5.473
10	Mutations, Recombination and Insertion in the Evolution of 2019-nCoV	7.150
11	The architecture of SARS-CoV-2 transcriptome	9.577
12	Routes for COVID-19 importation in Brazil	7.624
13	Spatio-temporal propagation of COVID-19 pandemics	7.462
14	Presence of SARS-Coronavirus-2 in sewage	10.080
15	Work-related Covid-19 transmission	9.856
16	COVID-19 is an emergent disease of aging	6.918
17	Identification of super-transmitters of SARS-CoV-2	16.268
18	*COVID-19 spreading: a model	8.260
19	AI334 and AQ806 antibodies recognize the spike S protein from SARS-CoV-2 by ELISA	9.338
20	Placental pathology in COVID-19	8.308
21	Metamorphosis of COVID-19 Pandemic	11.073
22	Are we #stayinghome to Flatten the Curve?	8.388
23	Cytokine biomarkers of COVID-19	9.935
24	Stability of SARS-CoV-2 Phylogenies	12.348
25	Hypokalemia in Patients with COVID-19	8.513
26	Surveillance testing of SARS-CoV-2	8.335
27	From predictions to prescriptions: A data-driven response to COVID-19	10.656
28	Are men dying more than women by COVID-19?	7.455
29	Cold sensitivity of the SARS-CoV-2 spike ectodomain	11.843
30	Phylogeny of the COVID-19 Virus SARS-CoV-2 by Compression	7.128
31	CAR Macrophages for SARS-CoV-2 Immunotherapy	8.481

*False positive

Based on the embedded space of titles, we used the following ways to identify years that might be novel compared to the past:

Strategy Y1. Find the mean title vector for each year, and trace the path from year to year. The longer paths may represent a significant change between adjacent years.

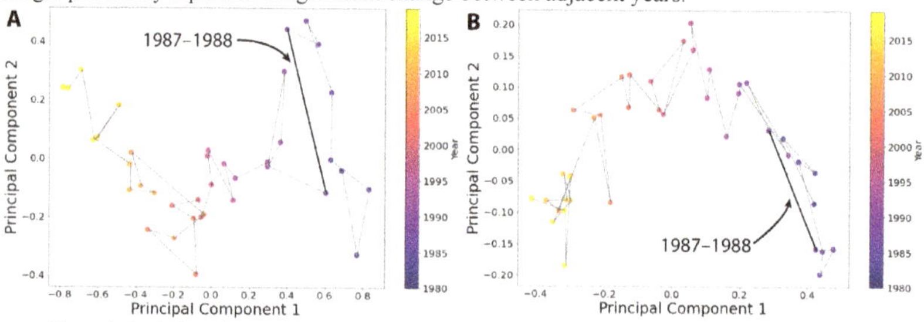

Figure 5. Depiction of strategy Y1: PCA projection of the mean embedded vector from papers on leprosy. Left: embedded titles; Right: embedded abstracts.

Strategy Y2. Estimate the proportion of titles in a given year that are located far away from any title in the preceding time period. A large change in the proportion of such titles in a given year may suggest a new and growing area of research.

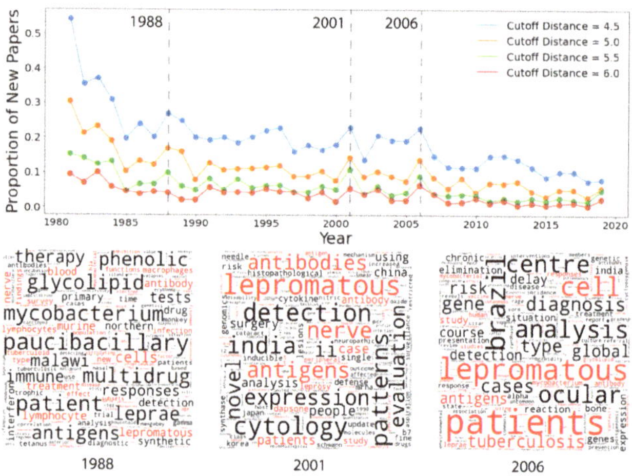

Figure 6. Top: Strategy Y2 depicts the proportion of 'novel' papers each year compared to all past publications. Bottom: Differential word clouds of terms within titles containing the word "leprosy" at times detected by Y2.

A low-dimensional projection of strategy Y1 is shown in Figure 5. The long path between 1987 and 1988 is in alignment with previous results (changepoint analysis, Strategy T2) in this paper regarding the literature on multidrug therapy in leprosy. Based on strategy Y2, Fig. 6 shows the proportion of potentially novel titles at each time-point, and corresponding differential word clouds to indicate terms possibly indicative of new areas.

While the title might be the most succinct 'sentence' representation of the topic of a paper, a rhetorical or terse title may fail to convey the essence of a paper. We therefore attempted to embed entire abstracts as an alternative version of strategy Y1. While this shows a possibly less noisy path from year to year, the variance (range of values on axes) decreases so that the semantic 'hops' from year to year become smaller (Fig. 5). Results from strategies such as Figure 5 and 6 could be used to calibrate and determine the thresholds for measures of novelty (projected path lengths or significant proportion of novel papers) to be indicative of novelty in the recent past.

4. Conclusions

We have presented and illustrated approaches to the detection of semantic changepoints within a set of research publications. Admittedly, a novel paper is not synonymous with a high impact paper. Nor is it synonymous with a novel conclusion. False negatives are also possible. For instance, analogous sentences differing in only one term (e.g., a highly effective new therapeutic intervention instead of an older marginally effective one) may have similar embeddings, especially within long titles. More specialized embedding schemes that are domain and problem specific may be necessary. Changepoint analysis is based on acute differences, and therefore less able to detect steady growth in a new area over a longer time period. As proof of concept, we have focused on a subset of the publications for each disease. For detection of truly novel papers, multiple databases will need to be considered together with sophisticated ontology and machine-based querying to maximize recall without sacrificing precision. The approach can be potentially enhanced by using named entity recognition approaches for more insightful analysis of the underlying reasons for an observed changepoint. A more detailed modeling of content based on incorporating topic modeling and/or the analysis of full text journal papers can help provide more granular changepoints and corresponding interpretations. Ultimately, the approach presented in this paper could potentially be incorporated into a real time system for the detection of novel information as it appears. Clinicians could potentially use it to find new vistas in their specific disciplines, especially in the context of hitherto incurable diseases. Researchers could find nascent topics worth expanding into.

5. Supplementary Information

Software to carry out the analysis on a given set of papers, and a larger set of detailed results for several of the infectious diseases on the WHO list of neglected tropical diseases are available at: https://github.com/pdddinakar/SemanticChangepointDetection.

6. Acknowledgements

We would like to thank Larry Hunter for insightful feedback on the ideas presented in this paper.

References

1. Blei, D. M., Ng, A. Y. & Jordan, M. I. Latent dirichlet allocation. *J. Mach. Learn. Res.* **3**, 993–1022 (2003).
2. Boyack, K. W. *et al.* Clustering more than two million biomedical publications: Comparing the accuracies of nine text-based similarity approaches. *PloS One* **6**, e18029 (2011).
3. Chen, C. CiteSpace II: Detecting and visualizing emerging trends and transient patterns in scientific literature. *J. Am. Soc. Inf. Sci. Technol.* **57**, 359–377 (2006).
4. Jun, S.-P., Yoo, H. S. & Choi, S. Ten years of research change using Google Trends: From the perspective of big data utilizations and applications. *Technol. Forecast. Soc. Change* **130**, 69–87 (2018).
5. Batagelj, V. Efficient algorithms for citation network analysis. *ArXiv Prepr. Cs0309023* (2003).
6. Wiersinga, W. J., Rhodes, A., Cheng, A. C., Peacock, S. J. & Prescott, H. C. Pathophysiology, Transmission, Diagnosis, and Treatment of Coronavirus Disease 2019 (COVID-19): A Review. *JAMA* (2020) doi:10.1001/jama.2020.12839.
7. WHO | Global Leprosy Strategy 2016–2020: Accelerating towards a leprosy-free world. *WHO* http://www.who.int/lep/resources/9789290225096/en/.
8. bioRxiv COVID-19 SARS-CoV-2 preprints from medRxiv and bioRxiv. https://connect.bioRxiv.org/relate/content/181.
9. Hunter, J. D. Matplotlib: A 2D Graphics Environment. *Comput. Sci. Eng.* **9**, 90–95 (2007).
10. Pedregosa, F. *et al.* Scikit-learn: Machine Learning in Python. *J. Mach. Learn. Res.* **12**, 2825–2830 (2011).
11. Fearnhead, P. Exact and efficient Bayesian inference for multiple changepoint problems. (2006).
12. Adams, R. P. & MacKay, D. J. C. Bayesian Online Changepoint Detection. *ArXiv07103742 Stat* (2007).
13. Xuan, X. & Murphy, K. Modeling changing dependency structure in multivariate time series. in *Proceedings of the 24th international conference on Machine learning* 1055–1062 (Association for Computing Machinery, 2007). doi:10.1145/1273496.1273629.
14. Bird, S., Klein, E. & Loper, E. *Natural language processing with Python.* (O'Reilly, 2009).
15. Mueller, A. *Python Word Cloud Package.* (2020).
16. Chen, Q., Peng, Y. & Lu, Z. BioSentVec: creating sentence embeddings for biomedical texts. in *2019 IEEE International Conference on Healthcare Informatics (ICHI)* 1–5 (2019). doi:10.1109/ICHI.2019.8904728.
17. WHO fact sheet on leprosy. https://www.who.int/news-room/fact-sheets/detail/leprosy.
18. Pennington, J., Socher, R. & Manning, C. D. Glove: Global vectors for word representation. in *Proceedings of the 2014 conference on empirical methods in natural language processing (EMNLP)* 1532–1543 (2014).
19. Mikolov, T., Chen, K., Corrado, G. & Dean, J. Efficient Estimation of Word Representations in Vector Space. *ArXiv13013781 Cs* (2013).
20. Zhang, Y., Chen, Q., Yang, Z., Lin, H. & Lu, Z. BioWordVec, improving biomedical word embeddings with subword information and MeSH. *Sci. Data* **6**, 52 (2019).
21. Allot, A. *et al.* LitSense: making sense of biomedical literature at sentence level. *Nucleic Acids Res.* **47**, W594–W599 (2019).

TreeFix-TP: Phylogenetic Error-Correction for Infectious Disease Transmission Network Inference

Samuel Sledzieski, Chengchen Zhang, Ion Mandoiu, and Mukul S. Bansal†

Department of Computer Science and Engineering, University of Connecticut
Storrs, CT 06269, USA

†*Corresponding Author: mukul.bansal@uconn.edu*

Many existing methods for estimation of infectious disease transmission networks use a phylogeny of the infecting strains as the basis for transmission network inference, and accurate network inference relies on accuracy of this underlying evolutionary history. However, phylogenetic reconstruction can be highly error prone and more sophisticated methods can fail to scale to larger outbreaks, negatively impacting downstream transmission network inference.

We introduce a new method, TreeFix-TP, for accurate and scalable reconstruction of transmission phylogenies based on an error-correction framework. Our method uses intra-host strain diversity and host information to balance a parsimonious evaluation of the implied transmission network with statistical hypothesis testing on sequence data likelihood. The reconstructed tree minimizes the number of required disease transmissions while being as well supported by sequence data as the maximum likelihood phylogeny. Using a simulation framework for viral transmission and evolution and real data from ten HCV outbreaks, we demonstrate that error-correction with TreeFix-TP improves phylogenetic accuracy and outbreak source detection. Our results show that using TreeFix-TP can lead to significant improvement in transmission phylogeny inference and that its performance is robust to variations in transmission and evolutionary parameters. TreeFix-TP is freely available open-source from `https://compbio.engr.uconn.edu/software/treefix-tp/`.

Keywords: phylogeny reconstruction, transmission network inference, infectious disease, computational epidemiology

1. Background

The study of infectious disease has benefited greatly from advances in computational molecular epidemiology. The efficacy of public health efforts to combat the spread of these pathogens has rapidly expanded as technology improves – most notably, the onset of powerful high throughput or next-generation sequencing (NGS) methods has provided molecular epidemiologists with the ability to quickly and cheaply sequence the genomes of the infecting strains (viral or bacterial)[1] which in turn has opened the door for computational analysis of these sequences and of disease transmission. By understanding disease transmission, those investigating a disease can more effectively combat its spread. Computational methods for molecular epidemiology have had a positive impact on public health in a number of cases,[2,3] and continue to be widely used for the study of infectious disease transmission,[4] including for the ongoing COVID-19 pandemic (e.g., through the popular `https://nextstrain.org/ncov/global` web portal).

Transmission network inference is a challenging computational problem, which has been reflected in the number of new methods developed for understanding disease transmission, especially that of rapidly-evolving RNA viruses.[5–10] A key challenge with studying the transmission of rapidly evolving RNA and retroviruses[11] is that they exist in the host as "clouds" of closely related sequences. These strain variants are referred to as *quasispecies* by virologists,[12–16] and the resulting genetic diversity of the strains circulating within a host has important implications for efficiency of virus transmission, virulence, disease progression, drug/vaccine resistance, etc..[17–21] The advent of next-generation sequencing technologies, has revolutionized the study of quasispecies, but most existing transmission network inference methods are unable to make use of the ability to sequence multiple distinct strain sequences per host. However, methods that explicitly consider multiple strain sequence per host have recently started to be developed; such methods include Phyloscanner,[7] SharpTNI,[22] and TNet.[23]

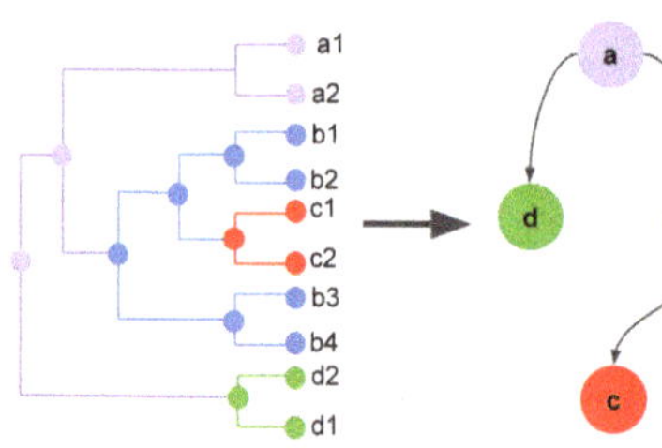

Fig. 1. **Phylogeny-based transmission network inference:** In this figure, internal nodes of the phylogenetic tree on the left are labeled by one of hosts a, b, c, or d, represented here by the different colors. This labeling of internal nodes causes some of the edges in the tree to have different labels at their two end points, and such edges represent transmission edges in the final transmission network. In the figure we see transitions from a to b, a to d, and b to c, yielding the transmission network shown on the right.

Some of the most powerful and widely used techniques for transmission network inference, including Phyloscanner,[7] SharpTNI,[22] and TNet,[23] are based on computing and using phylogenies of the infecting strains.[5–8,24] We refer to these strain phylogenies as *transmission phylogenies*. These phylogeny-based methods infer transmission networks through a host assignment for each node of the transmission phylogeny, where this phylogeny is either first constructed independently or is co-estimated along with the host assignment. Leaves of the transmission phylogeny are labeled corresponding to the host from which they are sampled, and an ancestral host assignment is then inferred for each node/edge of the phylogeny. This ancestral host assignment defines a transmission network, where transmission is inferred along any edge connecting two nodes labeled with different hosts. In the case of a rooted phylogeny, this coloring also confers direction of transmission, where the host for the ancestral sequence along a transmission edge is considered to be the source of the transmission, and the host of the child sequence is considered to be the recipient. This is illustrated in Figure 1.

Two of the most widely-used methods for inference of transmission phylogenies are BEAST[25] and RAxML.[26] For instance, among existing transmission inference methods, TransPhylo[6] uses BEAST to infer a transmission phylogeny, while Phyloscanner[7] uses RAxML. BEAST uses Markov Chain Monte Carlo (MCMC) to estimate phylogenies and evolutionary parameters for several sophisticated models of evolution. Because the models implemented are highly complex, BEAST is prohibitively slow for use on anything other than small data sets. As a result, more scalable, but slightly less accurate, maximum likelihood based methods, such as the state-of-the-art RAxML method,[26] are often used in practice for inferring transmission phylogenies. There are also several methods which address transmission phylogeny reconstruction specifically from a transmission perspective, and use transmission information to inform phylogenetic inference. These methods often perform co-estimation of both the transmission phylogeny and network, and often model within-host evolution. BEASTlier[5] and Phybreak[8] both use Bayesian inference for co-estimation of transmission phylogeny and network, and so run into the same scalability issues as BEAST. Thus, even though accurate reconstruction of the transmission phylogeny has a direct impact on transmission network inference, all existing phylogenetic inference methods for transmission phylogenies are either prohibitively slow and unscalable or suffer from poor inference accuracy. Furthermore, none of these existing phylogenetic inference methods can take advantage of the information provided by multiple sequences from each infected host.

In this work, we introduce *TreeFix-TP*, a new method for reconstructing transmission phylogenies that is as scalable as RAxML but significantly more accurate. TreeFix-TP improves the accuracy of infectious disease transmission phylogenies using an error-correction approach. Specifically, TreeFix-TP leverages both sequence and host information to reconstruct more accurate phylogenies than maximum likelihood on its own by minimizing the number of inter-host transmissions while maintaining statistical support. Similar error correction approaches have been successfully used for reconstruction of gene trees;[27,28] however, these previous methods are based on leveraging a known species phylogeny to error-correct and improve gene trees, and they are therefore inapplicable to the current setting where the goal is to reconstruct the strain tree itself (analogous to the species tree). We address this problem by leveraging intra-host strain diversity and defining a fitness function based on minimizing the number of inter-host transmissions implied by the underlying phylogeny.

In this study, we compare the phylogenetic reconstruction accuracy of Treefix-TP to RAxML.[26] We show that TreeFix-TP reconstructs significantly more accurate transmission phylogenies than RAxML, and is robust to variations in transmission model, sequence length, rate of evolution, and number of viruses. Furthermore, we demonstrate the use of TreeFix-TP for improving source detection in 10 real-world HCV outbreaks.

2. Methods

2.1. *Minimizing inter-host transmissions*

The availability of multiple strain sequences from each host provides valuable additional information that can be used to improve the inference of transmission phylogenies. Consider an ideal evolutionary scenario with a complete transmission bottleneck and no re-infection. In such a scenario, all sequences sampled from the same host should form a single monophyletic clade. For N hosts, this ideal case would result in a coloring with N single-color sub-graphs and would imply $N - 1$ transmissions. Deviations from this ideal would be reflected in the transmission phylogeny and imply a few additional transmissions. Thus, when multiple strain sequences are available from each host, a biologically meaningful criterion for estimating the "correctness" of a transmission phylogeny is to minimize the number of implied inter-host transmissions. Note that the problem of computing the minimum number of implied inter-host transmissions on a given transmission phylogeny is equivalent to the well-known small parsimony problem in phylogenetics and can be solved very efficiently.[29] By minimizing the number of inter-host transmissions implied by a candidate phylogeny, and carefully avoiding over-fitting, we can improve the accuracy of a given phylogeny.

2.2. *Description of TreeFix-TP*

The input for TreeFix-TP is a multiple sequence alignment of infectious disease sequences, a maximum likelihood phylogeny constructed on the infectious disease sequences, and a mapping from all sequences to known hosts. TreeFix-TP aims to find the transmission phylogeny which is well supported by sequence data and has the minimum transmission cost. Using the maximum likelihood phylogeny as a starting point, we perform iterative local searches and evaluate each candidate tree using a statistical likelihood test and an evaluation of the transmission cost. Candidate phylogenies which are statistically equivalent to the maximum likelihood phylogeny, and with a lower transmission cost, are accepted and set as the starting point for the next local search iteration.

TreeFix-TP uses the Shimodaira-Hasegawa (SH) statistical likelihood test[30] to determine sequence support for a given phylogeny. This test considers two trees, in our case the maximum likelihood phylogeny and a candidate phylogeny, with the null hypothesis that the two trees are equally supported by the sequence data. The null hypothesis is rejected at a significance level α which can be defined by the user. If the null hypothesis fails to be rejected, the two trees are considered to be statistically equivalent

The transmission cost for a candidate phylogeny is calculated by solving an instance of the small parsimony problem using Fitch's algorithm.[29] The states at the leaves of a candidate phylogeny are the hosts from which each sequence is known to be sampled. Fitch's algorithm, then, calculates the minimum number of state changes required to generate the given phylogeny, which corresponds to minimizing the number of inter-host transmissions. In this case, we are concerned only with the cost of a candidate and not the internal assignments of hosts, so only the upward pass of Fitch's algorithm is performed. Full details of the algorithm and efficient implementation can be found in Section S1 in the Supplementary Material.

2.3. *Evaluation using simulated data sets*

2.3.1. *Data set generation*

To evaluate the performance of TreeFix-TP, we generated a number of simulated data sets across a variety of parameters and developed a testing pipeline to compare TreeFix-TP with RAxML (see Figure 2). Our simulated viral data sets were generated using FAVITES,[31] a recently developed framework for simultaneous simulation of viral transmission networks, phylogenetic trees, and sequences.

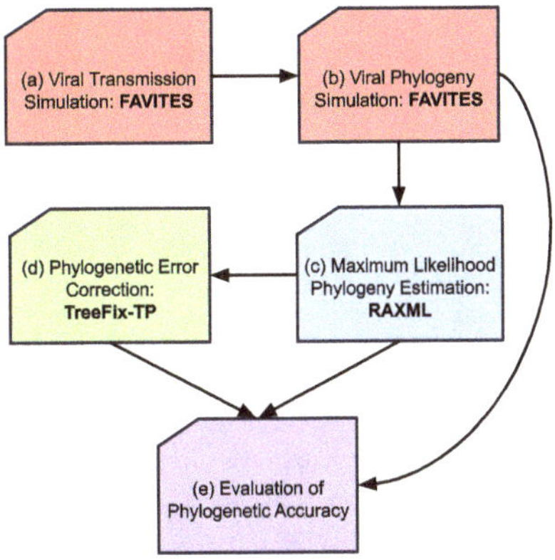

Fig. 2. **TreeFix-TP Testing Pipeline:** To evaluate TreeFix-TP, we first used FAVITES to generate a transmission network (a) and ground truth viral phylogeny (b). Maximum-likelihood phylogenies were then reconstructed from sequences using RAxML (c), and were error corrected with TreeFix-TP (d). The RAxML and TreeFix-TP phylogenies were compared using RF distance, as described in Section 2.3.2 (e).

A contact network was generated using the Barabasi-Albert model[32] with 1000 individuals each with 100 outgoing edges preferentially attached to high-degree nodes. One host was randomly selected to be the source of the infection. Transmission was simulated for a predefined amount of time, or until all hosts were recovered under one of two different compartmental models, either Susceptible-Exposed-Infected-Recovered (SEIR) or Susceptible-Infected-Recovered (SIR).[33] These models are parameterized by transition rates β, λ, and δ, where β is the rate of transition from susceptible to exposed in the SEIR model or susceptible to infected in the SIR model, λ is the rate of transition from exposed to infected (only in the SEIR model), and δ is the rate of recovery for infected individuals. In our simulation, we had four categories of data sets with variations on infection rate β to explore the effect of transmission model on reconstruction accuracy. λ and δ were set according to the infection rate. These parameters can be found in Supplementary Table S2.

Due to the simulation of latent periods, data sets generated under the SEIR model tend to exhibit an outbreak structure, where one high-degree individual infects several of its neighbors, followed by a period of low infection. When one of the newly-infected neighbors becomes infectious, another outbreak occurs. This is contrary to the SIR model, which tends to have a more periodic pattern of disease transmission. In addition to varying the transmission model, we simulated data sets with different rates of infection and recovery. This resulted in four categories of simulation with infection rates of 0.015, 0.003, 0.01, and 0.01 respectively. We group SEIR (0.015) and SEIR (0.01) together, and group SIR (0.003) and SIR (0.01) together since there was no significant difference between the transmission model parameter settings.

These transmission network parameter settings were generally based on the defaults suggested by FAVITES, with some adjustments as necessary for preventing the occurrence of long edges separating sequences from different hosts.

Internal evolution of the virus in infected hosts was simulated under a logistic-growth coalescent model. Each internal phylogeny was connected according to transmission to form a full transmission phylogeny. The branch lengths of this phylogeny were scaled to simulate different rates of sequence evolution. Sequences were simulated using the GTR + Γ model starting with a real HCV viral sequence from HCV outbreak data at the root (discussed in more detail in Section 3.2). The GTR rate matrix and gamma parameter were determined by applying RAxML to estimate parameters and construct a phylogeny for real sequences from an HCV outbreak. Thus, the simulated sequences are designed to reflect real rapidly-evolving RNA viral sequences.

By default, we simulated sequences of length 1000 nucleotides and sampled 10 sequences per infected host. We scaled the branch lengths of the phylogeny by 0.25 on data sets where the SEIR and SIR (0.003) models were used, and by 1.5 on data sets where the SIR (0.01) model was used. These scale factors were chosen so that the height of the tree would be approximately ten expected mutations per-site. We varied the sequence length, number of sequences per host, and mutation rate to quantify the robustness of TreeFix-TP to variance in sequence evolution. The list of all transmission and evolution simulation parameters can be found in Supplementary Table S2. For each of the four categories, we tested the effects of varying sequence length, number of samples per host, and scale factor, varying one of these parameters at a time from the default setting. Specifically, we simulated sequences of length 250, 500, and 1000, sampled 5, 10, and 20 sequences, and scaled the tree by double or half the default. Including the default setting, this resulted in 7 distinct parameterizations per category, or 28 total. Full specifications of parameters for each variation can also be found in Supplementary Table S2.

For each set of simulation parameters, we simulated 20 different data sets for a total of 560 simulated data sets. RAxML and TreeFix-TP were limited to 8GB of memory and 10 days, and due to these limitations we were able to reconstruct phylogenies using TreeFix-TP for 486 of these data sets. Of the 74 runs which did not complete, the simulated trees had an average of 733.43 leaves. For the 486 simulated data sets on which we obtained results, we had between 35 and 630 sequences, with an average of 223.41 leaves. The average number of transmissions was 22, and 95% of data sets had between 7 and 49 transmissions. Of the data sets for which we obtained results, only 6 had more than 60 transmissions.

2.3.2. *Evaluating reconstruction accuracy*

The accuracy of the reconstructed phylogeny was evaluated by calculating the Robinson-Foulds distance[34] between the true evolutionary history from the simulated data and both the maximum likelihood tree reconstructed by RAxML and the error-corrected tree reconstructed by TreeFix-TP. We calculated the average RF distances, normalized by the maximum possible RF distance (number of internal edges). We calculated the *RF percent decrease* as follows: Given simulated tree S, maximum likelihood tree R, and TreeFix-TP tree T, RF percent decrease is given by $100 \times (RF(S,R) - RF(S,T))/RF(S,R)$. We calculated p-values using the one-tailed Wilcoxon Signed-Rank test implemented in Scipy 1.3.1. Additionally, we looked at the minimum transmission cost implied by the RAxML and TreeFix-TP trees. The cost of the TreeFix-TP tree is guaranteed to be no greater than that of the RAxML tree, but it is valuable to see by how much the transmission cost is decreased and the relationship between transmission cost and Robinson-Foulds distance. Note that we did not compare reconstruction accuracy against BEAST[25] since it is not scalable to data set sizes used in this study.

3. Results

3.1. *Phylogenetic error correction results*

For baseline evaluation, we compared the phylogenies reconstructed by TreeFix-TP and RAxML on 35 data sets corresponding to the SEIR transmission model, sequence length 1000, 10 sequences per host, and a mutation rate of 0.25. Among these trials, 48.6% of the data sets showed a decrease in RF distance with TreeFix-TP, while 42.86% saw no improvement and 8.57% saw an increase. The average RF percent decrease for trees which improved was 14.6%, and as high as 46.154%, while the average RF percent increase for those trees that got worse was only 3.644%. In every run where there was no improvement, the maximum likelihood tree generated with RAxML implied exactly as many or only one more transmission than the true number of transmissions, so the ability for TreeFix-TP to correct errors by minimizing transmission was limited. Across all 35 data sets, the average normalized RF distance of trees reconstructed with RAxML was 0.152, while trees reconstructed with TreeFix-TP had an average normalized RF distance of 0.137 ($p = 0.0003$, Wilcoxon Signed-Rank). The overall average RF percent decrease was 9.99%.

We also evaluated 32 data sets corresponding to the SIR transmission model, sequence length 1000, 10 sequences per host, and a mutation rate of either 1.5 or 0.25 (aggregated over both transmission rate categories). The average normalized RF distance of trees constructed with RAxML was 0.103, while trees reconstructed with TreeFix-TP had an average normalized RF distance 0.098 ($p = 0.006$, Wilcoxon Signed-Rank). The magnitude of improvement is impacted by the large number of no-change error corrections. Specifically, under the SIR model of transmission, 68.75% of runs had no-change, while 28.13% showed a decrease in RF distance, and the remaining 3.13% showed an increase. The overall average RF percent decrease was 4.36%, but those which improved had an average RF percent decrease of 14.116%, and as high as 28.57%. For those which got worse, the average RF percent increase was 9.8%. A comparison of these results across the SEIR and SIR transmission models suggests that error correction might be more effective under a model of transmission that includes a latent period, which results in transmissions patterns which more closely reflect outbreaks.

Impact of varying sequence length To evaluate the robustness of TreeFix-TP to the amount of sequence information available, we varied sequence length from the base 1000 nucleotides to 250 and 500 nucleotides (Figure 3a). Under the SEIR model, we found that TreeFix-TP continued to improve the accuracy of phylogenetic reconstruction with shorter sequence lengths, and that sequence length didn't seem to have a large effect on the ability

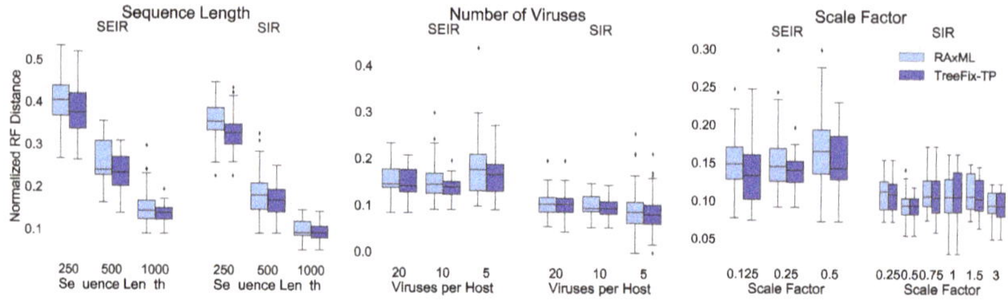

Fig. 3. **Robustness of phylogeny reconstruction to different parameters:** Normalized Robinson-Foulds (RF) distance from the simulated phylogeny for reconstructions with both RAxML and TreeFix-TP under a variety of settings. TreeFix-TP reconstructs the most accurate trees across all data sets. **(a)** RF distance for varied sequence lengths. Trees are in general more accurate with longer sequences, and TreeFix-TP improves upon RAxML to a greater extent with shorter sequences. **(b)** RF distance for varied numbers of viruses sampled from each host. TreeFix-TP has the largest improvement when fewer viruses are sampled. **(c)** RF distance across multiple different scale factors. TreeFix-TP reconstructed the most accurate phylogenies with all scale factors.

of error correction to improved the accuracy of the phylogeny. At sequence length 1000, the average normalized RF distance decreased by 9.99% from 0.152 to 0.137 after error correction ($p = 0.0003$, Wilcoxon Signed-Rank). At length 500, this was a decrease of 11.03% from 0.264 to 0.235 ($p = 1e-5$, Wilcoxon Signed-Rank). At sequence length 250, the average RF distance decreased by an average of 5.68% from 0.403 to 0.380 ($p = 6e-5$, Wilcoxon Signed-Rank). As expected, the absolute error rate increases sharply, for both RAxML and TreeFix-TP, as sequence length decreases.

Under the SIR model, we found the error correction continued to have an impact at all sequence lengths, and that error correction was more effective at shorter sequence lengths. With sequence length of 1000, the average RF distance decreased by 4.36% (0.103 to 0.099 normalized RF, $p = 0.006$, Wilcoxon Signed-Rank). At length 500, there was a 7.65% decrease (0.187 to 0.172 normalized RF, $p = 0.0001$, Wilcoxon Signed-Rank), and at length 250 there was a 7.59% decrease (0.357 to 0.330 normalized RF, ($p = 9e-5$, Wilcoxon Signed-Rank). Under this model, error correction seems to be more effective with shorter sequences, likely because longer sequences contain more information which allows maximum likelihood methods to reconstruct a relatively accurate tree before any error correction occurs.

Impact of varying number of viruses We observed the effect of sampling different numbers of viruses from each infected host, from the default of 10 to 5 and 20 viral sequence samples (Figure 3b). TreeFix-TP reconstructed more accurate phylogenies in each case, with the largest overall improvement occurring for trees with 5 sequences from each host. Under the SEIR model, with 20 viruses, there was an average RF distance decrease of 2.78% (0.152 to 0.148 normalized RF, $p = 0.018$, Wilcoxon Signed-Rank). With 10 and 5 viruses, there were larger decreases of 9.99% and 10.18% respectively (0.152 to 0.137 and 0.183 to 0.164 normalized RF, $p = 0.0003$, $p = 8e-5$ Wilcoxon Signed-Rank). Under the SIR model, with 20 viruses, there was an decrease in average RF distance of only 1.56% (0.108 to 0.106 normalized RF, $p = 0.072$, Wilcoxon Signed-Rank). This decrease was 4.36% with 10 viruses and 6.52% with 5 viruses (0.103 to 0.099 and 0.096 to 0.089 normalized RF, $p = 0.006$, $p = 0.013$ Wilcoxon Signed-Rank). Error correction seems to be more effective with fewer viruses, which matches the intuition about sequence length - that more sequence data leads to originally accurate phylogenies, and less potential for error correction.

Impact of varying scale factor We found that TreeFix-TP is also robust to various rates of sequence evolution (Figure 3c). Under the SEIR model of evolution, scale factors of 0.125, 0.25, and 0.5 resulted in a decrease in average RF distance by 6.64%, 9.99%, and 9.76% respectively (0.151 to 0.141, 0.152 to 0.137, 0.168 to 0.152 normalized RF, $p = 0.004, 0.0003, 0.0006$ Wilcoxon Signed-Rank). Under the SIR model, we used two different sets of scale factors dependent on the disease transmission parameters. Aggregated across SIR (0.003) and SIR (0.01), we tested scale factors of 0.25, 0.5, 0.75, 1, 1.5, and 3. These scale factors had average RF percent decreases of 3.05%, 2.87%, 2.02%, 0.07%, 5.97%, and 1.20% (0.111 to 0.108, 0.095 to 0.093, 0.111 to 0.109, 0.1043 to 0.1042, 0.113 to 0.106, and 0.095 to 0.094 normalized RF, $p = 0.045, 0.054, 0.034, 0.327, 0.021, 0.250$). As expected, the overall RF distances tended to be larger for very small and very large scale factors, which indicates that a reasonable rate of evolution is important to overall phylogenetic reconstruction accuracy, but plays less of an impact on error correction.

3.2. *Source recovery in HCV outbreaks*

We also evaluated the impact of using TreeFix-TP on real data sets of HCV outbreaks made available by the CDC.[9] In total, there are 10 different data sets, each representing a separate HCV outbreak. Each of these outbreak data sets contains between 2 and 19 infected hosts and a few dozen to a few hundred strain sequences. For each of these 10 outbreaks, the source host of the outbreak is known (through the CDC's epidemiological efforts). We used a simple phylogenetic pipeline to infer a source for each of these 10 data sets as follows: We constructed phylogenetic trees using RAxML and TreeFix-TP and rooted them using two of the most widely used rooting methods, balanced rooting (implemented in RAxML[26]) and midpoint rooting.[35,36] We then used Sankoff's algorithm for the small parsimony problem[37] to label the internal nodes of these phylogenies with hosts and report the host assignment at the root as the inferred source of that outbreak. (Note that Phyloscanner also uses Sankoff's algorithm to label internal nodes of the phylogeny, but we chose not to use Phyloscanner directly because it is very conservative in its host assignments and often leaves nodes unlabeled.) Using the RAxML trees, the source was correctly identified in 6 (balanced rooting) and 7 (midpoint rooting) of the 10 outbreaks. In contrast, the trees reconstructed by TreeFix-TP correctly identified the source in 8 out of the 10 outbreaks with both rooting strategies. Furthermore, the outbreaks correctly identified by RAxML were a strict subset of those identified by TreeFix-TP.

3.3. *Running time and scalability*

Using its default number of iterations (5000) TreeFix-TP required an average of approximately 37 hours for each run, but this running time varied depending on the number of tips and length of sequence. TreeFix-TP took less than an hour and a half for trees of 50-60 tips, but upwards of 200 hours for trees with more than 500 tips and 1000 nucleotide-length sequences. On average, runs took fewer than 9 minutes per tip, and scaled linearly in tree size, number of hosts, and sequence length.

4. Discussion and Conclusions

In this paper, we have introduced a new method, TreeFix-TP, for more accurate and scalable reconstruction of infectious disease transmission phylogenies when multiple strain sequences are sampled from each infected host, and demonstrated its impact on phylogenetic inference and outbreak source detection. TreeFix-TP uses an error-correction approach where it seeks to improve a given maximum-likelihood phylogeny of the infecting strains by using additional information about which host each strain was sampled from and balancing it with sequence-only likelihood using a statistical hypothesis testing framework. As our experimental results show, TreeFix-TP consistently reconstructs more accurate phylogenies than the state-of-the-art maximum-likelihood phylogeny inference method RAxML. We also demonstrate how TreeFix-TP can be used to augment existing phylogeny-based pipelines for transmission network inference by error correcting the phylogenies before they are used for network inference or outbreak source detection.

Going forward, it would be worthwhile to develop even more advanced, yet scalable, methods for construction of transmission phylogenies. As our experimental results show, even though the absolute error rate of TreeFix-TP phylogenies is often significantly lower than that of RAxML trees, this absolute error rate still remains quite high overall even after error correction. This is partly because the ability of TreeFix-TP to error-correct depends on the number of different hosts represented in the phylogeny, rather than on the size of the tree itself. In the future, it may be possible to use additional information about within-host strain evolution to further improve transmission phylogeny inference.

Acknowledgments

The authors wish to thank Dr. Pavel Skums (Georgia State University) and the Centers for Disease Control for sharing their HCV outbreak data. This work was supported in part by NSF award CCF 1618347 to IM and MSB.

Authors' Contributions

SS contributed to the theoretical results, implemented the software, performed the experimental study, analyzed the results, and contributed to the writing of the manuscript. CZ contributed to initial project development and conducted the experimental analysis on real data. IM helped supervise the research and contributed to writing the manuscript. MSB conceived the research project, supervised the research, and contributed to the writing of the manuscript.

Supplementary Material

Supplementary material can be found at:
`https://compbio.engr.uconn.edu/treefix-tp_supplement/`

References

1. S. C. Shuster, Next-generatino sequencing transforms today's biology, *Nature* **5** (dec 2007).
2. A. Grulich, A. Pinto, A. Kelleher, D. Cooper, P. Keen, F. Di Giallonardo, C. Cooper and B. Telfer, A10 Using the molecular epidemiology of HIV transmission in New South Wales to inform public health response: Assessing the representativeness of linked phylogenetic data, *Virus Evolution* **4** (04 2018).
3. D. Clutter, R. W. Shafer, S.-Y. Rhee, W. J. Fessel, D. Klein, S. Slome, B. A. Pinsky, J. L. Marcus, L. Hurley, M. J. Silverberg and S. L. Kosakovsky Pond, Trends in the Molecular Epidemiology and Genetic Mechanisms of Transmitted Human Immunodeficiency Virus Type 1 Drug Resistance in a Large US Clinic Population, *Clinical Infectious Diseases* **68**, 213 (05 2018).
4. E. O. Romero-Severson, I. Bulla and T. Leitner, Phylogenetically resolving epidemiologic linkage, *Proceedings of the National Academy of Sciences* **113**, 2690 (mar 2016).
5. M. Hall, M. Woolhouse and A. Rambaut, Epidemic Reconstruction in a Phylogenetics Framework: Transmission Trees as Partitions of the Node Set, *PLoS Computational Biology* **11**, p. e1004613 (dec 2015).
6. X. Didelot, C. Fraser, J. Gardy, C. Colijn and H. Malik, Genomic infectious disease epidemiology in partially sampled and ongoing outbreaks, *Molecular Biology and Evolution* **34**, 997 (jan 2017).
7. C. Wymant, M. Hall, O. Ratmann, D. Bonsall, T. Golubchik, M. de Cesare, A. Gall, M. Cornelissen and C. Fraser, PHYLOSCANNER: Inferring Transmission from Within- and Between-Host Pathogen Genetic Diversity, *Molecular Biology And Evolution* **35**, 719 (mar 2017).
8. D. Klinkenberg, J. A. Backer, X. Didelot, C. Colijn and J. Wallinga, *Simultaneous inference of phylogenetic and transmission trees in infectious disease outbreaks* (PLoS, 2017).
9. P. Skums, A. Zelikovsky, R. Singh, W. Gussler, Z. Dimitrova, S. Knyazev, I. Mandric, S. Ramachandran, D. Campo, D. Jha, L. Bunimovich, E. Costenbader, C. Sexton, S. O'Connor, G. L. Xia and Y. Khudyakov, QUENTIN: Reconstruction of disease transmissions from viral quasispecies genomic data, *Bioinformatics* **34**, 163 (jun 2018).
10. S. L. Kosakovsky Pond, S. Weaver, A. J. Leigh Brown and J. O. Wertheim, HIV-TRACE (TRAnsmission Cluster Engine): a Tool for Large Scale Molecular Epidemiology of HIV-1 and Other Rapidly Evolving Pathogens, *Molecular Biology and Evolution* **35**, 1812 (01 2018).
11. J. W. Drake and J. J. Holland, Mutation rates among RNA viruses, *Proceedings of the National Academy of Sciences of the United States of America* **96**, 13910 (1999).
12. E. Domingo and J. Holland, RNA virus mutations and fitness for survival, *Annu Rev Microbiol* **51**, 151 (1997).
13. E. Domingo, M.-S. E., F. Sobrino, J. de la Torre, A. Portela, J. Ortin, C. Lopez-Galindez, P. Perez-Brena, N. Villanueva and R. Najera, The quasispecies (extremely heterogeneous) nature of viral rna genome populations: biological relevance – review, *Gene* **40**, 1 (1985).
14. M. E. M, J. McCaskill and P. Schuster, The molecular quasi-species, *Adv Chem Phys* **75**, 149 (1989).
15. M. Martell, J. Esteban, J. Quer, J. Genesca, A. Weiner, R. Esteban, J. Guardia and J. Gomez, Hepatitis c virus (hcv) circulates as a population of different but closely related genomes: quasispecies nature of hcv genome distribution, *Journal of Virology* **66**, 3225 (1992).
16. D. Steinhauer and J. Holland, Rapid evolution of rna viruses, *Annual Review of Microbiology* **41**, 409 (1987).
17. N. Beerenwinkel, T. Sing, T. Lengauer, J. Rahnenfuehrer and K. R. et al., Computational meth-

ods for the design of effective therapies against drug resistant HIV strains, *Bioinformatics* **21**, 3943 (2005).
18. N. G. Douek DC, Kwong PD, The rational design of an AIDS vaccine., *Cell* **124**, 677 (2006).
19. B. Gaschen, J. Taylor, K. Yusim, B. Foley and F. G. et al., Diversity considerations in HIV-1 vaccine selection, *Science* **296**, 2354 (2002).
20. J. Holland, J. de la Torre and D. Steinhauer, Rna virus populations as quasispecies, *Current Topics in Microbiology and Immunology* **176**, 1 (1992).
21. S.-Y. Rhee, T. Liu, S. Holmes and R. Shafer, HIV-1 subtype B protease and reverse transcriptase amino acid covariation, *PLoS Comput Biol* **3**, p. e87 (2007).
22. P. Sashittal and M. El-Kebir, SharpTNI: Counting and Sampling Parsimonious Transmission Networks under a Weak Bottleneck, *bioRxiv* (2019).
23. S. Dhar, C. Zhang, I. Mandoiu and M. S. Bansal, Tnet: Phylogeny-based inference of disease transmission networks using within-host strain diversity, in *Bioinformatics Research and Applications*, eds. Z. Cai, I. Mandoiu, G. Narasimhan and P. Skums (Springer Nature, 2020)
24. E. M. Volz, K. Koelle and T. Bedford, Viral Phylodynamics, *PLoS Computational Biology* **9**, p. e1002947 (mar 2013).
25. A. J. Drummond and A. Rambaut, BEAST: Bayesian evolutionary analysis by sampling trees, *BMC Evolutionary Biology* **7**, p. 214 (nov 2007).
26. A. Stamatakis, RAxML version 8: a tool for phylogenetic analysis and post-analysis of large phylogenies, *Bioinformatics* **30**, 1312 (may 2014).
27. Y.-C. Wu, M. D. Rasmussen, M. S. Bansal and M. Kellis, TreeFix: Statistically Informed Gene Tree Error Correction Using Species Trees, *Systematic Biology* **62**, 110 (jan 2013).
28. M. S. Bansal, Y. C. Wu, E. J. Alm and M. Kellis, Improved gene tree error correction in the presence of horizontal gene transfer, *Bioinformatics* **31**, 1211 (apr 2015).
29. W. Fitch, Towards defining the course of evolution: minimum change for a specified tree topology, *Syst. Zool.* **20**, 406 (1971).
30. H. Shimodaira and M. Hasegawa, Multiple comparisons of log-likelihoods with applications to phylogenetic inference, *Molecular Biology and Evolution* **16**, p. 1114 (1999).
31. N. Moshiri, J. O. Wertheim, M. Ragonnet-Cronin and S. Mirarab, FAVITES: simultaneous simulation of transmission networks, phylogenetic trees and sequences, *Bioinformatics* **35** (11 2018).
32. R. Albert and A.-L. Barabási, Statistical mechanics of complex networks, *Rev. Mod. Phys.* **74**, 47 (Jan 2002).
33. W. O. Kermack, A. G. McKendrick and G. T. Walker, A contribution to the mathematical theory of epidemics, *Proceedings of the Royal Society of London. Series A, Containing Papers of a Mathematical and Physical Character* **115**, 700 (1927).
34. D. F. Robinson and L. R. Foulds, Comparison of phylogenetic trees, *Math. Biosci.* **53**, 131 (feb 1981).
35. J. S. Farris, Estimating phylogenetic trees from distance matrices, *The American Naturalist* **106**, 645 (1972).
36. D. Swofford, G. Olsen, P. Waddell and D. Hillis, Phylogenetic inference, in *Molecular systematics*, eds. D. Hillis, C. Moritz and e. B. Mabl (Sinauer Associates, 1996) pp. 407–514.
37. D. Sankoff, Minimal mutation trees of sequences, *SIAM Journal on Applied Mathematics* **28**, 35 (1975).

SARS-CoV-2 Drug Discovery based on Intrinsically Disordered Regions

Anish Mudide
Phillips Exeter Academy
20 Main Street, Exeter, NH 03833, USA
Email: amudide@gmail.com

Gil Alterovitz
Biomedical Cybernetics Laboratory, Brigham and Women's Hospital and Harvard Medical School
Department of Veterans Affairs, National Artificial Intelligence Institute
25 Shattuck Street, Boston, MA 02115, USA
Email: ga@alum.mit.edu

Severe acute respiratory syndrome coronavirus 2 (SARS-CoV-2), a close relative of SARS-CoV-1, causes coronavirus disease 2019 (COVID-19), which, at the time of writing, has spread to over 19.9 million people worldwide. In this work, we aim to discover drugs capable of inhibiting SARS-CoV-2 through interaction modeling and statistical methods. Currently, many drug discovery approaches follow the typical protein structure-function paradigm, designing drugs to bind to fixed three-dimensional structures. However, in recent years such approaches have failed to address drug resistance and limit the set of possible drug targets and candidates. For these reasons we instead focus on targeting protein regions that lack a stable structure, known as intrinsically disordered regions (IDRs). Such regions are essential to numerous biological pathways that contribute to the virulence of various viruses. In this work, we discover eleven new SARS-CoV-2 drug candidates targeting IDRs and provide further evidence for the involvement of IDRs in viral processes such as enzymatic peptide cleavage while demonstrating the efficacy of our unique docking approach.

1. Introduction

IDRs lack a fixed three-dimensional structure, and instead fold dynamically into a set of continuous conformations based on surrounding conditions [1]. This allows IDRs to have a wide range of binding partners, and as a result, serve significant roles in critical biological processes such as cell signaling and transcription [2-3]. Moreover, certain short IDRs known as molecular recognition features (MoRFs) are essential for initiating protein-protein interactions (PPIs) [4]. For over a decade now, it has been clear that IDRs are functionally important to and incredibly abundant in proteins implicated across the disease spectrum [5].

While IDRs are not incredibly common in the SARS-CoV-2 proteome, the IDRs that do exist contribute greatly to the functioning and overall virulence of the pathogen [6-7]. In fact, nearly all SARS-CoV-2 proteins are predicted to have MoRFs, highly suggestive of the importance of IDRs in PPI networks [7]. SARS-CoV-2 IDRs therefore serve as promising drug targets for antiviral drug discovery.

Of the 27 mature viral proteins within the SARS-CoV-2 proteome, the majority of current drug discovery research is largely focused on three main targets: the RNA polymerase, the Papain-like protease, and the 3C-like protease (3CLpro) [8-9]. The 3CLpro's main role is to cleave the polyproteins into functional parts [10]. While all three targets are disordered [7], in this study we focus on the CoV-2 3CLpro since it is highly similar (96% sequence identity) to its CoV-1 relative, for which an abundance of bioassay data is available [10]. In particular, we concentrate our efforts on the N-terminally short IDR (residues 1-6; see footnote 'b') predicted to be a MoRF in both CoV-1 and CoV-2 [7]. A drug capable of binding to this IDR could thereby inhibit PPIs within the virus.

Our approach to drug discovery consists of two major steps. First, we compute binding affinities between the CoV-2 3CLpro IDR and over 1400 ligands from the NCI Diversity Set III through a unique docking procedure. While older docking procedures focus on targeting structured protein pockets [11], in this study we account for the wide range of IDR conformations through the allowance of residue side chain rotation as well as through ensemble docking. High binding affinities are a key first indicator of drug potential since they imply a great attractive force toward the receptor and demonstrate that the binding energy can be used to alter the receptor structure. We discovered over 60 ligands with binding affinities of -8.0 kcal/mol or better. However, drug discovery approaches relying solely on docking often fail to produce seriously meaningful results, and expert opinion suggests the cross-verification of results using distinct techniques [12-13]. Thus, in the second step of our approach, we validate and filter our results using a statistical model. The results of bioassay AID 1706, which screens over 290,000 compounds for inhibition of CoV-1 3CLpro-mediated peptide cleavage [14], are used to train a message passing neural network (MPNN) to distinguish between positive (3CLpro inhibiting) and negative (non-inhibiting) compounds. Due to the high similarity between the two CoV 3CLpros, such a model is likely to make meaningful predictions relevant to CoV-2 3CLpro inhibition [10, 15]. This model is then used to predict activity scores for each of the previously docked ligands. We show a correlation between activity scores and binding affinity, suggesting the efficacy of our docking approach. Moreover, we combine the results of our steps to determine 11 new CoV-2 drug candidates, many of which show antibiotic or antiviral properties. Figure 1 summarizes the process.

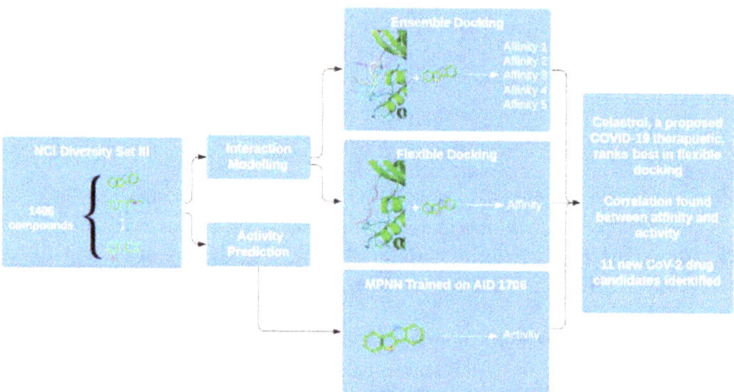

Fig. 1. Drug discovery flowchart.

2. Methods

2.1. *Molecular docking*[a]

2.1.1. *Data collection*

The three-dimensional structures of all 27 mature viral proteins were predicted by the Oak Ridge National Laboratory (ORNL) with a workflow consisting of X-ray crystallography results, homology modeling, and disorder prediction among other techniques. In particular, the structure of the monomeric form of 3CLpro was obtained directly from ORNL's COVID-19 site.[b]

In this study, we use the National Cancer Institute (NCI) Diversity Set III as our ligand dataset. Diversity sets are constrained such that no two ligands can be overly similar to one another, resulting in heterogeneity. A single SDF file was retrieved from NCI's website[c] describing the structures of each of the ligands in the set.

2.1.2. *Data preprocessing*

AutoDockTools was used to prepare and preprocess the PDB file for the 3CLpro before docking. Water molecules were removed, polar hydrogen atoms were added, and Kollman charges were added to the entire structure. The structure was then saved as a PDBQT file.

The ligands were extracted from the SDF file into individual PDB files. Then, the prepare_ligand function from the AutoDock Flexible Receptor (ADFR) suite[d] was used to preprocess each of these ligand files, generating PDBQT files ready for docking.

2.1.3. *Target file generation*

The protein-ligand docking software used in this study is AutoDock Flexible Receptor (ADFR). ADFR requires at least two parameters to be passed: the protein receptor, specified by a target file, and the ligand, specified by a PDBQT file. Target files specify the docking box size and position, calculated binding pockets, residue side chains to be made flexible, affinity maps, as well as other meta-data. AutoGridFR was used to generate such a target file for the 3CLpro. In particular, the docking box was specified to enclose residues 1-9, and residues within the IDR (1-6) were specified as having flexible side chains. Additionally, AutoSite 1.0 was used to generate ligand binding pockets through a clustering algorithm that groups high affinity points into disjoint "fills." Fills with high scores in close proximity to the disordered region were chosen to be targeted during docking. Figure 2 graphically summarizes the parameters chosen for target file generation.

[a] Our code, data and results are available at https://github.com/Biomedical-Cybernetics-Lab2/IDR-SARS-CoV-2.

[b] https://compsysbio.ornl.gov/covid-19/covid-19-structome/.

[c] https://wiki.nci.nih.gov/display/NCIDTPdata/Compound+Sets.

[d] https://ccsb.scripps.edu/adfr/.

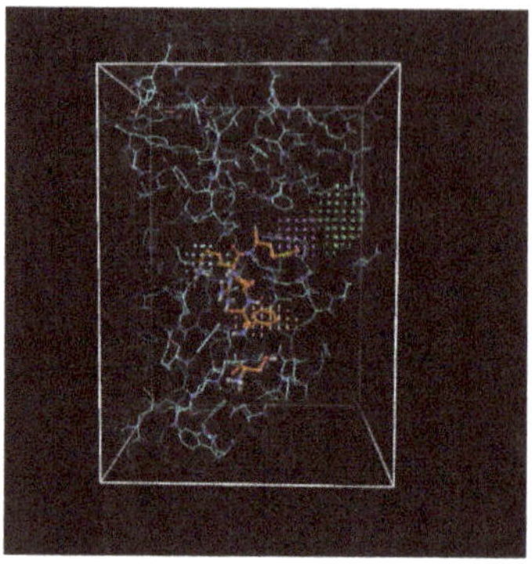

Fig. 2. Orange residues are part of the IDR and specified as flexible. The docking box is shown in white. Fills chosen are shown in purple, light blue, green and orange.

2.1.4. *Flexible docking*

ADFR is unique from other protein-ligand docking software in that it can handle both ligand and receptor flexibility. As a result, ADFR is of incredible use when performing IDR-related docking. ADFR employs a genetic algorithm (GA) to find the best docked position of a given ligand. For each protein-ligand pair, the GA is run several times in case the GA converges to local rather than global optima. Moreover, the user can specify both how many runs are executed as well as an upper bound for the number of times the scoring function is called per run. This allows us to drastically cut down on compute time by potentially terminating searches before they converge. The default values for the number of GA runs and the maximum number of score evaluations are 50 and 2.5 million respectively; in this study, at least initially, we modify these parameters to 7 runs with at most 28,000 evaluations each. Docking is performed with these parameters for 1405 distinct ligands from the NCI Diversity Set III, and results are compiled.

2.1.5. *Ensemble docking*

In our pursuit of simulating the conformational flexibility of the IDR for accurate drug discovery, we also utilize ensemble docking. In this approach, we generate many possible conformations of the IDR, and dock each ligand onto each possible conformation. In this study, we generate conformations by treating the IDR as a loop of the protein. Loop modelling implemented by MODELLER[e] is then used to generate five likely IDR conformations. We then repeat the processes outlined in the above sections: we preprocess each newly generated PDB file, generate a target file for each, and perform docking on each conformation-ligand pair using ADFR. Figure 3 illustrates how the five different conformations of the IDR compare to each other. After docking is complete, results are compiled.

[e] https://salilab.org/modeller/.

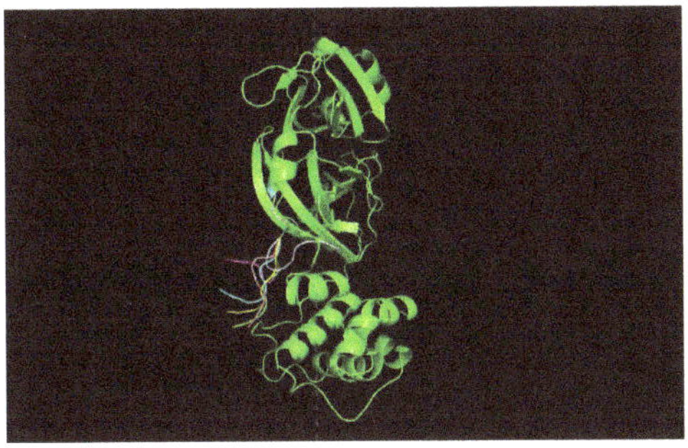

Fig. 3. Five conformations of the 3CLpro IDR superimposed onto the original structure.

2.2. *Statistical model*

2.2.1. *Chemprop*

Chemprop[f] is a freely available implementation of a message passing neural network. Such models are designed to predict properties of graph-based inputs. In the case of molecular property prediction, molecules are transformed to graphs by treating atoms as nodes and the bonds between atoms as edges. Using this representation, a feature vector is generated through a learned algorithm that aggregates chemical features within the graph. This vector is then passed to a typical feed-forward neural network [16]. For our purposes, this neural network outputs a real value between 0 and 1 representing the model's confidence that a certain molecule has a desired binary property.

2.2.2. *Data and training*

Our aim was to train a MPNN model to predict whether a molecule can inhibit the CoV-2 3CLpro *in vitro* to further validate and filter our results from molecular docking. We make use of the results from bioassay AID 1706, which screens over 290,000 compounds for inhibition of CoV-1 3CLpro peptide cleavage, to train such a model. Concretely, the bioassay screens for cleavage inhibition by attaching a fluorescent compound and a quencher to opposite sides of a 3CLpro substrate. A compound can then be classified as active or inactive since fluorescence increases if and only if cleavage occurs [14]. Due to the high similarity between the two CoV 3CLpros, a model trained on CoV-1 data is likely to make meaningful predictions relevant to CoV-2 3CLpro inhibition. Each training example in the dataset[g] consists of one feature (the SMILES string of the compound) and one label (a binary output; 1 for inhibition, 0 for no inhibition). Just 405 of the compounds are classified as positive (label = 1), whereas the other 290,321 compounds are negative (label = 0). To

[f] https://github.com/chemprop/chemprop.

[g] Retrieved from https://github.com/yangkevin2/coronavirus_data.

account for this imbalance between positive and negative data points in the training set, an equal number of positives and negatives are used in each batch during training. Furthermore, additional features generated by RDKit are appended to the feature vector generated before being passed into the neural network during training and predicting. Once trained, the model achieves a test ROC AUC of .739. We then apply the model to predict activity scores for each of the previously docked ligands.

3. Results

3.1. *Interaction modelling*

The binding affinities of over 1400 ligands with the proposed IDR target were analyzed *in silico* using molecular docking. We first simulated IDR conformational flexibility by allowing IDR residue side chains to rotate while searching for the optimal ligand pose. With this docking procedure, 57 ligands were found to have binding affinities of -8.0 kcal/mol or better. Considering that we terminated the docking searches before convergence by bounding the maximum number of score evaluations, their true binding affinities are likely to exceed -8.0 kcal/mol. Therefore, we deemed all 57 ligands as ideal drug candidates. Table 1 summarizes these results of this first docking procedure.

Table 1. Binding affinity results from flexible docking (abridged)

Molecule (NSC)	Binding Affinity (kcal/mol)
70931	-9.8
177862	-9.7
16437	-9.3
96541	-9.1
117987	-8.8
45527	-8.8
...	...

With a binding affinity of -9.8 kcal/mol, the top molecule found is NSC-70931, also known as the triterpenoid named celastrol. Celastrol displays antiviral properties against influenza A virus as well as dengue virus in mice [17-18]. In fact, celastrol has already been suggested as an anti-inflammatory therapeutic for the lethal pneumonia stage of COVID-19 [19]. These results indicate the potential of our first docking method.

When we reran the docking of celastrol onto the 3CLpro IDR with the default parameters mentioned above, the search converged and found a pose with an improved docking score of -11.4 kcal/mol (shown in Figure 4). This further solidifies our claim that the binding affinities presented in this study are likely sub-optimal.

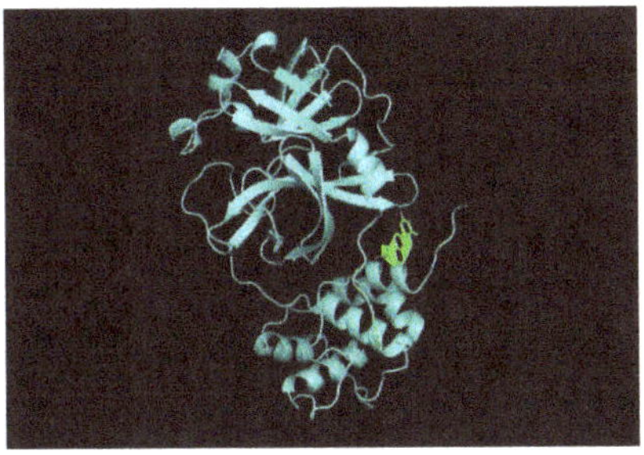

Fig. 4. Docked pose of celastrol (-11.4 kcal/mol) after search converged.

We then simulated IDR conformational flexibility using a different approach known as ensemble docking. Concretely, each ligand was docked onto five distinct conformations of the IDR generated by loop modelling techniques. These five binding affinities were retrieved, but only the highest of the five was used to compare ligands with each other. With this docking procedure, 49 ligands were found to have highest binding affinities of -8.0 kcal/mol or better. Table 2 summarizes the results of this second docking procedure.

Table 2. Binding affinity results from ensemble docking (abridged)

Molecule (NSC)	Best Binding Affinity (kcal/mol)
166259	-9.3
37641	-9.1
121868	-9.1
727038	-9.1
117987	-8.7
70931	-8.6
...	...

The top molecule found is NSC-166259, a close relative of succinic acid found to have a highest binding affinity of -9.3 kcal/mol with conformation 2 of the IDR. NSC-166259 displays anticancer properties, showing activity in human tumor cell bioassays. Upon closer inspection of NSC-166259's docked pose, it becomes apparent that NSC-166259 interacts with the receptor at two sites: residue 126 as well as residue 3, which is within the IDR (see Figure 5). This confirms the notion that our docking approach can find ligand poses that interact directly with the IDR.

Finally, given the current need for efficient drug discovery through repurposing, a set of well-known compounds such as danazol, genistein and estramustine found to perform well in both docking procedures are listed along with their modern uses in Table 3.

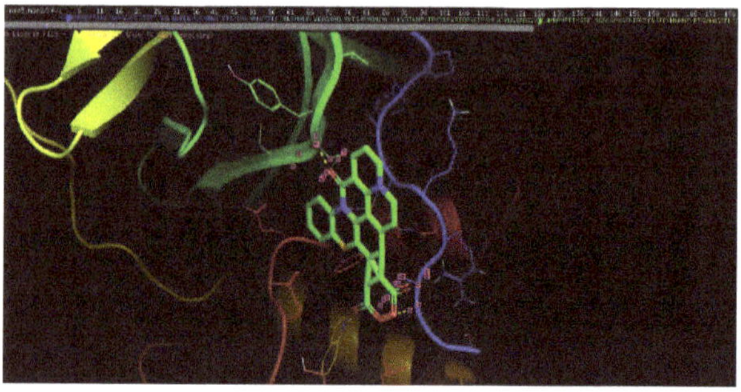

Fig. 5. NSC-166259 interacting with IDR.

Table 3. Drug repurposing candidates with their uses and additional information.

Drug	Pharmacological Use	Additional Information
Celastrol	Inflammation, cancer (lung, prostate)	Suppresses NF-kB signaling
Danazol	Fibrocystic breast disease, endometriosis	Targets estrogen receptor alpha
Estramustine	Cancer (prostate)	Targets estrogen receptor alpha/beta
Camptothecin	Cancer	Targets topoisomerase
Genistein	Cardiovascular risk, cancer	Targets estrogen receptor alpha/beta
Benzbromarone	Heart failure, chronic kidney disease	Targets cytochrome P450 2C9

3.2. *Activity prediction*

The goal of this work is to find CoV-2 3CLpro inhibitors by concentrating our efforts on the IDR/MoRF present at the N-terminus. The first step in this effort using molecular docking yielded promising results; however, in general, docking approaches need to be cross verified by a different method. Thus, our next goal was to create a statistical model capable of predicting *in vitro* inhibition of our protein target to filter and provide further evidence for our docking results. Such a model would be capable of making predictions many orders of magnitude faster than standard bioassays. Here, we train a model to predict whether a compound can inhibit 3CLpro-mediated peptide cleavage.

Due to the scarcity of CoV-2 data, we train our model using CoV-1 3CLpro peptide cleavage inhibition data from bioassay AID 1706. The model structure chosen is a MPNN implemented by Chemprop. Our model achieves a test ROC AUC of .739 (80% train, 10% validation, 10% test).

We then apply the trained model to predict activity scores for each of the previously docked ligands. A total of 11 ligands (see Table 4) are identified as having both high affinity (absolute affinity $\geq$ 7.9 kcal/mol) as well as high activity ($\geq$ 0.8). These ligands have high probabilities of binding to the IDR, having enough binding energy to deform the 3CLpro, and inhibiting peptide cleavage. Therefore, we deem these 11 ligands promising drug candidates. Furthermore, known use cases for these 11 ligands include orthopoxviruses, foot-and-mouth disease virus, human tumors, and malaria. We are currently investigating a potential collaboration to validate the efficacy of these 11 new drug candidates *in vitro*.

Table 4. Top 11 drug candidates in terms of affinity and activity.

Molecule (NSC)	Activity	Affinity	Active Bioassays
16437	.859	-9.3	Foot-and-mouth disease (FMD) virus
117987	.872	-8.8	
601359	.855	-8.4	Melanoma cell line, Malaria
13294	.825	-8.4	
127133	.908	-8.3	
61610	.823	-8.2	Malaria
107582	.877	-8.1	
128606	.920	-8.0	
211490	.808	-8.0	Hepatitis C virus, Human cytomegalovirus
679525	.894	-8.0	Orthopoxviruses, FMD virus
204232	.800	-7.9	DNA Polymerase Beta

We also investigate the possible link between 3CLpro cleavage inhibition and IDR binding affinity. A scatter plot of the binding affinities and activity scores for each of the 1405 docked ligands is shown in Figure 6. The correlation coefficient r measuring the strength and direction of the linear relationship between the two variables is computed to be 0.38, suggesting a weak to slightly moderate correlation. This means that higher binding affinities to the IDR of the CoV-2 3CLpro weakly/moderately correlate with higher rates of cleavage inhibition. This suggests that the IDR/MoRF of the CoV-2 3CLpro is involved in the peptide cleavage process. As a matter of fact, it is well known that the dimerization of 3CLpro that develops its active site involves our targeted IDR [7]. Therefore, since our method realizes this relationship, it suggests that targeting the IDR in the monomeric form is an effective way of finding 3CLpro peptide cleavage inhibitors. This also could suggest that our approach of cross verifying docking results with statistical models could be used to hypothesize other biological relationships key to drug discovery in the future. In Figure 7, we show the distribution of the IDR binding affinities of known CoV-1 3CLpro inhibitors from bioassay AID 1706, and in Figure 8 we show the same distribution for the NCI Diversity Set III. We find the average binding affinity of CoV-1 3CLpro inhibitors to be -6.74 kcal/mol, which is above the typical threshold for choosing possible drug candidates, whereas the average for the NCI Diversity Set III, which we assume to be a representation of the drug-like ligand space, is just -5.93 kcal/mol. Consequently, the distributions indicate that the average 3CLpro inhibitor falls within the top 23.5% of all ligands in terms of binding affinity to the IDR of 3CLpro, further supporting our previous claims. Furthermore, it is possible that the correlation between the IDR and cleavage inhibition is higher than mentioned above but is dampened in our data since the MPNN was trained on *in vitro* results, but high binding affinities do not always correspond to *in vitro* binding.

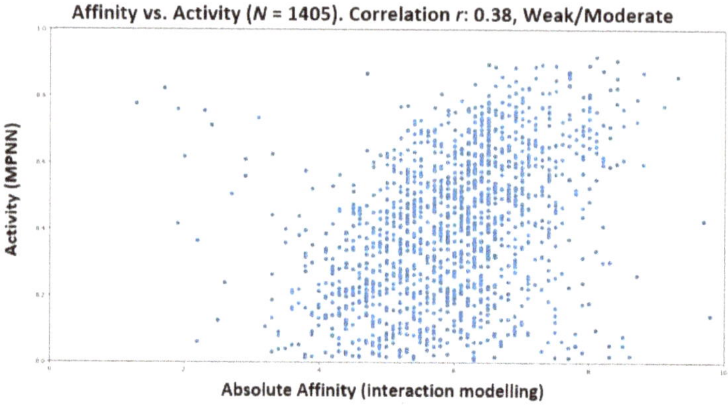

Fig. 6. Affinity versus activity

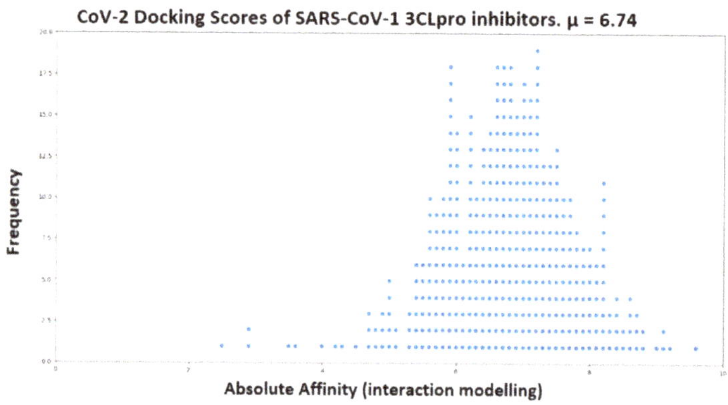

Fig. 7. Distribution of 3CLpro inhibitor binding affinities.

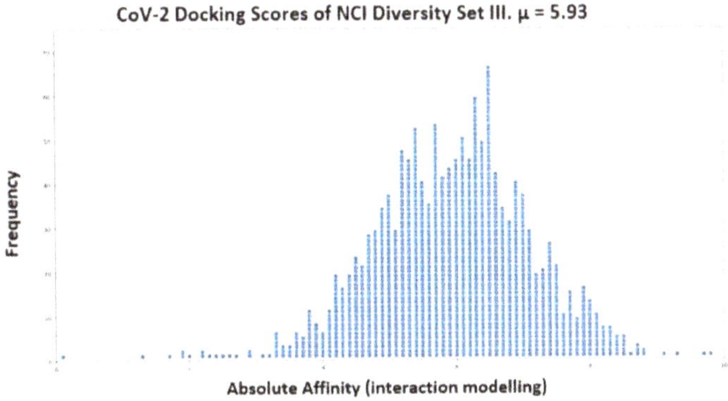

Fig. 8. Distribution of NCI Diversity Set III binding affinities.

4. Conclusion

Currently, there are no widely approved CoV-2 antivirals or vaccines available. Given the infectious and fatal nature of COVID-19, there exists a dire need for immediate drug discovery research. In this work, we make advancements by specifically focusing on targeting disordered protein regions. We demonstrate how these IDRs can be targeted through molecular docking and illustrate how results can be verified in a multi-faceted approach. Ultimately, we identify 11 new drug candidates with high binding and activity scores, along with known antiviral properties. In the future we would like to validate our results *in vitro* as well as further explore the IDR interactions within the SARS-CoV-2 proteome through MoRF mimicry.

5. Acknowledgements

This research was undertaken as part of the MIT-PRIMES program. Ning Xie and Ling Teng of the Biomedical Cybernetics Lab provided frequent support, feedback, and organization.

References

1. Wright, P. E., & Dyson, H. J. (2015). Intrinsically disordered proteins in cellular signalling and regulation. *Nature reviews. Molecular cell biology*, *16*(1), 18–29.
2. Wright, P. E., & Dyson, H. J. (1999). Intrinsically unstructured proteins: re-assessing the protein structure-function paradigm. *Journal of molecular biology*, *293*(2), 321-331.
3. Rhoades, E. (2018). *Intrinsically Disordered Proteins*. Academic Press.
4. Mohan, A., Oldfield, C. J., Radivojac, P., Vacic, V., Cortese, M. S., Dunker, A. K., & Uversky, V. N. (2006). Analysis of molecular recognition features (MoRFs). *Journal of molecular biology*, *362*(5), 1043-1059.

5. Uversky, V. N., Oldfield, C. J., & Dunker, A. K. (2008). Intrinsically disordered proteins in human diseases: introducing the D2 concept. *Annu. Rev. Biophys.*, *37*, 215-246.
6. Chang, C. K., Hou, M. H., Chang, C. F., Hsiao, C. D., & Huang, T. H. (2014). The SARS coronavirus nucleocapsid protein--forms and functions. *Antiviral research*, *103*, 39–50.
7. Giri, R., Bhardwaj, T., Shegane, M., Gehi, B. R., Kumar, P., Gadhave, K., ... & Uversky, V. N. (2020). When Darkness Becomes a Ray of Light in the Dark Times: Understanding the COVID-19 via the Comparative Analysis of the Dark Proteomes of SARS-CoV-2, Human SARS and Bat SARS-Like Coronaviruses. *bioRxiv*.
8. Wang, R., Hozumi, Y., Yin, C., & Wei, G. W. (2020). Decoding SARS-CoV-2 Transmission and Evolution and Ramifications for COVID-19 Diagnosis, Vaccine, and Medicine. *Journal of chemical information and modeling*, acs.jcim.0c00501. Advance online publication.
9. Joshi, S., Joshi, M., & Degani, M. S. (2020). Tackling SARS-CoV-2: proposed targets and repurposed drugs. *Future medicinal chemistry*, 10.4155/fmc-2020-0147. Advance online publication.
10. Chen, Y. W., Yiu, C. B., & Wong, K. Y. (2020). Prediction of the SARS-CoV-2 (2019-nCoV) 3C-like protease (3CL ^{pro}) structure: virtual screening reveals velpatasvir, ledipasvir, and other drug repurposing candidates. *F1000Research*, *9*, 129.
11. Antunes, D. A., Devaurs, D., & Kavraki, L. E. (2015). Understanding the challenges of protein flexibility in drug design. *Expert opinion on drug discovery*, *10*(12), 1301-1313.
12. Thafar, M., Raies, A. B., Albaradei, S., Essack, M., & Bajic, V. B. (2019). Comparison Study of Computational Prediction Tools for Drug-Target Binding Affinities. *Frontiers in Chemistry*, *7*.
13. Kairys, V., Baranauskiene, L., Kazlauskiene, M., Matulis, D., & Kazlauskas, E. (2019). Binding affinity in drug design: experimental and computational techniques. *Expert opinion on drug discovery*, *14*(8), 755–768.
14. National Center for Biotechnology Information (2020). PubChem Bioassay Record for AID 1706, Source: The Scripps Research Institute Molecular Screening Center.
15. Suárez, D., & Díaz, N. (2020). SARS-CoV-2 Main Protease: A Molecular Dynamics Study. Journal of chemical information and modeling.
16. Yang, K., Swanson, K., Jin, W., Coley, C., Eiden, P., Gao, H., ... & Palmer, A. (2019). Analyzing learned molecular representations for property prediction. *Journal of chemical information and modeling*, *59*(8), 3370-3388.
17. Khalili, N., Karimi, A., Moradi, M. T., & Shirzad, H. (2018). In vitro immunomodulatory activity of celastrol against influenza A virus infection. *Immunopharmacology and Immunotoxicology*, *40*(3), 250-255.
18. Yu, J. S., Tseng, C. K., Lin, C. K., Hsu, Y. C., Wu, Y. H., Hsieh, C. L., & Lee, J. C. (2017). Celastrol inhibits dengue virus replication via up-regulating type I interferon and downstream interferon-stimulated responses. *Antiviral research*, *137*, 49-57.
19. Habtemariam, S., Nabavi, S. F., Berindan-Neagoe, I., Cismaru, C. A., Izadi, M., Sureda, A., & Nabavi, S. M. (2020). Should we try the antiinflammatory natural product, celastrol, for COVID-19?. *Phytotherapy Research*.

Feasibility of the vaccine development for SARS-CoV-2 and other viruses using the shell disorder analysis

Gerard Kian-Meng Goh,[1,*] A. Keith Dunker,[2] James A. Foster,[3] and Vladimir N. Uversky[4]

[1]*Goh's BioComputing, Singapore 548957, Republic of Singapore (gohsbiocomputing@yahoo.com)*

[2]*Center for Computational Biology and Bioinformatics, Department of Biochemistry and Molecular Biology, Indiana University School of Medicine, 410 W. 10th St, HS5000, Indianapolis, IN46202, USA*

[3]*Department of Biological Sciences University of Idaho, Moscow, ID 83843, USA*

[4]*Department of Molecular Medicine,Morsani College of Medicine, University of South Florida, Tampa, FL, USA*

Several related viral shell disorder (disorder of shell proteins of viruses) models were built using a disorder predictor via AI. The parent model detected the presence of high levels of disorder at the outer shell in viruses, for which vaccines are not available. Another model found correlations between inner shell disorder and viral virulence. A third model was able to positively correlate the levels of respiratory transmission of coronaviruses (CoVs). These models are linked together by the fact that they have uncovered two novel immune evading strategies employed by the various viruses. The first involve the use of highly disordered "shape-shifting" outer shell to prevent antibodies from binding tightly to the virus thus leading to vaccine failure. The second usually involves a more disordered inner shell that provides for more efficient binding in the rapid replication of viral particles before any host immune response. This "Trojan horse" immune evasion often backfires on the virus, when the viral load becomes too great at a vital organ, which leads to death of the host. Just as such virulence entails the viral load to exceed at a vital organ, a minimal viral load in the saliva/mucus is necessary for respiratory transmission to be feasible. As for the SARS-CoV-2, no high levels of disorder can be detected at the outer shell membrane (M) protein, but some evidence of correlation between virulence and inner shell (nucleocapsid, N) disorder has been observed. This suggests that not only the development of vaccine for SARS-CoV-2, unlike HIV, HSV and HCV, is feasible but its attenuated vaccine strain can either be found in nature or generated by genetically modifying N.

Keywords: SARS; COVID; disorder; Coronavirus; HIV; vaccine; virulence; viral shell.

1. Introduction

1.1. *SARS-COV-2 Vaccine*

Since its outbreak in December 2019, a dangerous coronavirus (CoV), severe acute respiratory syndrome CoV-2 (SARS-CoV2), causing Coronavirus Disease (COVID-19) has spread rampantly with the dire consequences including large numbers of deaths and morbidities [1]. The SARS-CoV-2 spread has been so severe that many believe that it could only be kept in check with the discovery and availability of effective vaccines. While the successes in the discovery of vaccines for a large variety of viruses, including classical viruses, such as smallpox and rabies viruses, provide grounds for greater hope, optimism and inspiration toward the discovery of COVID-19 vaccines, there are also nightmare scenarios, in cases, such as HIV (human immunodeficiency virus, HCV (hepatitis C virus), and HSV (herpes simplex virus), for which no vaccine has been found despite searches that span close to 40, 30, and 100 years, respectively. The polio vaccine development itself took 30-40 years, but those years were before the era of powerful modern molecular technology. It would therefore be unfair to make such comparison [3,5]. A question is then: Will the search for a SARS-CoV-2 vaccine be a nightmare as seen in HIV, or will it be a spectacular success, as in the case of rabies and smallpox? We shall see that the shell disorder analysis has much to say in this regard.

1.2. *Shell disorder analysis of HIV and other viruses*

In 2008, we reported that the use of artificial intelligence (AI) found some strange features in the outer shell (matrix) of the HIV-1, which was found to be very disordered [6]. We were not able to detect this feature in any other virus, despite a search among of a somewhat wide variety of unrelated viruses, such as influenza virus, rabies virus and the HIV's cousin, EIAV (equine infectious anemia virus) [6]. In subsequent years, similar levels of disorder can be found in very few other viruses, including HSV and HCV [2]. Both viruses are associated with sexual transmission, and no effective vaccine has been found for both. These cannot be explained by current the standard textbook paradigm [2].

In these and similar studies, the levels of protein intrinsic disorder [7-12] were measured for proteins constituting shell of each analyzed virus using a neural network-based predictor PONDR® VLXT [13,14]. This algorithm predicts the intrinsic disorder predisposition of each residue in a · protein. A convenient yardstick to measure the level of disorder in a protein is PID (percentage of disorder). In the case of HIV-1, the matrix PID reaches the high level of 70% [2].

1.3. *Spinoff projects including coronaviruses: Shell disorder and modes of transmission*

Following the success of the HIV shell project, several spinoff projects based on the similar ideology were initiated. One of these spinoffs was the coronavirus project. Before the MERS-CoV outbreak in 2012, a shell disorder model was built to predict the mode of transmission of this virus based mostly on the levels of intrinsic disorder in its inner shell (N, Nucleocapsid) but also partly taking into account the disorder status of the outer shell (membrane, M) [15]. When the PIDs of M and N were measured for the variety of CoVs, the viruses were clustered into three groups mainly based on their N PID values. Those with the highest PIDs are those with higher respiratory but lower fecal-oral transmission potentials. Those with intermediate levels of N PIDs are the CoVs predicted to have intermediate levels of both respiratory and fecal-oral transmission potentials. The model was developed using knowledge of the behaviors of animal CoVs, particularly porcine CoVs from the veterinary community [3] and was later further validated using multivariate analysis [3,15,16].

In this model, SARS-CoV was placed in group B that contains CoVs with intermediate - respiratory and fecal-oral transmission potential and the results of the model were published in 2012 [15]. Upon the MERS-CoV (Middle Eastern respiratory syndrome CoV) outbreak, the characterization and classification of MERS-CoV had to wait until the time when the genome or proteome sequences of this virus became available. However, as soon as this information was released, it was used to analyze MERS-CoV. This analysis revealed that MERS-CoV belongs to the group C that contains CoVs with higher fecal-oral but lower respiratory transmission potentials [16]. The results further validate the reliability of the model as clinical data of MERS-CoV do show that it is not easily transmissible among humans via respiratory routes and is associated with camels, which are often farmed and thus allow for greater fecal-oral transmission [17].

Yet another opportunity to validate the shell disorder model came with the COVID-19 outbreak. Again, the model was consistent with existing datae and placed SARS-CoV-2 in the same category as SARS-CoV; i.e., CoVs with the intermediate levels of both respiratory and fecal-oral transmission potentials [18]. There was, however, something noticeably odd about this virus. Analysis showed that with the exception for HCoV-HKU1, SARS-CoV-2 had the hardest outer shell (lowest M_{PID}) in our fairly wide variety of CoVs analyzed then. This means that SARS-CoV-2 is likely to resist the anti-microbial enzymes found in the saliva and mucus of the host and is also likely to survive longer in non-physiological environments [19,20]. Further search has found that such hardness is associated with burrowing animals, such as rabbits and pangolins that are in contact with buried fecal materials [21,22]. Furthermore, clinical studies have shown that COVID-19 patients shed large amounts of SARS-CoV-2 viral particles, which are far exceeding levels of viral particles shed by infected with SARS-CoV [23]. The hardness of the outer shell and the ability of the virus to resist the anti-microbial enzymes in body fluids could account for these observations

1.4. *Yet another spinoff: Correlations between the inner shell disorder and virulence*

Yet another spinoff from the parent HIV vaccine mystery project is the discovery of a correlations between the inner shell disorder and virulence in fairly diverse set of related and unrelated viruses including Nipah virus (NiV), flaviviruses, Dengue virus (DENV), and Ebola virus (EBOV) [24-28]. In this paper, we will address the feasibility of SARS-CoV-2 vaccine development based on the shell disorder models and discuss the evolutionary aspects of SARS-CoV-2 and other viruses.

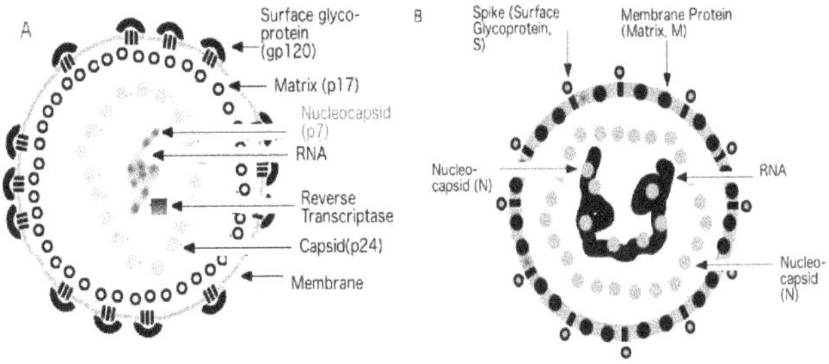

Figure 1. Virion Physiology A) HIV B) Coronavirus (CoV). (Figures reproduced with the permission of Gerard K. M. Goh 2017)

2. Results

2.1. Clustering of CoV based mainly on N_{PID}

As already mentioned, the CoV shell disordered model clustered CoVs into three statistically identifiable groups (ANOVA $p < 0.01$, **Table 1**), which correlated positively with the levels of respiratory transmission but negatively with the levels of fecal-oral transmission potentials. While the main contributing independent variable is the N_{PID} (r^2=0.77, $p < 0.01$), a slight increase in the correlation coefficient can be seen when both M_{PID} and N_{PID} are used as independent variables (r^2=0.80, $p < 0.01$). This implies that M_{PID} does contribute to the model even if slightly. **Figure 1** provides schematic virion physiology, with HIV and CoV as examples[3,4]. The inner and outer shells of CoV is the N and M proteins, respectively, as seen in **Figure 1B**.

Table 1. Categorization of coronaviruses by mainly N PID to predict levels of respiratory and fecal-oral transmission potentials ($p< 0.001$, $r^2 = 0.8$).

Shell Disorder Group	Coronavirus	Accession Code (M Proteins)[a]	Accession Code (N Proteins)[a]	M_{PID}	N_{PID}	Remarks
A	HCoV-229E	P15422(U)	P15130(U)	23	56	Higher levels of respiratory transmission lower levels of fecal-oral transmission
	IBV(Avian)	P69606(U)	Q8JMI6(U)	10	56	
B	Bovine	P69704(U)	Q8V432(U)	7.8	53.1	Intermediate levels of respiratory and fecal-oral transmission
	Rabbit	H9AA37(U)	H9AA59(U)	5.7	52.2	
	PEDV (Porcine)	P59771(U)	Q07499(U)	8	51	
	Canine (Resp.)	A3E2F6(U)	A3E2F7(U)	7	50.5	
	HCoV-OC43	Q4VID2(U)	P33469(U)	7	51	
	SARS-CoV	P59596(U)	P59595(U)	8.6	50.2	
	HCoV-NL63	Q6Q1R9(U)	Q6Q1R8(U)	11	49	
	SARS-Cov-2	P0DTC5(U)	P0DTC9(U)	5.9	48.2	
	Bats[b]	A3EXD6(U)	Q3LZX4(U)	11.2±5.3	47.7±0.9	
C	MHV(Murine)	Q9JEB4(U)	P03416(U)	8	46.8	Lower levels of respiratory transmission higher levels of fecal-oral transmission
	Pangolin-CoV[c]	QIA428617(G)	QIA48630(G)	5.6±0.9	46.6±1.6	
	MERS-CoV	K0BU37(U)	K0BVN3(U)	9.1	44.3	
	TGEV(Porcine)	P09175(U)	P04134(U)	14	43	
	Canine (Ent.)	B8RIR2(U)	Q04700(U)	8	40	
	HCoV-HKU1	Q14EA7(U)	Q0ZME3(U)	4.5	37.4	

[a]UniProt(U): https://www.uniProt.org; Genbank-NCBI(G): https://www.ncbi.nlm.nih.gov/protein
[b]3 out of 4 bat-CoVs are in group B. Note: Large standard deviation can be seen for N_{PID} as denoted by "±"
[c]2 out of 3 pangolin-CoVs are in group C. One is almost identical to SARS-CoV-2 in N PID. All samples were found to have high sequence similarities to the corresponding proteins found in SARS-CoV-2

2.2 *Outer shell disorder is an indicator for the presence or absence of effective vaccines*

While disorder at the inner shell is correlated with the mode of transmission, high outer shell disorder is associated with difficulties in finding effective vaccines. We shall see later that disorder at the inner shell (and sometimes at the outer shell as well) correlates with virulence. **Tables 2-3** summarize the disorder of the different shells of a variety of related and non-related viruses. There are no effective vaccines for HIV, HCV and HSV, which have abnormally high outer shell disorder. Conversely, viruses for which effective vaccines are available have ordered outer shells. These include rabies, yellow fever virus, smallpox virus and rotavirus [2-4,29]. The poliovirus has a capsid that is made up of a complex of several proteins, which are all relatively ordered. As aforementioned, the effective vaccines for polio have been available since the 1950s [4,5].

Table 2. Viruses and their descriptions. UniProt (http://www.uniprot.org) accession codes for shell proteins are given.

Virus	Virus Type, Tansmission	Outer Shell, Proteins (UniProt Accession)*	Intermediate Shell	Inner Shell
EIAV[a]	Retroviridae (RNA) Insect	Matrix, p15 (P69732)	Capsid, p26 (P69732)	Nucleocapsid, p11(P69732)
FIV[b]	Retroviridae, Fights, Blood contacts	Matrix, p15 (P16087)	Capsid, p24 (P16087)	Nucleocapsid. p13 (P16087)
HIV-1	Retroviridae Sexual	Matrix, p17 (P03348)	Capsid, p24 (P03348)	Nucleocapsid, p7(p17)

HIV-2	Retroviridae Sexual, Bite	Matrix, p17 (P04584)	Capsid, p24 (P04584)	Nucleocapsid, p7(P04584)
Variola/ Smallpox[c]	Poxviridae (DNA) Inhalation	Membrane, C9L(Q76U97), A14(P33839), F5(P33865)		Core, VP8(Q0N570), 4A(Q0N532), 4B(Q0N539)
Rabies	Rhabdopviridae (RNA) Bites	Matrix, M(P25224)		Nucleocapsid, N(P151979)
Poliovirus	Picornaviridae (RNA) Fecal-Oral		Capsid, VP1-4 (P03302)	
Yellow Fever (YFV)	Flaviviridae (RNA) Insect	Membrane, M (P03314)	Capsid, C (P03314)	
Rotavirus	Reoviridae (RNA) Fecal-oral	Outer Capsid , VP7 (P21285)	Capsid, VP6 (P03530)	Capsid, VP2 (P12472)
SARS-CoV-2	Coronavirus (RNA)Respiratory, Fecal-oral	Membrane (A3EXD6)		Nucleocapsid (Q3LXZ4)
Hepatitis C (HCV)	Flaviviridae (RNA) Sexual			Core, p19, p21 (P26663)
Herpes Simplex Virus-2 (HSV-2)[c]		Tegument, VP22-UL49 (A74K33), VP1/2-UL36 (I1UYK0), VP13/14-UL47 (A7LK25), VP16-UL48 (P68335), US3 (P13287)	Capsid, VP5 (p89442)	

[a]Equine Infectious Anemia Virus (EIAV)
[b]Feline Immunodeficiency Virus (FIV)
[c]Only major shell proteins are considered

2.3. *A disordered outer shell provides an immune evasion tactic: Viral shapeshifting*

Evidently, as seen in **Tables 2-3**, some viruses like HIV evade the immune system via their disordered outer shell. The question is then: How do viruses do it? **Figure 2** summarizes the mechanism of immune evasion as seen in the case of HIV and its disordered matrix. The disordered matrix allows for motions that increase the movements of the surface glycoproteins such that the antibodies are not able to bind firmly to the virus. In the case of HIV, HIV antibodies can easily be found but neutralizing antibodies are difficult to find [2,3]. There are obviously various degrees of vaccine failures that depends on the level of outer shell disorder as we shall see in the case of FIV.

Table 3. Disorder levels (PID) of shell proteins. PIDs are arranged according proteins as stated in Table 2. Effective vaccines have been discovered for EIAV, rabies, polio, smallpox and rotavirus.

Virus	PID of Outer Shell	PID of Intermediate Shell	PID of Inner Shell
EIAV	13±0.1	29±0.1	26±0.1
FIV	53.3±2	38.8±2.7	64.5+11.8
HIV-1, SIVcpz	56.5±10.8	44.5±2.6	39.5±3.0
HIV-2, SIVmac	51.5±2.5	26.6±2.9	46.5±0.1
Smallpox*	13±0.1 8±0.1 15±0.1		19±0.1 4±0.1 12±0.1
Rabies	25.8±1.4	21.5±0.8	
Poliovirus*		34±3.8 15.12±6.1 31.3±3.6 27±0.1	

Yellow Fever (YFV)	35.2±0.9	74.3±0.9		
Rotavirus	12.9±1	9.8±1.4	19±1.7	
SARS-CoV-2	5.9±0.1		48.2±0.9	
HCV*			52.5±0.5	48.5±0.5
HSV-2	61±2 50±0.1 38±1 39±0.6 37±0.1		18±0.1	

*The standard error is denoted by "±"

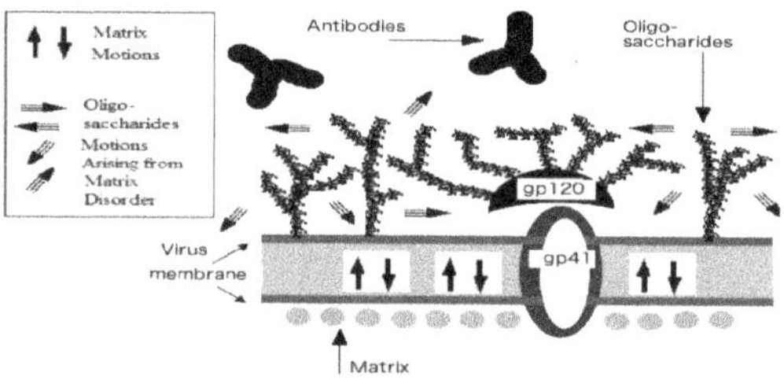

Figure 2. Viral "shape-shifting" immune evasion. Highly disordered matrix allows for greater motions at the glycoprotein that will prevent the antibodies from binding tightly to the virus. (Figure reproduced with the permission of Gerard K. M. Goh 2017)

2.4. *SARS-CoV-2: Exceptionally hard shell (low M_{PID}) associated with burrowing animals and buried feces*

While **Table 2** shows that it is difficult to find vaccines for viruses with extremely high levels of disorder in outer shell, **Figure 3A** reiterates that even though virtually all CoVs have relatively hard outer shell (for them, the fecal-oral transmission is generally an important route) SARS-CoV-2 has an exceptionally hard outer shell ie low M_{PID}. As a matter of fact, it has one of the hardest outer shells (among all the CoVs). A later search for CoVs with harder (low disorder) M protein came up with rabbit-CoV and pangolin-CoVs (see **Table 1**) [21,22]. There were obvious associations with burrowing animals that are likely in contact with buried feces. It should also be noted that while pangolin-CoVs were found to be closely related to SARS-CoV-2, the rabbit-CoV is not. Independent but parallel evolutions with similar evolutionary pressures are implied in the case of SARS-CoV-2 pangolin-CoVs, and rabbit-CoV [21,22].

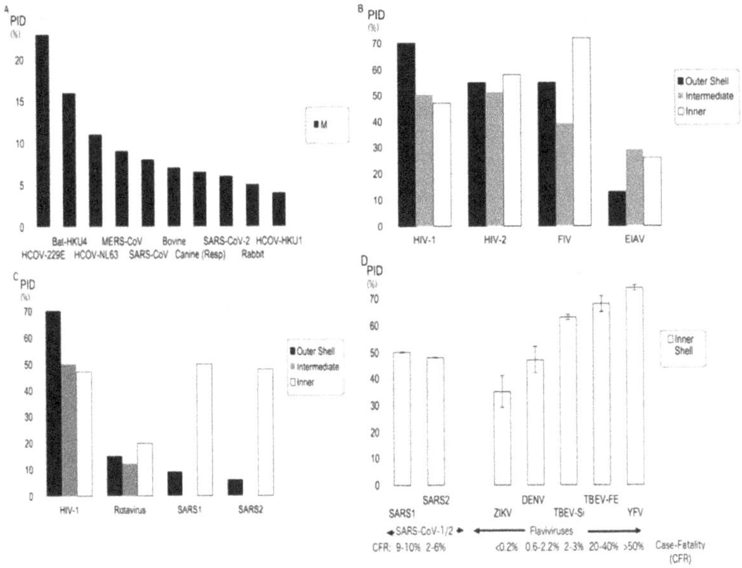

Figure 3. Quantification of shell disorder in different viruses. A) PIDs of M by CoV. SARS-CoV-2 has among the hardest M ie lowest M_{PID}. B) Shell disorder of selected retroviruses. C) Comparative shell disorder. SARS-CoV-2 (SARS2). 2003 SARS-CoV (SAR1). D) Inner shell disorder *vs*. virulence. Flavivirus capsid, C (inner shell. Case-fatality rate (CFR)

2.5. Behavior of the animal hosts matters in the evolutions of the viruses: EIAV vs. HIV

While it has been seen that higher outer shell disorder and the absence of effective vaccines are not just seen in retroviruses but also in other viruses, such as HCV and HSV as shown in **Table 2**, **Figure 3B** illustrates that not all retroviruses have high matrix disorder or lack effective vaccines, EIAV has ordered shells at all levels, and an effective vaccine for this virus was discovered in 1973 or before [30]. It should also be noted that HIV is predominantly sexually transmitted, whereas EIAV is transmitted by blood sucking horseflies, which hold its blood meal in their mouthpiece where the virus is exposed to the insect's saliva [2]. Unlike HIV, the EIAV needs the hard (low disorder) shells to protect itself against destructive effects arising from the anti-microbial enzymes found in the saliva. The "viral shape-shifting" immune evasive characteristic found in HIV is therefore absence in viruses, such as EIAV and rabies viruses. Experimental observations of outer shell disorder and the resulting immune evasion have been made [31-34].

2.6. Feasibility of developing attenuated vaccine strains for SARS-CoV-2

We have seen that SARS-CoV-2 looks nothing like the viruses, for which the effective vaccines are unavailable. Furthermore, it has one of the hardest outer shells among CoVs (**Figure 3A**). **Figure 3D** suggests that the attenuated vaccine strains can be obtained by lowering the disorder levels in its inner shell; i.e., N_{PID}. Flaviviruses are used here only as an example of the evidence of correlation [24.26]. While only correlations between flavivirus inner shell disorder and virulence are shown, a variety of other viruses have been found to have such characteristics. These include NiV, EBOV and DENVs [24-28]. SARS-CoV-2 and the 2003 SARS-CoV do also provide hints of such correlation by having respective N_{PID} of 48% and 50%, while also having CFR of 2-6% and 9-10%, as seen in **Figure 3C**. The HIV and other viral shape-shifters exhibit a unique immune evading tactic as seen in **Figure 2**, SARS-CoV1/2, NiV, EBOV, and flavivirus use a different tactic, "Trojan Horse", where the virus would replicate rapidly upon infection before the host immune system even recognizes its presence [3].

3. Discussion

3.1. Links between respiratory transmission, N (Inner shell) disorder, and virulence: Viral load in body fluids vs. vital organs

A puzzle arises when we inspect the data shown in **Table 1** and **Figure 3D**. In fact, **Table 1** tells us that there is a strong positive correlation between the inner shell disorder and respiratory transmission potentials of CoV, whereas **Figure 3D** reports positive correlations between inner shell disorder and virulence. What is then the connection between these two types of correlation? They are connected, because they are related to viral loads at different parts of the body. In order for respiratory transmission to be feasible, a minimal level of viral load in the saliva or mucus has to be attained. Similarly, death often occurs when the viral load in a specific vital organ exceeds a minimal threshold. What could then account for the discrepancy between viral loads in body fluids and vital organs? One answer is related to the ability of the virus to resist the anti-microbial enzymes found in saliva and mucus. Given this and the anti-microbial resistant hard (low disorder) outer shell of SARS-CoV-2, the significance of the observation that SARS-CoV-2 sheds high amount of viral particles without increase in virulence, when compared with SARS-CoV is reiterated.

3.2. Greater disorder in the inner shell proteins provide means for the more efficient replication of viral particles

It has to be kept in mind that inner shell proteins have varied but similar functions across virus species [4]. This accounts for the observation of correlations between inner shell disorder and virulence across virus species. For instance, N proteins of CoVs are involved in assembly of various viral proteins near or at the endoplasmic reticulum (ER) and Golgi apparatus in preparation for packaging [35,36]]. Similarly, the C protein precursors of flaviviruses move towards the ER and bind to its membrane, where interactions of other viral proteins take place in the assembling of viral particles [4]. In the case of EBOV, the NP (nucleoprotein) helps forming a tube-like structure that assists in the transportation of viral proteins to the ER [37]. As for the NiV, the N protein binds to both L and P proteins to form the RNA polymerase [38], which is responsible for the viral RNA replication. The inner shell proteins therefore play important roles in the rapid replication of the viral particles. As we can see, instances of protein-protein/DNA/RNA interactions taking place are aplenty. The greater the disorder, the more efficient the inner shell protein is able to play its role in the replication of the virus as disorder provides for more effective protein-protein/DNA/RNA binding [7,18,39].

3.3 Two modes of immune evasion: "Trojan Horse" (inner shell disorder) and "viral shape-shifting" (outer shell disorder)

We have just described "Trojan Horse" immune evading strategy, where the virus replicates rapidly via inner shell disorder, before the host immune system could even recognize it. Oftentimes, such strategy backfires on the virus especially when the viral load overwhelms viral organs and thus killing the host. We have also described the other immune evading strategy, "viral shape-shifting" in HIV as manifested in the outer shell disorder (**Figure 2**), there is evidence that HIV actually employs both strategies. This is complicated by the fact that the matrix (outer shell) assumes many of the roles that inner shell proteins of other viruses would normally have. These roles include embedding the proteins into the host membrane and assembling the viral proteins [4]. Evidence of HIV's adoption of such strategy can be seen by the fact over 90% patients infected by HIV-1 dies within two years of infection but the onset of symptoms (AIDS) for HIV-2 and FIV may take may years if at all [6,40,41]. Unsurprisingly, the maximal matrix PIDs of HIV-1, HIV-2 and FIV are 70%, 55% and 55% respectively (**Figure 3B**).

3.4. FIV, HIV-1 and HIV-2: Similarities and differences

As it was already mentioned, the "viral shape-shifting" immune evading strategy requires high outer shell disorder, as seen in HIV, HSV and HCV. This, too, presents an enigma, as HIV-1, HIV-2, and FIV have different degrees of outer shell disorder (70%, 55%, 55%, see **Figure 3B**). HIV-1 is spread globally via sexual transmission. While HIV-2 is also predominantly sexually transmitted, it is mainly found in parts of Africa near its rainforest reservoir of the monkey, sooty mungabey, which replenishes the virus by bites and other human interactions [41]. Similarly, FIV is predominantly spread through fights and subsequent blood contacts [40]. These could explain the discrepancies in levels of matrix disorder between HIV-1 and FIV/HIV-2.

3.5. *FIV vaccine enigma: Questionable efficacy*

While there are apparent differences in evolution and matrix disorder of EIAV, HIV-1, and FIV, there are also differences in the successes with respect to the search for their vaccines. The search for an HIV vaccine has been ongoing for nearly 40 years with abysmal failures, but an effective vaccine for its horse cousin, EIAV, was discovered in 1973 or before [30]. It was tested on 60 million horses and was able to deflect an oncoming pandemic. Needless to say, the matrix PID of EIAV is 13%, compared to the HIV-1 M_{PID} of 70%. A more enigmatic story can be found in the case of FIV. A ray of hope came in 2002, when a FIV vaccine became commercially available. It initially boasted of 82% efficacy against the FIV subtype A, but was later shown to be totally ineffective against strains from countries, such as United Kingdom. It was shown to provide only 58% protection for cats in Australia [42]. Finally, the vaccine was later withdrawn from the market in USA and Canada partly because of its questionable efficacy [40]. This is consistent with our data showing that FIV (matrix PID = 55%), like HIV-2, has more moderate matrix disorder than HIV-1.

4. Conclusions

4.1. Development of the SARS-CoV-2 vaccine is feasible and vaccine strains can be found in nature

The nightmare scenario in the frantic search for the SARS-CoV-2 vaccine would be that it all ends up like the search for the HIV vaccine or, even worse, FIV vaccine. However, we can all have a sigh of relief as the shell disorder models are unable to detect any similarities between SARS-CoV-2, HIV-1 and FIV in terms of the outer shell disorder or peculiarities of evolution with respect to their modes of transmission. The outer shell disorder of SARS-CoV-2 and SARS-CoV resembles more viruses, like rotavirus, for which there are effective vaccines. Unlike rotavirus that is solely reliant on fecal-oral routes, SARS-CoV-2 has a somewhat disordered inner shell. The presence of such feature is necessary for most viruses with respiratory transmission potentials, because, as already explained, a minimal viral load in the mucus or saliva is required for transmission. The higher inner shell disorder also provides means for the "Trojan horse" immune evasion. Because of this, strategies involving attenuation of the SARS-CoV-2 by creating N with greater order levels can be contrived. In fact, a previous study has suggested that a SARS-CoV-2 precursor could have entered the human population via pangolins in 2017 or before as an attenuated mild virus as a result of the peculiarities of the behaviors of pangolins [21,22]. Therefore, the disorder analysis not only suggests that vaccine development for SARS-CoV-2 is viable but also points out that the attenuated vaccine strains can already exist in nature.

5. Materials and Methods

As already mentioned, protein intrinsic disorder is an important concept that can be used for various analyses of proteins. It basically refers to the protein regions or entire proteins that have no unique 3D structures. Disorder in proteins plays an array of significant roles, such as the recognition of binding sites [7-12]. AI has been successfully employed to recognize disordered regions. For instance, PONDR® VLXT (www.pondr.com) [11,12,43] deploys neural networks to recognize such regions, as it was trained using known disordered and ordered sequences. PONDR® VLXT has been successfully used in the study of structural proteins of a variety of viruses including HIV, HSV, HCV, NiV, EBOV, 1918 HIN1 influenza A virus, CoVs, DENV, and several flaviviruses, e.g., Yellow fever virus (YFV), and Zika virus (ZIKV) [1-3,6,15,16,21,22,24-28,44]]. The reason that PONDR® VLXT is highly suitable for the studies of structural proteins of viruses has to do with its sensitivity in detecting local sites for potential protein-protein/DNA/RNA/lipid interactions [45]. A useful ratio used in this study is PID (Percentage of Intrinsic Disorder), which is defined as the number of residues predicted to be disordered divided by the total number of residues in the protein and multiplied by 100%. The value of this parameter provides an estimate of the extent of disorder in the protein of interest. A relational database was built using MYSQL, JDBC and JAVA. A JAVA program imports the sequence and disorder information into the database [44]. Sequence information are obtained from UniProt (http://www.uniProt.org) Multivariate analyses were done using R-statistical package [46].

References

1 G. K. Goh, et al. (2020) Biomolecules. 10: 331.
2. G. K. Goh, et al. (2019) Biomolecules. 9: 178.
3. G. K. Goh, Viral shapeshifters: Strange behaviors of HIV and other viruses (Simplicity, 2017).
4. N. H. Acheson, Fundamentals of molecular virology (Wiley, 2007).
5. D. M. Oshinsky, Polio: An American story (Oxford University Press, 2005).
6. G. K. Goh, et al. (2008) Vir J. 5: 126.
7. P.E. Wright, et al. (1999) J Mol Bio. 9293: 321-331.
8. V.N. Uversky, et al. (2000) Proteins 41: 415-427.
9. A.K. Dunker, et al. (2000) J Mol Graph Model 19: 26-59.
10. P. Tompa. (2002) Trend Biochem Sci. 27: 527–533.
11. A. Hatos, et al (2020) Nucleic Acids Res. 48(D):D269-D276.
12. A.K. Dunker et al (2015) Semin Cell Dev Bio. 37:44-55.
13. E. Garner, et al. (1999) Genome Inf. 10: 41–50.
14. P. Romero, et al. (2001) Proteins. 42 38–48.
15. G. K. Goh, et al. (2012) J Pathog. 2012: 738590.
16. G. K. Goh, et al. (2013) PloS Curr. 5.
17. M. Ferguson, et al. (2014) Lancet Infect Dis. 14: 93–94.
18. G. K. Goh, et al. (2020) Microb Pathog. 144: 104177.
19.. A. M. Cole, et al. (1999) Inf immun. 67: 3267–3275.
20. D. Malamud, et al. (2011) Adv Dent Res. 23: 34–37.
21. G. K. Goh, et al. (2020) Preprints. 2020060327.
22. G.K. Goh et al (2020) J. Proteome Research. 2020.
23. R. Wolfel, et al. (2020) Nature. 581:465-469.

24. G. K. Goh, et al. (2016) Mol Biosyt. 12: 1881–1891.
25. G. K. Goh, et al. (2015) Mol Biosyt. 11: 2227–2344.
26. G. K. Goh, et al. (2019) Biomolecules. 9: e710.
27. G. K. Goh, et al. (2020) Microb Pathog. 141: 103976.
28. G. K. Goh, et al. (2020) Preprints. 2020050116.
29. J. Angel, et al. (2007) Nat Rev Microbiol. 5: 529-39.
30. H. Wang, et al. (2017) Oncotarget. 9: 1356-64.
31. N. Kol et al (2007) Biophys J. 92:1777-1783.
32. H. Pang et al (2013) Retrovirology. 10:4.
33. Y. Ohori et al (2014) Biochim Biophys Acta. 1844:520-526.
34. F. Caccuri et al (2104) J Vir. 88:5706-5717.
35. S. Lu et al (2020) BioRxiv. 2020.07.31.228023.
36. T. McBride, et al. (2014) Viruses. 6: 2991-2018.
37. S. Wantanabe, et al. (2006) J. Vir. 80: 3743-51.
38. J. Habchi, et al. (2012) Mol Biosyst. 8: 69-81.
39. M. Macossay-Castillo, et al. (2019) J Mol Bio. 431: 1650-70.
40. M. Scherk, et al. (2012) J Fel Med Surg. 15: 785–808..
41. J. Goudsmit, Viral sex: Nature of AIDS (Oxford University Press, 1997).
42. B. Sahay, et al. (2018) Viruses. 10: 277.
43. X. Li, et al. (1999) Genome Inform Ser Workshop Genome Inform. 10: 30–40.
44. G. K. Goh, et al. (2008) BMC Genomics. 9 Suppl 2: S4.
45. C.J. Oldsfield et al (2005) Biochemistry. 44:12454-12470.
46. R Core Team. (2019) A language and environment for statistical computing.

Protein sequence models for prediction and comparative analysis of the SARS-CoV-2 –human interactome

Meghana Kshirsagar[†], Nure Tasnina[*], Michael D. Ward[‡], Jeffrey N. Law[*], T. M. Murali[*], Juan M. Lavista Ferres[†], Gregory R. Bowman[‡], Judith Klein-Seetharaman[a]

[†]*Microsoft, AI for Good Research Lab, Redmond, WA, USA*

‡*Dept. of Biochemistry & Molecular Biophysics, Washington Univ., St. Louis, MO, USA*

[a]*Colorado School of Mines, Initiative for AI in Bio and Health, Golden, CO, USA*

[*]*Dept. of Computer Science, Virginia Tech, Blacksburg, VA, USA*

Viruses such as the novel coronavirus, SARS-CoV-2, that is wreaking havoc on the world, depend on interactions of its own proteins with those of the human host cells. Relatively small changes in sequence such as between SARS-CoV and SARS-CoV-2 can dramatically change clinical phenotypes of the virus, including transmission rates and severity of the disease. On the other hand, highly dissimilar virus families such as *Coronaviridae*, *Ebola*, and *HIV* have overlap in functions. In this work we aim to analyze the role of protein sequence in the binding of SARS-CoV-2 virus proteins towards human proteins and compare it to that of the above other viruses. We build supervised machine learning models, using Generalized Additive Models to predict interactions based on sequence features and find that our models perform well with an AUC-PR of 0.65 in a class-skew of 1:10. Analysis of the novel predictions using an independent dataset showed statistically significant enrichment. We further map the importance of specific amino-acid sequence features in predicting binding and summarize what combinations of sequences from the virus and the host is correlated with an interaction. By analyzing the sequence-based embeddings of the interactomes from different viruses and clustering them together we find some functionally similar proteins from different viruses. For example, `vif` protein from *HIV-1*, `vp24` from *Ebola* and `orf3b` from SARS-CoV all function as interferon antagonists. Furthermore, we can differentiate the functions of similar viruses, for example `orf3a`'s interactions are more diverged than `orf7b` interactions when comparing SARS-CoV and SARS-CoV-2.

Keywords: protein interaction prediction; SARS-CoV-2; SARS-CoV; generalized additive models ; protein sequence

1. Introduction

Disease-causing pathogens such as viruses introduce their proteins into the host cells where they interact with the host's proteins enabling the virus to replicate inside the host. These interactions between pathogen and host proteins are key to understanding infectious diseases. The experimental discovery of protein-protein interactions (PPI) in general, and including those between host and pathogen, involves biochemical and biophysical methods, most frequently on a large scale using yeast two-hybrid (Y2H) assays and co-immunoprecipitation (co-IP) usually combined with mass spectrometry, but also many others usually applied at smaller scales such as co-crystallization or surface plasmon resonance. Computational techniques complement laboratory-based methods by predicting highly probable PPIs. Supervised machine learning based methods use the known interactions as training data and formulate the interaction prediction problem in a classification setting.[1–3]

For a newly emerged virus such as SARS-CoV-2, the type of information that is most easily obtained is genome sequence information. Within the first few weeks of its discovery, thousands of DNA sequences had been deposited. The much more complex task of discovering the interactome took a few months of the pandemic and the first global interactome study was published in Gordon et al.[4] A sequence based PPI prediction approach, which can use protein sequences derived from the viral DNA sequence, can thus be very informative in the initial stages of understanding a new virus. The rationale behind a sequence-based approach is that the amino-acid sequences of proteins determine its structure and consequently its function in the organism. By using amino-acid sequences of the two proteins of interest as inputs to a model, we can capture the dependence between their individual structural properties, their functions and their binding affinities. Towards this, we make the following contributions:

- We present an *interpretable* model for SARS-CoV-2 – human PPI prediction using only sequence-based features and evaluate these models on various metrics. We show that the performance of our interpretable model on SARS-CoV-2 PPI prediction, is better than that of Random Forests (which have been popular in prior work) and a deep learning approach that uses a Transformer based architecture for modeling protein sequences
- We analyze the interactomes from a sequence perspective, within SARS-CoV-2 and in comparison to other viruses and find interesting observations
- We validate predictions from our model using an additional recently published dataset from Stukalov et al.[5]

2. Methods

Given a virus-human protein interaction represented as the tuple: (p_v, p_h), we model the joint dependence of both the virus protein p_v and human protein p_h's sequences on the output variable, explicitly in the form of sequence feature level interactions. Towards this, we use a non-linear model GA^{2}M (Lou et al.[6]), which extends traditional Generalized Additive Models (GAMs) by incorporating higher-order feature interactions.

The standard GAM model is a generalized linear model in which the predictor depends linearly on unknown smooth functions f_i of some input covariates x_i. It has the following form: $g(E[y]) = \sum_{i\in[1,\ldots,d]} f_i(x_i)$, where d is the number of features or covariates, y is the output variable for an input, g is the link function (for instance: log). Here f_i is a linear function over the ith feature of example x.

2.1. *Generalized Additive Models with interactions (GA^{2}M)*

While GAMs usually model the dependent variable as a sum of univariate terms, GA^{2}M permits interactions and consists of univariate and a small number of pairwise interaction terms between pairs of features:

$$g(E[y]) = \sum_i f_i(x_i) + \sum_{i,j\in[1,\dots,d],i\neq j} f_{ij}(x_i, x_j)$$

Here i, j are indices over the set of all features. In Section 3.2 we describe our feature set in detail. To represent each virus-human PPI example (p_v, p_h), we concatenate the protein sequence features of both p_v and p_h to get a single feature vector of dimension d.

Since GA^2M only include one- and two-dimensional components, these components can be visualized and interpreted which has been difficult with neural networks. Lou et al.[6] propose an algorithm to learn GA^2M models that learn non-linear functions (trees) for every univariate and bivariate term, with pairs of features for the latter being selected by efficiently ranking all possible pairs of features as candidates and choosing the top k, where k is a hyper-parameter.

3. Gold Standard Interaction Datasets

We consider the following datasets (details in Table 1) in various experimental settings.

(1) a set of human proteins that physically interact with SARS-CoV-2 in human embryonic kidney cells (HEK293) based on affinity-purification mass spectrometry[4]
(2) a multi-level proteomics study[5] of SARS-CoV and SARS-CoV-2 proteins that also involves an affinity-purification mass spectrometry-based binding study but carried out in a human lung epithelial cell line (A549)
(3) Virus-human interactions data for other viruses was downloaded from VirHostNet[7a]

Unlike the interactions reported in the first mass spectrometry study,[4] the data from the second study[5] has homologous PPI within each dataset as well as several interologs between SARS-CoV and SARS-CoV-2. We downloaded the sequences for *Ebola* and *HIV-1* proteins from UniprotKB and those for SARS-CoV and SARS-CoV-2 from the respective publications' supplementary materials.

Table 1. Dataset characteristics

Virus and source	Interactions	Human proteins	Virus proteins
SARS-CoV-2 (Gordon et al.[4])	332	332	28
SARS-CoV (Stukalov et al.[5])	711	624	24
SARS-CoV-2 (Stukalov et al.[5])	1089	882	22
SARS-CoV (VirHostNet)[7]	141	122	23
Ebola (VirHostNet)[7]	221	221	7
HIV-1 (VirHostNet)[7]	618	583	8

[a] `http://virhostnet.prabi.fr/`

3.1. *Dealing with the lack of negative examples*

Due to the way protein interaction studies are designed, it is not possible to identify non-binding proteins: we cannot rule out interactions between baits and preys that are not co-purified in an affinity purification experiment, for instance. In order to build supervised machine learning models from PPI data, negative datasets comprising pairs of proteins that are unlikely to interact are constructed using heuristics such as (a) randomly selecting pairs of proteins from the set of all possible protein pairs,[8] which has ≈600,000 pairs (b) considering two proteins that do not co-locate within a cell. An approach using (b) is infeasible when considering cross-species protein interactions and also has a bias towards functionally dissimilar proteins. Other negative sets are manually curated in databases like Negatome [b], which are based on known protein domain properties such as hydrophobicity and derived from observational studies that note specific protein domains' lack of affinity towards certain other domains. While this data adjusts the bias mentioned above, it does not contain protein domains or families of many viruses, in particular none from *Coronaviridae*.

We found that using the set of 6,532 non-interacting pairs from Negatome resulted in models that were discriminating virus proteins from other proteins (AUC-PR of 0.98) due to the lack of virus proteins in the negative class. The negatives generated by approach (a) do not have this issue or the functional bias discussed above. Hence we randomly sample the requisite number of negatives from a combination of Negatome and the heuristic in (a).

Choice of class skew: We sample negatives at various positive to negative class-skews: balanced, 1:5 meaning we sample five times as many negatives as the number of positives, 1:10, 1:20 and 1:50. Using a balanced set of positives and negatives results in a biased model that has many false positives whereas using a large class-skew (1:50) that represents our prior that most pairs of proteins are unlikely to interact results in a model that captures the properties of the random protein pairs rather than the positive class (which is out-numbered). We analyzed the ranking of positives from the validation data (using the metric Precision @ 10% Recall) to decide the class-skew, which we treat as a global hyper-parameter. We found 1:10 to be the optimal setting that lead to the best Precision @ 10% Recall.

3.2. *Features*

We derived amino acid sequence k-mer features: consisting of the normalized frequency of 1-mers, 2-mers and 3-mers in the protein sequence. In addition to the above, we also derive conjoint triad features.[9] This approach first partitions the twenty amino acids into seven classes based on their dipoles and the volumes of the side chains. Trimers are represented using the classes of amino acids; hence trimers with amino acids belonging to the same classes, such as ART and VKS, are treated identically. There are 7^3 such tri-mers owing to the 7 classes. The `protr`[c] package was used for generating the conjoint triad features and the `fasta2matrix`[d]

[b] `http://mips.helmholtz-muenchen.de/proj/ppi/negatome/`

[c] `https://cran.r-project.org/web/packages/protr/vignettes/protr.html#46_conjoint_triad_descriptors`

[d] `https://noble.gs.washington.edu/proj/nucsvm/fasta2matrix.py`

utility was used to generate other k-mer features. For each virus-host protein pair, we concatenated the feature vectors of the individual proteins. Therefore, each virus-host protein pair had a feature vector of length 17,526 ($20 + 20^2 + 20^3 + 7^3$ from each protein).

Feature selection: The implementation of GA^{2}M that we use[e] does not scale well beyond a few thousand features because the number of pairs of features to consider is very large and the computational complexity of the feature-pair ranking algorithm.[6] To reduce the number of feature-pairs to consider, we select the top 2500 tri-mers in a feature selection step that builds a linear model on other virus-human interactions. This reduces the number of features in our model to ≈7000 features ($20 + 20^2 + 7^3 + 2500$ features per protein to be precise).

4. Experiments

We train various supervised machine learning models on these datasets to explore the strengths and weaknesses of each approach and illustrate that our method of choice, namely GA^{2}M perform well while giving us interpretability. We compare GA^{2}M with Random Forests, which have been popular in prior work on protein-sequence based prediction and a deep learning embeddings based approach, TAPE.[10]

4.1. *TAPE: Transformer based model for protein sequences*

We use the Unirep model[f] from the TAPE repository[10] which was pretrained on masked language modeling of 31 million protein sequences using a Transformer architecture derived from BERT. This model takes as input, a protein, in the form of its amino acid sequence $x = (x_1, \ldots x_n)$, where n is the length of the protein sequence and outputs a sequence of continuous embeddings $y = (y_1 \ldots y_n)$. The architecture comprises 12 encoder layers, each of which includes multiple attention heads. Intuitively, attention weights define the influence of every token on the next layer's representation for the current token.

To derive TAPE-based embeddings, we apply a `UniRep babbler-1900` model on all protein sequences in our dataset, which gives us 1900 dimensional embeddings for each protein in two modes: `pooled` and `avg` where the former incorporates the temporal aspect of the input sequence and the latter averages over the per-position embeddings. We concatenate the embeddings from the virus and human proteins to get a 3800 dimensional embedding. We trained two types of supervised models using these as features: Logistic Regression and Random Forests. We found no significant difference in the performance from either and show results from the Logistic Regression based models due to computational efficiency. For the embeddings, we found the setting `avg` worked better probably because it captures homology better. The hyper-parameters of all algorithms were trained using nested cross-validation and grid-search over various values. For GA^{2}M, the main hyper-parameter is the number of interaction terms k for which we tried the following values: $0, 10, 50, 100, 200, 500$. We observed that the performance improved until $k = 100$ and then got worse with higher k. We choose $k = 50$ to trade-off computational speed against a small drop in accuracy.

[e] `https://github.com/interpretml/interpret`
[f] `https://github.com/songlab-cal/tape`

5. Results

5.1. *Prediction performance and validation of predicted interactions*

In Fig 1 we show the predictive performance of all approaches in a 5-fold cross validation setup, for a class-skew of 1:10, where each experiment was repeated 20 times, each time with a different set of negative examples. The reported numbers show the mean (horizontal line in the bar) and standard deviation of the metrics. The GA^2M model has an AUC-PR of 0.67 on predicting SARS-CoV-2-human PPI and 0.59 on predicting SARS-CoV-human PPI. The results from the TAPE embeddings are similar to that of Random Forests on SARS-CoV-2-human PPI possibly due to the small scale of PPI data.

To evaluate our models further, we score the set of all possible SARS-CoV-2-human protein-pairs: let us call this set U comprising of 29 x $\approx 21,000 = 609,000$ for ≈21,000 `reviewed` proteins from UniprotKB, and validate these scores using the more recently published PPI from Stukalov et al.[5] Towards this, we first train 100 different models on the gold standard dataset from Gordon et al.[4] by using the 332 positives and sampling a random set of negatives from the unlabeled protein pairs for each of the 100 runs. Since the predictions from a single model are likely to have a bias dependent on the exact set of negatives used, we train 100 different models and apply each of them on the set U. The score for each example from U is averaged over the scores from the 100 different models.

After excluding the gold-standard PPI from the set U, we found that 10,211 examples crossed the classifier score threshold of 0.5. Suppl. Table S1 shows the 28 predictions from this list of 10,211 which appear as experimentally determined interactions in.[5]We performed Fisher's exact test to evaluate the statistical significance of this observation (i.e the probability of seeing 28 of 1089 interactions if 10,211 pairs of proteins are sampled from 609,840 pairs) and obtained a p-value of 0.014. Since pull-down/mass spectrometry based methods are prone to false negatives because of technical limitations in the technology, it is likely that additional pairs within the 10,211 highly ranked predictions are also interacting.

5.2. *Enrichment analysis of predicted human binding partners*

Our models predicted 113 unique human proteins to have at least one interaction with a SARS-CoV-2 protein having a score larger than 0.9. We used Fisher's exact test to determine the enrichment of Gene Ontology (GO) biological processes and cellular components in this set of proteins. We considered terms with Benjamini-Hochberg corrected p-value ≤ 0.01. To remove the redundancy resulting from the parent-child relationships in the GO, we used REVIGO[11] to simplify the sets of enriched terms. REVIGO forms groups of highly similar GO terms using a clustering algorithm (which is similar to hierarchical clustering) and then chooses one representative for each cluster while ensuring that no two representatives are more similar than a user-provided cutoff. We used SimRel[12] as the semantic similarity measure and 0.7 as the cutoff. We now discuss some key enriched GO cellular components. The full set of enriched cellular components and biological processes is available in the supplementary materials.

The GO cellular component "actin cytoskeleton" was significantly enriched (p-value 1.74×10^{-14}) in the predicted human binding proteins. Many viruses use and modify the

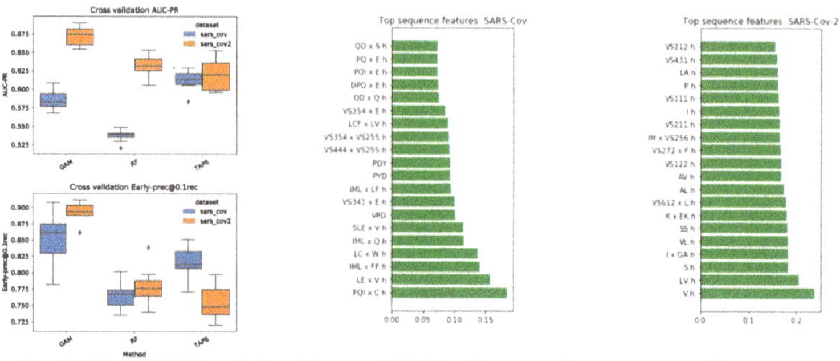

Fig. 1. **(left)** AUC-PR and Precision at 10% Recall averaged over 20 runs for a class skew of 1:10. **(center)** Sequence features relevant to predicting interactions between SARS-CoV and human proteins and **(right)** SARS-CoV-2 and human proteins. Statistics obtained by averaging feature weight from 20 models. Feature names with no suffix are from the virus protein and the suffix '`.h`' refers to that feature from the human protein. Pairwise interaction features are shown as: f_1 x f_2, for instance: `I x GA.h` refers to an interaction between feature `I` from the virus protein and `GA` from the human protein. Features with a prefix of `VS` are conjoint triad features. `VS612` represents a trimer that contains amino-acids from classes 6, 1, and 2. See Fig 3 for the mapping of these classes.

host cell's actin cytoskeleton at different stages of their life cycle including entry, replication, egress, and infection of new cells.[13] In uninfected host cells, viral particles bind to cellular receptors associated with actin filaments in order to travel along filopodia and reach entry sites where endocytosis occurs.[13] Filopodial extensions also act as bridges between infected to uninfected cells to transport virus particles.[13] A global phosphoproteomic analysis[14] of SARS-CoV-2 infection in Caco-2 cells found that the virus induced substantial increase in filopodial protrusions. The authors hypothesized that induction of filopodia might be crucial for egress of SARS-CoV-2 and/or its spread from one cell to another within epithelial monolayers.

The GO cellular component "kinesin complex" was significantly enriched (p-value 1.28×10^{-8}). Kinesins are a family of motor proteins that play an important role in the replication and spread of different viruses by mediating their long distance movement in the microtubule transport system.[15] Our predictions suggest that SARS-CoV-2 may also use kinesins for transport within infected host cells.

6. Discussion

6.1. *Visualizing the virus-human interactions*

Fig. 2(left) shows the embedding of the PPI datasets from Stukalov et al[5] in comparison to HIV, Ebola and SARS. PCA was used for dimensionality reduction from 17,526 features to 100 dimensions, followed by t-SNE to visualize the embedding. The interactive versions of these figures are available in our repository [g], where the user can hover over each entry and

[g] `https://github.com/meghana-kshirsagar/sars_ppi/blob/master/allviruses_plot.html`

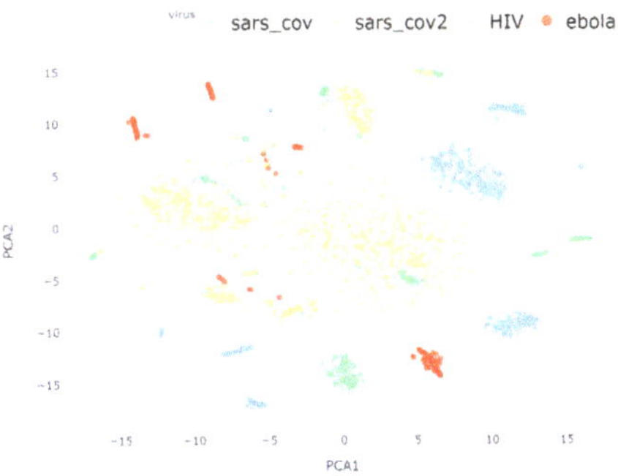

Fig. 2. Embedding of the SARS-CoV and SARS-CoV-2 PPI from Stukalov et al[5] jointly with the *Ebola* and *HIV-1* PPI described in Table 1. Each dot represents a virus-human PPI, colored by the virus species (details in Section 6.1). The large cluster of overlapping yellow and green points at the center shows the interologs between SARS-CoV and SARS-CoV-2.

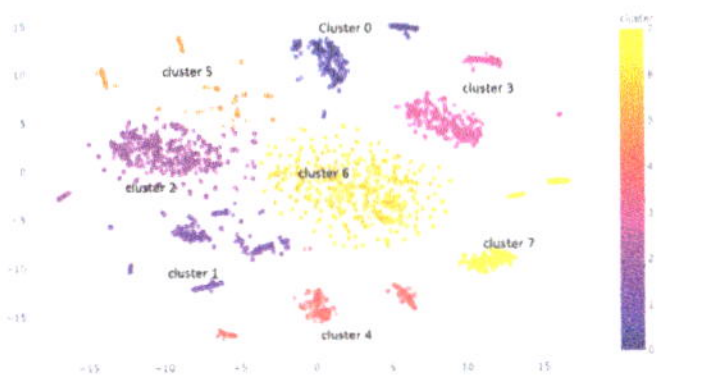

Class	1	2	3	4	5	6	7
Amino acids	Ala, Gly, Val	Ile, Leu, Phe, Pro	Tyr, Met, Thr, Ser	His, Asn, Gln, Trp	Arg, Lys	Asp, Glu	Cys

Fig. 3. **(top)** K-means clustering of the dimensionality reduced data from the left panel. Each dot is a virus-human PPI coloured by the cluster it was assigned to by the k-means algorithm. **(bottom)** The seven amino-acid classes used in the conjoint triad features; details of the properties used in their classification can be found in Shen et al.[9]

find the protein pair's identity. One can see that in both graphs, there are obvious clusters of interactions, some of which involve only proteins from a single type of virus. In contrast, others show overlap with several viruses.

For further analysis of the PPI clusters, we apply k-means clustering on the 100-dimensional data obtained from PCA and colour the PPI based on which cluster they were assigned to. The result of k-means clustering is shown in Fig. 3 (right). There are 8 clusters, some of which we discuss here. Cluster 0 contains several visually distinct sub-clusters. On the right, there are mostly SARS `N` protein interactions, overlapping with SARS-CoV-2 `N` protein interactions, while those on the left are mostly `M` protein interactions. Cluster 1 includes sub-clusters for *HIV-1* `rev`, SARS `nsp6` (a protein with 4 transmembrane helices and a protease domain), as well as SARS and SARS-CoV-2 `E` and `orf7a` proteins. In the vicinity of the E protein interactions there are several *Ebola* `vp40` interactions as well as a subcluster of *HIV-1* `vpu` interactions. The close proximity of all four viruses implies that there may be commonality in the functions of these interactions. Indeed, in SARS the `M`, `E`, and `N` proteins are required for efficient assembly, trafficking, and release of virus-like particles, as evidenced by the need for co-expression of both `E` and `N` proteins with `M` protein.[16] This is remarkably similar to what has been observed in *Ebola*, where expression of `vp40` alone in mammalian cells induces the production of virus particles with a density similar to that of virions but proper particles require co-expression of `vp40` and `GP`.[17] How do the `nsp6` and `orf7a` proteins fit into this process? While it is known that `nsp6` is involved in autophagy (it limits autophagosome diameter), the proximity to the SARS/SARS-CoV-2 E protein interactions and the *Ebola* `vp40` interactions suggest that there is a connection to virion formation. Unclear is also the role of the *HIV-1* accessory protein `vpu`, and this proximity may shed light on its function.

Cluster 2 contains a small subcluster on the left, composed mostly with `orf9b` SARS, and a few `orf9b` SARS-CoV-2 interactions, but the majority of this cluster are `orf3` interactions from SARS-CoV-2, lined with some on the top and the bottom of `orf3a` from SARS. Cluster 4 contains three subclusters, left: *HIV-1* `vif` interactions, middle: SARS `orf3b` interactions and right: *Ebola* `vp24` interactions. The functions of `vif` are not well understood, but for `vp24` and `orf3b` it is clear that they act as IFN antagonists,[18] although the two proteins don't share any detectable sequence similarity. Furthermore, the `vif` protein in another virus, the caprine arthritis encephalitis virus, appears to be an interferon antagonist as well.[19] This cluster is a particularly strong validation for the concept that the PPI network that a virus protein engages in defines its functions and provides a novel way to identify functional similarity where sequence and structure similarity is not detectable.

Cluster 6 is a large cluster that contains only `orf7b` from SARS and SARS-CoV-2. Clusters 0, 2 and 6 are the ones most unique to the coronaviruses but with different levels of similarity within. It has been speculated that the differences between `orf9b` in SARS and SARS-CoV-2 may contribute to the enhanced transmissibility of SARS-CoV-2, possibly due to increased ability to suppress the interferon response.[20] Finally, cluster 7 involves three subclusters, HIV tat, SARS `orf8`, and `orf8a`. `tat` activates RNA Polymerase II,[21] while the functions of orf8/a are not known.[22] Thus, it is tempting to speculate that there may be overlap in these functions with those of `tat` in HIV.

6.2. *Highly ranked sequence features*

Fig. 1 (center) shows the top-ranked features from SARS-CoV-human interactions and (right) SARS-CoV-2-human interactions. Single letters refer to amino acids in the k-mer, while those with a prefix `VS` refer to the conjoint triad feature with amino acid groups shown in Fig. 3 (bottom). An extension `.h` indicates that the feature refers to the human binding partner. One can clearly see that the top-ranked features for the two viruses are different in their detail (which supports that experimental observation that the sequence variations between the two viruses affect their PPIs[5]) but follow similar trends. For example, many of the features refer to hydrophilic amino acid combinations such as `QD`, `PQ`, `DPQ`, `QD` reflecting the fact that it is the water-exposed surfaces of proteins that engage in PPI interfaces. Furthermore, it is well established that the bulky aromatic, yet hydrophilic side-chain `Y` is often found as anchor residues in PPI interfaces. Thus, it is encouraging to find `PDY`, `PYD` and triad features involving class 3 amongst the top ranked features.

6.3. *Structural analysis*

Highly ranked sequence features from the model correspond to amino acid residues that form cryptic pockets. Cryptic pockets are cavities that form in protein structures due to thermal fluctuations in vivo, but are not observed in experimentally derived protein structures.[23] These pockets can expose functionally important residues to the surface of a protein and can also be used as targets for drug development.[24] A recent study performed molecular dynamics simulations on the majority of proteins in the SARS-CoV-2 proteome to sample the ensemble of structural poses that each protein adopts,[25] using a specialized algorithm to focus on sampling cryptic pockets.[26] The group curated a dataset indicating which residues are part of a cryptic pocket based on analysis using LIGSITE,[27] which performs a grid-based search for pockets, and exposons,[28] which identifies residues that have cooperative changes in their solvent exposure. Overlaying sequence features from the PPI model onto one of the SARS-CoV-2 proteins, Nonstructural protein 16 (`nsp16`), we find that the positions found significant by the model coincide with the location of 3 out of the 5 pockets. This protein is of particular interest since it has more pockets than any other protein in the dataset, and is an interesting drug target since it is known to be involved in evading the host immune response.[29]

7. Prior Work

Network analysis of SARS-CoV-2 has been carried out since the first SARS-CoV-2 related PPI dataset was deposited in BioRxiv on March 22, 2020.[4] The majority of analyses have focused on identifying targets for repurposing drugs,[30–32] and/or to better understand the molecular details underlying viral pathogenesis.[33,34] These network analysis papers use known human-human PPI to follow the paths from original human-virus pair into the human interactome. This network propagation approach has also been extended to include predicted human-human PPI.[35] A few groups have also looked at the prediction of new interactions between virus and human host proteins: PIPE[36] uses sequence-based PPI predictors PIPE4 and SPRINT to predict interactions for only 14 of the 29 SARS-CoV-2 proteins based on known PPI obtained from the VirusMentha database[37] which currently contains 5 SARS (not SARS-CoV-2) PPIs.

8. Conclusion

We developed a sequence-only based feature prediction model for interactions between SARS-CoV-2 and human proteins. Validation by an independent dataset showed significant enrichment of experimentally validated interactions in the highly-ranked predictions, strongly supporting the approach. The interpretability of our model also allows designing hypotheses toward disrupting these interactions, a crucial step in exploiting PPI prediction for antiviral drug discovery.

Supplementary material: Additional plots, tables, predicted PPI and enrichment analysis are available at: `https://github.com/meghana-kshirsagar/sars_ppi`

9. Acknowledgements

We would like to thank Mark Crovella and Simon Kasif for their help in discussions. This work was supported by National Science Foundation CISE grants 2031614 and 1940169 (to J.K-S.) and NSF grants DBI-1759858 and MCB-1817736 (to T.M.M.) and the Computational Tissue Engineering Graduate Education Program at Virginia Tech.

References

1. M. D. Dyer, T. Murali and B. W. Sobral, Supervised learning and prediction of physical interactions between human and HIV proteins, *Infection, Genetics and Evolution* **11**, 917 (2011).

2. M. Kshirsagar, K. Murugesan, J. G. Carbonell and J. Klein-Seetharaman, Multitask matrix completion for learning protein interactions across diseases, *Journal of Computational Biology* **24**, 501 (2017).
3. M. Chen, C. J.-T. Ju, G. Zhou, X. Chen, T. Zhang, K.-W. Chang, C. Zaniolo and W. Wang, Multifaceted protein–protein interaction prediction based on siamese residual rcnn, *Bioinformatics* **35**, i305 (2019).
4. D. E. Gordon, G. M. Jang, M. Bouhaddou, J. Xu, K. Obernier, K. M. White, M. J. O'Meara, V. V. Rezelj, J. Z. Guo, D. L. Swaney *et al.*, A SARS-CoV-2 protein interaction map reveals targets for drug repurposing, *Nature* , 1 (2020).
5. A. Stukalov, V. Girault, V. Grass, V. Bergant, O. Karayel, C. Urban, D. A. Haas, Y. Huang, L. Oubraham, A. Wang *et al.*, Multi-level proteomics reveals host-perturbation strategies of SARS-CoV-2 and SARS-CoV, *bioRxiv* (2020).
6. Y. Lou, R. Caruana, J. Gehrke and G. Hooker, Accurate intelligible models with pairwise interactions, *Proceedings of the 19th ACM SIGKDD international conference on Knowledge discovery and data mining* , 623 (2013).
7. T. Guirimand, S. Delmotte and V. Navratil, VirHostNet 2.0: surfing on the web of virus/host molecular interactions data, *Nucleic acids research* **43**, D583 (2015).
8. M. Kshirsagar, J. Carbonell and J. Klein-Seetharaman, Multitask learning for host–pathogen protein interactions, *Bioinformatics* **29**, i217 (2013).
9. J. Shen, J. Zhang, X. Luo, W. Zhu, K. Yu, K. Chen, Y. Li and H. Jiang, Predicting protein–protein interactions based only on sequences information, *Proceedings of the National Academy of Sciences* **104**, 4337 (2007).
10. R. Rao, N. Bhattacharya, N. Thomas, Y. Duan, X. Chen, J. Canny, P. Abbeel and Y. S. Song, Evaluating protein transfer learning with tape, *Advances in Neural Information Processing Systems* (2019).
11. F. Supek, M. Bošnjak, N. Škunca and T. Šmuc, REVIGO summarizes and visualizes long lists of gene ontology terms, *PloS one* **6**, p. e21800 (2011).
12. A. Schlicker, F. S. Domingues, J. Rahnenführer and T. Lengauer, A new measure for functional similarity of gene products based on Gene Ontology, *BMC bioinformatics* **7**, p. 302 (2006).
13. M. P. Taylor, O. O. Koyuncu and L. W. Enquist, Subversion of the actin cytoskeleton during viral infection, *Nature Reviews Microbiology* **9**, 427 (2011).
14. M. Bouhaddou, D. Memon, B. Meyer, K. M. White, V. V. Rezelj, M. C. Marrero, B. J. Polacco, J. E. Melnyk, S. Ulferts, R. M. Kaake *et al.*, The global phosphorylation landscape of sars-cov-2 infection, *Cell* **182**, 685 (2020).
15. M. P. Dodding and M. Way, Coupling viruses to dynein and kinesin-1, *The EMBO journal* **30**, 3527 (2011).
16. Y. Siu, K. Teoh, J. Lo, C. Chan, F. Kien, N. Escriou, S. Tsao, J. Nicholls, R. Altmeyer, J. Peiris *et al.*, The M, E, and N structural proteins of the severe acute respiratory syndrome coronavirus are required for efficient assembly, trafficking, and release of virus-like particles, *Journal of virology* **82**, 11318 (2008).
17. T. Noda, H. Sagara, E. Suzuki, A. Takada, H. Kida and Y. Kawaoka, Ebola virus VP40 drives the formation of virus-like filamentous particles along with GP, *Journal of virology* **76**, 4855 (2002).
18. A. P. Zhang, D. M. Abelson, Z. A. Bornholdt, T. Liu, V. L. Woods, Jr and E. O. Saphire, The ebolavirus VP24 interferon antagonist: know your enemy, *Virulence* **3**, 440 (2012).
19. Y. Fu, D. Lu, Y. Su, H. Chi, J. Wang and J. Huang, The vif protein of caprine arthritis encephalitis virus inhibits interferon production, *Archives of virology* (2020).
20. H. Jiang, H. Zhang, Q. Meng, J. Xie, Y. Li, H. Chen, Y. Zheng, X. Wang, H. Qi, J. Zhang *et al.*, SARS-CoV-2 orf9b suppresses type i interferon responses by targeting TOM70, *Cellular*

& *Molecular Immunology* , 1 (2020).
21. A. P. Rice, The HIV-1 tat protein: mechanism of action and target for hiv-1 cure strategies, *Current pharmaceutical design* **23**, 4098 (2017).
22. C.-T. Keng and Y.-J. Tan, Molecular and biochemical characterization of the sars-cov accessory proteins orf8a, orf8b and orf8ab, in *Molecular Biology of the SARS-Coronavirus*, (Springer, 2010) pp. 177–191.
23. C. R. Knoverek, G. K. Amarasinghe and G. R. Bowman, Advanced methods for accessing protein shape-shifting present new therapeutic opportunities, *Trends in Biochemical Sciences* (2019).
24. D. Beglov, D. R. Hall, A. E. Wakefield, L. Luo, K. N. Allen, D. Kozakov, A. Whitty and S. Vajda, Exploring the structural origins of cryptic sites on proteins, *Proceedings of the National Academy of Sciences* (2018).
25. M. I. Zimmerman, J. R. Porter, M. D. Ward, S. Singh, N. Vithani, A. Meller, U. L. Mallimadugula, C. E. Kuhn, J. H. Borowsky, R. P. Wiewiora, M. F. Hurley, A. M. Harbison, C. A. Fogarty, J. E. Coffland, E. Fadda, V. A. Voelz, J. D. Chodera and G. R. Bowman, Citizen scientists create an exascale computer to combat COVID-19, *bioRxiv* (2020).
26. M. I. Zimmerman and G. R. Bowman, FAST conformational searches by balancing exploration/exploitation trade-offs, *Journal of Chemical Theory and Computation* (2015).
27. M. Hendlich, F. Rippmann and G. Barnickel, LIGSITE: automatic and efficient detection of potential small molecule-binding sites in proteins, *Journal of Molecular Graphics and Modelling* (1997).
28. J. R. Porter, K. E. Moeder, C. A. Sibbald, M. I. Zimmerman, K. M. Hart, M. J. Greenberg and G. R. Bowman, Cooperative changes in solvent exposure identify cryptic pockets, switches, and allosteric coupling, *Biophysical Journal* (2018).
29. T. Viswanathan, S. Arya, S.-H. Chan, S. Qi, N. Dai, A. Misra, J.-G. Park, F. Oladunni, D. Kovalskyy, R. A. Hromas, L. Martinez-Sobrido and Y. K. Gupta, Structural basis of rna cap modification by SARS-CoV-2, *Nature Communications* (2020).
30. J. Bullock, A. S. Luccioni, K. H. Pham, C. S. N. L. Lam and M. Luengo-Oroz, Mapping the landscape of artificial intelligence applications against COVID-19, *arXiv* (2020).
31. J. N. Law, N. Tasnina, M. Kshirsagar, J. Klein-Seetharaman, M. Crovella, P. Rajagopalan, S. Kasif and T. Murali, Identifying human interactors of SARS-CoV-2 proteins and drug targets for COVID-19 using network-based label propagation, *bioRxiv* (2020).
32. Y. Zhou, Y. Hou, J. Shen, Y. Huang, W. Martin and F. Cheng, Network-based drug repurposing for novel coronavirus 2019-ncov/SARS-CoV-2, *Cell Discovery* **6** (2020).
33. N. Kumar, B. Mishra, A. Mehmood, M. Athar and S. Mukhtar, Integrative network biology framework elucidates molecular mechanisms of SARS-CoV-2 pathogenesis, *iScience* (2020).
34. D. Domingo-Fernandez, S. Baksi, B. Schultz, Y. Gadiya, R. Karki, T. Raschka, C. Ebeling, M. Hofmann-Apitius *et al.*, COVID-19 knowledge graph: a computable, multi-modal, cause-and-effect knowledge model of COVID-19 pathophysiology, *BioRxiv* (2020).
35. K. B. Karunakaran, N. Balakrishnan and M. K. Ganapatiraju, Interactome of SARS-CoV-2/ncov19 modulated host proteins presents clinically actionable targets for COVID-19, *Research Square* (2020).
36. K. Dick, K. K. Biggar and J. R. Green, Computational prediction of the comprehensive SARS-CoV-2 vs. human interactome to guide the esign of therapeutics, *bioRxiv* (2020).
37. A. Calderone, L. Licata and G. Cesareni, VirusMentha: a new resource for virus-host protein interactions, *Nucleic acids research* **43**, D588 (2015).

Computational Challenges and Artificial Intelligence in Precision Medicine

Olga Afanasiev[1], Joanne Berghout[2], Steven E. Brenner[3,4,5], Martha L. Bulyk[6,7], Dana C. Crawford[8,9], Jonathan H. Chen[10], Roxana Daneshjou[11], Łukasz Kidziński[12,*]

[1]Sutter Health - Palo Alto Medical Foundation, Palo Alto, California
[2]Department of Medicine, University of Arizona Health Science, Tucson, Arizona; currently: Rare Disease Research Unit, Pfizer Inc., Cambridge, Massachusetts
[3]Department of Bioengineering, University of California, Berkeley, California
[4]Department of Molecular & Cell Biology, University of California, Berkeley, California
[5]Department of Plant & Microbial Biology, University of California, Berkeley, California
[6]Department of Medicine, Brigham and Women's Hospital and Harvard Medical School, Boston, Massachusetts
[7]Department of Pathology, Brigham and Women's Hospital and Harvard Medical School, Boston, Massachusetts
[8]Department of Population and Quantitative Health Sciences, Case Western Reserve University, Cleveland, Ohio
[9]Department of Genetics and Genome Sciences, Case Western Reserve University, Cleveland, Ohio
[10]Department of Biomedical Informatics, Stanford University, Stanford, California
[11]Department of Biomedical Data Science, Stanford University, Stanford, California
[12]Department of Bioengineering, Stanford University, Stanford, California
[*]Corresponding author, e-mail: lukasz.kidzinski@stanford.edu

Continuously decreasing cost, speed and efficiency of DNA and RNA sequencing, coupled with advances in real-world sensing, storage of electronic health records, publicly available databases, and new data processing techniques enable precision medicine at unprecedented scale. Machine learning and artificial intelligence emerge naturally as tools for analyzing and summarizing data, supporting clinical decisions with data-driven insights and further unlocking genetically driven mechanisms underlying individualized risk. While these computational tools allow modeling of complex relations in large datasets, they pose new challenges especially because a patient's health is at stake. Due to an often black-box nature and high reliance on the training data, these new tools are prone to biases and most commonly provide correlational rather than causal insights. Results of these analyses have been difficult to validate, interpret, and explain to practitioners, and most genetic studies have struggled to encompass the full spectrum of human diversity. In this work, we summarize recent research trends in addressing these issues with examples from submissions to the "Computational Challenges and Artificial Intelligence in Precision Medicine" session at Pacific Symposium on Biocomputing 2021. We observe growing research interest in identifying biases, deriving causal and interpretable relations, tuning parameters of models for production, and using artificial intelligence for quality control. We expect further upsurge in work on interpretability and low-risk applications of advanced computational tools.

Keywords: artificial intelligence, augmented clinical decision making, bioinformatics, genomics, machine learning, personalized medicine, precision medicine, transcriptomics.

1. Introduction

High-volume genetic sequencing and 'omics data collection as well as increasingly accessible data streams from electronic health records (EHRs), clinical imaging, biobanks, wearables and more are opening up new vistas in biomedical and health data research. To integrate and/or identify meaningful insights from these large and typically noisy multi-dimensional data resources, the field has developed and applied novel computational tools, including many based on machine learning and artificial intelligence. Applied to genetics, these new methods have connected DNA variation to molecular functions and cellular perturbations, identified disease or patient subgroups and the biological processes driving these differences, suggested new therapeutic targets, and overall, dramatically increased our understanding of biomedicine. By integrating these datasets with rich clinical data, or developing algorithms to interpret, condense, or transform facets of these data into more interpretable modalities, much of the hidden information and patterns can be revealed and made useful for the practice of medicine.

2. Genomics and multi-omics data for precision medicine

An increasingly recognized problem for both health equity and methods development is the overrepresentation of European-ancestry participants in large-scale biobanks and omics resources[1]. Tools developed and trained on predominantly European-ancestry datasets have largely performed very poorly when generalized to more diverse populations, and this has serious bioethical, social, and scientific consequences that include missed insights, widened health disparities, and predictive inaccuracies[2,3]. In this PSB session, Singh et al. (2021) present a new method to analyze integrated DNA methylation, transcript expression, and sequence data in order to discover methylation-adjusted expression quantitative trait loci (eQTL) in cadaveric liver samples derived from donors of African American genetic ancestry[4]. Intersecting these data with cataloged genome-wide association study (GWAS) summary results presented several new genetic targets underlying GWAS loci in diseases that disproportionately impact African American populations. These targets had not been identified as candidates in previous work, and underscore the need for additional resources, methods development, and work in this area.

Along with better capturing the common variation present in humanity by expanding sample ascertainment to include historically excluded and currently underrepresented populations, one of the most compelling areas of research in precision medicine is to understand the functional impact of rare variation, and indeed, which rare variation is functional at all. Multiple tools have been generated to predict the functional consequences of protein coding variation, but only a few tools exist to analyze rare variation outside these coding regions. The problem is challenging due to incomplete annotation of functional regions and statistical limitations when considering ultra-rare variants that may appear uniquely within a dataset. Dong et al. (2021) have developed the AeQTL tool to identify rare heterogeneous variants that impact on levels of gene expression by aggregating rare variants according to user-specified regions and combining this genetic information with patient-matched transcriptomic data[5]. They applied their methods to breast cancer sample data and were able to discover associations between aggregated rare germline variants in cis exomic regions with the expression of BRCA1 and SLC25A39.

Moving closer to the clinic, pharmacogenomics has enormous capacity for clinical actionability by bringing genotype-data driven guidance to the task of selecting an appropriate maintenance dose for individual patients. Rapidly determining the correct dose effectively balances the risk of side effects or other adverse outcomes against patient benefit. McInnes and Altman (2021) conducted linear modeling analyses across more than 200,000 participants in the UK Biobank to interrogate the real-world, observational pharmacy evidence that patient genotype at pre-specified loci may influence the maintenance dose of certain drugs prescribed in practice by clinicians[6]. A significant genotype-drug dose relationship was observed across (i) those drugs with Clinical Implementation of Pharmacogenomics Consortium (CIPC) guidance[7], (ii) drugs with a relationship described in DrugBank but no formal practice guideline, and (iii) a discovery set, where six out of 561 tested drugs showed pharmacogenomic potential. They further leveraged the longitudinal nature of UK Biobank to identify associations to side-effects including the appearance of new diagnoses. While the existence of a genotype-dose relationship is an unsurprising result, it previously had not been demonstrated in real prescribing patterns at this scale, and clearly demonstrates how incorporating patient genotype can be an important advance to patient safety.

The fourth paper submitted to this track by Aoki and Ester (2021) presented another new computational tool designed to improve causal inference[8]. Finding relationships between genes and outcomes can lead to better understanding of biological pathways and processes. And so, correlating outcomes and genes is a natural screening tool. However, in purely observational studies, and particularly those with thousands of potential variables, we risk identifying non-causal relations, which are of lower importance for biological discovery or intervention. One approach for narrowing down research targets is to focus on causal relations rather than correlations. Aoki and colleagues have proposed the ParKCa framework which leverages a stacking ensemble meta-learner approach to combine outcomes of multiple causal discovery methods, exploit partially known causes, and predict new ones. They confirmed the efficacy of their approach through simulations and by using their analysis over a real-world dataset[9] to identify cancer driver genes.

3. Artificial intelligence in multi-modal datasets for clinical research and workflows

Massive amounts of clinical data require new methods for quality control, analysis, processing, validation, and deployment of algorithms. Data-driven algorithms are particularly scrutinized due to their black-box nature, and high reliance on the training dataset, resulting in overfitting and biases towards certain populations. As opposed to classical hand-crafted algorithms for which the entire processing pipeline can be diligently monitored, biases and errors cannot be easily removed from data-driven algorithms due to complicated relationships between millions of parameters automatically derived from the data.

Opportunities for novel computational tools in precision medicine are particularly emphasized by the COVID-19 pandemic. COVID-19 disease, caused by a highly infectious SARS-CoV-2[10], can be associated with severe pneumonia resulting in serious complications or death; these poor outcomes are more likely in patients with compromised immune systems due to other underlying conditions or age[11,12]. Rapid upsurge in the number of cases has exposed problems in healthcare systems across the globe. Moreover, restrictive measures for limiting the spread of the virus has led to the cancellation of face-to-face clinical visits for non-emergency visits. This situation has naturally resulted in the upsurge of telemedicine[13] and research on reading patient data automatically, with the intention of reducing the burden on clinicians. These developments among others are reflected in submissions and accepted papers to "Computational Challenges and Artificial Intelligence in Precision Medicine" session at Pacific Symposium on Biocomputing 2021.

Much of recent work in methods for clinical research has been focused on addressing these deployment issues, as it is exemplified by submissions and accepted papers to the "Computational tools and methods" track of this PSB 2021 session. First, for tuning models to certain populations[14], have analyzed optimization procedures for tuning model parameters. By analyzing a multi-study cohort of patients, they found that Bayesian optimization search is not more efficient than grid search and random sampling approaches despite prior evidence in literature based on single-study cohorts. Second, the bias towards the training set not only affects generalizability when we use it in different hospitals, but also results in varying behavior depending on demographics[15]. Third, data quality has fundamental importance not only for building medical machine learning tools, but also for clinical applications, particularly in telemedicine. Influx of telemedicine data due to COVID-19 motivated researchers to use deep learning for quality control of data[16].

3.1. *Optimization of genomic classifiers*

Machine learning and artificial intelligence allows researchers to identify relationships between patients' gene expression and their outcomes. These techniques can bring clinical benefits, but translation of research models into in-hospital deployment requires proper validation frameworks. While the research community focuses on accuracy metrics within a single cohort, practitioners often fail when attempting to deploy such models in practice.

Mayhew et al. (2021) addressed this problem by providing a framework for benchmarking solutions in the context of real-world deployment[14]. To that end, they built their models on a multi-study cohort of patients. They analyzed Bayesian optimization, grid search, and random search for hyperparameter optimization.

The authors illustrate an application of their framework on data on acute in-hospital infections with data coming from multiple studies. In contrast to previous research, they found that a Bayesian optimization framework was not more efficient and provided only marginal gains in performance of the final model. The study emphasized the need for deployment-centered benchmarking and validation on multi-study cohorts.

3.2. *Automatic reading of radiographic images*

Developments in computer vision, particularly in deep convolutional neural networks, have enabled a range of applications in medical imaging. Expert-level predictive models have been developed and published descriptions of algorithms capable of diagnosing skin cancer, brain cancer, lung lesions, or osteoarthritis progression from RGB camera photos, MRI sequences, X-ray, or CT scan input data.

Expert-level results can be achieved in a wide variety of use cases; however, applicability of these algorithms in practice is inhibited by any differences between the new real-world clinical data never seen by the model, and the datasets used for training. Additionally, even though demographics or ethnicity of patients are not explicitly expressed in radiographic images, there can be a bias in diagnostics, due to underrepresentation of certain groups or biased labels provided by clinicians.

In order to investigate the behavior of machine learning models as a function of demographics, ethnicity, or other patient data, one can look at a model's performance and analyze the True Positive Rate statistic of a model in different groups of interest. To that end, Sayyed-Kalantari et al. (2021) built a deep learning model for classifying chest X-rays, using multiple public chest X-ray datasets[15]. They trained a model with close to state-of-the-art performance and found that its accuracy depends on the patient's demographics, ethnicity, and insurance type. This discovery implied that validating the quality of the model across different populations should be one of the key quality checks for practitioners deploying a machine learning model in clinics. Without these kinds of quality checks, deep learning models may end up perpetuating biases rather than alleviating them.

3.3. *Quality control of images in telemedicine*

Other elements of clinical workflows can be addressed much more immediately than algorithmic imaging diagnostics, such as automating quality control. This is particularly important whenever images are collected by patients themselves (as in telemedicine) rather than by trained personnel under standardized hospital conditions. Systems for addressing quality control issues can bring immediate benefits to clinics and present low risk to patients while improving their care. Confident image assessment for clinicians, such as dermatologists, who are using telemedicine is challenging. Low quality images require extra time to read, causing additional delays related to retakes and extra reads. Vodrahalli et al. (2021) proposed a machine learning system for identifying low quality images automatically using a deep convolutional neural network classifier[16]. Their proof-of-concept algorithm could identify 50% of poor-quality images at a cost of only mislabeling 20% of the good quality images. Given a massive upsurge in telemedicine visits during the COVID-19 pandemic, this fraction could lead to significant time savings for hospitals and patients, as well as improved outcomes for time sensitive cases, such as malignant skin cancers. Moreover, these preliminary results could be further improved with better data and more thorough machine learning modelling.

4. Conclusion and future directions

Submissions to "Computational Challenges and Artificial Intelligence in Precision Medicine" session at Pacific Symposium on Biocomputing 2021 have revealed the growing importance and interest in decomposing components of black-box machine learning models, particularly for causality and for finding biases in data. Moreover, while automating the work of clinicians has always been a holy grail of artificial intelligence in medicine, papers in this session highlighted that there are more direct benefits of machine learning methods. Based on submissions to this session we expect further developments in interpretability of computational methods for precision medicine and more low-risk clinical applications, such as those motivated by COVID and post-COVID healthcare requirements.

5. Author contributions

All authors contributed equally to the PSB session. SEB was unable to review this manuscript for medical reasons.

References

1. Sirugo, G., Williams, S. M. & Tishkoff, S. A. The Missing Diversity in Human Genetic Studies. *Cell* **177**, 1080 (2019).
2. Obermeyer, Z., Powers, B., Vogeli, C. & Mullainathan, S. Dissecting racial bias in an algorithm used to manage the health of populations. *Science* **366**, 447–453 (2019).
3. Dias, R. & Torkamani, A. Artificial intelligence in clinical and genomic diagnostics. *Genome Med.* **11**, 70 (2019).
4. Singh, A., Zhong, Y., Nahlawi, L., Park, C.S., De, T., Alarcon, C., & Perera, M.A.. Incorporation of DNA methylation into eQTL mapping in African Americans. in *Pac Symp Biocomput* (2021).
5. Guanlan Dong, Michael C. Wendl, Bin Zhang, Li Ding, and Kuan-lin Huang. AeQTL: eQTL analysis using region-based aggregation of rare genomic variants. in *Pac Symp Biocomput* (2021).
6. McInnes, G., & Altman, R.B.. Drug Response Pharmacogenetics for 200,000 UK Biobank Participants. in *Pac Symp Biocomput* (2021).
7. Klein, M. E., Parvez, M. M. & Shin, J.-G. Clinical Implementation of Pharmacogenomics for Personalized Precision Medicine: Barriers and Solutions. *J. Pharm. Sci.* **106**, 2368–2379 (2017).
8. Aoki, R., & Ester, M. ParKCa: Causal Inference with Partially Known Causes. in *Pac Symp Biocomput* (2021).
9. Tomczak, K., Czerwińska, P. & Wiznerowicz, M. The Cancer Genome Atlas (TCGA): an immeasurable source of knowledge. *Contemp. Oncol.* **19**, A68–77 (2015).
10. Zhou, P. *et al.* A pneumonia outbreak associated with a new coronavirus of probable bat origin. *Nature* **579**, 270–273 (2020).
11. Chau, A. S. *et al.* The Longitudinal Immune Response to Coronavirus Disease 2019: Chasing the Cytokine Storm. *Arthritis Rheumatol* (2020) doi:10.1002/art.41526.
12. Gupta, S. *et al.* Factors Associated With Death in Critically Ill Patients With Coronavirus Disease 2019 in the US. *JAMA Intern. Med.* (2020) doi:10.1001/jamainternmed.2020.3596.
13. Alexander, G. C. *et al.* Use and Content of Primary Care Office-Based vs Telemedicine Care Visits During the COVID-19 Pandemic in the US. *JAMA Netw Open* **3**, e2021476 (2020).
14. Michael B. Mayhew, Elizabeth Tran, Kirindi Choi, Uros Midic, Roland Luethy, Nandita Damaraju, and Ljubomir Buturovic. Optimization of Genomic Classifiers for Clinical Deployment: Evaluation of Bayesian Optimization to Select Predictive Models of Acute Infection and In-Hospital Mortality. in *Pac Symp Biocomput* (2021).
15. Laleh Seyyed-Kalantari, Guanxiong Liu, Matthew McDermott, Irene Y. Chen, and Marzyeh Ghassemi. CheXclusion: Fairness gaps in deep chest X-ray classifiers. in *Pac Symp Biocomput* (2021).
16. Kailas Vodrahalli, Roxana Daneshjou, and James Zou. TeleQC: A Machine Learning Algorithm to Improve the Quality of Telehealth Photos. in *Pac Symp Biocomput* (2021).

AeQTL: eQTL analysis using region-based aggregation of rare genomic variants

Guanlan Dong[1], Michael C. Wendl[2], Bin Zhang[3], Li Ding[2] and Kuan-lin Huang[3,*]

[1]*Department of Biomedical Informatics, Harvard Medical School, Boston, MA 02115, USA*

[2]*Department of Medicine, McDonnel Genome Institute, Washington University in St. Louis, St. Louis, MO 63108, USA*

[3]*Department of Genetics and Genomic Sciences, Center for Transformative Disease Modeling, Icahn School of Medicine at Mount Sinai, New York, NY 10029, USA*

[*]*Corresponding Email: kuan-lin.huang@mssm.edu*

Concurrently available genomic and transcriptomic data from large cohorts provide opportunities to discover expression quantitative trait loci (eQTLs)—genetic variants associated with gene expression changes. However, the statistical power of detecting rare variant eQTLs is often limited and most existing eQTL tools are not compatible with sequence variant file formats. We have developed AeQTL (Aggregated eQTL), a software tool that performs eQTL analysis on variants aggregated according to user-specified regions and is designed to accommodate standard genomic files. AeQTL consistently yielded similar or higher powers for identifying rare variant eQTLs than single-variant tests. Using AeQTL, we discovered that aggregated rare germline truncations in *cis* exomic regions are significantly associated with the expression of *BRCA1* and *SLC25A39* in breast tumors. In a somatic mutation pan-cancer analysis, aggregated mutations of those predicted to be missense versus truncations were differentially associated with gene expressions of cancer drivers, and somatic truncation eQTLs were further identified as a new multi-omic classifier of oncogenes versus tumor-suppressor genes. AeQTL is easy to use and customize, allowing a broad application for discovering rare variants, including coding and noncoding variants, associated with gene expression. AeQTL is implemented in Python and the source code is freely available at https://github.com/Huang-lab/AeQTL under the MIT license.

Keywords: Gene expression; Sequencing; eQTL; Rare variants; Data integration.

1. Introduction

Advances in sequencing technologies have enabled the generation of large-scale disease cohorts with concurrently available genomic and transcriptomic data[1,2]. Samples with concurrent DNA- and RNA-sequencing (DNA-seq and RNA-seq) provide opportunities to discover expression quantitative trait loci (eQTLs), i.e. genetic variants associated with variations in gene expression[3]. Most existing eQTL tools focus on applying various statistical models to test for association between individual pairs of a variant and the associated gene expression[4–6]. However, for rare variants, the underlying power of the statistical testing is often limited and identifying eQTLs from rare variants remains a challenge[7].

Multiple methodologies and tools using aggregation strategies to group and identify rare variants associated with disease status have been developed[8–11], yet similar strategies have rarely been implemented for identifying eQTLs. In addition, these tools are not readily compatible with standard variant call files resulted from sequencing data, including VCFs/MAFs and RNA-seq data from large cohorts.

Here, we present AeQTL, a software tool that performs eQTL analysis on aggregated variants in specified genomic regions and is designed to accommodate standard file formats generated from sequencing data. Previous studies have found that rare germline variants are significantly enriched at both high and low extremes of gene expression in promoter regions[12]. Here, we show AeQTL's aggregation algorithm can increase the statistical power in order to discover rare variant eQTLs with a larger size of grouped carriers. Further, we demonstrate AeQTL's capacity in identifying both germline variants and somatic mutations associated with gene expression changes, which can help prioritize disease susceptibility genes or cancer driver genes. In sum, AeQTL offers a much-needed versatile multi-omics tool to integrate DNA-seq and RNA-seq data.

2. Methods

AeQTL implements standard eQTL analysis with user-defined variant-aggregation and its workflow is shown in Fig. 1. AeQTL requires three input files: an expression file, a genotype file, and a region file. The user can provide an additional covariate file for advanced analyses.

2.1. *Set up eQTL association tests*

The input region file (i.e. a BED file) is provided by the user to set up desired association tests between gene expressions and variants. Each line of the file contains a genomic region followed by one or more genes to be tested against. An association test will be set up between the expression level of each specified gene and aggregated variants in the genomic region. If no genes are specified, AeQTL by default will test each region against every gene's expression in the expression file in a *trans*-eQTL discovery mode. This user-constructed BED file allows flexibility in the design of eQTL analysis for testing both *cis*- and *trans*- eQTLs. We also provide a coding exomic region BED file on our Github page, which can be used for testing and exploratory purposes.

While all variants with matching samples in the expression file will be included in the tests, users can further restrict the aggregation by setting two optional thresholds: the number of mutated samples per region and the number of variants per region, in which case regions with samples or variants below the thresholds will be filtered out. Both thresholds are set to 1 by default.

2.2. *Aggregate variants and conduct regression analysis*

AeQTL aggregates variants by finding overlaps between variants and regions using the interval tree data structure, which is part of the bx-python package (https://github.com/bxlab/bx-python). We used a standalone wrapper of the interval tree (https://github.com/ccwang002/bx_interval_tree) for easier compilation. The interval tree is designed for fast intersect queries on one-dimensional intervals. Compared to other simple positional intersection methods, the interval tree has two major strengths: (1) it allows each interval to be annotated and the annotations will be preserved in queries; (2) the interval tree is implemented in Cython which is faster and more computationally efficient. AeQTL creates an interval tree for each chromosome. For each genomic region provided in the BED file, an interval is specified by the start and end positions, annotated by the region name, and added to the interval tree of its corresponding chromosome. Then, AeQTL finds the intervals that overlap with the given variant to extract its region name and aggregates variants of the same region. AeQTL accommodates different types of variants including single-nucleotide variants (SNVs), insertions, and deletions. After aggregation, AeQTL maps each region to samples and defines a regional mutation status by assigning a genotype "1" if a sample has any variants in this region and a genotype "0" otherwise.

For each tested gene in each region, AeQTL performs a linear regression analysis of RNA-seq gene expression e against regional genotype g:

$$e = \alpha + \beta g + \epsilon, \qquad \epsilon \sim i.i.d.\ N(0, \sigma^2).$$

The linear model is built using the ordinary least squares method with a residual term ϵ that follows a normal distribution with a mean of zero and a constant variance. AeQTL supports covariates c to be incorporated into the regression model:

$$e = \alpha + \beta_1 g + \beta_2 c + \epsilon,$$

which enables the model to account for clinical factors and population structures.

2.3. *Output intermediate mapped files and a result file with summary statistics*

AeQTL outputs mapped files with variant genotypes, expression, and covariates for each region, which can be readily routed into other aggregational statistical tests such as SKAT[8] to allow comparison. Notably, most of the other aggregational software do not allow common sequence file formats (i.e. VCF or MAF) and thus the intermediate files enable flexibility for users.

All the regression results are compiled in a summary file where AeQTL reports both p-values and coefficients of the intercept and all dependent variables, including regional genotype and covariates. To correct for multiple testing, AeQTL also reports adjusted p-values with false discovery rate (FDR) based on the Benjamini-Hochberg (BH) procedure.

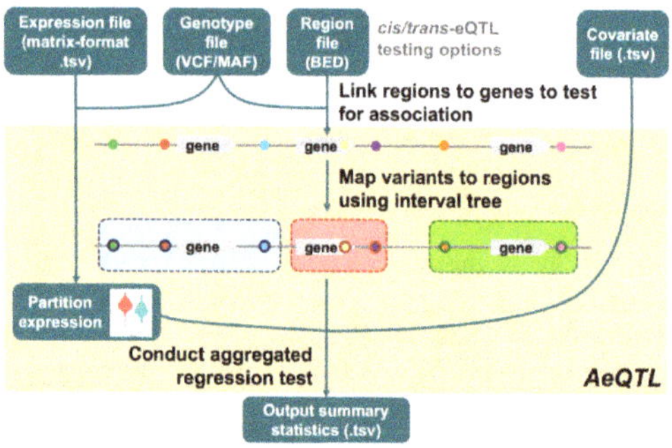

Fig. 1. AeQTL workflow. AeQTL links regions to gene expressions to set up the *cis*/*trans*-eQTL testing, partitions sample expression profiles based on aggregated variants in each region, conducts a linear regression test for each region-gene expression pair with optional covariates, and outputs a summary file.

3. Results

3.1. *AeQTL algorithm development and power simulation*

To demonstrate the aggregating effect on the statistical power of identifying rare variant eQTLs, we performed a simulation analysis using AeQTL. Based on VCF files and expression matrices of 10 rare variants (frequency = 0.1%, five were effective) with a series of sample sizes, we ran AeQTL on both single variants and grouped variants specified by BED files (Fig. 2a). Gene expression profiles were generated from a normal distribution with a mean of 20 and a standard deviation of 10, while effective variants had an effect size t ($t = -10$ or $t = -20$) from a normal distribution with a mean of t and a standard deviation of $| t/2 |$. For each sample size, power was calculated as the averaged value of 10,000 independent simulations.

Overall, statistical analysis of aggregated variants consistently demonstrated comparable or higher powers than individual variants. When the sample size was small, the powers of grouped and single variants were similarly low for both effect sizes. As the sample size increased, the powers under all testing conditions increased as expected. However, the powers of grouped variants increased noticeably faster than those of single variants. When effect size $= -20$, the increased power fold change provided by the AeQTL aggregation method was the most substantial within the sample size interval of 600 to 2,000. The power of grouped variants reached 97% with sample size = 2,000, while single variants required three times the sample size to reach a similar power. When effect size $= -10$, the power of grouped variants reached 95% with sample size = 6,000 and single variants did not reach the same power until sample size = 15,000. At a sample size of 5,000, the powers of all testing conditions except for single variants with effect size $= -10$ were higher than 90% and were saturated (> 99%) when the sample size reached 8,000.

3.2. *Germline eQTL detection*

We further tested AeQTL on rare germline truncations (minor allele frequency $\leq$ 0.05%) on chromosome 17 of the TCGA PanCanAtlas cohort[13]. We tested the hypothesis that rare truncations in cancer susceptibility genes are associated with their *cis*-expression in tumor samples. For the input BED file, we specified each of the genes on chromosome 17 as a region of interest and tested truncations in each gene region against the expression of its located gene. We used the level 3 TCGA RNA-seq gene expression data in RSEM[14] from breast invasive carcinoma (BRCA) patients and incorporated six covariates: age, gender, ethnicity, tumor stage, as well as the top two components from the principal component (PC) analysis on population structure (accounting for > 80% of the top 20 PCs). Because low gene expression levels would likely present technical noises, we filtered out genes with median expressions lower than log(2) (Fig. S1a). This germline analysis, which contained 1,071 samples, 3,150 variants, and 261 unique gene regions, took ~35 min on a Mac with a 2.3 GHz processor and 8 GB memory.

We visualized the distribution of adjusted genotype *p*-values on a QQ-plot (Fig. 2b). The expression of *BRCA1* was significantly associated with aggregated rare truncations in the *BRCA1* exomic region (P = 0.033) in the BRCA cohort, demonstrating that AeQTL could efficiently identify grouped genotype-expression association. In addition, *SLC25A39* (P = 0.030) was among the top-ranked genes whose expressions were negatively associated with the aggregated rare truncations in their regions. We also carried out a sensitivity analysis of adjusted genotype *p*-values against region sizes, which suggested no significant correlation between the two, indicating the lack of false-discovery from large genes ($r_s = 0.16$, P = 0.18, Fig. S1b).

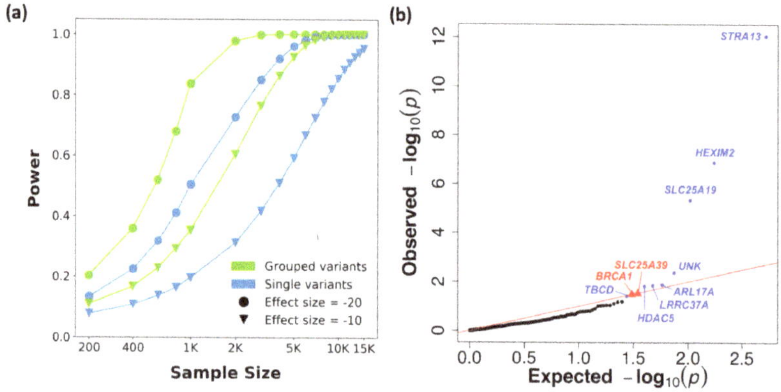

Fig. 2. (a) Power simulation on eQTL analyses using rare variants. The statistical powers of AeQTL (grouped variants; in green) and single-variant testing (in blue) are compared under different sample sizes. (b) QQ-plot of -log10 adjusted genotype *p*-values from rare germline truncations on chromosome 17 in breast cancer patients. The red diagonal line is the expected value. *BRCA1* and *SLC25A39* are marked in red triangles. Other genes showing significant associations (P < 0.05) are also labeled and marked in blue.

3.3. *Somatic eQTL detection*

Aside from germline variants, we also tested AeQTL on somatic truncations and missense mutations across 32 cancer types of the TCGA PanCancer cohort[15]. Similar to the germline eQTL detection, the input BED file contained coding sequence positions of all genes where each gene region was tested against the expression of itself in a *cis*-expression pattern. We used the TCGA PanCancer RNA-seq data and incorporated seven covariates: age, gender, ethnicity, the same top two PCs on population structure as in the germline analysis, cancer subtype, and whether the patient showed an onset age ≤ 50 years old. A separate AeQTL run was performed for each variant type in each cancer type, and all the output summary files for each variant type were compiled together for multiple testing correction using FDR to generate the final pan-cancer output.

We interrogated a subset of 299 genes that were reported as likely driver genes by Bailey et al.[16] and extracted 23,849 truncations and 11,966 missense mutations located in these genes from 8,639 samples. AeQTL identified 243 gene-cancer pairs with truncations and 77 gene-cancer pairs with missense mutations that were significantly associated with their respective gene expressions (FDR < 0.05, Fig. 3). The total and unique variant sites used in the analysis are summarized in Table S1. The top-ranked gene-cancer pairs with truncations include the *MET* proto-oncogene from brain lower grade glioma (LGG), the calcium channel gene *CACNA1A* from lung adenocarcinoma (LUAD), and *TP53* from BRCA; the top-ranked gene-cancer pairs with missense mutations include *JAK2* from stomach adenocarcinoma (STAD), *TP53* from lung squamous cell carcinoma (LUSC), and *FGFR3* from bladder urothelial carcinoma (BLCA).

To demonstrate the computational capacity of AeQTL, we expanded the analysis to the entire dataset, including 335,866 truncations and ~2 million missense mutations from 10,208 samples. AeQTL identified 1,179 gene-cancer pairs with truncations and 3,241 gene-cancer pairs with missense mutations significantly associated with their respective gene expressions (FDR < 0.05).

For significant gene-cancer pairs with truncations, 156 overlapped with the likely driver genes. For significant gene-cancer pairs with missense mutations, 115 overlapped with the likely driver genes. Interestingly, we also identified many top-ranked genes that were not previously identified drivers by TCGA PanCanAtlas driver project[16]. The top-ranked somatic eQTL genes with truncations include *OR8D1* in LUSC, *SOX10* in head and neck squamous cell carcinoma (HNSC), and *PSG7* in kidney renal clear cell carcinoma (KIRC). The top-ranked somatic eQTL genes with missense mutations include *USP29* in cholangiocarcinoma (CHOL) and *AMELX*, *CNTN5*, and *OR1L3* in lymphoid neoplasm diffuse large B-cell lymphoma (DLBC). Multiple recent reports highlight the functionality of the "long-tail driver genes" found with lesser mutations in multiple cancer types[17–21]. These somatic eQTL genes and their expression-associated mutations represent new candidates that warrant further investigations.

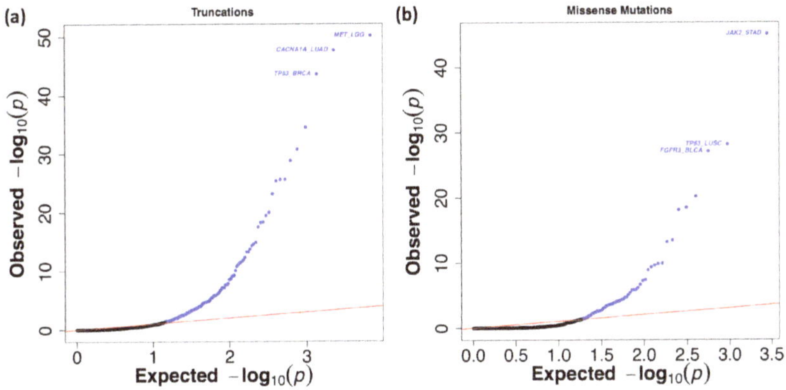

Fig. 3. QQ-plots of -log10 adjusted genotype *p*-values from somatic truncations (a) and missense mutations (b) on likely driver genes in 32 cancer types. The red diagonal line is the expected value. Gene-cancer pairs showing significant associations (P < 0.05) are marked in blue and the top three ranked pairs are labeled.

3.4. *eQTL patterns of oncogenes and tumor suppressor genes*

Cancer driver genes, depending on their mutated cancer type and pathway context, can be subclassified into oncogenes and tumor suppressor genes (TSGs). But most existing methods to classify oncogenes and TSGs leveraged cohort-level mutation data[22–25] that lack considerations of their downstream consequences. To understand whether eQTL patterns could capture the distinction between oncogenes and TSGs, we further investigated the significant genes classified as oncogene or TSG from Bailey et al.'s DNA mutation-based study[16].

In the likely-driver-gene subset analysis, the genotype coefficients of truncations showed a strong association with their respective predicted classifications of oncogenes or TSGs. Genes predicted to be oncogenes or possible oncogenes had larger positive genotype coefficients while genes predicted to be TSGs or possible TSGs had larger negative genotype coefficients (Fig. 4a), demonstrating a polarized pattern of how truncations in oncogenes versus TSGs may affect their respective genes' expression in opposite directions. Moreover, we performed a receiver operating characteristic (ROC) analysis evaluating how well genotype coefficients could predict the labels of driver genes. The analyses yielded an area under the curve (AUC) of 86.3% (Fig. 4c), suggesting the potential of using somatic truncations eQTL patterns to distinguish between oncogenes and TSGs. In comparison, such a pattern was not recapitulated in the genotype coefficients of missense mutations, where both oncogene and TSG mutations were associated with increased gene expressions (Fig. 4b). Overall, genotype-expression analyses revealed distinct eQTL patterns associated with missenses versus truncations and oncogenes versus TSGs in cancer drivers.

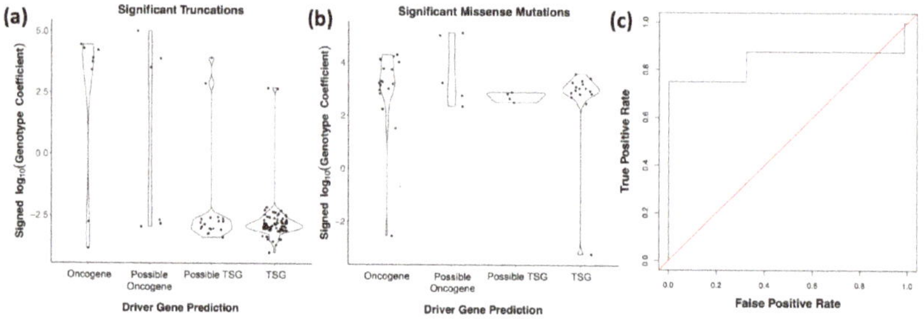

Fig. 4. Violin plots of signed log10 genotype coefficients from significant somatic truncations (a) and missense mutations (b) on likely driver genes in 32 cancer types ($P < 0.05$). The driver gene predictions are obtained from Bailey et al. and genes with no predictions are filtered out. (c) ROC curve of significant somatic truncations labeled as either "Oncogene" or "TSG" (AUC = 0.863). Genes labeled as "Possible Oncogene" or "Possible TSG" are filtered out.

3.5. *Comparison with existing variant-aggregation methods*

We used the intermediate mapped files from TCGA somatic likely driver subset to test two of the most popular variant-aggregation methods, SKAT[8] and SKAT-O[9], and performed the same multiple testing correction based on the BH procedure using FDR. For each gene-cancer pair, we analyzed the difference between the adjusted p-value from AeQTL and the adjusted p-values from SKAT and SKAT-O. The majority of the adjusted p-values from AeQTL were lower than the ones from SKAT and SKAT-O (median difference for truncations = -0.076 (SKAT) and -0.017 (SKAT-O), median difference for missense = -0.012 (SKAT) and -0.010 (SKAT-O), Fig. S2). For both truncations and missenses, SKAT-O identified more significant associations than SKAT while recapturing the ones identified by SKAT. This is not surprising since SKAT-O leverages both SKAT and burden test and implements a small-sample adjustment procedure, which should work well with somatic data. Notably, AeQTL was able to identify more significant truncation associations than SKAT-O and 40 out of the 243 associations were unique to AeQTL (Fig. S3a). On the other hand, SKAT-O identified more significant missense associations than AeQTL (Fig. S3b). This is possibly due to SKAT-O's better compatibility with scenarios where only a fraction of variants show functionalities and potentially different directionalities. Further, neither SKAT nor SKAT-O provides a regression coefficient for regional genotype, which makes it difficult to understand the direction of variant's effect on gene expression and make discoveries such as the polarized eQTL patterns of oncogenes and TSGs.

Most existing variant-aggregation methods are designed to conduct association tests on quantitative traits, most notably for SNP-array genotype data. While each gene expression value can be considered as a continuous trait for analyses using these methods, few of those readily accommodate sequencing data formats such as large VCFs/MAFs and expression matrices from cohorts. To address this challenge, AeQTL can complement the existing methods since it provides intermediate mapped files which can be routed into other aggregational statistical tests based on the users' preference and hypothesis. We believe such user-friendly functionality would be essential to help the field adopt aggregated eQTL testing from sequencing data.

4. Discussion

AeQTL increases the power of eQTL detection by aggregating variants in a defined genomic region. We have applied AeQTL to both synthetic and real datasets. The synthetic dataset demonstrated that variant aggregation consistently yielded similar or higher powers for rare variant eQTL detection. For real datasets, we used rare germline truncations in breast cancer to showcase that AeQTL can efficiently identify significant associations between grouped variants and gene expressions. Furthermore, we applied AeQTL to somatic mutations in a pan-cancer dataset and identified top-ranked gene-cancer pairs that were significantly associated with either truncations or missense mutations in their respective gene regions.

To facilitate users' adoption of AeQTL, we also provide input files to conduct analyses using MAF datasets, as used by TCGA PanCancer somatic mutation data[15]. The application procedure is described in detail and included as an example on Github.

AeQTL is easy to use and customize. Out of the three required input files (region, variant, and expression files), both variant and expression files can be directly taken by AeQTL without any complicated reformatting or pre-processing, while the user-constructed region file allows great flexibility for setting up association tests. Moreover, we provide the exome BED file used in our TCGA analyses on the Github page so that users can easily explore the tool in the *cis*-eQTL mode.

The simplicity of AeQTL's method design means that it can be broadly applied to datasets without imposing on them excessive assumptions or limitations. We have demonstrated AeQTL's promising performance when applied to cancer datasets. However, with more genomic and transcriptomic data being collected and made available in other fields such as neurodegenerative diseases and psychiatric diseases, we believe AeQTL will contribute to multiple areas of study. Aside from research, another important application of AeQTL is in educational settings. From processing standard sequencing data formats, to building classic regression models, and to producing FDR-controlled outputs, AeQTL has a clear and simple workflow that can facilitate the learning process of eQTL analysis.

There are a few aspects of the method that may be improved. First, a potential downside of having a simple method that suits more datasets is that the aggregated genotype of each region is not weighted. Having unweighted variants does not necessarily lead to worse performance, since the underlying mechanism is often unknown and having preset weights may actually confound the results. Nevertheless, we would like to offer more options for users in cases where there are known variations in the magnitudes of effect for certain variants. We plan to introduce more optional settings such as an annotated variant file with a scaling factor, either specified by the user or generated using other algorithms.

Traditional methods to classify oncogenes or tumor suppressors rely on algorithms considering only DNA-mutation patterns or functional curation[22–25]. Herein, we present truncation eQTL patterns revealed by AeQTL as a potential new method to distinguish oncogenes (elevated expression) from tumor suppressors (reduced expression). In TSGs, truncations including nonsense variants or frameshift variants may introduce early stop-codons that likely have led to nonsense-mediated decay (NMD), thus abolishing gene transcripts. In contrast, oncogene truncations show a higher frequency of inframe indels[16], albeit the mechanisms through which they are associated with higher gene expression warrant further investigation.

With increasingly available cohorts of matched genomic (e.g. whole-genome sequencing) and transcriptomic (e.g. RNA-seq) data, we expect that the robust and versatile AeQTL tool can be applied broadly for discovering rare coding and noncoding variants associated with gene expression.

5. Acknowledgement

The authors wish to acknowledge The Cancer Genome Atlas and its participating patients and families that generously contributed the data. This work was supported by Mount Sinai seed fund to KH.

6. Appendix

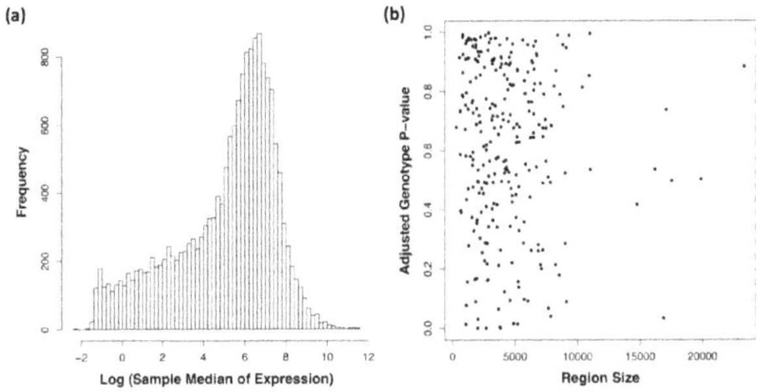

Fig. S1. (a) Histogram of the log-transformed sample median gene expression. (b) Sensitivity analysis of whether genomic region size affects eQTL detection. A randomly scattered pattern is shown when adjusted genotype p-values are plotted against region sizes. The Spearman correlation test also shows no significant correlation ($r_s = 0.16$, $P = 0.18$).

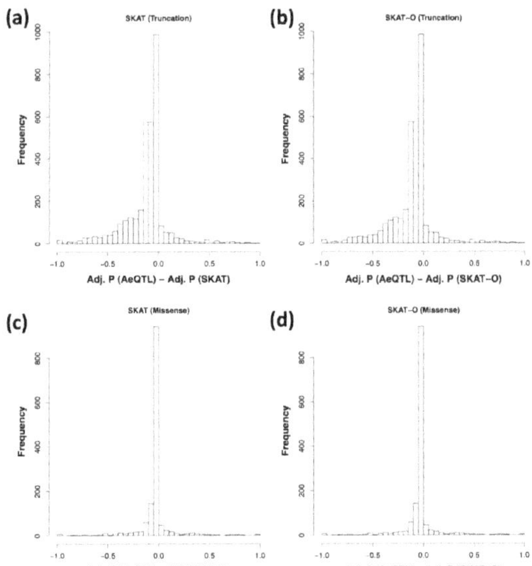

Fig. S2. Histograms of the differences between adjusted p-values from AeQTL and (a) SKAT for truncations, (b) SKAT-O for truncations, (c) SKAT for missense mutations, (d) SKAT-O for missense mutations in TCGA somatic likely-driver-subset analysis.

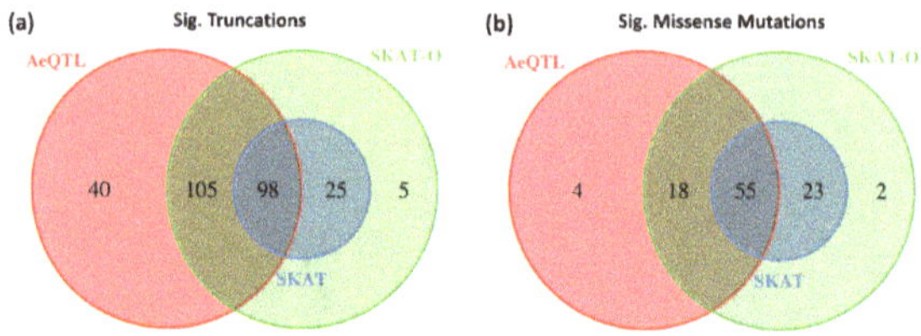

Fig. S3. Venn diagrams of significant associations identified by AeQTL, SKAT, and SKAT-O for (a) truncations and (b) missense mutations in TCGA somatic likely-driver-subset analysis.

Table S1. Summary of the number of variant sites used in TCGA somatic likely-driver-subset analysis.

	Likely Driver Genes	Likely Driver Genes (sig.)
Unique Truncations	15430	4190
Total Truncations	18124	5311
Unique Missense Mutations	5233	1364
Total Missense Mutations	9882	3059

References

1. Ding L, Bailey MH, Porta-Pardo E, et al. Perspective on Oncogenic Processes at the End of the Beginning of Cancer Genomics. *Cell*. 2018;173(2):305-320.e10. doi:10.1016/j.cell.2018.03.033
2. Ardlie KG, DeLuca DS, Segrè A V., et al. The Genotype-Tissue Expression (GTEx) pilot analysis: Multitissue gene regulation in humans. *Science (80-)*. 2015;348(6235):648-660. doi:10.1126/science.1262110
3. Zhu Z, Zhang F, Hu H, et al. Integration of summary data from GWAS and eQTL studies predicts complex trait gene targets. *Nat Genet*. 2016;48(5):481-487. doi:10.1038/ng.3538
4. Shabalin AA. Matrix eQTL: Ultra fast eQTL analysis via large matrix operations. *Bioinformatics*. 2012;28(10):1353-1358. doi:10.1093/bioinformatics/bts163
5. Purcell S, Neale B, Todd-Brown K, et al. PLINK: A tool set for whole-genome association and population-based linkage analyses. *Am J Hum Genet*. 2007;81(3):559-575. doi:10.1086/519795
6. Abecasis GR, Cherny SS, Cookson WO, Cardon LR. Merlin — Rapid analysis of dense genetic maps using sparse gene flow trees. *Nat Genet*. 2002;30(1):97-101. doi:10.1038/ng786
7. Battle A, Mostafavi S, Zhu X, et al. Characterizing the genetic basis of transcriptome diversity through RNA-sequencing of 922 individuals. *Genome Res*. 2014;24(1):14-24. doi:10.1101/gr.155192.113
8. Wu MC, Lee S, Cai T, Li Y, Boehnke M, Lin X. Rare-variant association testing for sequencing data with the sequence kernel association test. *Am J Hum Genet*. 2011;89(1):82-93. doi:10.1016/j.ajhg.2011.05.029
9. Lee S, Wu MC, Lin X. Optimal tests for rare variant effects in sequencing association

studies. *Biostatistics*. 2012;13(4):762-775. doi:10.1093/biostatistics/kxs014
10. Asimit JL, Day-williams AG, Morris AP, Zeggini E. Europe PMC Funders Group Europe PMC Funders Author Manuscripts ARIEL and AMELIA : Testing for an Accumulation of Rare Variants Using Next-Generation Sequencing Data. 2012;73(2):84-94. doi:10.1159/000336982.ARIEL
11. Han F, Pan W. A data-adaptive sum test for disease association with multiple common or rare variants. *Hum Hered*. 2010;70(1):42-54. doi:10.1159/000288704
12. Zhao J, Akinsanmi I, Arafat D, et al. A Burden of Rare Variants Associated with Extremes of Gene Expression in Human Peripheral Blood. *Am J Hum Genet*. 2016;98(2):299-309. doi:10.1016/j.ajhg.2015.12.023
13. Huang K lin, Mashl RJ, Wu Y, et al. Pathogenic Germline Variants in 10,389 Adult Cancers. *Cell*. 2018;173(2):355-370.e14. doi:10.1016/j.cell.2018.03.039
14. Li B, Dewey CN. RSEM: Accurate transcript quantification from RNA-Seq data with or without a reference genome. *BMC Bioinformatics*. 2011;12. doi:10.1186/1471-2105-12-323
15. Ellrott K, Bailey MH, Saksena G, et al. Scalable Open Science Approach for Mutation Calling of Tumor Exomes Using Multiple Genomic Pipelines. *Cell Syst*. 2018;6(3):271-281.e7. doi:10.1016/j.cels.2018.03.002
16. Bailey MH, Tokheim C, Porta-Pardo E, et al. Comprehensive Characterization of Cancer Driver Genes and Mutations. *Cell*. 2018;173(2):371-385.e18. doi:10.1016/j.cell.2018.02.060
17. Armenia J, Wankowicz SAM, Liu D, et al. The long tail of oncogenic drivers in prostate cancer. *Nat Genet*. 2018;50(5):645-651. doi:10.1038/s41588-018-0078-z
18. Leiserson MDM, Vandin F, Wu HT, et al. Pan-cancer network analysis identifies combinations of rare somatic mutations across pathways and protein complexes. *Nat Genet*. 2015;47(2):106-114. doi:10.1038/ng.3168
19. Elman JS, Ni TK, Mengwasser KE, et al. Identification of FUBP1 as a Long Tail Cancer Driver and Widespread Regulator of Tumor Suppressor and Oncogene Alternative Splicing. *Cell Rep*. 2019;28(13):3435-3449.e5. doi:10.1016/j.celrep.2019.08.060
20. Loganathan SK, Schleicher K, Malik A, et al. Rare driver mutations in head and neck squamous cell carcinomas converge on NOTCH signaling. *Science*. 2020;367(6483):1264-1269. doi:10.1126/science.aax0902
21. Gao J, Chang MT, Johnsen HC, et al. 3D clusters of somatic mutations in cancer reveal numerous rare mutations as functional targets. *Genome Med*. 2017;9(1):1-13. doi:10.1186/s13073-016-0393-x
22. Vogelstein B, Papadopoulos N, Velculescu VE, Zhou S, Diaz LA, Kinzler KW. Cancer genome landscapes. *Science (80-)*. 2013;340(6127):1546-1558. doi:10.1126/science.1235122
23. Tokheim CJ, Papadopoulos N, Kinzler KW, Vogelstein B, Karchin R. Evaluating the evaluation of cancer driver genes. *Proc Natl Acad Sci U S A*. 2016;113(50):14330-14335. doi:10.1073/pnas.1616440113
24. Kumar RD, Searleman AC, Swamidass SJ, Griffith OL, Bose R. Statistically identifying tumor suppressors and oncogenes from pan-cancer genome-sequencing data. *Bioinformatics*. 2015;31(22):3561-3568. doi:10.1093/bioinformatics/btv430
25. Collier O, Stoven V, Vert JP. LOTUS: A single- And multitask machine learning algorithm for the prediction of cancer driver genes. *PLoS Comput Biol*. 2019;15(9):1-27. doi:10.1371/journal.pcbi.1007381

Drug Response Pharmacogenetics for 200,000 UK Biobank Participants[1]

Gregory McInnes
Biomedical Informatics, Stanford University, 450 Serra Mall
Stanford, CA 94305, United States of America
Email: gmcinnes@stanford.edu

Russ B Altman
Departments of Bioengineering, Genetics, Medicine, Biomedical Data Science, Stanford University, 450 Serra Mall
Stanford, CA 94305, United States of America
Email: russ.altman@stanford.edu

Pharmacogenetics studies how genetic variation leads to variability in drug response. Guidelines for selecting the right drug and right dose for patients based on their genetics are clinically effective, but are widely unused. For some drugs, the normal clinical decision making process may lead to the optimal dose of a drug that minimizes side effects and maximizes effectiveness. Without measurements of genotype, physicians and patients may adjust dosage in a manner that reflects the underlying genetics. The emergence of genetic data linked to longitudinal clinical data in large biobanks offers an opportunity to confirm known pharmacogenetic interactions as well as discover novel associations by investigating outcomes from normal clinical practice. Here we use the UK Biobank to search for pharmacogenetic interactions among 200 drugs and 9 genes among 200,000 participants. We identify associations between pharmacogene phenotypes and drug maintenance dose as well as differential drug response phenotypes. We find support for several known drug-gene associations as well as novel pharmacogenetic interactions.

Keywords: Pharmacogenetics, Pharmacogenomics, Statistical Analysis, Biobank, UK Biobank

1. Introduction

Pharmacogenetics promises to revolutionize patient care by offering personalized drug selection and dosage based on an individual's genetics[1]. Variations in the genes that encode proteins involved in drug pharmacokinetics and pharmacodynamics are known to lead to interindividual heterogeneity in drug response and can greatly affect clinical outcome. Dosage guidelines have been developed by organizations such as the Clinical Implementation of Pharmacogenetics

---

[1] G.M. is supported by the Big Data to Knowledge (BD2K) from the National Institutes of Health (T32 LM012409). R.B.A is supported by NIH/National Institute of General Medical Sciences PharmGKB resource (U24HG010615) and NIH GM102365. RBA is supported by the Chan Zuckerberg Biohub.

Consortium (CPIC; cpicpgx.org) to aid physicians in incorporating pharmacogenetics into their practice, however the adoption of pharmacogenetics by practicing physicians has not lived up to the optimism in the field[2,3].

Doctor's may not directly be using pharmacogenetics to inform practice, but genetics influences how patients respond to drugs nonetheless. Some drugs, such as warfarin, have a narrow therapeutic index and blood concentration of the drug must be frequently measured to ensure patient safety[4]. The ultimate dose at which the patient achieves the appropriate, stable blood concentration of the drug is the maintenance dose. For warfarin, this dose is strongly influenced by genetic factors such including variations in the metabolizing enzymes CYP2C9 and CYP4F2, as well as the drug target VKORC1.

In other instances genetic variation may lead patients to be at higher risk for side effects. The frequently prescribed drug simvastatin has well known pharmacogenetic interactions with *SLCO1B1* that can lead to simvastatin-induced myopathy[5]. While this is a rare side effect, individuals with poor functioning SLCO1B1 are at higher risk for simvastatin-induced myopathy. CPIC guidelines for simvastatin recommend that individuals with poor functioning SLCO1B1 take a reduced simvastatin dose or a different drug altogether.

Numerous pharmacogenetic drug-gene relationships have been discovered, but most pharmacogenetic studies are small and narrowly focused. The use of electronic health record and biobank scale data as a means for pharmacogenetic discovery and validation of known relationships has been proposed, but until recently databases linking clinical data with genetic data for a large number of patients were unavailable[1,6]. Biobanks offer an opportunity to retrospectively assess known drug-gene relationships in a clinical setting as well as offer the opportunity to discover new drug-gene associations. Biobanks and electronic health records have been used to perform targeted association studies between genomics and response to individual drugs[7] as well as characterize frequency of pharmacogenetic alleles in populations[8,9], but studies of drug response across a large number of drugs have not yet been performed.

The UK Biobank has been widely used to perform genome-wide association studies on a wide variety of traits, but it also includes primary care data from the United Kingdom's National Health System[10]. This dataset offers longitudinal, structured clinical data for more than 220,000 participants that includes diagnoses, laboratory tests, and prescription data. This dataset offers a unique opportunity to identify associations between drug response phenotypes and genetics. Here we present a retrospective pharmacogenetic analysis linking drug exposure for 200 drugs to clinical outcome using the UK Biobank primary care data. We focus on two types of clinical outcomes of interest: maintenance dose and differential drug response.

2. Methods

2.1. *Pharmacogenetic Allele Calling*

We investigated drug-gene relationships for nine important pharmacogenes in the UK Biobank for 222,114 participants using primary care data from the National Health System, provided by the UK Biobank[10]. The pharmacogenetic alleles used in this study were derived from a previously reported procedure, described here in brief[8]. We used imputed genotypes from the Axiom Biobank Array released by the UK Biobank[11]. We included nine genes in our analysis: *CYP2B6, CYP2C19, CYP2C9, CYP2D6, CYP3A5, CYP4F2, SLCO1B1, TPMT,* and *UGT1A1*. The proteins encoded by these genes play critical roles in drug pharmacokinetics and each is included in a CPIC dosing guideline for a drug. We assigned pharmacogenetic phenotypes for each gene using PGxPOP, a tool designed for high throughput mapping of pharmacogenetic alleles and phenotypes (https://github.com/PharmGKB/PGxPOP). The analysis was limited to individuals of European descent. This included participants who self reported as European and were confirmed as European using principal component analysis.

2.2. *Drug Dosage Association with Pharmacogenetics*

Drugs used in this study were derived from the PharmGKB curated drug list (https://www.pharmgkb.org/downloads, drugs.zip)[12]. For each drug, we extracted prescription information from the UK Biobank primary care prescription data by matching the drug name and brand names in the prescription data. Dosage information and drug quantity was extracted using regular expressions that searched within the drug description. We excluded combination therapies from the analysis.

We calculated maintenance dose by determining the average milligrams of drug per day for the last five prescriptions of each drug. This was done by calculating the total milligrams of drug administered for a single prescription divided by the number of days until the next prescription. We then averaged the milligrams of drug per day over the five most recent prescriptions. Prescriptions with a quantity outside two standard deviations from the mean quantity across all participants for that drug were excluded. Subjects were required to receive a minimum of five prescriptions to be included in the analysis. We required drugs to have a minimum of 50 subjects with a maintenance dose to be included in the analysis.

We divided the analysis of maintenance dose associations into three groups of drug-gene pairs. First, we investigated the relationship between drug-gene pairs that have an existing CPIC guideline. This indicates a strong level of evidence of a relationship between a drug and a gene. Second, we investigated drug-gene pairs which have some level of evidence in PharmGKB, but no existing CPIC guideline. These pairs still have some prior evidence indicating an association, but not enough to develop a dosage guideline. Third, we investigated all other drug-gene pairs where an interaction is indicated in DrugBank[13]. These pairs have no prior evidence of a pharmacogenetic association. Data was grouped within each gene by predicted phenotype. For example, for *CYP2C9* participants were put into bins by metabolizer class (normal metabolizers (NM), intermediate metabolizers (IM), and poor metabolizers (PM). Phenotype groups with less than ten participants for a drug are excluded from analysis.

Association between maintenance dose and pharmacogenetic phenotypes was tested for 200 drugs using two types of non-parametric statistical association tests. We used both a Kruskal-Wallis one-way analysis of variance and Jonckheere-Terpstra trend tests to test for associations between each drug and gene pair. Both types of tests are necessary to detect various relationships between dosage and genetics. First, the Kruskal-Wallis test was used to identify any pharmacogenetic phenotype (e.g. CYP2C9 PMs) that have a significant difference in the dosage from other metabolizer classes. Second, Jonckheere-Terpstra tests for an ordered relationship in ranked groups. This is a natural fit for pharmacogenetic phenotypes since there is an inherent order in function which may lead to a linear relationship with dosage (e.g. NM > IM > PM). Resulting p-values are adjusted using a Bonferroni correction. We used a covariate-adjusted dose as the response variable for each test. To do this we fit a linear regression model to the dosage using several covariates: age (at time of last prescription), sex, BMI, genotyping array, and the first for principal components of a principal component analysis (PCA) using genotype data (UK Biobank Data-Field 22009).

We tested the impact of the intronic *CYP2C19* variant rs3814637 on warfarin dose. We used a two-sided Jonckheere-Terpstra test on the allele dosage against the warfarin maintenance dose. Allele dosage was determined as the sum of the alternate alleles for rs3814637.

2.3. *Differential Drug Response Phenotype Association*

In a separate analysis, we tested the relationship between pharmacogenes and drug response for all drugs using diagnosis codes in primary care data. We sought to identify pharmacogenomic phenotypes that would lead to a differential drug response phenotype, for example, instances where poor metabolizers have an increased risk of developing some side effect compared to normal metabolizers. For each drug included in the dosage analysis we identified all diagnoses in the primary care data in the year following the first exposure to the drug. Diagnosis codes in the primary care data are provided as Read Codes (version 2 and version 3), we mapped the Read Codes to ICD-10 codes including only the first three digits (the chapter and first two numerals). ICD-10 codes from chapters V, W, X, Y, and Z were excluded from analysis. Codes were required to have at least 100 events per drug to be included in the analysis. Diagnosis codes may represent the primary disease indication for the drug, side effects, comorbidities, or other unrelated events.

We used logistic regression to test the association between gene phenotypes and ICD-10 code incidence for each drug. This was set up using a binary indicator as the response variable and a one-hot encoding of gene phenotype. We included age (at time of first prescription), sex, genotyping array, and the first four principal components from a genotype PCA as covariates.

We evaluated three tiers of drug-gene relationship, as in the maintenance dose analysis. Drug-gene pairs with CPIC guidelines, drug-gene pairs with any level of evidence in PharmGKB but no CPIC guideline, and an exploratory analysis. For the exploratory analysis of side effect relationships we limited our search to drugs known to interact with *CYP2C9*, *CYP2C19*, and *CYP2D6*, as indicated by DrugBank. These genes were selected because they are promiscuous metabolizing enzymes with well defined pharmacogenetics.

3. Results

The pharmacogenetic analyses presented here included a total of 201,498 participants, after removing 20,615 participants not of European descent. More than 57 million prescriptions are contained within the primary care data, an average of 262 prescriptions per participant. Our initial drug list included 3,358 drugs. Of this, 200 were found in the UK Biobank prescription data with sufficient counts to be included in subsequent analysis.

3.1. *Drug Dosage Association with Pharmacogenetics*

We sought to evaluate methods for testing the relationship between maintenance dose and pharmacogenes at a biobank scale. We performed this analysis using three groups of drug-gene pairs. Of the drugs with CPIC guidance for any of the nine genes queried, there were 24 that had the minimum of 50 participants for whom a maintenance dose could be calculated. We find that nine of the drug-gene pairs have a significant difference in the dosage across gene phenotypes (Kruskal-Wallis or Jonckeere-Terpstra $p < 0.05$, Table 1). We do not adjust for multiple tests because these are known relationships not discoveries. Warfarin and *CYP2C9* phenotypes had the most significant relationship ($p \cong 0$, Jonckeere-Terpstra). The remaining twenty drug-gene pairs did not have a significant relationship between maintenance dose and gene phenotype.

Table 1. Drug-gene dose relationship results. Drug-gene pairs are presented in three groups: drugs with CPIC guidelines, without guidelines but PharmGKB evidence, and novel associations. Level of Evidence represents the maximum level of evidence for the drug-gene relationship in PharmGKB. p-values with a * are significant at $p <= 8.6 \times 10^{-6}$, bonferroni adjusted. Test indicates which type of test achieved the p-value shown (JT=Jonckheere-Terpstra, KW=Kruskal Wallis). Only results with a standard error less than 0.2 are included.

Group	Drug	Gene	Level of Evidence	# Samples	Test	p-value
CPIC guidance	warfarin	CYP2C9	1A	6,409	JT	0.00E+00
	phenytoin	CYP2C9	1A	459	KW	1.04E-05
	azathioprine	TPMT	1A	799	KW	9.13E-03
	imipramine	CYP2C19	2A	348	JT	1.10E-23
	lansoprazole	CYP2C19	2A	2,793	JT	2.52E-02
	pantoprazole	CYP2C19	3	114	JT	2.56E-02
	simvastatin	SLCO1B1	1A	34,611	KW	3.52E-02
	warfarin	CYP4F2	1A	4,559	KW	3.69E-02
	paroxetine	CYP2D6	1A	2,804	KW	4.22E-02
No guidance	warfarin	CYP2C19	3	6,410	KW	2.22E-14
	nicotine	CYP2B6	3	391	JT	6.38E-04
Novel associations	cyclosporine	CYP2C19	NA	166	JT	1.87E-05*
	rabeprazole	CYP2C9	NA	223	JT	4.55E-05*

We then investigated association between maintenance dose and gene phenotype for drug-gene pairs with any level of evidence in PharmGKB but no CPIC guideline. We found two drug-gene pairs with a p-value less than 0.05 for either the Kruskal-Wallis test or Joncheere-Terpstra trend test (Table 1). The most significant was the Kruskal-Wallis test for warfarin and CYP2C19 phenotype. Investigating the dose relationship with phenotype reveals that CYP2C19 normal metabolizers have a decreased maintenance dose compared to the other CYP2C19 metabolizer classes (Figure 1, second row, first column). We followed up on this finding by interrogating the association between rs3814637 and warfarin maintenance dose.

The intronic variant rs3814637 within *CYP2C19* has been previously reported to be associated with warfarin response[14–16]. This variant is contained within several *CYP2C19* star alleles: *CYP2C19*1.004*, *CYP2C19*1.005*, and *CYP2C19*15.001*, all of which are normal functioning alleles. We observed that normal metabolizers had an average daily dose of 4.8 mg (compared to 5.3 mg for the other metabolizer classes). We then tested the association between rs3814637 and warfarin maintenance dose. We find a significant relationship between rs3814637 dosage and warfarin maintenance dose ($p <= 1.0 \times 10^{-46}$, two-sided Jonckheere-Terpstra, Fig. 2).

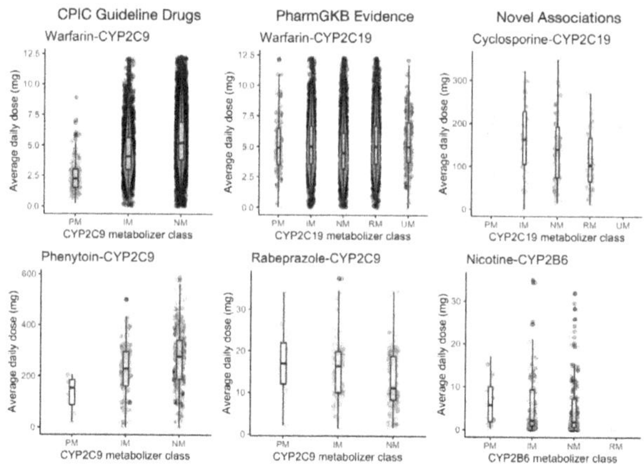

Figure 1. Box plots of maintenance dose for most significant drug-gene pairs. The top two most significant pairs are shown for each group (columns). Enzyme metabolizer classes are represented along the x-axis and the distribution of maintenance dose along the y-axis.

We then analyzed the relationship between maintenance dose and gene phenotype for drug-gene pairs that had no previous indication of a pharmacogenetic relationship but are known to interact. We tested 581 drug-gene pairs and found two significant relationships between dose and gene phenotype: cyclosporine and *CYP2C19*, and nicotine and *CYP2B6* ($p < 8.6 \times 10^{-6}$, Jonckheere-Terpstra, bonferroni adjusted, table 3: Novel associations).

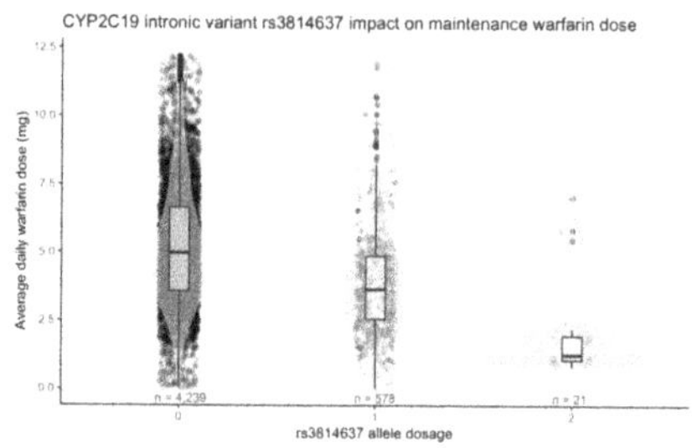

Fig. 2. CYP2C19 intronic variant rs3814637 has a strong influence on warfarin maintenance dose. The x-axis indicates the alternate allele dosage. The y-axis is the maintenance dose.

3.2. *Differential Drug Response Phenotype Association*

We investigated the degree to which adverse drug reactions related to pharmacogenetics could be discovered by performing a statistical analysis of pharmacogene phenotypes and coded medical events within a one year window following the first administration of a drug. We again evaluated three drug-gene groups starting with drug-gene pairs with CPIC guidelines (Table 2, CPIC Guidance Group). The most significant side effect is a decreased incidence of herpes zoster diagnoses among CYP2C19 intermediate metabolizers ($p <= 8.76 \times 10^{-5}$).

Table 2. Drug-gene side effect relationship results. Associations are presented in three groups: drug-gene pairs with CPIC guidelines, pairs with no guidelines but evidence in PharmGKB, and novel associations. Phenotype is the gene phenotype (IM: Intermediate Metabolizer, PM: Poor Metabolizer, RM: Rapid Metabolizer, UM: Ultrarapid Metabolizer, IF: Increased Function, PF: Poor Function). Odds ratio is the odds ratio relative to normal metabolizer or normal function alleles. * indicates significance with Bonferroni adjusted p-value threshold of 1.0×10^{-5}. Only results with a standard error less than 0.2 are included.

Group	Drug	Gene	Level of Evidence	Phenotype	ICD-10	Code definition	Odds ratio	p-value
CPIC Guidance	citalopram	CYP2C19	1A	IM	B02	Herpes zoster	0.53	8.76E-05
	simvastatin	SLCO1B1	1A	IF	M65	Synovitis and tenosynovitis	1.82	1.42E-04
	amitriptyline	CYP2C19	1A	RM	R53	Malaise and fatigue	1.55	1.74E-04
	amitriptyline	CYP2C19	1A	UM	J30	Vasomotor and allergic rhinitis	1.94	2.75E-04
	codeine	CYP2D6	1A	PM	A52	Late syphilis	1.78	3.30E-04
	ibuprofen	CYP2C9	1A	PM	E13	Other specified diabetes mellitus	2.00	4.90E-04
	clopidogrel	CYP2C19	1A	RM	B08	Viral infections characterized by skin and mucous membrane lesions	0.59	5.17E-04
	tamoxifen	CYP2D6	1A	IM	C50	Malignant neoplasm of breast	0.62	6.98E-04
	simvastatin	SLCO1B1	1A	PF	M79	Unspecified soft tissue disorders	1.49	7.46E-04
	simvastatin	SLCO1B1	1A	DF	M65	Synovitis and tenosynovitis	1.79	7.75E-04
No Guidance	citalopram	CYP2D6	3	IM	J45	Asthma	1.44	9.13E-05
	citalopram	CYP2D6	3	IM	I50	Heart failure	1.56	1.12E-04
	simvastatin	CYP2C9	3	PM	J01	Acute sinusitis	1.74	1.56E-04
	citalopram	CYP2D6	3	IM	J64	Unspecified pneumoconiosis	1.56	5.74E-04
	propranolol	CYP2D6	4	IM	O86	Other puerperal infections	1.85	6.38E-04
Novel associations	diazepam	CYP2C9	NA	PM	M19	Osteoarthritis	2.33	4.52E-06*
	zopiclone	CYP2C9	NA	IM	H91	Unspecified hearing loss	2.20	1.73E-05
	loratadine	CYP2D6	NA	IM	M16	Osteoarthritis of hip	1.98	1.20E-04
	tramadol	CYP2B6	NA	PM	H61	Disorders of external ear	1.95	1.86E-04
	quinine	SLCO1B1	NA	IF	N39	Disorders of urinary system	1.95	1.87E-04

Next we looked to see if there are any differential drug response phenotypes enriched among drug-gene pairs with any level of evidence but no CPIC guideline. The top five results are shown in Table 2 under "No Guidance". We find several phenotypes enriched among CYP2D6 intermediate metabolizers taking citalopram, including respiratory issues and heart failure. We also find an increased risk of sinus infections among CYP2C9 poor metabolizers on simvastatin, and an increased risk of puerperal infections among CYP2D6 intermediate metabolizers on propranolol.

We interrogated all other drugs known to be metabolized by CYP2C9, CYP2C19, or CYP2D6 for differential drug response phenotypes. This resulted in 4,806 independent association tests across 81 drugs. After multiple hypothesis corrections one side effect was significantly associated with a drug-gene pair: increased incidence of osteoarthritis in CYP2C9 poor metabolizers after taking diazepam. We show the top five results from the exploratory analysis in Table 2.

4. Discussion

Biobanks offer a powerful solution for enabling the study of relationships between drugs and genes. Large datasets linking genetic and longitudinal clinical data are becoming more broadly available and allow interrogation of the relationship between drug response and pharmacogenetic phenotypes. Here we derived drug phenotypes in the form of maintenance dose and differential drug response phenotypes for more than 200,000 participants across 200 drugs in the UK Biobank and tested their association with well established pharmacogenetic phenotypes for nine genes.

Pharmacogenetic testing is not yet common practice, but for some drugs the standard clinical procedures used to determine maintenance dose are influenced by genetics. We find evidence to support existing pharmacogenetic associations with maintenance dose. Among 24 drugs with CPIC guidance in our study we find evidence for a genetic influence on maintenance dose for nine drugs. For the remaining pairs with guidance, it is possible we are not likely to observe an association with maintenance dose because efficacy is difficult to measure or side effects are rare. Among drugs with any prior evidence of a pharmacogenetic relationship but no CPIC dosage guideline we find that maintenance dose supports the association for two drug-gene pairs. Most notably, carriers of the *CYP2C19* intronic variant rs3814637 have a significantly decreased warfarin maintenance dose. The causal mechanism through which this effect occurs is unclear, and this variant itself may not be causal, rather in linkage disequilibrium with a causal variant. In GTEx, rs3814637 is associated with increased expression of *CYP2C9* (the gene typically associated with warfarin response) in several tissues, although importantly not in liver. There is a gap in the amount of warfarin dosing variability that can be explained by genetics among individuals of African descent[17]. rs3814637 has nearly twice the allele frequency in the

African population as it does in the European population (11.6% vs 6.7%)[18]. Although this study focuses on Europeans, this variant may explain some of the missing heritability of warfarin response among Africans, but further study is needed to confirm this relationship.

We discovered potential novel pharmacogenetic associations with maintenance dose for two drugs: cyclosporine with *CYP2C19*, and nicotine with *CYP2B6*. Both drugs are known to be metabolized by their respective associated enzymes, however there is no prior literature evidence suggesting a pharmacogenetic relationship. For both drugs, we find a decreasing association between dose and metabolizer class of their associated enzymes, where individuals with higher rates of metabolism tend to be on lower doses.

Our analysis of differential drug response phenotypes reveals associations with side effects among drug-gene pairs. This analysis is limited due to the large number of tests requiring a strict multiple hypothesis testing threshold, but produces interesting hypotheses. At first glance many of the differential phenotype associations seem unlikely, but literature evidence exists for many of the findings. For example, the most significant association among drugs with CPIC guidelines was a decreased incidence of herpes zoster among CYP2C19 intermediate metabolizers compared to CYP2C19 normal metabolizers treated with citalopram. However, two previous studies have demonstrated that SSRIs can lead to increased resistance to herpes[19,20]. CYP2C19 intermediate metabolizers have an increased blood concentration of citalopram and may have an increased resistance to a herpes infection. We also find CYP2C19 rapid metabolizers on clopidogrel have a decreased risk of viral skin lesions compared to CYP2C19 normal metabolizers. There is evidence that clopidogrel may inhibit viral clearance[21]. It may be possible that CYP2C19 rapid metabolizers have a lower concentration of clopidogrel and therefore the degree to which they are able to fight off viral infections is higher than that of CYP2C19 normal metabolizers. The most significant association is between CYP2C9 poor metabolizers on diazepam having an increased incidence of osteoarthritis. There is no literature that suggests osteoarthritis may be a side effect of diazepam, although there are studies that suggest diazepam could be used to treat pain as a result of rheumatoid arthritis. Without further evidence we cannot say whether this relationship results from pharmacogenetics and not a correlation with the drug indication or a statistical artifact.

This work has several limitations. First, we use pharmacogenetic alleles called from data imputed from genotyping arrays. We previously reported limitations in accuracy of the ability to accurately call alleles in several pharmacogenes from imputed data, notably in *CYP2D6*[8]. The lack of structural variants in the dataset in addition to the inability to call rare variants may lead to inaccurate prediction of CYP2D6 phenotypes. Second, we broadly apply our maintenance dose algorithm to drugs in the UK Biobank. While this is effective for some drugs, better clinical end points may provide an improved representation of patient response. For example, a dose response curve may provide more fine grained insight into individual response and yield better insight into the genetics of drug response. It is challenging to broadly define response across drugs from numerous classes with varying indications and therapeutic indices. Even a single drug can be used for different indications and may require different doses to treat each indication. Additionally, this approach will miss patients who take a drug once and experience side effects that lead them to immediately switch drugs. No catch-all definition will suffice, but maintenance dose does reveal insight into patient response. Third, the data we used to define drug usage is in the form of prescription orders. We do not know whether the prescriptions were filled or if the patient took the drug as prescribed. Finally, we do not provide any clinical validation of the predictions presented here; further followup is needed.

Biobanks are an immense resource that allow for pharmacogenetic association testing at an unprecedented scale. Longitudinal clinical data is critical to be able to define drug response phenotypes in order to accurately assess patient response to treatment and ultimately test genetic associations. As access to biobanks continue to expand and more data is available, the ability to perform pharmacogenetic studies at large scale will increase. We believe that these resources offer a promising avenue for discovery and will further advance the field of pharmacogenetics.

5. Acknowledgments

This research has been conducted using the UK Biobank Resource under Application Number 33722. We thank all the participants in the UK Biobank study. Most of the computing for this project was performed on the Sherlock cluster. We would like to thank Stanford University, the PharmGKB resource (NIH HG010615), and the Stanford Research Computing Center for providing the computational resources that contributed to these research results. Thank you to Adam Lavertu who helped develop the ideas that led to this work. The Genotype-Tissue Expression (GTEx) Project was supported by the Common Fund of the Office of the Director of the National Institutes of Health, and by NCI, NHGRI, NHLBI, NIDA, NIMH, and NINDS. The data used for the analyses described in this manuscript were obtained from the GTEx Portal on 10/01/20.

References

1. Lavertu, A. *et al.* Pharmacogenomics and big genomic data: from lab to clinic and back again. *Hum. Mol. Genet.* **27**, R72–R78 (2018).
2. Relling, M. V. & Klein, T. E. CPIC: Clinical Pharmacogenetics Implementation Consortium of the Pharmacogenomics Research Network. *Clin. Pharmacol. Ther.* **89**, 464–467 (2011).

3. Krebs, K. & Milani, L. Translating pharmacogenomics into clinical decisions: do not let the perfect be the enemy of the good. *Hum. Genomics* **13**, 39 (2019).
4. International Warfarin Pharmacogenetics Consortium *et al.* Estimation of the warfarin dose with clinical and pharmacogenetic data. *N. Engl. J. Med.* **360**, 753–764 (2009).
5. Ramsey, L. B. *et al.* The clinical pharmacogenetics implementation consortium guideline for SLCO1B1 and simvastatin-induced myopathy: 2014 update. *Clin. Pharmacol. Ther.* **96**, 423–428 (2014).
6. Wilke, R. A. *et al.* The emerging role of electronic medical records in pharmacogenomics. *Clin. Pharmacol. Ther.* **89**, 379–386 (2011).
7. Wei, W.-Q. *et al.* Characterization of statin dose response in electronic medical records. *Clin. Pharmacol. Ther.* **95**, 331–338 (2014).
8. McInnes, G. *et al.* Pharmacogenetics at scale: An analysis of the UK Biobank. 2020.05.30.125583 (2020) doi:10.1101/2020.05.30.125583.
9. Reisberg, S. *et al.* Translating genotype data of 44,000 biobank participants into clinical pharmacogenetic recommendations: challenges and solutions. *Genet. Med.* (2018) doi:10.1038/s41436-018-0337-5.
10. Sudlow, C. *et al.* UK biobank: an open access resource for identifying the causes of a wide range of complex diseases of middle and old age. *PLoS Med.* **12**, e1001779 (2015).
11. Bycroft, C. *et al.* Genome-wide genetic data on ~500,000 UK Biobank participants. 166298 (2017) doi:10.1101/166298.
12. Whirl-Carrillo, M. *et al.* Pharmacogenomics knowledge for personalized medicine. *Clin. Pharmacol. Ther.* **92**, 414–417 (2012).
13. Wishart, D. S. *et al.* DrugBank 5.0: a major update to the DrugBank database for 2018. *Nucleic Acids Res.* **46**, D1074–D1082 (2018).
14. Lane, S. *et al.* The population pharmacokinetics of R- and S-warfarin: effect of genetic and clinical factors. *Br. J. Clin. Pharmacol.* **73**, 66–76 (2012).
15. Liang, Y. *et al.* Association of genetic polymorphisms with warfarin dose requirements in Chinese patients. *Genet. Test. Mol. Biomarkers* **17**, 932–936 (2013).
16. Jorgensen, A. L. *et al.* Genetic and environmental factors determining clinical outcomes and cost of warfarin therapy: a prospective study. *Pharmacogenet. Genomics* **19**, 800–812 (2009).
17. Perera, M. A. *et al.* Genetic variants associated with warfarin dose in African-American individuals: a genome-wide association study. *Lancet* **382**, 790–796 (2013).
18. Karczewski, K. J. *et al.* The mutational constraint spectrum quantified from variation in 141,456 humans. *Nature* **581**, 434–443 (2020).
19. Irwin, M. R. *et al.* Major depressive disorder and immunity to varicella-zoster virus in the elderly. *Brain Behav. Immun.* **25**, 759–766 (2011).
20. Irwin, M. R. *et al.* Varicella zoster virus-specific immune responses to a herpes zoster vaccine in elderly recipients with major depression and the impact of antidepressant medications. *Clin. Infect. Dis.* **56**, 1085–1093 (2013).
21. Iannacone, M., Sitia, G., Narvaiza, I., Ruggeri, Z. M. & Guidotti, L. G. Antiplatelet drug therapy moderates immune-mediated liver disease and inhibits viral clearance in mice infected with a replication-deficient adenovirus. *Clin. Vaccine Immunol.* **14**, 1532–1535 (2007).

ParKCa: Causal Inference with Partially Known Causes

Raquel Aoki[†] and Martin Ester
School of Computing Science, Simon Fraser University
Burnaby, Canada
[†]*E-mail: raoki@sfu.ca*

Methods for causal inference from observational data are an alternative for scenarios where collecting counterfactual data or realizing a randomized experiment is not possible. Our proposed method ParKCA combines the results of several causal inference methods to learn new causes in applications with some known causes and many potential causes. We validate ParKCA in two Genome-wide association studies, one real-world and one simulated dataset. Our results show that ParKCA can infer more causes than existing methods.

Keywords: Causality, Precision Medicine

1. Introduction

The vision of precision medicine is the development of prevention and treatment strategies that take individual variability into account.[1] Precision medicine promises to allow more precise diagnosis, prognosis, and treatment of patients, based on their individual data. In the context of precision medicine, it is important to understand the leading causes of the outcome of interest, which can be achieved using causal discovery methods. Drug response and adversarial drug reactions are examples of an outcome of interest, and OMICS data record potential causes and confounders.

The gold standard of causal inference is based on experimental design, with randomized trials and control groups, which is not always available due to the lack of the full experiment and counterfactual data, either because it is too expensive or impossible to collect. Therefore, there is a rise of more data-driven methods, based on observational data, to either perform causal discovery or estimate treatment effects.[2–4] Furthermore, some applications, such as Driver Gene Discovery,[5] have a few causes that are well known. Most of the existing methods use these only to evaluate or to eliminate edges on constraint-based causal discovery methods.

To handle computational biology (CB) applications with thousands of treatments, unobserved confounders, and partially known causes, we propose ParKCa: a method that uses the few known causes to learn new causes through the combination of causal discovery methods. The intuition is that ParKCa will learn how to identify causes based on the outputs of the other methods (similar to ensemble learning) and a few known examples. ParKCa has many advantages. First, leveraging several methods instead of using a single one can minimize biases and highlight patterns common across the methods. Second, it allows the use of known causes to help identify new causes. Finally, it also allows the combinations of several datasets that share the same set of possible causes but might differ in the datatype or set of rows.

The proposed method ParKCa is validated in a simulated dataset and on the driver gene discovery application. The existence of associations between cancer and specific genes is well accepted in the precision medicine field. However, the human body has more than 20,000 genes, and not all genes mutations lead to cancer.[5] Hence, the challenge here is to recognize those genes that are associated with cancer spreading from the original site to other areas of the body (metastasis) through causal inference. These genes are known as *driver genes* and play an important role in cancer prevention and treatment.

There are many challenges around driver gene discovery. The progress in sequencing technology and the lower cost of collecting genetic information allowed the creation of datasets such as The Cancer Genome Atlas (TCGA). However, the number of columns (genes) is often much larger than the number of rows (patients), which poses a challenge for machine learning models. The partially known dependence between genes due to pathways is also a challenge. Pathways are sets of genes where the alteration or mutation in one gene can cause changes in other genes that share the same pathway.[6] Additionally, some elements that cause cancer might not be included in the dataset. Examples of attributes not observed are the structured clinical information about the patient, such as their lab results and lifestyle. Finally, the lack of a well-defined training set is also a challenge that makes the evaluation of results tricky. There is no 'true' list of driver genes to evaluate the quality of the machine learning models.[7–9]

Table 1. Toy example of the transposed input data (on the left) and output data (on the right). Note that in the input data we have Y and in the outcome data, the *known causes*.

	Gene 1	...	Gene V	Y
Patient 1	7.39	...	1.60	0
...	...	...	...	...
Patient J	3.25	...	2.73	1

$\rightarrow$

	L_1	L_2	Known Cause
Gene 1	-1.2	-2.4	1
...	...	...	..
Gene V	0	12.3	0

A toy example is shown in Table 1. The goal is to learn which genes among V genes are causally associated with a phenotype Y from a dataset with J patients as in the left side of Table 1. The input data represents the gene expression of patients, and it is used to fit the level 0 models L_1 and L_2. The right side of Table 1 shows the output data, constructed with the learners' output. We add to the output an attribute with the partially known causes.

The main contributions of this paper are as follows:

- We introduce the problem of causal discovery from observational data with partially known causes. We are the first ones to formalize it as a stacking problem.
- We propose ParKCa, a flexible method that learns new causes from the outputs of causal discovery methods and from partially known causes.
- ParKCa is validated on a real-world TCGA dataset for identifying genes that are potential causes of cancer metastases and on simulated genomic datasets.

2. Related Work

Our work combines several research areas:

Causality: Motivated by the need for models that are more robust, reproducible, and easier to explain, causality has received a lot of attention. Constraint-based and score-based causal discovery methods, such as PC-algorithm,[10] fast PC-algorithm,[11] FCI,[12] RFCI,[13] and fGES,[14] are still largely used. Their main goal is to recover the causal structure that fits the observed data. However, these methods have a poor performance on large dimensional datasets and/or assume causal sufficiency, suppositions that fail on most of the computational biology (CB) applications.

Deep learning models are making significant contributions to the estimation of treatment effects in the past years.[2,3,15] BART[16] uses the Conditional Average Treatment Effect (CATE) to estimate the treatment effects has been successful in many applications. Finally, the Deconfounder Algorithm (DA),[4] combines probabilistic factor models and outcome models to estimate causal effects. Considering the challenge of learning causes from several datasets, Tillman and Spirtes[17] proposed a method to learn equivalence classes from multiple datasets. A limitation of this work, however, is the lack of scalability to large dimensional datasets.

Ensembles: Our work is based on stacking ensemble.[18] Its main idea is to use several learners models whose outputs are combined and used as input for fitting a meta-model. The idea of using an ensemble approach to calculate causal effects is not new.[19,20] Instead of using ensemble learning to make more accurate causal effect estimates, our work focuses on using ensemble learning to discover new causes. We adopted commonly used meta-learner models (Logistic Regression, Random Forest, Neural Networks, and others), and we also explore PU-learning classification models,[21–23] a sub-class of semi-supervised learning. Our method performs a classification task with the meta-learner on the level 1 data $D^1_{V \times L}$ and a new variable that encodes the *known causes*. The labels are $Y^1_v = 1$ for well-known causes and $Y^1_v = 0$ for non-causal or unknown causes. In other words, the classification learns from positive (known causes) and unlabeled (not causal or unknown causes) data.

Driver Gene Discovery (real-world dataset): Spurious correlations or associations between genes and metastasis are common. Therefore, the challenge lies in identifying those genes that are true causes (*driver genes*) of the underlying condition, not just associated with it. In our real-world dataset, we want to find genes that contribute to cancer metastasis development. In this condition, cancer spreads from the original site to other areas. Previous methods that explored this application are: MuSiC,[7] OncodriveFM,[24] ActiveDriver,[25] TUSON,[26] OncodriveCLUST,[27] MutsigCV,[28] OncodriveFML,[29] 20/20+ (`https://github.com/KarchinLab/2020plus`), and others.[6,8] The first challenge of this application is the large number of genes (possible causes) along with the small sample size and the known (and unknown) dependencies among genes, which adds a certain complexity to the problem. The existence of confounders, some possible to be observed (such as clinical information), others not (such as family history or lifestyle)[5] poses another challenge. Finally, the limited and biased list of known driver genes[30] also needs some attention.[9] This list, here referred to as Cancer Gene Census (CGC), is the gold-standard of driver genes currently available.

3. The ParKCa Method

ParKCa deals with causal discovery from a stacking ensemble perspective with some adaptations. Typically, each causal discovery method is estimated individually, and their results compared. In ParKCa, we use the causal discovery methods' outputs as a classifier's features to learn how these methods agree to identify new causes based on a few known causes used as examples. According to the stacking nomenclature, the causal discovery methods are our learners, and the classification model is our meta-learner, as shown in Figure 1.

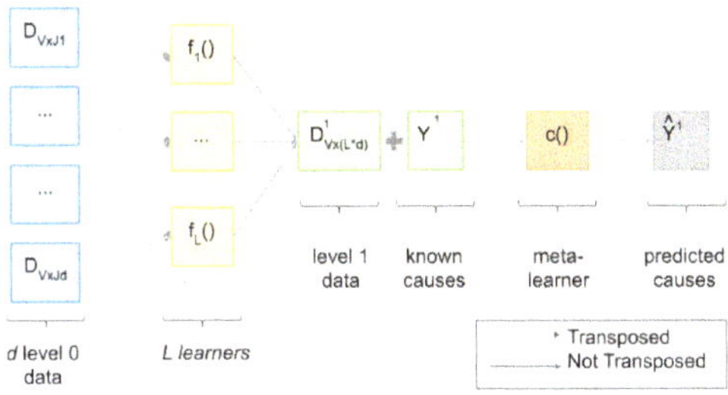

Fig. 1. Illustration of the ParKCa method. From d datasets/subsets with J_d columns (examples) and V rows (possible causes), $L \times d$ outputs are extracted using L level 0 models. These outputs are aggregated in a single dataset $D^1_{V\times(L*d)}$. In this step, we also add the partially known causes Y^1 and fit a meta-learner model to predict new potential causes.

Compared to a standard stacking model, the first modification required by our approach is how we feed the learners. Unlike stacking used as a predictive model, where the goal is to maximize the accuracy of the predictions, we focus on the features (potential causes). Therefore, a learners f_L receives a $transpose(D^0_{V\times J})$ as input data. The outcome of interest Y^0_J can be easily added using the transposed level 0 data. In our real-world dataset, Y^0 encodes if the patient had cancer metastasis or not. The learners'output is one value per feature $v \in \{1, ..., V\}$. The second modification is how we create level 1 data. Traditionally, to avoid data leaking and overfitting, stacking learning models use cross-validation to make the predictions used on level 1 data. The same is not possible in our approach: subsets of the possible causes would violate assumptions, such as causal sufficiency. Instead, we use bootstrapping on the transposed level 0 data. By doing so, we can decrease biases and test the significance of coefficients when suitable.

All assumptions that the learners make, such as discrete treatments, also need to be satisfied. ParKCa requires the number of possible causes V to be sufficiently large to be able to fit the meta-learner. As an ensemble method, ParKCa requires sufficient diversity among the learners. We check the diversity with the averaged Q statistics[31] over all pairs of classifiers. Considering two learners, f_i and f_j, and the number of True Positives (TP), True Negatives (TN), False Positives (FP) and False Negatives (FN) from a confusion matrix between the two learners, the Q Statistic is $Q_{i,j} = \frac{TP\times TN - FP\times FN}{TP\times TN + FP\times FN}$, and $Q_{i,j} \in [-1, 1]$. The diversity increases when the learners commit errors in different objects, resulting in a negative or close to zero $Q_{i,j}$. For L learners, the average Q is defined as:

$$Q_{av} = \frac{2}{L(L-1)} \sum_{i=1}^{L-1} \sum_{j=i+1}^{L} Q_{i,j} \tag{1}$$

3.1. *Learners*

ParKCa starts by fitting the learners, also called level 0 models. As Figure 1 shows, the $transpose(D^0_{V\times J})$ is the input of the learners $f_l, \forall l \in [1, L]$. The level 0 models of ParKCA are causal discovery methods or models that estimate the treatment effect, and their output concatenated is the level 1 data $D^1_{V\times L}$. The learners employed must have all their assumptions satisfied for the validity of the results. The Deconfounder Algorithm (DA),[4] for example, requires a predictive check of its latent variables. Therefore, it is necessary to verify if the factor model passes the predictive check.

Defining $\phi_{(v,l)}$ as the outcome from learner f_l and potential cause v, the value $\phi_{(v,l)}$ can be continuous or discrete depending on the outputs of the learners. The outcomes $\phi_{(v,l)} \forall v \in V$ and $\forall l \in L$ aggregated form the level 1 data $D^1_{V\times L}$. Optionally, one might choose to set all non-causal variables to 0. In this case, if v is a non-causal variable according to method f_l, then $D^1_{v\times l} = 0$, else, $D^1_{v\times l} = \phi_{(v,l)}$.

Some methods, such as models that estimate treatment effects, might benefit from using bootstrapping to decrease biases and, when suitable, perform a statistical test for $H_0 : \phi_{(v,l)} = 0$ or $H_1 : \phi_{(v,l)} \neq 0$. To use bootstrap, we suggest the following steps:

(1) Take B samples of size $J' = \lfloor 0.9 * J \rfloor$ from $transpose(D^0_{J\times V})$ and fit the learner f_l in each sample $D^0_{J'\times V}$ saving the estimated outputs $\phi_{(v,l),b}$; then, set $D^1[v,l] = \frac{1}{B}\sum_{b=1}^{B} \phi_{(v,l),b}$ (Strong Law of Large Numbers)

(2) *(Optional)* Apply a two-tailed test to check the hypothesis test with the sample $\{\phi_{(v,l),1}, ..., \phi_{(v,l),B}\}$. If p-value ≤ 0.05, reject H_0 and set $D^1[v,l] = \frac{1}{B}\sum_{b=1}^{B} \phi_{(v,l),b}$, else 0.

According to the Strong Law of Large Numbers, let $\{\phi_{(v,l),1}, ..., \phi_{(v,l),B}\}$ be independent identically distributed random variables with $E|\phi_{(v,l),b}| < \infty$, then the average of the samples converges to the true mean when B is sufficient large.

ParKCa assumes that the number of possible causes V is sufficiently large to fit a meta-learner. Therefore, the learners must be robust to large datasets. A few examples are the RFCI[13] and fGES[14] work well in applications where there are no unobserved confounders; the PC-algorithm fast[11] is robust to unobserved confounders, but its performance in large datasets is poor; the DA[32] and CEVAE[3] are suitable to applications with unobserved confounders.

To validate ParKCa, in the experiments we worked with three methods: the Deconfounder Algorithm (DA)[a],[4] BART,[16] and CEVAE.[3] The main idea behind DA is to learn latent features as a substitute for unobserved confounders. Then, use the data augmented with the latent

[a]There is an ongoing discussion about DA, with some recent criticism[33,34] and extra clarification[35] presented. ParKCa assumes that the original work[4] is correct. However, in case the reader is uncomfortable with the use of this method, we recommend to replace it with other suitable learner.

variables to make the causal inference through an outcome model. The use of proxies to replace true confounders in causal inference analysis[36] will also be employed on the BART model. BART makes data interventions to estimate the conditional average treatment effect (CATE). For each possible cause $v \in \{1, ..., V\}$:

$$CATE_v = E[Y|X = x, do(X_v = a), Z] - E[Y|X = x, do(X_v = 0), Z] \quad (2)$$

where $X_v = 0$ represents the intervention component on the observed data $X_v = a$ and Z the estimated proxies. Finally, CEVAE infers causal effects from observational data and is robust to unobserved confounders. Based on Variational Autoencoders (VAE), it tries to simultaneously discover the hidden confounders and infer how they affect the treatment and output.

The learners of our experiments were selected to satisfy the requirements and assumptions of the application. We would like to emphasize that ParKCa is not limited to these learners.

3.2. *Meta-learner*

The level 1 dataset $D^1_{V\times L}$ records the outputs of L learners for V possible causes. If multiple level 0 datasets are being used, then the format is $D^1_{V\times L*d}$, where d is the number of level 0 datasets (see Figure 1). The prior knowledge about known causes is added as a new attribute Y^1, where $Y^1_v = 1$ if v is a known cause, and 0 otherwise. Note that, unless all the possible causes are known, some true causes will be labeled as 0. ParKCa uses binary classification models as meta-learners. The level 1 data contains only positive or unlabeled examples, so we tested PU-learning classification models and compared their results with traditional classification models (Logistic Regression (LR), Random Forest (RF), and Neural Network (NN)).

The PU-learning model Adapter-PU[22] uses a traditional probabilistic classifier $c^o(X)$ such that $c^o(X) = p(Y = 1|X)$ is as close as possible. This method assumes that the labeled positive examples are randomly selected among all true positive examples. Unbiased PU (UPU)[23] is another PU-learning model adopted. UPU is a convex classification method that aims to cancel the bias from the unlabeled data being a mix of positive and negative examples by using a loss function for positive examples and another loss function for unlabeled examples.

The traditional binary classification models consider all unlabeled examples as negatives examples or non-causal variables, which can add bias and noise to the predictions. A majority vote ensemble from the methods described above is also used. Finally, a random model is also compared. The random model assigns the labels 1 and 0 according to the proportion of 1's and 0's in the training data.

4. Experiments

We performed experiments to validate our method on two Genome-wide association studies (GWAS) datasets, a real-world dataset, and a simulated dataset.

Real-world dataset: We use The Cancer Genome Atlas Program (TCGA) dataset, which has available the gene expression (RNA-seq) of patients with cancer. The data pre-processing is described in the Supplemental Material A.1, and the level 0 dataset from this application has 7066 genes and 2854 patients, of which 1039 (36%) have metastases. From the 7066 genes,

681 (9%) are known driver genes.[30] These known driver genes are our positive examples in the meta-learner models. For this application, we also worked with multiple datasets considering their clinical information, such as gender and cancer type.

Simulated datasets: We simulated GWAS data following the scheme described by Wang and Blei[4] and Song et al.,[37] illustrated with more details in the Supplemental Material A.2. Single nucleotide polymorphisms (SNPs), the most common type of genetic variation among people, is the datatype adopted. We simulated 10 independent datasets, with 5000 individuals and 10000 SNPs, and confounders. 10% of these SNPs were set to be causal of a binary trait.

To validate our method, we first evaluate the learners adopted. Then, we check if ParKCa indeed contributes to detecting more causes. We also verify for the simulated dataset if ParKCa can be used to make better estimates of the treatment effect. Finally, as an extra analysis, we compare our results in the real-world dataset with the state-of-art methods in driver gene discovery. We say 'extra' because these methods are not standard causal discovery methods but aim to solve the same problem.

Evaluation of the learners: We adopted DA with the probabilistic PCA[38] as a factor model and Logistic Regression with the elastic net as an outcome model. In our experiments, increasing the number of the latent variables did not improve the results (See Supplemental Material B.1); thus, we adopted $k = 15$. The DA models passed the predictive check with $k = 15$ (average p-value= 0.7181 at real-world dataset and 0.5381 on the simulated dataset).

The ROC curves of the learners on level 0 data were also evaluated (Supplemental Material B.2). To construct the ROC curves, we split the transposed dataset $D^0_{V\times J}$ into training (67%) and testing set (33%). Thus, each set has all the possible causes and a subset of the samples. Using the models fitted on the training set, we predicted the outputs for the testing set, which is the outcome of interest at level 0. We used 12 learners on the real-world dataset: BART + 3 datasets (all patients, female and male patients), and DA + 9 datasets (all patients, female and male patients, and six groups of patients with same cancer type). All level 0 models had an excellent performance, except for 4 DA models constructed using datasets based on cancer type. Therefore, we removed the outputs obtained through datasets based on cancer types ESCA, LIHC, PAAD, and SARC from the level 1 dataset. In the simulation study, we repeated the experiment ten times, once for each simulated dataset, and reported the rate of True Positives and False Positives using either CEVAE or DA for each repetition. All learners had a good performance in predicting the outcome of the level 0 datasets. The CEVAE's convergence plots indicate that the model converges after 40 epochs on average. Some randomly selected convergence plots are shown in the Supplemental Material B.3.

The last step of the learners' evaluation is checking the diversity among the final level 0 models measured by Q_{av}. Negative values of Q_{av} indicate diversity because the learners are committing errors on different objects. The learners used on the real-world dataset about driver gene discovery has diversity equal to -0.013, and the average diversity of the learners used with the simulated dataset was -0.051, from which we conclude there is diversity among the learners adopted in our experiments.

Learners *versus* Meta-Learners: This comparison is the core of our validation process. Here, we investigate if adding an extra layer in ParCKa, the meta-learner, contributes to discovering more causes. Therefore, we compare the learners and the meta-learners capacity of identifying causal variables. The learners are used individually to predict if a variable is causal according to their metrics and definitions. These predictions are compared with the meta-learners' predictions $\hat{Y}^1$.

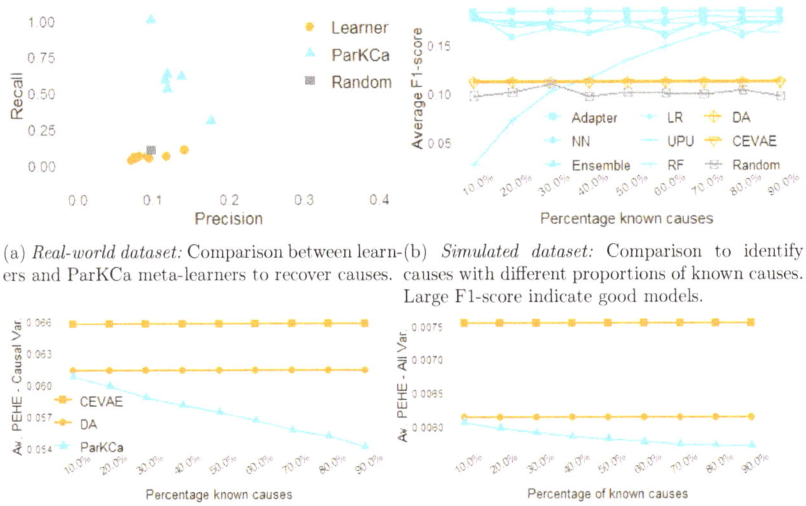

(a) *Real-world dataset:* Comparison between learners and ParKCa meta-learners to recover causes.

(b) *Simulated dataset:* Comparison to identify causes with different proportions of known causes. Large F1-score indicate good models.

(c) *Simulated dataset:* PEHE on the level 1 test set. Small PEHE indicates good models.

Fig. 2. Causal discovery task evaluation. The learners consider only the original data; ParKCA meta-learners use the learners' outputs and partially known causes to identify more causes.

The DA considers variables significantly different from 0 in the outcome model as a causal variable. BART and CEVAE do not have a similar metric, and they only provide treatment effect estimates. One option is using the bootstrap method explained earlier to fit a confidence interval and check if the treatment effect estimated is significantly different from zero. In our experiments, however, we obtained better performance by assigning the 10% largest estimated treatment effects as causal variables and setting the other variable as non-causal. The learners are compared individually against 6 meta-learners (UPU, Adapted-PU, Logistic Regression - LR, Random Forest - RF, Neural Network - NN, Ensemble - E), and a random model.

To perform a fair comparison between learners and meta-learners, we split the level 1 data into training and testing sets. We calculated the precision and recall using only the predicted values for the testing set, and the list of known causes as ground truth. The results for the real-world data are shown in Figure 2a. While meta-learners and learners models have similar precision, meta-learners tend to have much better recall then learners. Overall,

ParKCa has fewer causal variables undetected (False Negatives) than the learners. Figure 2b shows the average F1-score on the simulated datasets versus the proportion of known causes. The proportion of known causes represents how much of the true causes ParCKa has access: if ParKCa knows 40% of the causes, during the training phase, we randomly label 40% of the true causes as 1 and the other 60% as 0, to replicate what we usually encounter in the real-world.We observe that even when ParKCa can access to only a small proportion of true causes, it performs better than the existing baselines, which are independent of the percentage of known causes. The meta-learners, except for the RF, have a higher average F1-score even when only 10% of the causes are known. These results validate our claim that the stacking approach used by ParKCa can identify more causes then existing methods when some causes are known. All meta-learners, except the RF, seem to be independent of the proportion of known causes on the testing set, which points out another ParKCa's quality: even a small portion of known causes can produce better results than traditional methods. These results sustain our claim that the stacking approach (ParKCa) is capable of identifying more causes than isolated causal methods (learners).

Treatment Effect Estimates: We investigate a secondary result of the learners used on the simulated datasets, the estimation of treatment effect. We compare the Precision in Estimation of Heterogeneous Effect (PEHE)[3,16] of our approach against that of the learners used in dependency from the percentage of known causes. $PEHE = \frac{1}{N}\sum_{i=1}^{N}((y_{i1} - y_{i0}) - (\hat{y}_{i1} - \hat{y}_{i0}))$, where y_{i1} and y_{i0} are the true treatment effects, and $\hat{y}_{i1}$ and $\hat{y}_{i0}$ are the estimated treatment effects. This scenario requires known treatment effect estimates, which are hardly ever available in real-world applications. However, this experiment can easily be performed on a simulated dataset, and its results collaborate with our claim that ParKCa finds better results than the learners. We adopted a simple linear regression model as a meta-learner, and we split the level 1 data into training and testing sets. We compared the PEHE of the meta-learner with CEVAE and DA on the test set. Figure 2c shows the PEHE for the causal variables in the left and for all variables on the right. The average PEHE for the ParKCa meta-learner is similar to that of DA when only 10% of the causes are known; however, it decreases when more causes are known in both plots. These results point out an alternative use of ParKCa for treatment effect estimation.

Comparison between ParKCa and other baselines: We compared our results from the real-world dataset with reported results from eight driver gene discovery methods analyzed using the Cancer Gene Census (CGC).[9] The baselines are MutsigCV, ActiveDriver (AD), MuSiC, OncodriveCLUST (ODC), OncodriveFM (ODFM), OncodriveFML (ODFML), TUSON, and 20/20+. Their approaches vary from analysis of somatic point mutations, mutation significance, functional impact and clusters of somatic mutations, and Random Forest of previous driver genes methods. These baselines are not considered causal discovery methods, which is why we did not use them as learners, but are strong methods that try to solve the same problem on the real-world dataset. We remind the reader that ParKCA takes partial knowledge of causal genes and genomic data as input, while the compared methods only have genomic data as input, but use multiple and more sophisticated types of genomic data. We used the results reported by Tokheim et al.[9]

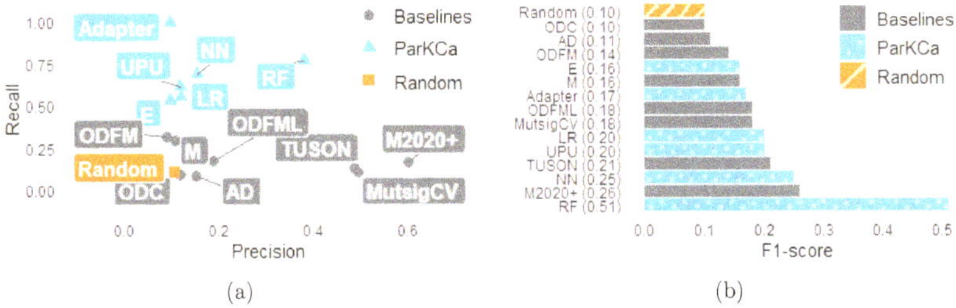

Fig. 3. *(Real-world dataset)* Comparison between driver gene discovery baselines and ParCKa. The goal is to predict driver genes (causal variables) correctly. Large Recall, Precision, and F1-score indicate good models. The score reported is from the full real-world dataset.

It is important to point out that the choice of what is considered a good driver gene discovery model is also an open question. Large recall and small precision indicate models that can recover many known driver genes at the cost of a high rate of False Positives (FP). One can interpret this as a bad model because the true driver genes are lost in the middle of the FP, while others might think that this is an indication of a larger number of unknown driver genes yet to be discovered and explored. On the other hand, high precision and low recall indicate models good at identifying certain driver genes; however, they fail to identify a broader range of them, reflected by a large number of False Negatives (FN). The F1-score summarizes these measures by giving the same importance to both of them. Figure 3a indicates that ParKCA meta-learners have a larger recall (0.69 on average) and smaller precision (0.16 on average). On the other hand, the baselines have lower recall (0.17 on average) and larger precision (0.28 on average). Figure 3b shows that ParKCA with RF has the largest F1-score, which is almost the double of the largest F1-score from the baseline methods. Overall, ParKCa has competitive results when compared to existing driver gene discovery methods.

5. Discussion and Conclusion

Our proposed method ParKCa demonstrated excellent results in the experiments. For small percentages of known causes, Adapter-PU was the meta-learner with the best performance. Furthermore, there was almost no difference between PU models and traditional classification models when the percentage of known causes was above 70%. If the unlabeled examples are mostly negative, the contribution that PU methods can make is limited, and PU classification reduces to traditional binary classification.

While our simulations show the efficacy of our method, we highlight that in practice, the results crucially depend on the list of known causes. If this list is comprehensive and includes causes with diverse behaviors, ParKCa will likely succeed in its task. On the other hand, if the list is biased towards certain characteristics, our method might only identify causes with behavior similar to the known examples. Furthermore, ParKCa performance also relies on the assumptions of the level 0 learner's being met.

In conclusion, we believe our proposed method ParKCa makes important contributions to the causal discovery and causal inference. ParKCa exploits partial knowledge of causes, is flexible, robust, easy to use, and demonstrated promising results on a real-life dataset and in simulations. After narrowing down from thousands or millions of potential causes, the causes detected by ParKCa can be better explored and confirmed thought rigorous laboratory studies. Improvements in causal discovery and causal inference methods that work on CB applications can bring many other benefits beyond the development of sophisticated causal discovery techniques. A successful approach has the potential to significantly improve therapy recommendations, working towards the goal of precision medicine to provide the "right drug at the right dose to the right patient".[1]

Supplemental Material:

`sites.google.com/view/raquelaoki/publications/parkca-supplemental-material`
Code: `https://github.com/raquelaoki/ParKCa`

References

1. F. S. Collins and H. Varmus, A new initiative on precision medicine, *New England journal of medicine* **372**, 793 (2015).
2. *Learning representations for counterfactual inference* 2016.
3. *Causal effect inference with deep latent-variable models* 2017.
4. Y. Wang and D. M. Blei, The blessings of multiple causes, *Journal of the American Statistical Association* (2019).
5. M. R. Stratton, P. J. Campbell and P. A. Futreal, The cancer genome, *Nature* **458**, p. 719 (2009).
6. F. Vandin, E. Upfal and B. J. Raphael, De novo discovery of mutated driver pathways in cancer, *Genome research* **22**, 375 (2012).
7. N. D. Dees, Q. Zhang, C. Kandoth, M. C. Wendl, W. Schierding, D. C. Koboldt, T. B. Mooney, M. B. Callaway, D. Dooling, E. R. Mardis *et al.*, Music: identifying mutational significance in cancer genomes, *Genome research* **22**, 1589 (2012).
8. M. P. Schroeder, C. Rubio-Perez, D. Tamborero, A. Gonzalez-Perez and N. Lopez-Bigas, Oncodriverole classifies cancer driver genes in loss of function and activating mode of action, *Bioinformatics* **30**, i549 (2014).
9. C. J. Tokheim, N. Papadopoulos, K. W. Kinzler, B. Vogelstein and R. Karchin, Evaluating the evaluation of cancer driver genes, *Proceedings of the National Academy of Sciences* (2016).
10. P. Spirtes and C. Glymour, An algorithm for fast recovery of sparse causal graphs, *Social science computer review* **9**, 62 (1991).
11. T. Le, T. Hoang, J. Li, L. Liu, H. Liu and S. Hu, A fast pc algorithm for high dimensional causal discovery with multi-core pcs, *IEEE/ACM transactions on computational biology and bioinformatics* (2016).
12. P. Spirtes, C. N. Glymour, R. Scheines, D. Heckerman, C. Meek, G. Cooper and T. Richardson, *Causation, prediction, and search* (MIT press, 2000).
13. D. Colombo, M. H. Maathuis, M. Kalisch and T. S. Richardson, Learning high-dimensional directed acyclic graphs with latent and selection variables, *The Annals of Statistics* , 294 (2012).
14. J. Ramsey, M. Glymour, R. Sanchez-Romero and C. Glymour, A million variables and more: the fast greedy equivalence search algorithm for learning high-dimensional graphical causal models, with an application to functional magnetic resonance images, *International journal of data*

science and analytics **3**, 121 (2017).
15. *Adapting neural networks for the estimation of treatment effects* 2019.
16. J. L. Hill, Bayesian nonparametric modeling for causal inference, *Journal of Computational and Graphical Statistics* **20**, 217 (2011).
17. *Learning equivalence classes of acyclic models with latent and selection variables from multiple datasets with overlapping variables* 2011.
18. D. H. Wolpert, Stacked generalization, *Neural networks* **5**, 241 (1992).
19. M. S. Schuler and S. Rose, Targeted maximum likelihood estimation for causal inference in observational studies, *American journal of epidemiology* **185**, 65 (2017).
20. S. R. Künzel, J. S. Sekhon, P. J. Bickel and B. Yu, Metalearners for estimating heterogeneous treatment effects using machine learning, *Proceedings of the national academy of sciences* (2019).
21. IEEE, *Building text classifiers using positive and unlabeled examples* 2003.
22. *Learning classifiers from only positive and unlabeled data* 2008.
23. *Convex formulation for learning from positive and unlabeled data* 2015.
24. A. Gonzalez-Perez and N. Lopez-Bigas, Functional impact bias reveals cancer drivers, *Nucleic acids research* **40**, e169 (2012).
25. J. Reimand and G. D. Bader, Systematic analysis of somatic mutations in phosphorylation signaling predicts novel cancer drivers, *Molecular systems biology* **9** (2013).
26. T. Davoli, A. W. Xu, K. E. Mengwasser, L. M. Sack, J. C. Yoon, P. J. Park and S. J. Elledge, Cumulative haploinsufficiency and triplosensitivity drive aneuploidy patterns and shape the cancer genome, *Cell* **155**, 948 (2013).
27. D. Tamborero, A. Gonzalez-Perez and N. Lopez-Bigas, Oncodriveclust: exploiting the positional clustering of somatic mutations to identify cancer genes, *Bioinformatics* **29**, 2238 (2013).
28. M. S. Lawrence, P. Stojanov, C. H. Mermel, J. T. Robinson, L. A. Garraway, T. R. Golub, M. Meyerson, S. B. Gabriel, E. S. Lander and G. Getz, Discovery and saturation analysis of cancer genes across 21 tumour types, *Nature* **505**, 495 (2014).
29. L. Mularoni, R. Sabarinathan, J. Deu-Pons, A. Gonzalez-Perez and N. López-Bigas, Oncodrivefml: a general framework to identify coding and non-coding regions with cancer driver mutations, *Genome biology* **17**, p. 128 (2016).
30. P. A. Futreal, L. Coin, M. Marshall, T. Down, T. Hubbard, R. Wooster, N. Rahman and M. R. Stratton, A census of human cancer genes, *Nature reviews cancer* **4**, 177 (2004).
31. L. I. Kuncheva and C. J. Whitaker, Measures of diversity in classifier ensembles and their relationship with the ensemble accuracy, *Machine learning* **51**, 181 (2003).
32. S. Wager and S. Athey, Estimation and inference of heterogeneous treatment effects using random forests, *Journal of the American Statistical Association* **113**, 1228 (2018).
33. E. L. Ogburn, I. Shpitser and E. J. T. Tchetgen, Comment on "blessings of multiple causes", *Journal of the American Statistical Association* **114**, 1611 (2019).
34. A. D'Amour, Comment: Reflections on the deconfounder, *Journal of the American Statistical Association* **114**, 1597 (2019).
35. Y. Wang and D. M. Blei, The blessings of multiple causes: A reply to ogburn et al.(2019), *arXiv preprint arXiv:1910.07320* (2019).
36. *Causal inference with noisy and missing covariates via matrix factorization* 2018.
37. M. Song, W. Hao and J. D. Storey, Testing for genetic associations in arbitrarily structured populations, *Nature genetics* **47**, p. 550 (2015).
38. M. E. Tipping and C. M. Bishop, Probabilistic principal component analysis, *Journal of the Royal Statistical Society: Series B (Statistical Methodology)* **61**, 611 (1999).

Optimization of Genomic Classifiers for Clinical Deployment: Evaluation of Bayesian Optimization to Select Predictive Models of Acute Infection and In-Hospital Mortality*

Michael B. Mayhew†, Elizabeth Tran, Kirindi Choi, Uros Midic, Roland Luethy, Nandita Damaraju and Ljubomir Buturovic

Inflammatix, Inc.
Burlingame, California 94010, USA
†*E-mail: mmayhew@inflammatix.com*
www.inflammatix.com

Acute infection, if not rapidly and accurately detected, can lead to sepsis, organ failure and even death. Current detection of acute infection as well as assessment of a patient's severity of illness are imperfect. Characterization of a patient's immune response by quantifying expression levels of specific genes from blood represents a potentially more timely and precise means of accomplishing both tasks. Machine learning methods provide a platform to leverage this *host response* for development of deployment-ready classification models. Prioritization of promising classifiers is dependent, in part, on hyperparameter optimization for which a number of approaches including grid search, random sampling and Bayesian optimization have been shown to be effective. We compare HO approaches for the development of diagnostic classifiers of acute infection and in-hospital mortality from gene expression of 29 diagnostic markers. We take a deployment-centered approach to our comprehensive analysis, accounting for heterogeneity in our multi-study patient cohort with our choices of dataset partitioning and hyperparameter optimization objective as well as assessing selected classifiers in external (as well as internal) validation. We find that classifiers selected by Bayesian optimization for in-hospital mortality can outperform those selected by grid search or random sampling. However, in contrast to previous research: 1) Bayesian optimization is not more efficient in selecting classifiers in all instances compared to grid search or random sampling-based methods and 2) we note marginal gains in classifier performance in only specific circumstances when using a common variant of Bayesian optimization (i.e. automatic relevance determination). Our analysis highlights the need for further practical, deployment-centered benchmarking of HO approaches in the healthcare context.

Keywords: hyperparameter optimization; Bayesian optimization; acute infection; sepsis; disease severity; mortality; classification; molecular diagnostics; genomics.

1. Introduction

Patient lives depend on the swiftness and accuracy of 1) assessment of the severity of their illness and 2) detection of acute infection (when present). The COVID-19 pandemic has put this fact into stark relief. Currently, clinicians determine severity of illness by computing scores

*Supplementary material can be found at `https://arxiv.org/abs/2003.12310`

(e.g. SOFA[1]) based on patient physiological features associated with the risk of adverse events (e.g. in-hospital mortality, organ failure). Similarly, detection of acute infection generally involves evaluation of symptoms (e.g. cough, runny nose, fever) as well as laboratory tests for the presence of specific pathogens. However, these methods provide superficial and imprecise measures of patient illness. Recent work has highlighted the potential of using gene expression measurements from patient blood to detect the presence and type of infection to which the patient is responding[2–5] as well as the patient's severity of illness.[6]

Coupled with these host response signatures, advances in machine learning (ML) provide a platform for the development of robust, diagnostic classifiers of acute infection status (e.g. bacterial or viral) and in-hospital mortality from gene expression. An important step in this development is optimization of the classifier's hyperparameters (e.g. penalty coefficient in a LASSO logistic regression, learning rates for gradient descent). Hyperparameter optimization begins with specification of a search space and proceeds by generating a user-specified number of hyperparameter configurations, training the classifier models given by each configuration, and evaluating the performance of the trained classifier in *internal validation*. Internal validation performance is typically assessed either on a separate validation/tuning dataset or by cross-validation. Configurations are then ranked by this performance, with the top configuration selected and retained for *external validation* (application to a held-out dataset).

Multiple HO approaches have been proposed. For classifiers with relatively small hyperparameter spaces (e.g. support vector machines), optimizing over a pre-defined grid of hyperparameter values (grid search; GS) has proven effective. More recent work has shown that optimization by randomly sampling (RS) hyperparameter configurations can lead to better coverage of high-dimensional hyperparameter spaces and potentially better classifier performance.[7] Bayesian optimization (BO) is a global optimization procedure that has also proven effective for hyperparameter optimization in classical[8–12] and biomedical[13–16] ML applications. In BO, one uses a model (commonly a Gaussian process (GP)[17]) to approximate the objective function one wants to optimize; for hyperparameter optimization, the objective function maps from hyperparameter configurations to the internal validation performance of their corresponding classifiers. In contrast to GS/RS, BO proceeds by sequentially evaluating configurations with each newly visited configuration used to update the model of the objective function.

In this work, we compare GS/RS and BO methods for hyperparameter optimization of gene expression-based diagnostic classifiers for two clinical tasks: 1) detection of acute infection and 2) prediction of mortality within 30 days of hospitalization. We optimize and train three different types of classifiers using gene expression features from 29 diagnostic markers in a multi-study cohort of 3413 patient samples for acute infection detection (3288 for 30-day mortality prediction). Patient samples were assayed on a variety of technical platforms and collected from a range of geographical regions, healthcare settings, and disease contexts. Our extensive analysis evaluates the BO approach, in particular, under a range of computational budgets and optimization settings. Crucially, beyond assessing and comparing the performance of top classifiers in internal validation, we further evaluate top models selected by all HO approaches in a multi-cohort external validation set comprising nearly 300 patients profiled by a targeted diagnostic instrument (NanoString). Our analysis provides important insights for diagnostic classifier development using genomic data, and, more generally, about the implementation and practical usage of HO methods in healthcare.

2. Related Work

Previous studies comparing HO approaches in the ML community have demonstrated that BO can select promising classifiers more efficiently (with fewer evaluations of hyperparameter configurations) than GS/RS methods.[8–12,15,16,18] However, these studies have focused on internal validation performance and on benchmark datasets whose composition and handling (i.e. partitioning into training-validation-test splits) doesn't necessarily reflect characteristics of healthcare settings (i.e. smaller, structured, and more heterogeneous datasets; high propensity for models to be applied to out-of-distribution samples at test time[19]).

Bayesian optimization has also found recent success in genomics and biomedical applications.[20–22] Ghassemi et al.[13] compare multiple HO approaches, including BO, for tuning parameters of the multi-scale entropy of heart rate time series to aid mortality prediction among sepsis patients. Colopy et al.[14] analyzed RS and BO methods for optimization of patient-specific GP regression models used in vital-sign forecasting. A study by Nishio et al.[15] evaluated both RBF SVM and XGBoost classifiers tuned by either RS or BO for detection of lung cancer from nodule CT scans. Borgli et al.[16] evaluated BO for tuning and transfer learning of pre-trained convolutional neural networks to detect gastrointestinal conditions from images. Again, however, these studies only reported either internal validation performance or performance on a test set partitioned from a full, relatively small and homogeneous (e.g. collected from a single hospital) dataset, making conclusions difficult to draw about the generalizability of selected models in other segments of the deployment population. Moreover, these studies focused on: 1) no more than two classifier types, 2) a narrow range of settings for BO, and 3) physiological or image data. To our knowledge, no studies have evaluated the external validation performance of selected models, an important pre-requisite for eventual model deployment. In addition, no comparison of HO approaches has yet been attempted for development of diagnostic classifiers using genomic data.

3. Methods

3.1. *Cohort & Feature Description*

To build our datasets, we combined gene expression data from public sources and in-house clinical studies designed for research in diagnosing acute infections and sepsis. We collected the publicly available studies from the NCBI GEO and EMBL-EBI ArrayExpress databases using a systematic search.[2] The public studies were profiled using a variety of different technical platforms (e.g. mostly microarrays). Samples from the in-house clinical studies were profiled on the NanoString nCounter platform using a custom codeset for 29 diagnostic genes of interest. All included studies consisted of samples from our target population: both adult and pediatric patients from diverse geographical regions and clinical settings. Each included study had measurements taken from patient blood for all 29 markers. To account for heterogeneity across studies, we performed co-normalization (see[5] and the Supplement).

The features we used in our analyses were based on the expression values of 29 genes previously found to accurately discriminate three different aspects of acute infection: 1) viral vs. bacterial infection (7 genes),[3] 2) infection vs. non-infectious inflammation (11 genes),[2] and 3) high vs. low risk of 30-day mortality (11 genes).[6] Building on our previous work,[5] we computed both the geometric means and arithmetic means of these six groups of genes, producing 12 features. We optimized and trained our classifiers on the combination of these 12 features and the expression values of all 29 genes (41 features in total). Labels for one of three classes of the acute infection detection or BVN task (**B**acterial infection, **V**iral infection, or **N**on-infectious inflammation) were determined differently for each of the training and validation studies depending on available data. For training set studies, we used the labels provided by each study, deferring to each study's criteria for adjudication which may have involved multi-clinician adjudication with or without positive pathogen identification or positive pathogen identification alone. When BVN adjudications were not directly provided by the study, we assigned class labels based on available pathogen test results from the study metadata/manuscripts. For validation data, one study was adjudicated by a panel of clinicians using all available clinical data (including pathogen test results) while all other validation studies were labeled by us using only pathogen test results. Non-infected determinations did not include healthy controls. Binary indicator labels of whether a patient died within 30 days of hospitalization were derived from study metadata (when available) and the associated study's manuscripts.

For both tasks, we separated studies into a training set and an external validation set. For the BVN task, the training set consisted of 43 studies (profiled outside Inflammatix) and 3413 patients (1087 with bacterial infection, 1244 with viral infection, and 1082 non-infected). The BVN external validation set consisted of six studies (profiled by Inflammatix) and 293 patients (153 with bacterial infection, 106 with viral infection, and 34 non-infected). For the mortality task, the training set consisted of 33 studies (profiled outside Inflammatix) and 3288 patients (175 30-day mortality events) while the mortality external validation set comprised four studies (profiled by Inflammatix) and 348 patients (80 30-day mortality events). A description of the publicly available studies in our training set appears in Supplementary Table 1.

3.2. *Grouped cross-validation*

Previous analyses by our group[5] suggested that alternative cross-validation strategies were preferable over conventional k-fold cross-validation (CV) for identifying classifiers able to generalize across heterogeneous patient populations. We use 5-fold grouped CV (full studies are allocated to one and only one of five folds) to rank and select hyperparameter configurations from GS/RS methods and as an objective function in BO.

3.3. *Classifier types and performance assessment*

We evaluated three types of classification models: 1) support vector machines with a radial basis function (RBF) kernel, 2) XGBoost (XGB[23]) and 3) multi-layer perceptrons (MLP). MLP models were trained with the Adam optimizer[24] with mini-batch size fixed at 128.

For the BVN task, we ranked and selected models based on multi-class AUC (mAUC).[25] For the mortality task, we selected models by binary AUC but report both AUC and average precision to account for class imbalance. To determine performance of models in grouped 5-

fold CV, we pooled the model's predicted probabilities for each fold and computed the relevant metric from the pooled probabilities. The top-performing hyperparameter configuration was then trained on the full training set and applied to the external validation set. We computed external validation performance for these top models using their predicted probabilities for the validation samples. We computed 95% bootstrap confidence intervals for differences in classification performance by sampling predicted probabilities with replacement 5000 times (using the same set of bootstrap sample IDs for both sets of predicted probabilities in the comparison), computing the relevant performance metric on each bootstrap sample, computing the difference between performance metrics for each bootstrap sample in a given comparison, and reporting the 2.5th and 97.5th quantiles of the 5000 differences.

3.4. *Hyperparameter optimization details*

For RBF SVM, we conduct a grid search over configurations of the cost, C, and bandwidth hyperparameters, γ. C values ranged from 1e-03 to 2.15 and γ values ranged from 1.12e-04 to 10. We generated RS samples for XGBoost and MLP uniformly and independently of one another from pre-specified ranges or from grids (Suppl. Tables 2 and 3).

For BO, the objective function maps from hyperparameter configurations to 5-fold grouped CV performance of the corresponding classifiers. The two main components of BO are: 1) a model that approximates the objective function, and 2) an acquisition function to propose the next configuration to visit. We use a GP regression model with Gaussian noise to approximate the objective function. To initialize construction of the objective function, we uniformly and independently sample configurations (either 5 or 25) from the hyperparameter space.

We investigate both the expected improvement and upper confidence bound acquisition functions. We use both standard and automatic relevance determination (ARD) forms of the Matern5/2 covariance function in BO's GP model of the objective (further details in Supplement). We also perform BO in the hyperparameters' native scales (*original* space) or in which continuous and discrete hyperparameter dimensions are searched in the continuous range 0 to 1 and transformed back to their native scales prior to their evaluation (*transformed*).

4. Results

We compared BO and GS/RS approaches for hyperparameter optimization of three types of classifiers for two clinical tasks. For the BVN task, we sought classifiers that could achieve high performance in predicting whether a patient had a bacterial or viral infection or was showing a non-infectious inflammatory response. For the mortality task, we sought high-performing classifiers of mortality events within 30 days of hospital admission. Though we considered BO at two initialization budgets (5 and 25 configurations), we did not see substantial differences in performance between classifiers with 5 and 25 initial configurations (Suppl. Table 4, Suppl. Figs. 3-6). We focus on BO results with 25 initial configurations and the expected improvement acquisition function for the remainder of this work (results for all runs in Supplement).

General comparison of classifier performance across tasks and HO approaches Across both tasks and HO approaches, we note distinct performance characteristics of the selected classifiers of each type. While RBF SVM classifiers performed similarly to the other two classifier types on the BVN task, they were the worst performers on the mortality task. XGB classifiers selected by either RS or BO demonstrated competitive performance in both tasks and were remarkably consistent in their performance regardless of the number of hyperparameter configurations evaluated for HO. MLPs achieved the highest internal and external validation performance for both acute infection detection and mortality prediction (Table 1), suggesting potential benefits of learning latent features (hidden layers) for these tasks. We also find that, despite the considerable class imbalance in the mortality task, all classifier types selected by AUC still demonstrated average precision considerably higher than the respective baselines for internal ($\frac{175}{3288} \approx 0.053$) and external ($\frac{80}{348} \approx 0.230$) validation.

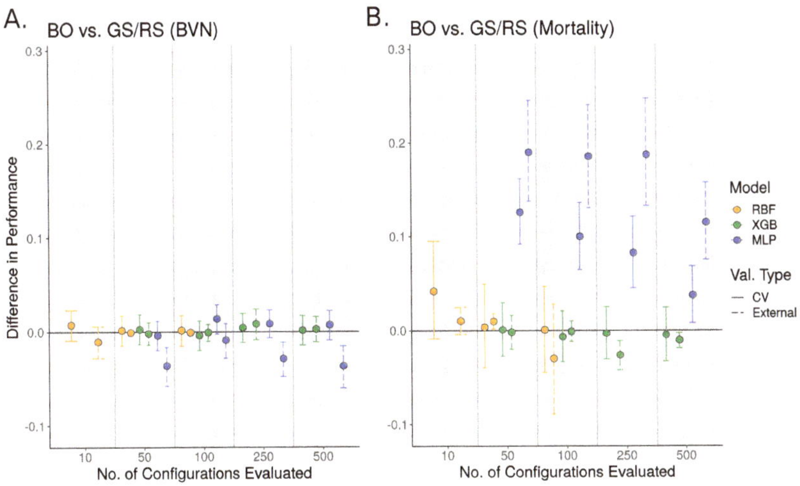

Fig. 1: **Differences in classification performance of models selected by either BO or GS/RS using BO evaluation budgets.** Performance differences greater than 0 on the BVN (A; mAUC) and mortality (B; AUC) tasks indicate better performance for the BO-selected classifier. Classifiers were selected with the indicated number of hyperparameter configurations evaluated. Automatic relevance determination was not enabled for BO. Points represent observed differences while error bars represent 95% bootstrap confidence intervals.

Evaluation of BO- and GS/RS-selected classifiers at evaluation budgets typical of BO. Previous studies have shown that BO can select promising classifiers more efficiently than GS/RS methods. Surprisingly, we find that at smaller numbers of configurations evaluated (more typical of BO), classifiers selected by GS/RS showed similar or better performance in both internal and external validation (Table 1 and Figs. 1) when compared with corresponding BO-selected classifiers. We observed similar trends when using the upper confidence

Table 1: Grouped 5-fold CV and external validation (Val.) performance of selected classifiers for the BVN and mortality tasks. BO results used the EI acquisition function and 25 initialization points. The ARD column indicates whether automatic relevance determination was enabled (Y/N) in BO's GP model of the objective function. **Bold** numbers indicate the best performance for a column. BVN column shows performance in mAUC; mortality column shows AUC performance with average precision in parentheses. *Grid specified only 4757 configurations.

Model	HO Type	No. of Evals.	ARD	BVN CV	BVN Val.	Mortality CV	Mortality Val.
RBF	GS	10	-	0.808	0.862	0.758 (0.182)	0.736 (0.375)
	GS	50	-	0.814	0.853	0.797 (0.169)	0.739 (0.372)
	GS	100	-	0.814	0.853	0.800 (0.192)	0.782 (0.533)
	GS	250	-	0.814	0.853	0.801 (0.191)	0.749 (0.386)
	GS	500	-	0.815	0.853	0.801 (0.191)	0.749 (0.386)
	GS	1000	-	0.815	0.853	0.839 (0.225)	0.708 (0.444)
	GS	5000*	-	0.815	0.853	0.839 (0.225)	0.708 (0.444)
	BO	10	Y	0.811	0.788	0.800 (0.190)	0.747 (0.383)
	BO	10	N	0.815	0.851	0.800 (0.187)	0.746 (0.381)
	BO	50	Y	0.816	0.852	0.801 (0.196)	0.752 (0.389)
	BO	50	N	0.816	0.852	0.801 (0.194)	0.749 (0.385)
	BO	100	Y	0.816	0.852	0.800 (0.197)	0.753 (0.392)
	BO	100	N	0.816	0.852	0.801 (0.196)	0.752 (0.389)
XGB	RS	50	-	0.809	0.830	0.880 (0.315)	0.819 (0.542)
	RS	100	-	0.813	0.827	0.885 (0.288)	0.819 (0.526)
	RS	250	-	0.812	0.826	0.885 (0.308)	0.829 (0.556)
	RS	500	-	0.810	0.829	0.885 (0.320)	0.826 (0.559)
	RS	1000	-	0.810	0.822	0.885 (0.311)	0.822 (0.552)
	RS	5000	-	0.813	0.830	0.888 (0.310)	0.823 (0.552)
	RS	25000	-	0.815	0.860	0.889 (0.303)	0.816 (0.532)
	BO	50	Y	0.818	0.865	0.887 (0.301)	0.814 (0.540)
	BO	50	N	0.812	0.828	0.881 (0.275)	0.817 (0.516)
	BO	100	Y	0.811	0.825	0.885 (0.314)	0.825 (0.559)
	BO	100	N	0.809	0.826	0.878 (0.288)	0.817 (0.521)
	BO	250	Y	0.818	0.865	0.886 (0.290)	0.826 (0.539)
	BO	250	N	0.816	0.834	0.882 (0.272)	0.802 (0.483)
	BO	500	Y	0.818	0.865	0.889 (0.346)	0.827 (0.591)
	BO	500	N	0.812	0.831	0.880 (0.313)	0.815 (0.538)
MLP	RS	50	-	0.818	0.860	0.763 (0.121)	0.631 (0.288)
	RS	100	-	0.814	0.863	0.785 (0.156)	0.640 (0.301)
	RS	250	-	0.824	0.861	0.807 (0.211)	0.625 (0.366)
	RS	500	-	0.819	0.859	0.853 (0.240)	0.691 (0.401)
	RS	1000	-	0.835	**0.872**	0.809 (0.158)	0.637 (0.333)
	RS	5000	-	0.837	0.835	0.826 (0.249)	0.796 (0.546)
	RS	25000	-	**0.840**	0.856	0.859 (0.267)	0.743 (0.428)
	BO	50	Y	0.816	0.820	0.888 (0.340)	0.823 (0.554)
	BO	50	N	0.814	0.824	0.888 (0.290)	0.820 (0.564)
	BO	100	Y	0.822	0.845	0.886 (0.296)	**0.847** (0.631)
	BO	100	N	0.828	0.854	0.884 (0.292)	0.825 (0.577)
	BO	250	Y	0.817	0.848	0.890 (0.312)	0.842 (0.614)
	BO	250	N	0.832	0.832	0.889 (0.335)	0.812 (0.566)
	BO	500	Y	0.837	0.855	**0.894** (0.304)	0.835 (0.593)
	BO	500	N	0.826	0.822	0.890 (0.330)	0.806 (0.561)

bound acquisition function (Suppl. Figs. 7 and 8, Suppl. Table 5) or the transformed hyperparameter space (Suppl. Figs. 11 and 12, Suppl. Table 6). However, we do note two instances in which BO-selected classifiers exceeded performance of GS/RS-selected classifiers: 1) XGBoost classifiers in external validation for the BVN task and 2) MLP classifiers for the mortality task. While these instances support prior findings of BO's efficiency, our results also suggest that simply committing to a single HO approach could miss models that generalize well and that performance of selected classifiers will depend on the task and classifier type.

Evaluation of BO- and GS/RS-selected classifiers at evaluation budgets typical of GS/RS. In the previous analysis, we compared BO- and GS/RS-selected classifiers at evaluation budgets typical of BO (i.e. fewer configurations evaluated). In Figure 2, we compare BO-selected classifiers from their highest evaluation budgets (100 evaluations for RBF and 500 evaluations for XGB and MLP) to classifiers selected by GS/RS at larger evaluation budgets. Interestingly, we find that the BO-selected MLP classifiers for the mortality task continue to outperform their corresponding RS-selected counterparts, even with 25000 configurations evaluated for RS. Similarly, we find that BO-selected XGBoost classifiers exceed external validation performance of RS-selected classifiers on the BVN task up to an evaluation budget of 25000 configurations (though the differences do not persist at 25000 configurations). We observe these differences when conducting BO with the upper confidence bound acquisition function or with a transformed hyperparameter space (Suppl. Figs. 9, 10, 13 and 14). These results indicate the relative efficiency of BO in candidate classifier selection in these two instances but also illustrate the competitiveness of GS/RS-selected classifiers in our setting.

Assessment of effects on classifier performance of automatic relevance determination in BO. For high-dimensional hyperparameter spaces, some hyperparameters may have a greater impact on the model's generalization performance than others. Automatic relevance determination (ARD;[26]) in the GP model of BO's objective provides the means to estimate effects of variations in hyperparameter dimensions on the objective's value and has been used in multiple implementations of BO (e.g. Snoek et al., 2012[8] and BoTorch, `https://botorch.org/docs/models`). We directly compare the internal and external validation performance of classifiers selected by BO with and without ARD. In Figure 3, we find that enabling ARD seems to lead to comparable if not slightly better internal validation performance at higher evaluation budgets. Moreover, enabling ARD seems to improve external validation performance for both XGB (BVN task) and MLP classifiers (both tasks). In fact, the highest external validation performance by XGB classifiers on the BVN task is only achieved with ARD enabled (Table 1). However, these differences in performance are not as evident when using the upper confidence bound acquisition function (Suppl. Fig. 15) or conducting BO in the transformed hyperparameter space (Suppl. Fig. 16). Thus, ARD may not be necessary to select top-performing diagnostic classifiers for these two clinical tasks.

5. Discussion & Conclusions

In this analysis, we compared HO approaches for diagnostic classifier development to determine what approach (if any) led to improvements in: 1) external validation performance or

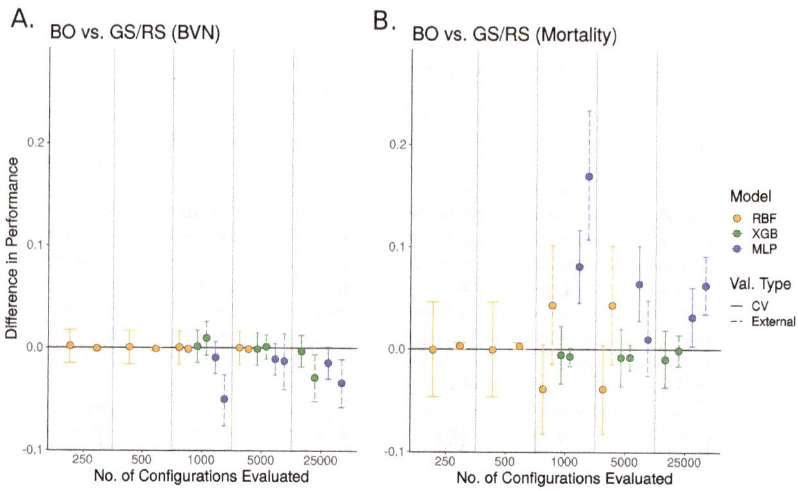

Fig. 2: **Differences in classification performance of models selected by either BO or GS/RS using GS/RS evaluation budgets.** Run settings and figure layout are the same as in Figure 1 except that here, indicated evaluation budgets apply to GS/RS-selected classifiers; BO-selected classifiers are taken from 100-evaluation (RBF) or 500-evaluation (XGB and MLP) runs.

2) computational efficiency. Consistent with previous findings, we found that BO was able to prioritize candidate classifiers for two tasks relevant to emergency and critical care with a fraction of the configurations evaluated using GS/RS. As embarrassingly parallel approaches like GS/RS can necessitate the use of commodity computing clusters, BO's efficiency makes the approach a potentially cost-effective solution. We also found that external validation performance of BO-selected MLPs for in-hospital mortality was consistently better across a range of HO evaluation budgets than that of GS/RS-selected classifiers, highlighting BO's potential to uncover diagnostic classifiers that generalize better to unseen patients.

However, and in contrast to previous comparisons of HO approaches, our analyses indicated that GS/RS methods could select classifiers for both tasks with evaluation budgets comparable to those used for BO. We also found mixed evidence in support of enabling ARD in the kernel of BO's GP model of the objective function. Thus, while we hoped we would uncover distinct and general differences between HO approaches in order to develop better guidelines about when (or even if) to use one approach over another, we did not identify such clear differences across tasks, classifier types, and optimization settings. Rather, our analysis suggests that both GS/RS and BO approaches should be investigated for classifier development.

We acknowledge limitations of our approach. For our RS runs, we sampled configurations

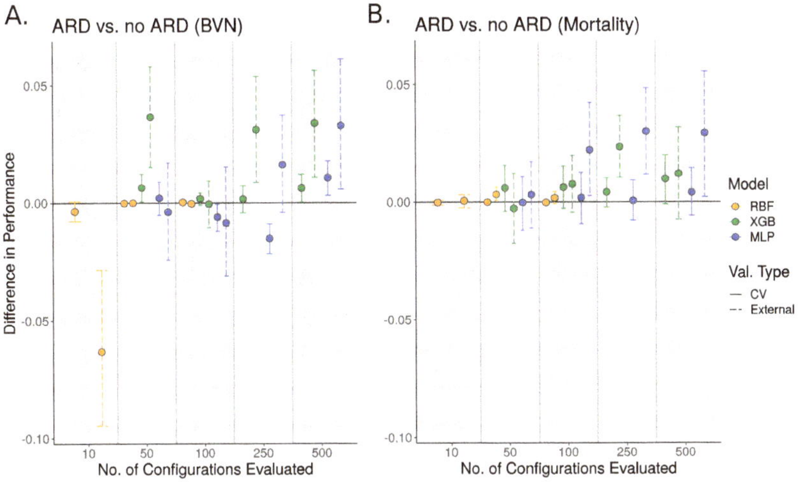

Fig. 3: **Differences in classification performance for BO-selected classifiers with or without automatic relevance determination (ARD) enabled.** Performance differences greater than 0 on the BVN (A; mAUC) and mortality (B; AUC) tasks indicate better performance for the classifier selected by ARD-enabled BO. Points represent observed differences while error bars represent 95% bootstrap confidence intervals.

uniformly and independently from pre-defined ranges or grids of values. Other random sampling approaches could've been used in which configurations are generated dependent on the values of previously generated configurations (e.g. Latin hypercube or low-discrepancy Sobol sequences) in order to encourage diversity of the resulting sample.[7] We felt that the similar performance we observed between BO and GS/RS-selected models using basic variants of GS/RS didn't necessarily justify further analysis with more sophisticated GS/RS variants. A second limitation is that we used a single set of features derived from a previously identified set of 29 gene expression markers. We chose these features based on previous analyses[5] and consistent with our goal of developing diagnostic classifiers from these specific markers for clinical deployment. We acknowledge our conclusions may not hold with other feature sets.

Throughout this work, we wanted our hyperparameter optimization to reflect our clinical deployment scenario: that classifiers would likely be evaluated on structured populations (e.g. from a given geographic region) not seen in training. A recent study by Google highlighted this challenge for deployment in healthcare: their AI system for breast cancer screening showed drops in predictive performance when trained on mammograms from the UK and applied to mammograms from the US.[27] However, our survey of ML studies comparing hyperparameter optimization approaches highlighted important differences from our setting in terms of dataset partitioning and, consequently, in the choice of internal validation-based objective function. For example, we found that ML studies primarily focused on larger ($N >\sim 100$k) datasets composed mainly of natural images. These benchmarks were often constructed (e.g. MNIST; `http://yann.lecun.com/exdb/mnist/`) to satisfy the assumption that the distribution of training and external validation samples are similar if not the same. Internal validation was then performed on subsets of these 'mixed' datasets, with samples from the same structured group in the full dataset appearing in both the training and validation set. However, as patient data is known to be heterogeneous due to biological differences as well as differences in geography, healthcare delivery, and assay technologies used, that assumption of distributional similarity between training and external validation samples is likely to be violated. Indeed, our recent work found that standard k-fold cross-validation gives optimistically biased estimates of generalization error in our setting,[5] breaking the group structure in left-out folds by randomly distributing patients from the same study into different cross-validation folds (akin to test set contamination). Consequently, in difference to the ML studies we reviewed, we opted for grouped 5-fold cross-validation as our objective function as well as evaluation of performance in external validation to aid model selection.

In conclusion, we find that both GS/RS and BO remain promising avenues for hyper-parameter optimization and represent key components in the development of more effective diagnostics for emergency and critical care.

References

1. A. E. Jones, S. Trzeciak and J. A. Kline, The Sequential Organ Failure Assessment score for predicting outcome in patients with severe sepsis and evidence of hypoperfusion at the time of emergency department presentation, *Critical care medicine* **37**, 1649 (May 2009), 19325482[pmid].
2. T. Sweeney, A. Shidham, H. R. Wong and P. Khatri, A comprehensive time-course–based multicohort analysis of sepsis and sterile inflammation reveals a robust diagnostic gene set, *Science Translational Medicine* **7** (2015).
3. T. Sweeney, H. R. Wong and P. Khatri, Robust classification of bacterial and viral infections via integrated host gene expression diagnostics, *Science Translational Medicine* **8** (2016).
4. T. Sweeney and P. Khatri, Benchmarking sepsis gene expression diagnostics using public data, *Critical care medicine* **45**, p. 1 (2017).
5. M. B. Mayhew, L. Buturovic, R. Luethy, U. Midic, A. R. Moore, J. A. Roque, B. D. Shaller, T. Asuni, D. Rawling, M. Remmel, K. Choi, J. Wacker, P. Khatri, A. J. Rogers and T. E. Sweeney, A generalizable 29-mrna neural-network classifier for acute bacterial and viral infections, *Nature Communications* **11**, p. 1177 (2020).
6. T. Sweeney, T. Perumal and R. e. a. Henao, A community approach to mortality prediction in sepsis via gene expression analysis, *Nat Commun* (2018).
7. J. Bergstra and Y. Bengio, Random search for hyper-parameter optimization, *Journal of Machine Learning Research* **13**, 281 (2012).
8. J. Snoek, H. Larochelle and R. P. Adams, Practical Bayesian optimization of machine learning algorithms, *In Advances in neural information processing systems* , 2951 (2012).
9. J. Snoek, O. Rippel, K. Swersky, R. Kiros, N. Satish, N. Sundaram, M. Patwary, M. Prabhat and R. Adams, Scalable Bayesian optimization using deep neural networks, in *International conference on machine learning*, 2015.
10. A. Klein, S. Falkner, S. Bartels, P. Hennig and F. Hutter, Fast Bayesian Optimization of Machine Learning Hyperparameters on Large Datasets, in *Proceedings of the 20th International Confer-*

ence on Artificial Intelligence and Statistics, eds. A. Singh and J. Zhu, Proceedings of Machine Learning Research, Vol. 54 (PMLR, Fort Lauderdale, FL, USA, 20–22 Apr 2017).
11. S. Falkner, A. Klein and F. Hutter, BOHB: Fast and Efficient Hyperparameter Optimization at Scale, in *ICML*, 2018.
12. A. Klein, Z. Dai, F. Hutter, N. Lawrence and J. Gonzalez, Meta-Surrogate Benchmarking for Hyperparameter Optimization, in *Advances in Neural Information Processing Systems 32*, eds. H. Wallach, H. Larochelle, A. Beygelzimer, F. dAlché-Buc, E. Fox and R. Garnett (Curran Associates, Inc., 2019) pp. 6270–6280.
13. M. Ghassemi, L. Lehman, J. Snoek and S. Nemati, Global optimization approaches for parameter tuning in biomedical signal processing: A focus on multi-scale entropy, in *Computing in Cardiology 2014*, Sep. 2014.
14. G. W. Colopy, S. J. Roberts and D. A. Clifton, Bayesian Optimization of Personalized Models for Patient Vital-Sign Monitoring, *IEEE Journal of Biomedical and Health Informatics* **22**, 301 (March 2018).
15. M. Nishio, M. Nishizawa, O. Sugiyama, R. Kojima, M. Yakami, T. Kuroda and K. Togashi, Computer-aided diagnosis of lung nodule using gradient tree boosting and Bayesian optimization, *PloS one* **13**, e0195875 (Apr 2018), 29672639[pmid].
16. R. J. Borgli, H. Kvale Stensland, M. A. Riegler and P. Halvorsen, Automatic Hyperparameter Optimization for Transfer Learning on Medical Image Datasets Using Bayesian Optimization, in *2019 13th International Symposium on Medical Information and Communication Technology (ISMICT)*, May 2019.
17. C. E. Rasmussen and C. K. I. Williams, *Gaussian Processes for Machine Learning* (The MIT Press, 2006).
18. J. Bergstra, D. Yamins and D. D. Cox, Making a science of model search: Hyperparameter optimization in hundreds of dimensions for vision architectures (2013).
19. S. Ben-David, J. Blitzer, K. Crammer and F. Pereira, Analysis of representations for domain adaptation, in *Advances in Neural Information Processing Systems 19*, eds. B. Schölkopf, J. C. Platt and T. Hoffman (MIT Press, 2007) pp. 137–144.
20. M. Thomas and R. Schwartz, A method for efficient Bayesian optimization of self-assembly systems from scattering data, *BMC Systems Biology* **12**, p. 65 (2018).
21. R. Tanaka and H. Iwata, Bayesian optimization for genomic selection: a method for discovering the best genotype among a large number of candidates., *Theor Appl Genet* **131**, 93 (2018).
22. S. Mao, Y. Jiang, E. B. Mathew and S. Kannan, BOAssembler: a Bayesian Optimization Framework to Improve RNA-Seq Assembly Performance (2019).
23. T. Chen and C. Guestrin, XGBoost: A Scalable Tree Boosting System, in *Proceedings of the 22nd ACM SIGKDD International Conference on Knowledge Discovery and Data Mining*, KDD '16 (ACM, New York, NY, USA, 2016).
24. D. P. Kingma and J. Ba, Adam: A Method for Stochastic Optimization (2014).
25. D. J. Hand and R. J. Till, A simple generalisation of the area under the ROC curve for multiple class classification problems, *Machine learning* **45**, 171 (2001).
26. R. M. Neal, *Bayesian Learning for Neural Networks* (Springer-Verlag, Berlin, Heidelberg, 1996).
27. S. M. McKinney, M. Sieniek, V. Godbole, J. Godwin, N. Antropova, H. Ashrafian, T. Back, M. Chesus, G. C. Corrado, A. Darzi, M. Etemadi, F. Garcia-Vicente, F. J. Gilbert, M. Halling-Brown, D. Hassabis, S. Jansen, A. Karthikesalingam, C. J. Kelly, D. King, J. R. Ledsam, D. Melnick, H. Mostofi, L. Peng, J. J. Reicher, B. Romera-Paredes, R. Sidebottom, M. Suleyman, D. Tse, K. C. Young, J. De Fauw and S. Shetty, International evaluation of an AI system for breast cancer screening, *Nature* **577**, 89 (2020).

TrueImage: A Machine Learning Algorithm to Improve the Quality of Telehealth Photos

Kailas Vodrahalli[1,†,*], Roxana Daneshjou[2,3,†,*], Roberto A Novoa[2,4], Albert Chiou[2], Justin M Ko[2], and James Zou[1,3,†]

[1]*Department of Electrical Engineering, Stanford University, Stanford, CA 94305, USA*

[2]*Department of Dermatology, Stanford University School of Medicine, Redwood City, CA 94063, USA*

[3]*Department of Biomedical Data Science, Stanford University School of Medicine, Stanford, CA 94305, USA*

[4]*Department of Pathology, Stanford University School of Medicine, Stanford, CA 94305, USA*

† *Correspondence can be addressed to: kailasv@stanford.edu, roxanad@stanford.edu, jamesz@stanford.edu*

* *These authors contributed equally*

Telehealth is an increasingly critical component of the health care ecosystem, especially due to the COVID-19 pandemic. Rapid adoption of telehealth has exposed limitations in the existing infrastructure. In this paper, we study and highlight photo quality as a major challenge in the telehealth workflow. We focus on teledermatology, where photo quality is particularly important; the framework proposed here can be generalized to other health domains. For telemedicine, dermatologists request that patients submit images of their lesions for assessment. However, these images are often of insufficient quality to make a clinical diagnosis since patients do not have experience taking clinical photos. A clinician has to manually triage poor quality images and request new images to be submitted, leading to wasted time for both the clinician and the patient. We propose an automated image assessment machine learning pipeline, TrueImage, to detect poor quality dermatology photos and to guide patients in taking better photos. Our experiments indicate that TrueImage can reject ~50% of the sub-par quality images, while retaining ~80% of good quality images patients send in, despite heterogeneity and limitations in the training data. These promising results suggest that our solution is feasible and can improve the quality of teledermatology care.

Keywords: Telemedicine; Teledermatology; Computer Vision; Image Quality Assessment.

1. Introduction

Due to the SARS-CoV-2 (COVID-19) pandemic, many hospitals have rapidly transitioned patient visits to video conference calls on a digital platform to limit exposure for both patients and healthcare workers. Although these digital visits have some limitations, they have recently accounted for more than 10% of all visits in the US, corresponding to more than an 10000%

increase since February 2020.[1]

The rapid adoption of telehealth has unearthed substantial challenges. For example, productive teledermatology visits require high clinical quality images of the area of concern; however, video call platforms do not have sufficient imaging resolution for diagnosis. In teledermatology, a clinician will often request patients to send in photos of their lesions or rash ahead of time. The clinician will use these images for assessing the patient's condition and use the digital platform of the visit to communicate with a patient rather than for making assessments.

Patients are guided on how to take photos of their lesions; see Figure 1 for standard guidelines. Despite these instructions, it is common for patients to take blurry images, images in poor lighting conditions (e.g., too much glare or too dark), or images that do not adequately show the lesion (e.g., taken from too far away). Prior assessments of image quality in dermatology are not applicable to real world teledermatology, as trained medical professionals took the photos in these studies.[2] So, we conducted an informal survey of dermatologists that suggests up to one fifth of all images sent in by patients could be of too low quality to be of use; see Table 1.

Due to this high percentage of low quality images, dermatologists or other staff members screen images prior to a visit and request a patient to retake an image when necessary. This process is time consuming and can take a similar amount of time as a regularly scheduled visit. Moreover, it is common for patients to send in images just prior to a visit leaving no time for image quality screening. When these images are low quality, the clinical visit is spent coaching the patient on retaking the photo rather than the clinical issue. Therefore, poor quality images can significantly disrupt a clinician's schedule and affect clinical care.

We propose an automated machine learning method for assessing dermatology image quality and giving concrete feedback for how to improve quality when necessary (e.g., "turn on camera flash" for dim lighting). We envision this solution as integrated into a smartphone application that can guide a patient through the process of taking an image with interactive, real-time feedback so that only high-quality photos are submitted for the televisit. This necessitates a computationally lightweight solution and motivates some of our design decisions.

We detail our prototype algorithm, TrueImage, and assess our method on a dataset of dermatology photos as a proof-of-concept. The method provides a quality score to quantify how suitable a photo is for teledermatology. The score enables clinicians to flexibly set a threshold for filtering photo quality – a more stringent threshold makes it more likely that the clinician works with high-quality images but could require more patients to retake photos. For example, we can reject ~50% of poor quality images at while retaining ~80% of good quality images; alternatively, we can reject ~10% of all poor quality images while retaining >95% of good quality images.

Contributions We identify photo quality as an important emerging challenge for telehealth, especially for dermatology. There has been relatively little work in this area. We develop an algorithm for automatic quality detection in dermatology images and to provide guidance to patients. This can potentially improve the clinical workflow and efficiency.

Photo tips:

Do not send multiple files with the exact same photo. If possible, avoid taking photos yourself and ask a family member, friend, or acquaintance to take them for you.

Send clear, helpful photos. We are unable to evaluate blurry photos.

Avoid shadows and use soft, indirect light. In low-light situations make sure the flash is turned on to avoid grainy or blurry images. In brighter light, you may not need flash. Experiment with the camera flash, it may help eliminate shadows but may also cause too much white out.

Do not take photographs in front of a window or other brightly lit background.

Do not use editing software for your photos prior to sending to your doctor.

All material uploaded to the patient portal will be viewable in your medical record.

Fig. 1. Example set of instructions given to patients at Stanford Health Care for how to take images for dermatology visits.

Table 1. Results of a survey we conducted asking dermatologists how often patients send in poor quality photos. Samples size is 37. Several responders reported poor image frequency as high as 1/2.

Frequency of poor quality photos per visit	1/5	1/10	1/20	1/50
Percent of survey response	78.4%	10.8%	5.4%	5.4%

2. Background

Dermatology has become an important application of machine learning research in recent years with the success of deep learning and the acquisition of large dermatology datasets. Much of this work is related to disease diagnosis[3,4] or lesion segmentation,[5] and most public data is taken using dermatoscopes, a special tool for magnifying lesions. However, as large-scale teledermatology is relatively new, little work has been done in solving problems specific to automatically assessing the quality of patient-taken images. There are several related problems that we detail here.

2.1. *Clinical Image Guidelines*

Photography for dermatology is commonly used in a clinical setting for both educational purposes and to track disease progression in patients. To ensure high quality photos, there are several guidelines that have been developed to counter the common issues that produce low quality photos in dermatology.[6] See Figure 2 for illustrative examples. These issues can be summarized as

1. Skin lesion area is blurry (Figure 2c).
2. Skin is discolored due to lighting conditions – this may be induced by a dim environment,

excessive shadows, excessive glare (e.g., due to camera flash), or the background reflecting tinted light (Figure 2b).

3. Skin lesion is cropped or image is taken from too far away (Figure 2d).
4. Image is distorted due to camera effects (e.g., fish-eye effect).
5. Background is distracting or patient is wearing distracting clothing and jewelry.
6. Image is taken at a poor orientation (e.g., a leg is photographed horizontally; a vertical photograph is preferable so that the entire frame is filled with the leg).

Items 5 and 6 above tend to be less critical for dermatologists to make an assessment. Empirically, we have noted that the most common issues in patient-taken images are items 1-3, with the ordering corresponding to their prevalence. So in this work, we focus on these 3 issues.

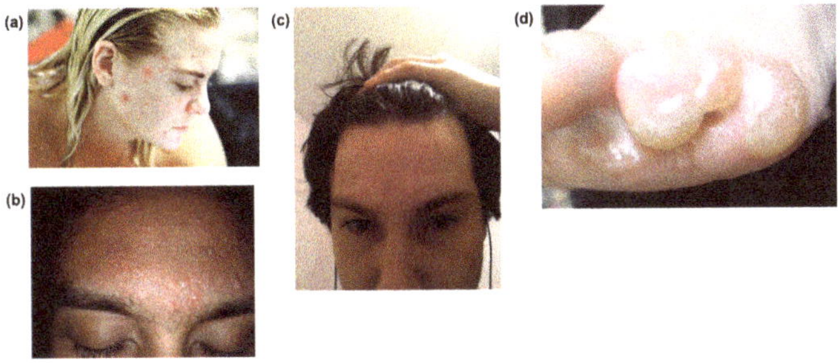

Fig. 2. Examples of poor quality images. (a) is a good quality image; the acne is in-focus, the lighting is not too dark and there is little glare. (b) has excessive glare and shadows on the skin. (c) is too blurry. Note that the lighting is also dim here; brighter lighting would likely also reduce issues with blur. (d) is cropped and zoomed in too close; the slight blur and glare exacerbates the problem.

2.2. *Image Quality Assessment*

Image quality assessment (IQA) methods attempt to measure the quality of digital images, where image quality is usually defined by human labelers. IQA methods can be split into two categories – full-reference IQA (FR-IQA) which require a high quality reference image and no-reference IQA (NR-IQA) which work given a single image; some techniques are adaptable to both settings.[7] Additionally, some methods are designed with respect to specific distortions like Gaussian blur, additive white noise, and JPEG compression artifacts,[8–10] while others are general purpose and adapt to the distortions present in a training dataset.[11] IQA has generally been studied for use on natural images, though some techniques have been adapted for detecting artifacts in MRI, CT, and ultrasonography[12] in clinical settings.

Due to the cost of labeling data (trained human labelers are required), most IQA datasets are small. However, deep learning methods have become prevalent recently and typically utilize data augmentation techniques like applying fixed distortions (e.g., Gaussian blur) to high quality images,[10] utilizing generative models like GANs,[13] or through leveraging larger image classification datasets for transfer learning.[14]

Classical methods also have good performance, in particular when a specific, known distortion is in consideration. Of particular interest to us, blur detection is a well-studied problem and efficient classical algorithms exist.[8,15] These algorithms generally rely on detecting the magnitude of low and high-frequency content in images – blurry images tend to have reduced high-frequency content.

2.3. *Semantic Segmentation*

Semantic segmentation is the problem of generating per-pixel class labels in an image. In our case, we are interested in a binary semantic segmentation problem of labeling skin and non-skin pixels in an image; segmentation is important to us as we are interested in the quality of *only* the parts of the image containing the lesion (e.g., the background can be blurry, but the lesion cannot).

Deep learning methods have dominated the field in recent years, with most modern methods relying on fully convolutional networks.[16,17] Methods like Mask RCNN[18] and YOLACT[19] layer a semantic segmentation module over an object detection framework to allow for instance level semantic segmentation. Other techniques use recurrent connections when multiple images (e.g., a video sequence) are available.[20]

Classical methods also exist and utilize a variety of approaches. Of particular interest to us, there are a large number of methods designed specifically for skin detection as part of gesture recognition or facial detection algorithms. Some of the most simple methods utilize decision trees learned from a dataset of skin pixels and classify each pixel independently;[21,22] other methods fit the distribution or skin colors using, for example, histogram-based techniques or a Gaussian mixture model and apply a threshold on the predicted probability to classify skin pixels.[23] More sophisticated methods apply additional steps to account for spatial information in the image.[24]

3. TrueImage Algorithm

Our algorithm can be described in 3 stages as shown in Figure 3. It consists of (1) semantic segmentation to identify skin regions, (2) feature generation, and (3) a quality classifier applied to these features. Our algorithm design is guided by two considerations/constraints:

1. We desire a computationally lightweight algorithm, due to the end-goal of having an interactive system deployable on older-generation smartphones.
2. As labeling data is costly, a more data-efficient algorithm is preferable.

These constraints motivated us to represent each photo by a relatively small number of interpretable features, which we explain in detail below. Additionally, the algorithm itself is more interpretable and so we can more easily ensure it is robust to various skin tones.

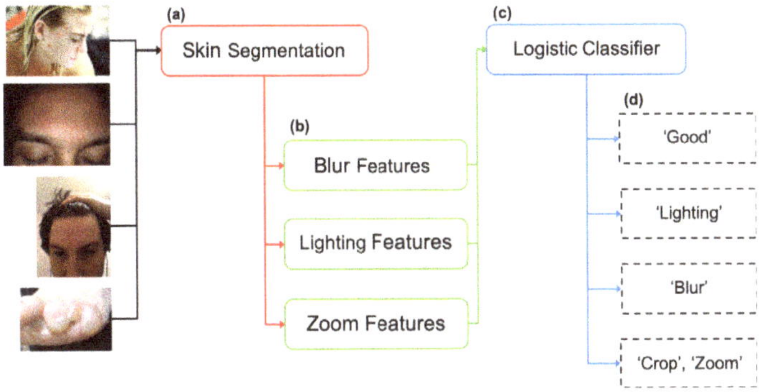

Fig. 3. Workflow schematic of the TrueImage dermatology image quality detection algorithm. (a) Image is input for skin segmentation. (b) After segmentation, the original image and segmentation mask are used for feature generation in three groups. Principal component analysis (PCA) is used to reduce the dimensionality of features within each feature group. (c) 4 classifiers are applied to the concatenated feature vectors, giving us (d) labels for reason(s) of poor image quality.

3.1. *Semantic Segmentation*

We use a per-pixel semantic segmentation algorithm to identify skin and lesion pixels. Each pixel is classified independently. Each pixel is transformed from RGB into both YCrCb and HSV color spaces, and the two representations are concatenated giving a vector in $\mathbb{R}^6$. We then use a Gaussian mixture model to assign the pixel a score corresponding to its likelihood of being "skin"; applying a threshold to the pixels scores gives us our semantic segmentation.

Additionally, we always consider border pixels to be non-skin. Empirically, we have seen that patient-taken images are generally well-centered. Thus, center cropping is a simple way of ensuring we are assessing the lesion area and not the surrounding skin or background clutter.

We also implement a simple, per-pixel lesion segmentation algorithm. The algorithm takes a pixel, transforms it to the LAB color space, and keeps the brightest pixels from each color channel. Then we compute the fraction of these brightest pixels located near the center of the image. The color channel with the highest fraction is used as the lesion mask; we additionally perform a bitwise-and operation with the skin segmentation mask to reduce false positives. An inspection of 15 images suggested that this algorithm detects red and dark patches well.

Although threshold-based algorithms are generally inferior to modern deep learning algorithms, our algorithm performs adequately in our setting, largely due to the constraints on our distribution of images (e.g., images are generally centered). Furthermore, since our end goal is a downstream task that uses the segmentation as an input, slight errors do not have significant impact.

3.2. *Features*

We consider 3 major reasons for poor quality images: blur, lighting conditions, and zoom. Although there are other reasons as noted in,[6] we have found empirically that these 3 are the most common issues in patient-taken images. We generate features designed to capture good-to-bad image variance for each of these issues.

3.2.1. *Blur Features*

Blur can be characterized by a higher presence of low-frequency components in an image as well as other effects like decreased color saturation. These effects can be measured by computation of the Fourier Transform on an image, through analysis of the image gradient (or higher order derivatives), or through various other means like looking at the color saturation.[8,15]

Subsequent to our semantic segmentation, we uniformly sample 100 32×32 pixel patches that contain at least 90% "skin" pixels. For each patch P, we compute

1. the magnitude of the high-pass filtered patch,

$$\frac{1}{32^2} \sum_{i,j \in [32]} 20 \ln \left(||F_h(P)_{i,j}|| \right),$$

   where F_h denotes the high-pass filter, and
2. the variance of the Laplacian of the patch,

$$\frac{1}{32^2} \sum_{i,j \in [32]} (L(x,y) - \mathbb{E}[L(x,y)])^2,$$

   where $L(x,y) = \frac{\partial^2 P}{\partial x^2} + \frac{\partial^2 P}{\partial y^2}$.

Subsequently, we summarize the patch-level distribution by computing the mean, median, max, min, and standard deviation of each feature across patches. This process results in 10 total features. We apply principal component analysis (PCA) to these features and keep the top 5 principal components.

3.2.2. *Lighting Features*

To detect poor lighting conditions, we use two types of features. We compute these features per patch, using the same 100 patches as for blur.

1. We transform the image to grayscale, G, and compute per each patch P_G (1) $underexposed := P_G[P_G < 50]$ and (2) $overexposed := P_G[P_G > 205]$. Color values are integers in the range $[0, 255]$. *underexposed* assess the amount of shadows and dim lighting in the image while *overexposed* assess the amount of glare. Note that both *underexposed* and *overexposed* are sets of values; we summarize each set by computing the median and the upper and lower quartiles. We summarize the patch-level distribution by computing the mean, median, max, min, and standard deviation across patches for each feature, resulting in 30 features.

2. We consider the probability distribution given by the Gaussian mixture model trained for skin segmentation for each patch. This distribution gives us information on the glare and shadow content, as well as discoloration in the skin due to poor lighting (e.g., a blue tinted light). We compute the median and lower and upper quartiles per patch, and subsequently compute the mean, median, max, min, and standard deviation across patches.

The above process results in 45 total features; we reduce this to 5 features using PCA.

3.2.3. *Zoom Features*

For assessing crop and zoom, we compute the ratio of skin to non-skin pixels and lesion to non-lesion detected inside the center cropped area. Images zoomed out too far will have fewer "skin" pixels near their center. Similarly, a high fraction of lesion pixels near the boundary of an image suggest that the lesion is not shown with adequate context.

Note that the relevance of these features are entirely dependent on the skin and lesion segmentation algorithms; to counteract deficiencies in our segmentation algorithm, we apply a more generous threshold for computing these features.

3.3. *Feature Aggregation*

We concatenate across feature groups and train four binary classifiers using logistic regression: one predicts whether an image is good quality, and the remaining three predict whether the image is blurry, has poor lighting, or is zoomed out / cropped respectively.

4. Training and Evaluation

4.1. *Skin Segmentation Model*

The Gaussian mixture model is trained to fit the distribution of skin pixels using the dataset from Bhatt et. al.[21] This dataset consists of roughly 50000 skin pixels sampled from 14,701 face images of a diverse set of individuals across age, gender, and ethnicity. Our model is fit using the standard expectation-maximization algorithm. Note that though this model is trained with pixels from face images, it is able to identify skin pixels in a generalizable manner.

4.2. *Logistic Classifiers*

To train our binary classifiers, we use four datasets. The first dataset was curated from Google Images using images licensed for free commercial use and contains images of various diseases and lesion types. A dermatologist added labels assessing image quality and giving reasons for poor quality. Note that the assigned labels should be interpreted as "too blurry to make an assessment" (if the assigned label is blurry). We augmented this dataset by applying one of two types of distortions to all good quality images. We split the dataset into training, validation, and test sets prior to augmentation so as to avoid data leakage.

1. (blur) Gaussian blur with randomly sized kernel.
2. (zoom/crop) Select random corner cropping of the image.

A dermatologist separately labeled some of the artificial images for use in our test set; only images the dermatologist deemed realistic were included for testing. See rows 1 and 2 ("Web" and "Web-Augmented") of Table 2 for a breakdown of the labels in this data.

We also used a dataset collected at Stanford Health Care. This dataset contains images taken by a dermatologist of various lesions observed in clinic. See row 3 ("Stanford") of Table 2 for details. We do not augment this dataset.

As these datasets have relatively few images of poor lighting conditions, we additionally collected our own poor lighting images by taking pictures of the authors and their families with their consent. We used this data only for training purposes. See row 4 ("Extra") of Table 2 for details. By aggregating across these four datasets from heterogeneous sources, we provide more representative photos to train and evaluate our approach.

Table 2. Image counts for our dermatology dataset. Some images have multiple labels. Images are randomly split into train, validation and test sets.

Data source	Total	Good	Blurry	Poor Lighting	Poor Zoom / Crop
Web	55	46	5	5	1
Web-Augmented	179	14	80	0	85
Stanford	99	86	5	7	2
Extra	29	0	0	29	0
All data	362	146	90	41	88

4.3. *Results on Dermatology Dataset*

After training our algorithm, we evaluated it on the test dataset described above. Our results are shown in Figure 4. There are four plots shown, one for each of the labels we generate: (a) good/poor quality, (b) blurry/not blurry, (c) good/poor lighting, (d) good/poor zoom or crop. We plot the receiver operating characteristic (ROC) curve.

Looking at Figure 4a, we notice that (1) we can reject ~50% of poor quality images while retaining ~80% of good quality images or (2) we can reject ~10% of poor quality images while retaining >95% of good quality images; this is particularly important as it suggests we can set TrueImage's parameters to reduce time spent by clinicians without adversely affecting patients. Looking at Figure 4b-d, we see that we can detect blur quite well, but can only detect poor lighting or crop moderately well. Lastly, we note that the result in Figure 4a is dependent on the distribution of poor quality images. Empirically, we have found that blurry images are the most common followed by poor lighting conditions.

Many of the "good" quality images are not actually good quality, but are adequate enough for clinical assessment. This label may vary between dermatologists, and the threshold parameter should be tuned per dermatologist and even per disease category (e.g., inflammatory lesions vs. non-inflammatory lesions). We envision that individual dermatologists or group practices will set their thresholds based on their preferences prior to sharing the application with patients; moreover, these settings would be adjustable by clinicians based on patient feedback. Some images are in a gray-area where their image quality rating is situation dependent (e.g., what context the dermatologist is viewing them in). The ~10% we can reject outright are in a set of images that are indisputably poor quality.

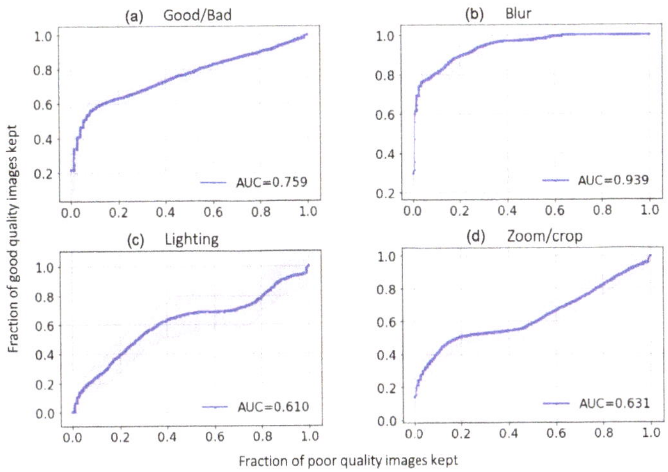

Fig. 4. Receiver operating characteristic (ROC) curve of TrueImage; 1 standard deviation confidence interval shaded in. Classifiers are for (a) Good/bad, (b) Blur, (c) Lighting, (d) Zoom/crop.

5. Discussion and Future Work

This work highlights photo quality as a significant and understudied challenge for teledermatology. We develop a novel, automated approach to detect poor quality dermatology photos.

We have also implemented an interactive user interface for TrueImage using gradio[25] shown in Figure 5, which will facilitate usage of TrueImage in clinical pilot studies. The eventual goal is to make an interactive interface run by the user to guide them in real-time in taking clinical photos. Our algorithm is computationally very efficient, and an unoptimized, single-threaded implementation takes about 1 second to run per photo using a standard laptop with a 1.8 GHz Intel Core i5 processor. This timing was computed by averaging over the runtime of 20 images. Because teledermatology is still quite new, there are limited publicly available datasets of patient photos annotated with quality measures. To address this challenge, we curated a dataset of 362 images from diverse sources and annotated the quality of each photo along with reasons for poor quality images. While our dataset is a good first step, creating larger datasets of dermatologist labeled photos would be a useful resource for telehealth research and can further improve the performance of TrueImage. Moreover, with a larger dataset, we hope to create more specific feedback for patients (e.g. instead of saying "poor lighting", stating "lighting is too dark" or "lighting is too bright"). Additionally, most of the images are of patients with lighter skin tones. A larger dataset with more diverse skin types is critical for TrueImage to be more broadly useful. However, our current results demonstrate that it is possible to robustly detect image quality in dermatology photos.

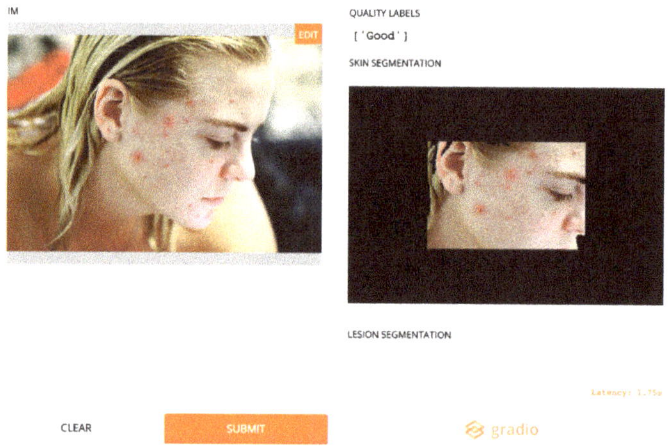

Fig. 5. User interface design using gradio.[25] Displays input image, skin segmentation, and quality classification.

References

1. A. Mehrotra, M. Chernew, D. Linetsky, H. Hatch and D. Cutler, The impact of the covid-19 pandemic on outpatient visits: Practices are adapting to the new normal (2020).
2. E. A. Krupinski, B. LeSueur, L. Ellsworth, N. Levine, R. Hansen, N. Silvis, P. Sarantopoulos, P. Hite, J. Wurzel, R. S. Weinstein *et al.*, Diagnostic accuracy and image quality using a digital camera for teledermatology, *Telemedicine Journal* **5**, 257 (1999).
3. A. Esteva, B. Kuprel, R. A. Novoa, J. Ko, S. M. Swetter, H. M. Blau and S. Thrun, Dermatologist-level classification of skin cancer with deep neural networks, *nature* **542**, 115 (2017).
4. S. S. Han, I. Park, S. E. Chang, W. Lim, M. S. Kim, G. H. Park, J. B. Chae, C. H. Huh and J.-I. Na, Augmented intelligence dermatology: Deep neural networks empower medical professionals in diagnosing skin cancer and predicting treatment options for 134 skin disorders, *Journal of Investigative Dermatology* (2020).
5. M. Goyal and M. H. Yap, Multi-class semantic segmentation of skin lesions via fully convolutional networks, *arXiv preprint arXiv:1711.10449* (2017).

6. L. Muraco, Improved medical photography: key tips for creating images of lasting value, *JAMA dermatology* **156**, 121 (2020).
7. S. Bosse, D. Maniry, K.-R. Müller, T. Wiegand and W. Samek, Deep neural networks for no-reference and full-reference image quality assessment, *IEEE Transactions on Image Processing* **27**, 206 (2017).
8. R. Liu, Z. Li and J. Jia, Image partial blur detection and classification, in *2008 IEEE conference on computer vision and pattern recognition*, (IEEE, 2008).
9. L. Li, W. Lin, X. Wang, G. Yang, K. Bahrami and A. C. Kot, No-reference image blur assessment based on discrete orthogonal moments, *IEEE transactions on cybernetics* **46**, 39 (2015).
10. X. Liu, J. van de Weijer and A. D. Bagdanov, Rankiqa: Learning from rankings for no-reference image quality assessment, in *Proceedings of the IEEE International Conference on Computer Vision*, (IEEE, 2017).
11. L. Zhang, L. Zhang and A. C. Bovik, A feature-enriched completely blind image quality evaluator, *IEEE Transactions on Image Processing* **24**, 2579 (2015).
12. L. S. Chow and R. Paramesran, Review of medical image quality assessment, *Biomedical signal processing and control* **27**, 145 (2016).
13. K.-Y. Lin and G. Wang, Hallucinated-iqa: No-reference image quality assessment via adversarial learning, in *Proceedings of the IEEE Conference on Computer Vision and Pattern Recognition*, (IEEE, 2018).
14. F. Gao, J. Yu, S. Zhu, Q. Huang and Q. Tian, Blind image quality prediction by exploiting multi-level deep representations, *Pattern Recognition* **81**, 432 (2018).
15. S. Pertuz, D. Puig and M. A. Garcia, Analysis of focus measure operators for shape-from-focus, *Pattern Recognition* **46**, 1415 (2013).
16. Y. Guo, Y. Liu, T. Georgiou and M. S. Lew, A review of semantic segmentation using deep neural networks, *International journal of multimedia information retrieval* **7**, 87 (2018).
17. I. Ulku and E. Akagunduz, A survey on deep learning-based architectures for semantic segmentation on 2d images, *arXiv preprint arXiv:1912.10230* (2019).
18. K. He, G. Gkioxari, P. Dollár and R. Girshick, Mask r-cnn, in *Proceedings of the IEEE international conference on computer vision*, (IEEE, 2017).
19. D. Bolya, C. Zhou, F. Xiao and Y. J. Lee, Yolact: Real-time instance segmentation, in *Proceedings of the IEEE international conference on computer vision*, (IEEE, 2019).
20. D. Nilsson and C. Sminchisescu, Semantic video segmentation by gated recurrent flow propagation, in *Proceedings of the IEEE Conference on Computer Vision and Pattern Recognition*, (IEEE, 2018).
21. R. B. Bhatt, G. Sharma, A. Dhall and S. Chaudhury, Efficient skin region segmentation using low complexity fuzzy decision tree model, in *2009 Annual IEEE India Conference*, (IEEE, 2009).
22. S. Kolkur, D. Kalbande, P. Shimpi, C. Bapat and J. Jatakia, Human skin detection using rgb, hsv and ycbcr color models, *arXiv preprint arXiv:1708.02694* (2017).
23. M. J. Jones and J. M. Rehg, Statistical color models with application to skin detection, *International Journal of Computer Vision* **46**, 81 (2002).
24. M. R. Mahmoodi, Fast and efficient skin detection for facial detection, *arXiv preprint arXiv:1701.05595* (2017).
25. A. Abid, A. Abdalla, A. Abid, D. Khan, A. Alfozan and J. Zou, An online platform for interactive feedback in biomedical machine learning, *Nature Machine Intelligence* **2**, 86 (2020).

Data used in this study Public data used in this study was released under a Wikimedia Commons-acceptable license (i.e., free to use for any purpose). The Stanford data comes from the Stanford Dermatology Clinic and is covered by IRB protocol 57673.

CheXclusion: Fairness gaps in deep chest X-ray classifiers

Laleh Seyyed-Kalantari[1,2*], Guanxiong Liu[1,2], Matthew McDermott[3], Irene Y. Chen[3], Marzyeh Ghassemi[1,2]

[1] *Computer Science, University of Toronto, Toronto, Ontario, Canada*

[2] *Vector Institute, Toronto, Ontario, Canada*

[3] *Electrical Engineering and Computer Science, Massachusetts Institute of Technology, Cambridge, MA USA*

Machine learning systems have received much attention recently for their ability to achieve expert-level performance on clinical tasks, particularly in medical imaging. Here, we examine the extent to which state-of-the-art deep learning classifiers trained to yield diagnostic labels from X-ray images are biased with respect to *protected attributes*. We train convolution neural networks to predict 14 diagnostic labels in 3 prominent public chest X-ray datasets: MIMIC-CXR, Chest-Xray8, CheXpert, as well as a multi-site aggregation of all those datasets. We evaluate the *TPR disparity* – the difference in true positive rates (TPR) – among different protected attributes such as patient sex, age, race, and insurance type as a proxy for socioeconomic status. We demonstrate that TPR disparities exist in the state-of-the-art classifiers in all datasets, for all clinical tasks, and all subgroups. A multi-source dataset corresponds to the smallest disparities, suggesting one way to reduce bias. We find that TPR disparities are not significantly correlated with a subgroup's proportional disease burden. As clinical models move from papers to products, we encourage clinical decision makers to carefully audit for algorithmic disparities prior to deployment. Our supplementary materials can be found at, `http://www.marzyehghassemi.com/chexclusion-supp-3/`.

Keywords: fairness, medical imaging, chest x-ray classifier, computer vision.

1. Introduction

Chest X-ray imaging is an important screening and diagnostic tool for several life-threatening diseases, but due to the shortage of radiologists, this screening tool cannot be used to treat all patients.[1,2] Deep-learning-based medical image classifiers are one potential solution, with significant prior work targeting chest X-rays specifically,[3,4] leveraging large-scale publicly available datasets,[3,5,6] and demonstrating radiologist-level accuracy in diagnostic classification.[6–8]

Despite the seemingly clear case for implementing AI-enabled diagnostic tools,[9] moving such methods from paper to practice require careful thought.[10] Models may exhibit disparities in performance across protected subgroups, and this could lead to different subgroups receiving different treatment.[11] During evaluation, machine learning algorithms usually optimize for, and

*Corresponding author email: `laleh@cs.toronto.edu`

report performance on, the general population rather than balancing accuracy on different subgroups. While some variance in performance is unavoidable, mitigating any systematic bias against protected subgroups may be desired or required in a deployable model.

In this paper, we examine whether state-of-the-art (SOTA) deep neural classifiers trained on large public medical imaging datasets are fair across different subgroups of *protected attributes*. We train classifiers on 3 large, public chest X-ray datasets: MIMIC-CXR,[5] CheXpert,[6] Chest-Xray8,[3] as well as an additional datasets formed of the aggregation of those three datasets on their shared labels. In each case, we implement chest X-ray pathology classifiers via a deep convolutional neural network (CNN) chest X-ray images as inputs, and optimize the multi-label probability of 14 diagnostic labels simultaneously. Because researchers have observed health disparities with respect to race,[12] sex,[13] age,[14] and socioeconomic status,[12] we extract structural data on race, sex, and age; we also use insurance type as an imperfect proxy[11] for socioeconomic status. To our knowledge, we are the first to examine whether SOTA chest X-ray pathology classifiers display systematic bias across race, age, and insurance type.

We analyze equality of opportunity[15] as our fairness metric based on the needs of the clinical diagnostic setting. In particular, we examine the differences in true positive rate (TPR) across different subgroups per attributes. A high TPR disparity indicates that sick members of a protected subgroup would *not* be given correct diagnoses—e.g., true positives—at the same rate as the general population, even in an algorithm with high overall accuracy.

We find three major findings: First, that there are indeed extensive patterns of bias in SOTA classifiers, shown in TPR disparities across datasets. Secondly, the disparity rate for most attributes/ datasets pairs is not significantly correlated with the subgroups' proportional disease membership. These findings suggest that underrepresented subgroups could be vulnerable to mistreatment in a systematic deployment, and that such vulnerability may not be addressable simply through increasing subgroup patient count. Lastly, we find that using the multi-source dataset which combines all the other datasets yields the lowest TPR disparities, suggesting using multi-source datasets may combat bias in the data collection process. As researchers increasingly apply artificial intelligence and machine learning to precision medicine, we hope that our work demonstrates how predictive models trained on large, well-balanced datasets can still yield disparate impact.

2. Background and Related Work

Fairness and Debiasing. Fairness in machine learning models is a topic of increasing attention, spanning sex bias in occupation classifiers,[16] racial bias in criminal defendant risk assessments algorithms,[17] and intersectional sex-racial bias in automated facial analysis.[18] Sources of bias arise in many different places along the classical machine learning pipeline. For example, input data may be biased, leaving supervised models vulnerable to labeling and cohort bias.[18] Minority groups may also be under-sampled, or the features collected may not be indicative of their trends.[19] There are several conflicting definitions of fairness, many of which are not simultaneously achievable.[20] The appropriate choice of a disparity metric is generally task dependent, but balancing error rates between different subgroups is a common consideration,[15,17] with equal accuracy across subgroups being a popular choice in medical settings.[21] In this work, we consider the equality of opportunity notion of fairness and evaluate the rate of correct diagnosis in patients across several protected attribute groups.

Ethical Algorithms in Health. Using machine learning algorithms to make decisions raises serious ethical concerns about risk of patient harm.[22] Notably, biases have already been demonstrated in several settings, including racial bias in the commercial risk score algorithms used in hospitals,[23] or an increased risk of electronic health record (EHR) miss-classification in patients with low socioeconomic status.[24] It is crucial that we actively consider fairness metrics when building models in systems that include human and structural biases.

Chest X-Ray Classification. With the releases of large public datasets like Chest-Xray8,[3] CheXpert,[6] and MIMIC-CXR,[5] many researchers have begun to train large deep neural network models for chest X-ray diagnosis.[4,6,8,25] Prior work[8] demonstrates a diagnostic classifier trained on Chest-Xray8 can achieve radiologist-level performance. Other work on CheXpert[6] reports high performance for five of their diagnostic labels. To our knowledge, however, no works have yet been published which examined whether any of these algorithms display systematic bias over age, race and insurance type (as a proxy of socialeconomic status).

3. Data

We use three public chest X-ray radiography datasets described in Table 1: MIMIC-CXR (CXR),[5] CheXpert (CXP),[6] Chest-Xray8 (NIH).[3] Images in CXR, CXP, and NIH are associated with 14 diagnostic labels (see Table 2). We combine all non-positive labels within CXR and CXP (including "negative", "not mentioned", or "uncertain") into an aggregate "negative" label for simplicity, equivalent to "U-zero" study of 'NaN' label in CXP. In CXR and CXP, one of the 14 labels is "No Finding", meaning no disease has been diagnosed for the image and all the other 13 labels are 0. Of the 14 total disease labels, only 8 are shared amongst all 3 datasets. Using these 8 labels, we define a multi-site dataset (ALL) that consists of the aggregation of all images in CXR, CXP, and NIH defined over this restricted label schema.

These datasets contain protected subgroup attributes, the full list of which includes sex (Male and Female), age (0-20, 20-40, 40-60, 60-80, and 80-), race (White, Black, Other, Asian, Hispanic, and Native) and insurance type (Medicare, Medicaid, and Other). These values are taken from the structured patient attributes. NIH, CXP, and ALL only have the patient sex and age, while CXR also has race and insurance type data (excluding around 100,000 images).

4. Methods

We implement CNN-based models to classify chest X-ray images into 14 diagnostic labels. We train separate models for CXR,[5] CXP,[6] NIH[3] and ALL and explore their fairness with respect to patient sex and age for all 4 datasets as well as race and insurance type for CXR.

4.1. *Models*

We initialize a 121-layer DenseNet[26] with pre-trained weights from ImageNet[27] and train multi-label models with a multi-label binary cross entropy loss. The 121-layer DenseNet was used as it produced the best results in prior studies[6,8] . We use a 80-10-10 train-validation-test

Table 1. Description of chest X-ray datasets, MIMIC-CXR (CXR),[5] CheXpert (CXP),[6] Chest-Xray8 (NIH),[3] and their aggregation on 8 shared labels (ALL). Here, the number of images, patients, view types, and the proportion of patients per subgroups of sex, age, race, and insurance type are presented. 'Frontal' and 'Latral' abbreviate frontal and lateral view, respectively. Native, Hispanic, and Black denote self-reported American Indian/Alaska Native, Hispanic/Latino, and Black/African American race respectively.

Subgroup	Attribute	**CXR**[5]	**CXP**[6]	**NIH**[3]	**ALL**
	# Images	371,858	223,648	112,120	707,626
	# Patients	65,079	64,740	30,805	129,819
	View	Frontal/Lateral	Frontal/Lateral	Frontal	Frontal/Lateral
sex	Female	47.83%	40.64%	43.51%	44.87%
	Male	52.17%	59.36%	56.49%	55.13%
Age	0-20	2.20%	0.87%	6.09%	2.40%
	20-40	19.51%	13.18%	25.96%	18.53%
	40-60	37.20%	31.00%	43.83%	36.29%
	60-80	34.12%	38.94%	23.11%	33.90%
	80+	6.96%	16.01%	1.01%	8.88%
Race	Asian	3.24%	—	—	—
	Black	18.59%	—	—	—
	Hispanic	6.41%	—	—	—
	Native	0.29%	—	—	—
	White	67.64%	—	—	—
	Other	3.83%	—	—	—
Insurance	Medicare	46.07%	—	—	—
	Medicaid	8.98%	—	—	—
	Other	44.95%	—	—	—

split with no patient shared across splits. We resize all images to 256 × 256 and normalize via the mean and standard deviation of the ImageNet dataset.[27] We apply center crop, random horizontal flip and random rotation, as some of the images maybe flipped or rotated within the dataset. The initial degree of random rotation is chosen by hyperparameter tuning. We use Adam[28] optimization with default parameters, and decrease the learning rate (LR) by a factor of 2 if the validation loss does not improve over three epochs; we stop learning if validation loss does not improve over 10 epochs. Thus the ultimate number of epochs for training each model is varied based on the early stop condition. For NIH, CXP and CXR we first tune models to get the highest average area under the receiver operating characteristic curve (AUC) over 14 labels by fine tuning the LR. For the best achieved model, we fine tune the degree of random rotation data augmentation from the set of 7, 10 and 15 and select the best model. Following this, best initial LR is 0.0005 for CXR and NIH where it is achieved as 0.0001 for CXP. Also, best initial degree for random rotation data augmentation is 10 for NIH and 15 for the CXR and CXP. For training on ALL, we use the majority vote of the best hyperparameters per individual dataset (e.g. 0.0005 initial LR and 15 degree random rotation). We then, fix the hyperparameters of the best model and train four extra models with the same hyperparameters but different random seeds between 0 to 100, per dataset. We report all the metrics based on the mean and 95% confidence intervals (CI) achieved over five studies per dataset. We choose batch size of 48 to use the maximum memory capacity of the GPU, for all datasets except NIH where we choose 32 similar to prior work.[8] The output of the network is an array of 14 numbers between 0 and 1 indicating the probability of each disease label. The binary prediction threshold per disease is chosen to maximize the F1 score measure on the validation dataset. We train models using a single NVIDIA GPU with 16G of memory in approximately 9, 20, 40, and 90 hours for NIH, CXP, CXR, and ALL, respectively.

4.2. *Classifier Disparity Evaluation*

Our primary measurement of bias is *TPR disparity*. For example, given a binary subgroup attribute such as sex (which in our data we classify as either 'male' or 'female'), we mimic prior work[16] and define the TPR disparity per diagnostic label i as simply the TPR of label i restricted to female patients minus that for male patients. More formally, letting g be the binary subgroup attribute, we define $\text{TPR}_{g,i} = P[\hat{Y}_i = y_i | G = g, Y_i = y_i]$, and the TPR disparity as, $\text{Gap}_{g,i} = \text{TPR}_{g,i} - \text{TPR}_{\neg g,i}$. For non-binary attributes $S_1, \ldots, S_N$, we use the difference between a subgroup's TPR and the median of all TPRs to define TPR disparity of the jth subgroup for the ith label as, $\text{Gap}_{S_j,i} = \text{TPR}_{S_j,i} - \text{Median}(\text{TPR}_{S_1,i}, .., \text{TPR}_{S_k,i})$.

5. Experiments

First, we demonstrate that the classifiers we train on all datasets reach near-SOTA level performance. This motivates using them to study fairness implications, as we can be confident any problematic disparities are not simply reflective of poor overall performance. Next, we explicitly test these classifiers for their implications on fairness. We target two investigations:

(1) **TPR disparity:** We quantify the TPR disparity per subgroup/disease for sex and age across all 4 datasets, and due to data availability for race and insurance type on CXR.
(2) **TPR disparity in proportion to membership:** We investigate the distribution of the positive patient proportion per subgroup S_j and label y_i (which is given by $P[S = S_j | Y = y_i]$) and the effect on TPR disparities. Prior work on chest X-ray diagnosis prediction has suggested data imbalance can explain sex TPR disparities[29] while work in other domains illustrates that disparities in small or vulnerable subgroups could be propagated if put into practice,[16,30] and these experiments are meant to probe that hypothesis.

6. Results

One potential reason that a model may be biased is because of poor performance, but we demonstrate that our models achieve near-SOTA classification performance. Table 2 shows overall performance numbers across all tasks and datasets. Though results have non-trivial

Table 2. The AUC for chest X-ray classifiers trained on CXP, CXR, NIH, and ALL averaged over 5 runs ±95%CI, where all runs have same hyperparameters but different random seed. ('Airspace Opacity'[5] and 'Lung Opacity'[6] denote the same label.)

Label (Abbr.)	CXR	CXP	NIH	ALL
Airspace Opacity (AO)	0.782 ± 0.002	0.747 ± 0.001	—	—
Atelectasis (A)	0.837 ± 0.001	0.717 ± 0.002	0.814 ± 0.004	0.808 ± 0.001
Cardiomegaly (Cd)	0.828 ± 0.002	0.855 ± 0.003	0.915 ± 0.002	0.856 ± 0.001
Consolidation (Co)	0.844 ± 0.001	0.734 ± 0.004	0.801 ± 0.005	0.805 ± 0.001
Edema (Ed)	0.904 ± 0.002	0.849 ± 0.001	0.915 ± 0.003	0.898 ± 0.001
Effusion (Ef)	0.933 ± 0.001	0.885 ± 0.001	0.875 ± 0.002	0.922 ± 0.001
Emphysema (Em)	—	—	0.897 ± 0.002	—
Enlarged Card (EC)	0.757 ± 0.003	0.668 ± 0.005	—	—
Fibrosis	—	—	0.788 ± 0.007	—
Fracture (Fr)	0.718 ± 0.007	0.790 ± 0.006	—	—
Hernia (H)	—	—	0.978 ± 0.004	—
Infiltration (In)	—	—	0.717 ± 0.004	—
Lung Lesion (LL)	0.772 ± 0.006	0.780 ± 0.005	—	—
Mas (M)	—	—	0.829 ± 0.006	—
Nodule (N)	—	—	0.779 ± 0.006	—
No Finding (NF)	0.868 ± 0.001	0.885 ± 0.001	—	0.890 ± 0.000
Pleural Thickening (PT)	—	—	0.813 ± 0.006	—
Pleural Other (PO)	0.848 ± 0.003	0.795 ± 0.004	—	—
Pneumonia (Pa)	0.748 ± 0.005	0.777 ± 0.003	0.759 ± 0.012	0.784 ± 0.001
Pneumothorax (Px)	0.903 ± 0.002	0.893 ± 0.002	0.879 ± 0.005	0.904 ± 0.002
Support Devices (SD)	0.927 ± 0.001	0.898 ± 0.001	—	—
Average	**0.834 ± 0.001**	**0.805±0.001**	**0.840 ± 0.001**	**0.859 ± 0.001**

variability, we show similar performance to the published SOTA of NIH,[8] the only dataset for which a published SOTA comparison exists for all labels. Note that the published results for CXP[6] are on a private, unreleased dataset of only 200 images and 5 labels. Our results for CXP are on a randomly sub-sampled test set of size 22,274 images, so the numbers for this dataset are not comparable to the published results there.

6.1. *TPR Disparities*

We calculate and identify TPR disparities and 95% CI across all labels, datasets and attributes. We see many instances of positive and negative disparities, which can denote bias for or against of a subgroup, here referred to *favorable* and *unfavorable* subgroups. As an illustrative example Fig. 1 shows the race TPR disparities distribution sorted by the the gap between least and most favorable subgroups per label. In a fair setting, all subgroups would have no appreciable TPR disparities, yielding a gap between least and most favorable subgroups within a label at '0'. Table 3 shows the summary of the disparities in all attributes and datasets. We note that the most frequent unfavorable subgroups are those with social disparities in the healthcare

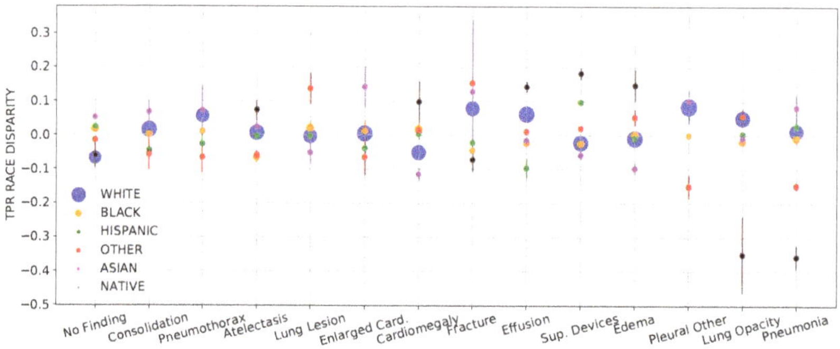

Fig. 1. The sorted distribution of TPR race disparity of CXR (y-axis) with label (x-axis). The scatter plot's circle area is proportional to group size. TPR disparities are averaged over five runs ±95%CI (shown with arrows). Hispanic patients are most unfavorable (highest count of negative TPR disparities, 9/13) whereas White patients are most favorable subgroup (9/13 zero or positive disparities). Labels 'No Finding' ('NF') and 'Pneumonia' ('Pa') have smallest (0.119) and largest (0.440) gap between least/most favorable subgroups. The average cross 14 labels gap is 0.226.

Table 3. Disparities overview over attributes and datasets. We average per label gaps between the least and most favorable subgroup's TPR disparities per attributes/datasets to obtain the average cross-label gap. The labels (full names on Table 2) that obtained the smallest and largest gaps are shown next to the average cross-label gap column, along with their gaps. We summarize and label in columns the most frequent "Unfavorable" and "Favorable" subgroups count, which are the ones that experience TPRs disparities below or above the zero gap line. See Section 6.1 for more details.

Attribute	**Dataset**	**Average Cross-Label Gap**			**Unfavorable**	**Favorable**
		Gap	**Lowest**	**Greatest**		
Sex	ALL	**0.045**	Ef:0.001	Pa:0.105	Female (4/7)	Male (4/7)
	CXP	0.062	Ed:0.000	Co:0.139	Female (7/13)	Male (7/13)
	CXR	0.072	Ed:0.011	EC:0.151	Female (10/13)	Male (10/13)
	NIH	0.190	M:0.001	Cd:0.393	Female (8/14)	Male (8/14)
Age	ALL	**0.215**	Ef:0.115	NF:0.444	0-20 (5/7)	40-60,60-80(5/7)
	CXR	0.245	SD:0.091	Cd:0.440	0-20, 20-40 (7/13)	60-80 (10/13)
	CXP	0.270	SD:0.084	NF:0.604	0-20, 20-40, 80- (7/13)	40-60 (8/13)
	NIH	0.413	In:0.188	Em:1.00	60-80 (7/14)	20-40 (9/14)
Race	CXR	0.226	NF:0.119	Pa:0.440	Hispanic (9/13)	White (9/13)
Insurance	CXR	0.100	SD:0.021	PO:0.190	Medicaid (10/13)	Other (10/13)

system, e.g., women and minorities, but no disease is consistently at the highest or lowest disparity. We show the average cross-label gap between and the labels of the least and most favorable subgroups per dataset and attributes. We count the number of time each subgroups experience negative disparities (unfavorable) and zero or positive disparities (favorable) across disease labels and report the most frequent unfavorable and favorable subgroups by count in Table 3. For CXP and CXR, we exclude "No Finding" label in the count (counts are out of 13) as we want to check negative bias in disease labels. Notably, the model trained on ALL has the smallest average cross- label gap between least/most favorable groups for sex and age.

6.2. *TPR Disparity Correlation with Membership Proportion*

We measure the Pearson correlation coefficients (r) between the TPR disparities and patients proportion per label across all subgroups/datasets. As we test multiple (33) hypotheses, (33 total comparisons amongst all protected attributes considered) with a desired significance level ($p < 0.05$), then based on Bonferroni correction,[31] the statistical significance level for each hypothesis is $p < 0.0015$ (0.05/33). The majority of correlation coefficients listed are positive, but the only statistically significant correlations are: race Other (r: 0.782, p: 0.0009) & age subgroups, 20-40 (r: 0.766, p: 0.0013), 60-80 (r: 0.787, p: 0.0008) and 80- (r:0.858 , p: 0.0000) in CXR, age group 60-80 (r: 0.853, p: 0.0001) in CXP, and age group 60-80 (r: 0.936, p: 0.0006) in ALL.

7. Summary and Discussion

We present a range of findings on the potential biases of deployed SOTA X-ray image classifiers over the sex, age, race and insurance type attributes on models trained on NIH, CXP and CXR. We focus on TPR disparities similar to prior work,[16] checking if the sick members of the different subgroups are given correct diagnosis at similar rates.

Our results demonstrate several main takeaways. First, all datasets and tasks display non-trivial TPR disparities. These disparities could pose serious barriers to effective deployment of these models and invite additional changes in either dataset design and/or modeling techniques to ensure more equitable models. Second, using a multi-source dataset leads to smaller TPR disparities, potentially due to removing bias in the data collection process. Third, while there is occasionally a proportionality between protected subgroup membership per label and TPR disparity, this relationship is not uniformly true across datasets and subgroups.

7.1. *Extensive Patterns of Bias*

We find that all datasets and tasks contain meaningful patterns of bias although no diseases are consistently at the highest or lowest disparity rates across all attributes and datasets. These disparities are present with respect to age and sex in all settings, with consistent subgroups (female, 0-20) showing consistently unfavorable outcomes. Note that in the case of the sex disparities, "female" patients are universally the least favored subgroup *despite* the fact that the proportion of female patients is only slightly less than male patients in all 4 datasets.

We also observe TPR disparities with respect to the patient race and insurance type in the CXR dataset. White patients, the majority, are the most favorable subgroup, where Hispanic patients are the most unfavorable. Additionally, bias exists against patients with Medicaid insurance, who are the minority population and are often from lower socioeconomic status. They are the most unfavorable subgroup with the model often providing incorrect diagnoses.

7.2. *Bias Reduction Through Multi-source Data*

Of the four datasets, the multi-source dataset led to the smallest disparities with respect to age and sex. Based on notions of generalizability in healthcare,[10,32] we hypothesize that this improvement stems from the combination of large datasets reducing data collection bias.

7.3. *Correlation Between TPR Disparities and Membership Proportion*

Although prior work has raised data imbalance as a potential cause of sex bias,[29] we observe TPR disparities are not often significantly correlated with disease membership. While we observe positive correlation between subgroups membership and TPR disparities, only 6 of 33 subgroups showed statistically significant correlation. By inspection, we identify diseases with the same patient proportion of a subgroup and completely different TPR disparities (e.g. 'Consolidation', 'Nodule' and 'Pneumothorax' in NIH have 45% Female, but the TPR disparities are in diverse range, -0.155, -0.079 and 0.047, respectively). Thus, having the same portion of images within all labels may not guarantee lack of bias.

7.4. *Discussion*

We identify subgroups that may experience more bias through the exploration of variance in TPR and FPR. Based on the equality of opportunity notion of fairness, a fair network should exhibit the same TPR on average among all subgroups regardless of how likely a subgroup may have a disease. Such an improvement would allow two patients with the same condition, but in different subgroups, to be diagnosed correctly and receive the same level of care. While we focused on some of the more obvious protected attributes, it is important to note that there are several other factors, subgroups, and attributes that we have not considered.

Identifying and eliminating disparities is particularly important as large datasets begin to be used by high-capacity neural models, but are based on highly skewed population, e.g., kidney injury prediction in a population that is 93.6% male.[33] While chest X-ray images datasets are not sex-skewed, we note that the age, race and insurance type attributes are highly unbalanced, e.g., 67.6% of patients are White, and only 8.98% are under Medicaid insurance. Subgroups with chronic underdiagnosis are those who experience more negative social determinants of health, specifically, women, minorities, and those of low socioeconomic status. Such patients may use healthcare services less than others. In some groups, such a dataset skew can increase the risk of miss-classification.[24]

Although "de-biasing" techniques[34,35] may reduce disparities, we should not ignore the important biases inherent in existent large public datasets. There are a number of reasons why dataset may induce disparities in algorithms, from imbalanced datasets to differences in statistical noise in each group (e.g. unmeasured predictive features) to differences in access to healthcare for patients of different groups.[12,19] For instance, an algorithm that can classify skin cancer[36] with high accuracy will not be able to generalize on different skin color if similar samples have not been represented enough in the trained dataset.[18] Intentionally adjusting the datasets to reduce disparities in to protect minorities and the subgroups with high disparities is one potential option in dataset creation, though our analyses suggest that dataset membership cannot always ameliorate bias.

With the great promise of advanced models for clinical care, we caution that advanced SOTA models must be carefully checked for such biases as those we have identified. Disparities in small or vulnerable subgroups could be propagated[30] within the development of machine learning models. This raises serious ethical concerns[22] about the equal accessibility to the required medical treatment. Usually the SOTA classifiers are trained to provides high AUC or accuracy on the general population. However we suggest additionally applying rigorous fairness analyses before deployment. Clear disclaimers about the dataset collection process and potential resulting algorithmic bias could improve model assessment for clinical use.

8. Limitations and Future Work

As SOTA deep learning diagnosis algorithms become more promising for medical screening, model bias investigation is essential. This work is a first step in quantifying these biases so that approaches for amelioration can be developed. However, important future work remains.

First, we note that across these models, our source of diagnostic labels for these images must be considered at best "silver" labels, as all currently existing public chest X-ray datasets use automatically deteremined labels based on natural language processing (NLP) techniques to extract labels from the radiology reports. These silver labels may be incorrect, in ways that could compound with observe biases or model errors, a risk that warrants further investigation. Additionally, we must consider the quality of the imaging devices themselves, the region of data collection, and the patient demographics at each hospital collection site. For instance, NIH was gathered from a hospital that covers more complicated cases, CXP contains more tertiary cases, and CXR was gathered from an emergency department, and prior literature has already shown that models are fully capable of taking advantage of such confounders.[32] These challenges may affect both the label quality,[37] and any patterns of bias in the labels, thereby affecting the resulting fairness metrics. Finally, exploration of existing de-biasing techniques, however limited, should also be undertaken over this modality to see if any of the problems we identified here can be resolved.

9. Conclusion

While the development and deployment of machine learning models in a clinical setting poses exciting opportunities, great care must be taken to understand how existing biases may be exacerbated and propagated. We show the TPR disparity of SOTA chest X-ray pathology classifiers trained on 4 datasets, (MIMIC-CXR, ChestX-ray8, CheXpert, and aggregation of those three on shared labels) across 14 diagnostic labels. We quantify the TPR disparity across datasets along sex, age, race and insurance type. Our results indicate that high-capacity models trained on large datasets do not provide equality of opportunity naturally, leading instead to potential disparities in care if deployed without modification.

Acknowledgment

We acknowledge the support of the Natural Sciences and Engineering Research Council of Canada (NSERC, funding number PDF-516984), Microsoft Research, CIFAR, NSERC Discovery Grant, and high performance computing platforms of Vector Institute. We also thank Dr. Alistair Johnson, Dr. Errol Colak and Grey Kuling for productive discussions.

References

1. A. Rimmer, Radiologist shortage leaves patient care at risk, warns royal college, *BMJ (Clinical research ed.)* **359**, p. j4683 (2017).
2. F. S Ali, S. G Harrington, S. B Kenned and S. Hussain, Diagnostic radiology in liberia: a country report, *Journal of Global Radiology* **1(2)** (2015).
3. X. Wang, Y. Peng, L. Lu, Z. Lu, M. Bagheri and R. M. Summers, ChestX-ray8: Hospital-Scale Chest X-Ray Database and Benchmarks on Weakly-Supervised Classification and Localization of Common Thorax Diseases, 2097 (2017).
4. L. Yao, E. Poblenz, D. Dagunts, B. Covington, D. Bernard and K. Lyman, Learning to diagnose from scratch by exploiting dependencies among labels, *arXiv:1710.10501* (2017).
5. A. E. W. Johnson, T. J. Pollard, S. J. Berkowitz, N. R. Greenbaum, M. P. Lungren, C.-y. Deng, R. G. Mark and S. Horng, MIMIC-CXR: A large publicly available database of labeled chest radiographs, *arXiv:1901.07042* (2019).
6. J. Irvin, P. Rajpurkar, M. Ko, Y. Yu, S. Ciurea-Ilcus, C. Chute, H. Marklund, B. Haghgoo, R. Ball, K. Shpanskaya, J. Seekins, D. A. Mong, S. S. Halabi, J. K. Sandberg, R. Jones, D. B. Larson, C. P. Langlotz, B. N. Patel, M. P. Lungren and A. Y. Ng, CheXpert: A Large Chest Radiograph Dataset with Uncertainty Labels and Expert Comparison, *arXiv:1901.07031* (2019).
7. P. Rajpurkar, J. Irvin, K. Zhu, B. Yang, H. Mehta, T. Duan, D. Ding, A. Bagul, C. Langlotz, K. Shpanskaya, M. Lungren and A. Ng, Chexnet: Radiologist-level pneumonia detection on chest x-rays with deep learning (2017).
8. P. Rajpurkar, J. Irvin, R. L. Ball, K. Zhu, B. Yang, H. Mehta, T. Duan, D. Ding, A. Bagul, C. P. Langlotz, B. N. Patel, K. W. Yeom, K. Shpanskaya, F. G. Blankenberg, J. Seekins, T. J. Amrhein, D. A. Mong, S. S. Halabi, E. J. Zucker, A. Y. Ng and M. P. Lungren, Deep learning for chest radiograph diagnosis: A retrospective comparison of the CheXNeXt algorithm to practicing radiologists, *PLOS Medicine* **15**, p. e1002686 (2018).
9. V. Institute, Thousands of images at the Radiologist's fingertips seeing the invisible (2019).
10. M. Ghassemi, T. Naumann, P. Schulam, A. L. Beam, I. Y. Chen and R. Ranganath, Practical guidance on artificial intelligence for health-care data, *The Lancet Digital Health* **1**, e157 (2019).
11. I. Y. Chen, P. Szolovits and M. Ghassemi, Can ai help reduce disparities in general medical and mental health care?, *AMA journal of ethics* **21**, 167 (2019).
12. I. Kawachi, N. Daniels and D. E. Robinson, Health disparities by race and class: why both matter, *Health Affairs* **24**, 343 (2005).
13. D. E. Hoffmann and A. J. Tarzian, The girl who cried pain: a bias against women in the treatment of pain, *The Journal of Law, Medicine & Ethics* **28**, 13 (2001).
14. J. Walter, A. Tufman, R. Holle and L. Schwarzkopf, "age matters"—german claims data indicate disparities in lung cancer care between elderly and young patients, *PloS one* **14**, p. e0217434 (2019).
15. M. Hardt, E. Price and N. Srebro, Equality of Opportunity in Supervised Learning, *NIPS'16* , 3323 (2016).
16. M. De-Arteaga, A. Romanov, H. Wallach, J. Chayes, C. Borgs, A. Chouldechova, S. Geyik, K. Kenthapadi and A. T. Kalai, Bias in bios: a case study of semantic representation bias in a

high-stakes setting (2019), Atlanta, GA.
17. A. Chouldechova, Fair prediction with disparate impact: A study of bias in recidivism prediction instruments, *Big data* **5**, 153 (2016).
18. J. Buolamwini and T. Gebru, Gender Shades: Intersectional Accuracy Disparities in Commercial Gender Classification, **81**, p. 15 (2018).
19. I. Chen, F. D. Johansson and D. Sontag, Why Is My Classifier Discriminatory?, in *NIPS'31*, 2018 pp. 3539–3550.
20. J. Kleinberg, S. Mullainathan and M. Raghavan, Inherent Trade-Offs in the Fair Determination of Risk Scores, *arXiv:1609.05807* (2016).
21. M. Srivastava, H. Heidari and A. Krause, Mathematical notions vs. human perception of fairness: a descriptive approach to fairness for machine learning, *arXiv preprint arXiv:1902.04783* (2019).
22. D. S. Char, N. H. Shah and D. Magnus, Implementing machine learning in health care — addressing ethical challenges, *New England Journal of Medicine* **378**, 981 (2018).
23. Z. Obermeyer and S. Mullainathan, Dissecting racial bias in an algorithm that guides health decisions for 70 million peoples, p. 89 (2019), Atlanta, GA.
24. M. A. Gianfrancesco, S. Tamang, J. Yazdany and G. Schmajuk, Potential biases in machine learning algorithms using electronic health record data., *JAMA internal medicine* **178**, 1544 (2018).
25. S. Akbarian, L. Seyyed-Kalantari, F. Khalvati and E. Dolatabadi, Evaluating knowledge transfer in neural network for medical images, *arXiv preprint arXiv:2008.13574* (2020).
26. G. Huang, Z. Liu, L. v. d. Maaten and K. Q. Weinberger, Densely Connected Convolutional Networks, 2261 (2017).
27. J. Deng, W. Dong, R. Socher, L.-J. Li, K. Li and L. Fei-Fei, ImageNet: A Large-Scale Hierarchical Image Database (2009).
28. D. P. Kingma and J. Ba, Adam: A method for stochastic optimization, *arXiv:1412.6980v9* (2017).
29. A. J. Larrazabal, N. Nieto, V. Peterson, D. H. Milone and E. Ferrante, Gender imbalance in medical imaging datasets produces biased classifiers for computer-aided diagnosis, *Proceedings of the National Academy of Sciences* **117**, 12592 (2020).
30. T. B. Hashimoto, M. Srivastava, H. Namkoong and P. Liang, Fairness Without Demographics in Repeated Loss Minimization, *arXiv:1806.08010* (2018).
31. R. G. J. Miller, *Simultaneous Statistical Inference* (Springer-Verlag New York, 1981).
32. J. R. Zech, M. A. Badgeley, M. Liu, A. B. Costa, J. J. Titano and E. K. Oermann, Confounding variables can degrade generalization performance of radiological deep learning models, *PLOS Medicine* **15**, p. e1002683 (2018).
33. N. Tomašev, X. Glorot, J. W. Rae, M. Zielinski, H. Askham, A. Saraiva, A. Mottram, C. Meyer, S. Ravuri, I. Protsyuk *et al.*, A clinically applicable approach to continuous prediction of future acute kidney injury, *Nature* **572**, p. 116 (2019).
34. A. Amini, A. P. Soleimany, W. Schwarting, S. N. Bhatia and D. Rus, Uncovering and Mitigating Algorithmic Bias through Learned Latent Structure, in *Proceedings of the 2019 AAAI/ACM Conference on AI, Ethics, and Society - AIES '19*, (ACM Press, Honolulu, HI, USA, 2019).
35. B. H. Zhang, B. Lemoine and M. Mitchell, Mitigating Unwanted Biases with Adversarial Learning, *arXiv:1801.07593* (2018).
36. A. Esteva, B. Kuprel, R. A. Novoa, J. Ko, S. M. Swetter, H. M. Blau and S. Thrun, Dermatologist-level classification of skin cancer with deep neural networks, *Nature* **542**, 115 (2017).
37. ~. Lukeoakdenrayner, Half a million x-rays! First impressions of the Stanford and MIT chest x-ray datasets (2019).

Incorporation of DNA methylation into eQTL mapping in African Americans

Anmol Singh, Yizhen Zhong, Layan Nahlawi, C. Sehwan Park, Tanima De, Cristina Alarcon, and Minoli A. Perera

Department of Pharmacology, Northwestern University Feinberg School of Medicine, Chicago, Illinois, United States of America
Email: minoli.perera@northwestern.edu

Epigenetics is a reversible molecular mechanism that plays a critical role in many developmental, adaptive, and disease processes. DNA methylation has been shown to regulate gene expression and the advent of high throughput technologies has made genome-wide DNA methylation analysis possible. We investigated the effect of DNA methylation on eQTL mapping (methylation-adjusted eQTLs), by incorporating DNA methylation as a SNP-based covariate in eQTL mapping in African American derived hepatocytes. We found that the addition of DNA methylation uncovered new eQTLs and eGenes. Previously discovered eQTLs were significantly altered by the addition of DNA methylation data suggesting that methylation may modulate the association of SNPs to gene expression. We found that methylation-adjusted eQTLs that were less significant compared to PC-adjusted eQTLs were enriched in lipoprotein measurements (FDR=0.0040), immune system disorders (FDR = 0.0042), and liver enzyme measurements (FDR=0.047), suggesting that DNA methylation modulates the genetic regulation of these phenotypes. Our methylation-adjusted eQTL analysis also uncovered novel SNP-gene pairs. For example, we found that the SNP, rs1332018, was associated to *GSTM3*. *GSTM3* expression has been linked to Hepatitis B which African Americans suffer from disproportionately. Our methylation-adjusted method adds new understanding to the genetic basis of complex diseases that disproportionally affect African Americans.

Keywords: genome-wide methylation, eQTLs, African Americans, Hepatocytes

1. Introduction

DNA methylation plays an important role in the regulation of gene expression which in turn affects many complex diseases and traits.[1] Integration of DNA methylation into expression Quantitative Trait Loci (eQTL) mapping, can be challenging as the addition of SNP-based covariates is computationally intensive and multi-omics datasets with matching samples are sparse.[2] Moreover, matching datasets in minority populations are nearly absent from public databases. DNA methylation patterns, in particular, are complex, vary greatly from sample to sample[3], and change with environmental exposures.[4] Therefore, DNA methylation studies can be hard to generalize. The advent of high throughput and next generation sequencing technologies, however, has made it possible for DNA methylation to be analyzed genome-wide.[4] Several investigators have previously integrated genome-wide sequencing data and DNA methylation to uncover SNPs that significantly associate to CpG methylation status, called methylation QTLs (meQTLs).[5, 6] These studies have found that DNA methylation plays a significant role in the onset of diseases and phenotypes such as obsessive-compulsive disorder and drug response.[5, 6] Most of these studies have been conducted in populations of European ancestry.

The African American population is widely underrepresented in genetic studies. In GWAS studies, only 19% of individuals are non-European and less than 5% are non-European and non-Asian.[7] While other eQTL mapping studies have used African American samples, the number of individuals have been very small, thus making them underpowered to adequately account for population specific variation. Furthermore, these studies did not account for DNA methylation as a SNP-based covariate.[7] In this study, we perform the first investigation of the effect of DNA methylation on eQTL mapping in African Americans and evaluate methylation-adjusted eQTL associations to complex diseases, phenotypic traits, and metabolic traits. These findings may explain the role DNA methylation plays in health disparities observed in African Americans.

2. Methods

2.1. *Cohort*

Fig. 1. Flowchart showing the study design and the methods used in each dataset.

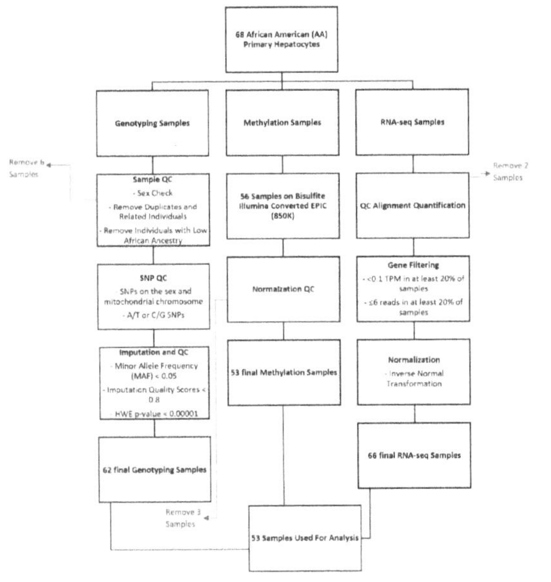

Sixty-eight African American hepatocyte cultures were acquired. After genotyping, DNA methylation quality control and RNA-sequencing quality control, 53 samples were used to conduct this analysis as shown in Fig. 1. Hepatocytes were either purchased from commercial companies (BioIVT, TRL, Life technologies, Corning, and Xenotech) or isolated from cadaveric livers using the same procedure described in Park et. al.[8] All genomic, transcriptomic and methylome data were gathered from the same hepatocyte samples.

2.2. *Genotyping, Imputation, and QC*

DNA was extracted from each hepatocyte culture using Gentra Puregene Blood kit (Qiagen) and all the DNA samples were bar coded for genotyping. The SNPs were genotyped using the Illumina Multi-Ethnic Genotyping array (MEGA) at the University of Chicago Functional Genomics Core using standard protocols. The outputs were then created by Genome Studio using a 0.15 GenCall score as the cutoff. PLINK[9] was then used to perform a sex check and to identify individuals with discordant sex information. The identity-by-descent method was used with a cut off score of 0.125 to identify duplicated or related individuals, where the cutoff score indicates third-degree relatedness. The following SNPs were excluded: SNPs on the sex and mitochondrial chromosome, A/T or C/G SNPs which may introduce flip-strand issues, SNPs with missing rate > 5% or failed Hardy-Weinberg equilibrium (HWE) tests ($p < .00001$), leaving 674,996 SNPs. Genotypes were phased using SHAPEIT and imputed with IMPUTE2. After imputations, SNPs were excluded for minor allele frequency < 0.05, imputation quality scores < 0.8, and HWE p-value < .00001, leaving 7,180,502 SNPs in the analysis.

2.3. *RNA-sequencing and QC*

Total RNA was extracted from each primary cell culture after three days of plating using the Qiagen RNeasy Plus mini-kit. Only the samples with RNA integrity number (RIN) score > 8 were sequenced. RNA-seq libraries were prepared using TruSeq RNA Sample Prep Kit, Set A (Illumina catalog # FC-122-1001) in accordance with the manufacturer's instructions. Illumina HiSeq 2500 and HiSeq 4000 machines were used to prepare the cDNA libraries sequence and. This resulted in 50 million reads per sample (single-end 50bp reads).

Quality of the raw reads from FASTQ files was assessed by FastQC (v0.11.2). A per base sequence quality threshold of > 20 across all bases was set for the fastq files. STAR 2.5[10] was used to align the reads to human Genome sequence GRCh38 and Comprehensive gene annotation (GENCODE version 25). Only uniquely mapped reads were retained and indexed by SAMTools 1.2.[11] To assess the nucleotide composition bias, GC content distribution and coverage skewness of the mapped reads read_NVC.py, read_GC.py and geneBody_coverage.py from RNA-SeQC (2.6.4)[12] were used. Lastly, Picard CollectRnaSeqMetrics was applied to evaluate the distribution of bases within transcripts. Fractions of nucleotides within specific genomic regions were measured and only samples with > 80% of bases aligned to exons and UTRs regions were retained for analysis.

2.4. *Gene expression quantification*

To quantify gene expression for the 17,992 genes used in the study a collapsed gene model was used, following the GTEx isoform collapsing procedure.[13] The reads were mapped to genes referenced with Comprehensive gene annotation (GENCODE version 25) to evaluate gene-level expression using RNA-SeQC.[12] The Bioconductor package, DESeq2 (version1.20.0)[14] was used to supply HTSeq[15] raw counts for the analysis of gene expression. DESeq2 was also used to perform principal component analysis (PCA). Using regularized log transformation, the counts were normalized. The two PC's used in the study, PC1 and PC2, were plotted to visualize the expression patterns of the samples and two samples with distinct expression patterns were excluded as outliers.

The gene expression was normalized by the trimmed mean of M-values normalization method (TMM), which was implemented in edgeR.[16] The TPM (transcript per million) was calculated by first normalizing the counts by gene length and then normalizing by read depth. The thresholds for gene expression values were set at < 0.1 TPM in at least 20% of samples and ≤ 6 reads in at least 20% of samples. Inverse normal transformation was used to normalize the expression values for each gene. The gene coordinates were remapped to hg19/GRCh 37 (GENCODE version 19) due to genotype imputation limitations.

2.5. *Methylation Sample Preparation and QC*

DNA was isolated from hepatocytes as described in Park et al.[8] As shown in Fig. 1, 56 of the hepatocyte samples produced sufficient bisulfite-converted DNA for analysis. The Illumina MethylationEPIC BeadChip microarray (San Diego, Ca, USA), consisting of approximately 850,000 probes[17] was used for methylation profiling from 56 AA hepatocytes that overlapped the samples used for gene expression analysis.

Methylation data QC and normalization was performed using the ChAMP R package (version 2.10.1)[18] as previously described in Park et.al.[8] This process removed: 9204 probes for any sample that did not have a detection p value <0.01, 1043 probes with a bead count <3 in at least 5% of samples, 49 probes that align to multiple locations as identified by Nordlund et al.[19], 2975 probes with no CG start sites, and 17,235 probes located on X and Y chromosomes. After QC, three samples were excluded, resulting in 53 samples remaining in the analysis.

2.6. *Methylation-adjusted eQTLs*

The R package Matrix eQTL[20] was used to determine the methylation site(s) that correspond to each SNP within a 2.5 kB window. CpG sites were then grouped together by SNP to determine the number of CpG sites on average at each SNP and to determine the pairwise correlation between CpG sites at each SNP. We used a weighted average based on the distance of the CpG site from the SNP to determine the methylation values for each SNP. If only one CpG site was linked to a SNP, then the weight of the CpG site would be:

$$\mathrm{w} = 1 - \left(\frac{d}{2500}\right) \tag{1}$$

, where d is the genomic distance (in base pairs) between the CpG site and the SNP and 2500 represents the 2.5kB window size used in this analysis. This weight would then be multiplied by the methylation value of the CpG site to get the normalized methylation value used in the analysis. This weighting system allowed proximal CpG sites to have a greater weight. If more than one CpG site was found within the 2.5kB region then each CpG site's weight, w_i, was calculated using equation (1) above and the final weight for each CpG site was calculated as:

$$w_f = \frac{w_i}{\sum_{k=1}^{N} w_k} \tag{2}$$

, where N is the total number of CpG sites that correspond to a particular SNP and $\sum_{k=1}^{N} w_k$ represents the sum of the initial weights of all the CpG sites that correspond to that SNP. This calculation ensures the sum of the final weights of all CpG sites corresponding to a single SNP are equal to one. The SNP-based methylation value was then calculated by:

$$\mathrm{M} = \sum_{f=1}^{N} w_f * m_f \tag{3}$$

, where M is the SNP-based methylation value and $\sum_{f=1}^{N} w_f * m_f$ represents the weighted sum of all of the methylation values for the CpG sites corresponding to that SNP. These averaged methylation values used as a SNP based covariate and eQTLs were mapped using the LAMatrix R package.[21] The methylation-adjusted eQTLs and PC-adjusted eQTLs were adjusted for sex, platform, batch, genotype-derived PCs 1 and 2, and 10 PEER variables estimated from normalized expression values as previously described in Zhong et.al.[7] The genotype-gene expression associations within a *cis* region (1 Mb around the gene) were tested. PC-adjusted eQTLs (mapped in the same hepatocyte cultures) were compared to methylation-adjusted eQTLs to investigate if changes in the eQTL statistical significance or change in effect size (Spearman's correlation).[21] All relevant data are within the manuscript are available from the GEO (GSE124076 and GSE147628).

2.7. *eQTL and GWAS overlap*

To understand how the methylation-adjusted eQTLs may explain the underlying mechanisms in GWAS findings, the method presented in Zhong et. al.[7] was used with some modifications. This included downloading the NHGRI/EBI GWAS Catalog file (v.1.0.2, 2019-03-22) and keeping only the associations that passed the genome-wide significant level ($p<5e-8$). Furthermore, the rsids were remapped from Build38 to Build37 using Ensembl API. The 1000 Genomes YRI population were used to extract all the variants in LD with the independent GWAS variants ($r^2 > 0.8$) and the traits of the corresponding GWAS hits were put into 17 groups which corresponded to ontology-based trait categories.[22] A false discovery rate (FDR) threshold of 0.05 was set as significant enrichment for an ontology. The methylation-adjusted eQTLs were split into three groups for this analysis: (i) eQTLs that were significant with PC-adjustment and increased in significance with methylation-adjustment, (ii) eQTLs that were not significant with PC-adjustment and became significant with methylation- adjustment (FDR<0.05), and (iii) eQTLs that were significant with PC-adjustment and became less significant with methylation-adjustment. These three groups of eQTLs were compared to the GWAS variants.

3. Results

Fifty-three African American hepatocyte samples were used in this analysis, with 28 (52.8%) males and 25 (47.2%) females. The age (mean ± standard deviation) of the cohort was 39 ± 18.4 years old. To account for methylation in this eQTL mapping analysis, LAMatrix was used.[21] Instead of incorporating local ancestry into the analysis as previously done[7], DNA methylation was used in its place. LAMatrix was chosen because the R package has increased power and controls false positives when gene expression differs by locus-specific covariate, such as methylation.[21]

3.1 Methylation-adjusted *eQTLs vs PC-adjusted eQTLs*

Out of the 7,180,502 total SNPs in the dataset, 2,494,181 SNPs had at least one CpG site within the 2.5 kB window, with an average of 3.08 CpG sites per SNP (ranging between of 1 to 95 CpG sites per SNP). We identified 2,296 eQTLs with methylation-adjustment at an FDR threshold of 0.05. To ascertain if any methylation-adjusted eQTLs resulted in the novel discovery of regulatory variation, we compared significant methylation-adjusted eQTLs (FDR<0.05) against significant PC-adjusted eQTLs (FDR<0.05). This comparison resulted in 308 unique methylation-adjusted eQTLs that were not found with PC-adjusted analysis, and 1,954 eQTLs which were common to both analyses. The remaining 19,567 found in PC-adjustment were not significant in this analysis (Fig.2A). The comparison revealed that there were 11,485 eQTLs that were significant with PC-adjustment and decreased in significance with methylation-adjustment and 50 eQTLs that were significant with PC-adjustment increased in significance with methylation-adjustment (Fig. 2B). We compared the effect size for methylation-adjusted eQTLs (all methylation-adjusted eQTLs and by the groups defined in Fig.2B) versus PC-adjusted eQTLs. All groups showed high correlation of effect size

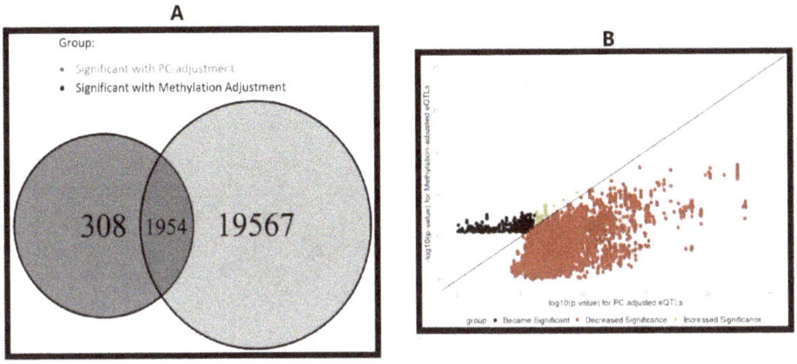

Fig. 2. Methylation-adjusted eQTLs as compared to PC-adjusted eQTLs

A) The Venn-Diagram showing the number of eQTLs that are significant with methylation-adjustment, significant with PC-adjustment, and significant in both analyses.

B) Comparison of the p-values of the 308 eQTLs that became significant with methylation-adjustment (black), 11,485 that were significant with PC-adjustment and decreased in significance with methylation-adjustment (red), and 50 that were significant with PC-adjustment and increased in significance with methylation-adjustment (gold).

(Spearman's correlation = 0.32-0.42, $p < 2.2e\text{-}6$, data not shown).

3.2 *GWAS Associations for Methylation-adjusted eQTLs*

We overlapped the methylation-adjusted eQTLs with SNPs in previously reported GWAS. Variants, from NHGRI-EBI GWAS catalog, or their tagging variants ($r^2 > 0.8$, 1000 Genomes YRI population) were used to determine the overlap with h the methylation-adjusted eQTLs. To analyze the effect of methylation even further, the methylation-adjusted eQTLs were broken into three groups: (i) eQTLs that are only significant with methylation-adjustment, (ii) eQTLs that were significant with PC-adjustment but became more significant with methylation-adjustment, and (iii) eQTLs that were significant with PC-adjustment but became less significant with methylation-adjustment. In total, there were 285 GWAS associations that intersect with methylation-adjusted eQTLs across the three groups.

3.2.1. *Group 1: eQTLs that were only significant with methylation-adjustment (N = 308)*

For eQTLs that were only significant with methylation-adjustment, 16 GWAS associations were found that intersected with these eQTLs. There was significant enrichment for digestive system disorders (FDR = 0.011), as shown in Fig. 3A. One of the eQTLs enriched for digestive system disorders, rs11546996, was associated with primary biliary cirrhosis.[23] Due to the intergenic location of rs11546996, the causal gene was reported as *SPIB* in this study. However, our analysis associated rs11546996 to *PNKP* (P-value = 1.05e-6, FDR = 0.026) thereby potentially identifying a new causal gene for primary biliary cirrhosis by accounting for methylation in eQTL mapping.

3.2.2. *Group 2: eQTLs that were significant with PC-adjustment and increased in significance with methylation-adjustment (N = 50)*

For eQTLs that were significant with PC-adjustment and increased in significance with methylation-adjustment, eight GWAS associations intersected with this group of eQTLs. No significantly enriched was found (Fig. 3B). This may be due to the very small number of eQTLs in this group.

3.2.3. *Group 3: eQTLs that were significant with PC-adjustment and decreased in significance with methylation-adjustment (N = 11,485)*

For eQTLs that were significant with PC-adjustment and decreased in significance with methylation-adjustment, 261 GWAS associations intersected with eQTLs in this group. There was significant enrichment for lipid or lipoprotein measurements (FDR = 0.0040), immune system disorders (FDR = 0.0042), and liver enzyme measurements (FDR = 0.047) (Fig. 3C). This suggests that these SNPs may be associated to susceptibility of disease, but that susceptibility may be modulated by DNA methylation.

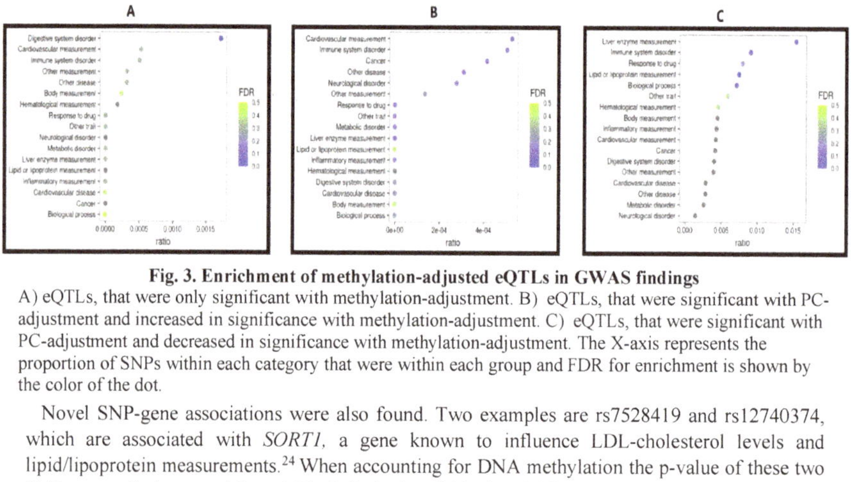

Fig. 3. Enrichment of methylation-adjusted eQTLs in GWAS findings
A) eQTLs, that were only significant with methylation-adjustment. B) eQTLs, that were significant with PC-adjustment and increased in significance with methylation-adjustment. C) eQTLs, that were significant with PC-adjustment and decreased in significance with methylation-adjustment. The X-axis represents the proportion of SNPs within each category that were within each group and FDR for enrichment is shown by the color of the dot.

Novel SNP-gene associations were also found. Two examples are rs7528419 and rs12740374, which are associated with *SORT1*, a gene known to influence LDL-cholesterol levels and lipid/lipoprotein measurements.[24] When accounting for DNA methylation the p-value of these two SNP-gene pairs increased from 1.99e-9 for both to 4.92e-8 and 1.25e-7, respectively. The FDR also increased from 3.01e-5 for both to 3.12e-3 and 6.08e-3, respectively. Furthermore, both SNPs had proportion of DNA methylation ranging from 0.12 to 0.33. Although these SNP-gene pairs remained significant with methylation-adjustment, their significance decreased dramatically indicating that methylation, near these SNPs, may play a role in the association between these SNPs to lipid phenotypes. This suggests that DNA methylation should be considered when assessing genomic risk of LDL-cholesterol levels and cholesterol-related diseases, such as myocardial infarction.

The methylation-adjusted eQTL, rs9296736 associated with the expression of *MLIP*, was previously found to be associated with liver enzyme measurements.[25] High levels of liver-enzymes in plasma are widely associated with an increased risk for developing many diseases including cirrhosis and cardiovascular disease.[25] This SNP-gene pair decreased in significance considerably when it was adjusted for methylation. The p-value and FDR of this methylation-adjusted eQTL went from 2.08e-9 and 3.13e-5 with PC-adjustment to 0.037 and 0.94, respectively, with methylation-adjustment. For this SNP-gene pair, rs9296736 was highly methylated, with proportion of DNA methylation ranging from 0.89 to 0.97. This result suggests that the association of rs9296736 to *MLIP* and liver enzyme measurements may depend on the DNA methylation landscape.

3.3. *Discovery of eGenes associated to disease traits using methylation-adjusted eQTL mapping*

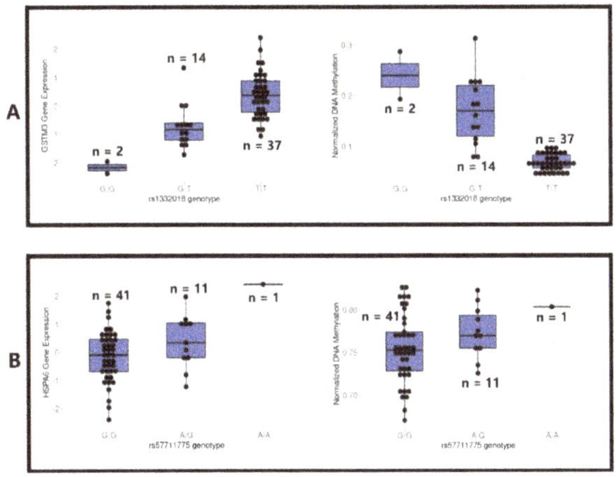

Fig. 4. Boxplots of genotype vs gene expression and DNA methylation for *GSTM3 and HSPA6.*

A) A significant increase in *GSTM3* gene expression (p = 1.1e-6) and a significant decrease in DNA methylation (p = 9.2e-13) are associated with rs1332018. The number of individuals (n) is shown for each genotype.

B) A significant increase in *HSPA6* gene expression (p=0.0099) and an increase in DNA methylation (p=0.20) are associated with rs57711775. The number of individuals (n) is shown for each genotype.

There were 179 eGenes found through methylation-adjusted eQTL mapping (FDR<0.05) of which 80 eGenes that were not significant with PC-adjustment. Two of these eGenes, *GSTM3* (FDR = 0.014) and *HSPA6* (FDR = 0.029), have been associated to disease traits such as Hepatitis B (HBV) for *GSTM3* and Hepatocellular Carcinoma (HCC) for both eGenes.[26-29] African Americans have a higher incidence and worse outcomes of HBV and HCC when compared to other demographics.[30, 31] Since these eGenes were not significant with PC-adjusted eQTL mapping, they may explain how methylation plays a role in the health disparities observed in African Americans. As shown in Fig.

4A, there is a significant assocication between rs1332018 genotype and *GSTM3* expression as well as rs1332018 genotype to DNA methylation. From this we can see that the T allele is associated with both increased gene expression and lower DNA methylation. A total of 18 CpG sites contributed to this association. As shown in Fig. 4B, there is also relationship between *HSPA6* gene expression and DNA methylation with the A allele associated with both increased gene expression and increased DNA methylation, though the later did not reach statistical significance. A total of 7 CpG sites contributed to this association.

4. Discussion

Through the integration of DNA methylation into eQTL mapping, we showed how methylation potentially plays a critical role in SNP-gene associations as well as the association of these eQTLs to diseases and metabolic traits. Our analysis was aided by using the computationally efficient R package, LAMatrix, which allows for the addition of a SNP based covariate to eQTL mapping. Additionally, our use of data from African Americans aided in the discovery of new regulatory variants as this population is more genetically diverse than European ancestry populations. Previous meQTLs studies have shown that SNPs can affect the methylation status of nearby CpGs, not only CpGs that overlap the SNP location. Shultz et. al. showed that SNPs within 0.2Mb of a CpG can significantly associate with methylation status.[32] We used a weighted approach which assumes that SNPs have a larger effect on closer CpG sites as previous studies showed a decrease in the association p-values of meQTLs with distance.[33, 34] In our analysis we accounted for methylation within a 2.5Kb window. Larger window sizes may be more appropriate, but as no previous study has incorporate methylation into eQTL mapping we took a conservative approach.

We found unique eGenes in our analysis that were not found by eQTL mapping with only PC-adjustment. Two of these eGenes, *GSTM3* and *HSPA6*, are associated with diseases such as HBV for *GSTM3* and HCC for both eGenes.[26-29] These are diseases that disproportionately affect African Americans.[30, 31] *GSTM3* has also been associated to oxidative stress and specifically several studies have found that epigenetic suppression of *GSTM3* in HBV-infected cells causes oxidative stress[27, 28], which can lead to HCC.[26] Furthermore, other studies showed that *GSTM3* expression was lowered with promoter hypermethylation[35] and in chemical-induced HCC.[36] This finding agrees with the previous studies mentioned, showing that epigenetic suppression of *GSTM3* leads to HCC in HBV-infected cells.[26-28] We found a significant inverse relationship between *GSTM3* expression and DNA methylation around rs1332018. This suggests that individuals with rs1332018 genotypes that have a lower *GSTM3* expression and higher methylation may be at a higher risk for HCC. *HSPA6* was also found to be overexpressed in human HCC tissues and a potential risk factor for HCC reccurence.[29] We found that the expression of *HSPA6* increased with methylation around rs57711775, which could mean that methylation potentially plays a role in upregulating *HSPA6*. Furthermore, the A allele of rs57711775 that is associated with higher *HSPA6* expression and higher methylation in our analysis. This SNP is not found in European ancestry populations according to the Ensembl database. Thus, we potentially elucidated a causal variant and risk allele for HCC specific to African Americans by incorporating methylation into this analysis. As has previously been reported, the direction of effect of DNA methylation is dependent on the location of methylation.[37, 38] Previous studies have shown that methylation within the transcriptional start site of the promoter is well known to repress gene expression while methylation within the gene body results in more variable expression.[37, 38] Therefore, both *GSTM3* expression and *HSPA6* expression may contribute to the onset of HCC in African Americans.

In our GWAS enrichment, we found a significant enrichment for digestive system disorders for the eQTLs significant only with methylation-adjustment and a significant enrichment for lipid or lipoprotein measurements, immune system disorders, and liver enzyme measurements for the eQTLs that are less significant with methylation-adjustment. Both immune-related phenotypes and lipid and lipoprotein measures differ by population and may contribute to disease disparities. Our findings suggest that methylation may play a role in these diseases. Further studies are needed to determine if DNA methylation around these specific SNPs and genes differ between populations. This analysis also revealed an interesting association with eQTLs that were only significant after methylation-adjustment. The SNP, rs11546996, a SNP associated with primary biliary cirrhosis, was a methylation-adjusted eQTL for *PNKP*. In a previous GWAS study, a causal SNP-gene association for primary biliary cirrhosis was found with rs11546996 and the causal gene was assumed to be *SPIB*, as it is the closest gene.[23] Since our study specifically looked at gene expression in hepatocytes, a tissue relevant for this disease, we may have found a potentially novel SNP-gene pair associated with primary biliary cirrhosis whose expression is regulated by both DNA methylation and gene variation. *PNKP* has also been associated with repairing DNA after damage from oxidative stress[39], so rs11546996 could be a SNP that effects this process.

There were several limitations to our study. First, we were only able to include 53 samples into this analysis and hence our analysis was underpowered. Second, we assessed DNA methylation with the Illumina EPIC array which is limited to the CpG sites chosen for the chip. Unmeasured DNA methylation may have effects on eQTLs that were not captured by our analysis. Third, our results, compared to the findings in the entire GWAS catalog, are only applicable to diseases in which hepatocytes play a key role. Our findings may not be generalizable to other cell or tissue types. Finally, we have assumed that DNA methylation closer to the SNP is more likely to influence eQTL mapping, however this may not always be the case. Additionally, we do not account for differences in effect size in our method. With greater meQTL analysis in relevant cell types and populations, we may be able to weight SNP/DNA methylation interactions more precisely as a SNP-based covariate.

In conclusion, this is the first study to explore the effect of DNA methylation in eQTL mapping in African Americans. The African American demographic is widely underrepresented in genetic studies and their greater genetic diversity may allow us to find novel SNP-gene pairs as well as population specific SNPs. Our findings can be used to understand how DNA methylation potentially plays a role in complex diseases, phenotypic traits, and metabolic traits in African Americans.

5. Acknowledgments

This work was supported by NIH National Institute on Minority Health and Health Disparities (NIMHD) Research Project 1R01MD009217-01 (R01).

References

1. Gamazon, E.R., et al., *Using an atlas of gene regulation across 44 human tissues to inform complex disease- and trait-associated variation.* Nature Genetics, 2018. **50**(7): p. 956-967.
2. Palsson, B. and K. Zengler, *The challenges of integrating multi-omic data sets.* Nature Chemical Biology, 2010. **6**(11): p. 787-789.
3. Laird, P.W., *Principles and challenges of genome-wide DNA methylation analysis.* Nature Reviews Genetics, 2010. **11**(3): p. 191-203.
4. He, Z., et al., *Role of genetic and environmental factors in DNA methylation of lipid metabolism.* Genes Dis, 2018. **5**(1): p. 9-15.
5. Bonder, M.J., et al., *Genetic and epigenetic regulation of gene expression in fetal and adult human livers.* BMC Genomics, 2014. **15**(1): p. 860.
6. Park, C.I., et al., *Reduced DNA methylation of the oxytocin receptor gene is associated with obsessive-compulsive disorder.* Clinical Epigenetics, 2020. **12**(1): p. 101.
7. Zhong, Y., et al., *Discovery of novel hepatocyte eQTLs in African Americans.* PLoS Genet, 2020. **16**(4): p. e1008662.
8. Park, C.S., et al., *Hepatocyte gene expression and DNA methylation as ancestry-dependent mechanisms in African Americans.* npj Genomic Medicine, 2019. **4**(1): p. 29.
9. Purcell, S., et al., *PLINK: a tool set for whole-genome association and population-based linkage analyses.* Am J Hum Genet, 2007. **81**(3): p. 559-75.
10. Dobin, A., et al., *STAR: ultrafast universal RNA-seq aligner.* Bioinformatics, 2013. **29**(1): p. 15-21.
11. Li, H., et al., *The Sequence Alignment/Map format and SAMtools.* Bioinformatics, 2009. **25**(16): p. 2078-9.
12. DeLuca, D.S., et al., *RNA-SeQC: RNA-seq metrics for quality control and process optimization.* Bioinformatics, 2012. **28**(11): p. 1530-2.
13. Aguet, F., et al., *Genetic effects on gene expression across human tissues.* Nature, 2017. **550**(7675): p. 204-213.
14. Love, M.I., W. Huber, and S. Anders, *Moderated estimation of fold change and dispersion for RNA-seq data with DESeq2.* Genome Biol, 2014. **15**(12): p. 550.
15. Anders, S., P.T. Pyl, and W. Huber, *HTSeq--a Python framework to work with high-throughput sequencing data.* (1367-4811 (Electronic)).
16. Robinson, M.D. and A. Oshlack, *A scaling normalization method for differential expression analysis of RNA-seq data.* Genome Biol, 2010. **11**(3): p. R25.
17. Bibikova, M., et al., *High density DNA methylation array with single CpG site resolution.* Genomics, 2011. **98**(4): p. 288-95.
18. Morris, T.J., et al., *ChAMP: 450k Chip Analysis Methylation Pipeline.* Bioinformatics, 2014. **30**(3): p. 428-30.
19. Nordlund, J., et al., *Genome-wide signatures of differential DNA methylation in pediatric acute lymphoblastic leukemia.* Genome Biol, 2013. **14**(9): p. r105.
20. Shabalin, A.A., *Matrix eQTL: ultra fast eQTL analysis via large matrix operations.* Bioinformatics, 2012. **28**(10): p. 1353-8.
21. Zhong, Y., M.A. Perera, and E.R. Gamazon, *On Using Local Ancestry to Characterize the Genetic Architecture of Human Traits: Genetic Regulation of Gene Expression in Multiethnic or Admixed Populations.* (1537-6605 (Electronic)).

22. Welter, D., et al., *The NHGRI GWAS Catalog, a curated resource of SNP-trait associations.* Nucleic Acids Res, 2014. **42**(Database issue): p. D1001-6.
23. Hirschfield, G.M., et al., *Association of primary biliary cirrhosis with variants in the CLEC16A, SOCS1, SPIB and SIAE immunomodulatory genes.* (1476-5470 (Electronic)).
24. Willer, C.J., et al., *Newly identified loci that influence lipid concentrations and risk of coronary artery disease.* (1546-1718 (Electronic)).
25. Chambers, J.C., et al., *Genome-wide association study identifies loci influencing concentrations of liver enzymes in plasma.* Nature Genetics, 2011. **43**(11): p. 1131-1138.
26. Ivanov, A.V., et al., *Oxidative stress, a trigger of hepatitis C and B virus-induced liver carcinogenesis.* Oncotarget, 2017. **8**(3): p. 3895-3932.
27. Qi, L., et al., *Methylation of the glutathione-S-transferase M3 gene promoter is associated with oxidative stress in acute-on-chronic hepatitis B liver failure.* (1349-3329 (Electronic)).
28. Sun, Y., et al., *GSTM3 reverses the resistance of hepatoma cells to radiation by regulating the expression of cell cycle/apoptosis-related molecules.* Oncol Lett, 2014. **8**(4): p. 1435-1440.
29. Yang, Z., et al., *Upregulation of heat shock proteins (HSPA12A, HSP90B1, HSPA4, HSPA5 and HSPA6) in tumour tissues is associated with poor outcomes from HBV-related early-stage hepatocellular carcinoma.* (1449-1907 (Electronic)).
30. Forde, K.A., *Ethnic Disparities in Chronic Hepatitis B Infection: African Americans and Hispanic Americans.* Current hepatology reports, 2017. **16**(2): p. 105-112.
31. Franco, R.A., et al., *Racial and Geographic Disparities in Hepatocellular Carcinoma Outcomes.* Am J Prev Med, 2018. **55**(5 Suppl 1): p. S40-S48.
32. Schulz, H., et al., *Genome-wide mapping of genetic determinants influencing DNA methylation and gene expression in human hippocampus.* (2041-1723 (Electronic)).
33. Banovich, N.E., et al., *Methylation QTLs are associated with coordinated changes in transcription factor binding, histone modifications, and gene expression levels.* PLoS Genet, 2014. **10**(9): p. e1004663.
34. McRae, A.F., et al., *Identification of 55,000 Replicated DNA Methylation QTL.* Sci Rep, 2018. **8**(1): p. 17605.
35. Peng, D.F., et al., *DNA hypermethylation regulates the expression of members of the Mu-class glutathione S-transferases and glutathione peroxidases in Barrett's adenocarcinoma.* Gut, 2009. **58**(1): p. 5-15.
36. Quiles-Perez, R., et al., *Inhibition of poly adenosine diphosphate-ribose polymerase decreases hepatocellular carcinoma growth by modulation of tumor-related gene expression.* Hepatology, 2010. **51**(1): p. 255-66.
37. Jiao, Y., M. Widschwendter, and A.E. Teschendorff, *A systems-level integrative framework for genome-wide DNA methylation and gene expression data identifies differential gene expression modules under epigenetic control.* Bioinformatics, 2014. **30**(16): p. 2360-6.
38. Yang, X., et al., *Gene body methylation can alter gene expression and is a therapeutic target in cancer.* Cancer cell, 2014. **26**(4): p. 577-590.
39. Jilani, A., et al., *Molecular cloning of the human gene, PNKP, encoding a polynucleotide kinase 3'-phosphatase and evidence for its role in repair of DNA strand breaks caused by oxidative damage.* J Biol Chem, 1999. **274**(34): p. 24176-86.

Innovative methodological approaches for data integration to derive patterns across diverse, large-scale biomedical datasets

Brett Beaulieu-Jones
Dept. of Biomedical Informatics, Harvard Medical School
10 Shattuck Street, 4th Floor
Boston, MA 02155
Email: brett_beaulieu-jones@hms.harvard.edu

Christian Darabos
Research, Teaching, and Learning at IT&C, Dartmouth College
37 Dewey Field Road
Hanover, NH 03755
Email: christian.darabos@dartmouth.edu

Dokyoon Kim
Dept. of Biostatistics and Epidemiology, Perelman School of Medicine at UPenn
3400 Civic Center Boulevard, Building 421
Philadelphia, PA 19104
Email: dokyoon.kim@pennmedicine.upenn.edu

Anurag Verma
Dept. of Genetics, Perelman School of Medicine at UPenn
3400 Civic Center Boulevard, Building 421
Philadelphia, PA 19104
Email: anuragv@upenn.edu

Shilpa Nadimpalli Kobren
Dept. of Biomedical Informatics, Harvard Medical School
10 Shattuck Street, 4th Floor
Boston, MA 02155
Email: shilpa_kobren@hms.harvard.edu

1 Introduction

"Biomedical data" refers to the increasingly large corpus of machine-mineable data encompassing two similar, yet pointedly distinct fields: biology and medicine. In recent years, experimental and technological advancements in these fields have resulted in an unprecedented diversity of molecular omics data and longitudinal health record data available for analysis (Lee et al., 2020; Mandel et al., 2016; Turro et al., 2020). Moreover, entirely new data sources such as social networking data, wearable technologies, and environmental measurements have emerged and are relevant indicators of phenomena observed across biology and medicine (Eagle et al., 2010; Le Goallec et al., 2020). Creative and sophisticated integration of these datasets promises the opportunity to further biological knowledge and understanding of disease and ultimately advance our ability to holistically detect and treat disease and improve patient care. However, challenges stemming from limited data quality and standardization, coupled with a dramatic increase in data size and required computational resources arise in pursuit of these goals.

Overcoming these inherent challenges to elucidate meaningful and relevant patterns from biomedical data requires integrating distinct data modalities and developing related methodological approaches (Lakhani et al., 2019). Data integration is necessitated by the noisiness, incompleteness, and/or other insufficiencies of information contained in any single biomedical data source when considered in isolation. Sometimes data is missing from certain sources in a biased manner as well. In other cases, labels assigned to data can be misleading or non-randomly incomplete. Additionally, emerging technologies are leading to more data that may not be amenable to traditional analysis approaches. Social media data, environmental data, wearable data, and patient-provided data, for instance, have become increasingly common in recent years, and each present unique challenges. Domain-specific knowledge and advanced technical processing are critical for properly integrating and deriving signals from these data.

Methodologically, it is critical to identify and understand the limits of labels assigned to biomedical data. For example, there are challenges in assigning levels of confidence or evidence to discoveries that do not have strong gold-standard truth assessments. In other instances, gold standard labels may be attainable through a costly and time-consuming process (e.g., clinical chart review). In biomedicine, multiple data sources are thus leveraged in practice to "build a case" that supports a hypothesis. For instance, genetic and medical imaging data can be examined in conjunction for improved pathology predictions (Pasco et al., 2011; Yu et al., 2016). Properly correcting for censoring challenges may require examining long term outcomes. Additionally, it may be necessary to impute data or otherwise account for its presence or absence. Building tools to visualize data or metadata may be helpful for human in-the-loop learning. Finally, it is critical to understand and mitigate sources of bias stemming from external factors or the data generation process, such as batch effects, institutional discrepancies in recording, and dataset shift.

Here, we highlight recent, innovative approaches utilizing new combinations of biomedical data sources to address previously intractable questions. We focus specifically on cutting-edge methods aimed at pattern discovery in biomedical data through novel pattern recognition or data integration. The research discussed here has two common themes: (i) using representation learning to model structures in data to enable biological or etiological understanding, and (ii) data integration with applications to cancer.

2 Understanding Biology by Modeling Structure and Processes with Machine Learning

In recent years, significant advances in learning representations have proven critical for modeling human biology and disease etiology processes with machine learning (Ching et al., 2018). These advances in knowledge representation can be applied to challenging questions such as modeling genomic and protein-protein interaction patterns in cancer to understand dysregulation patterns (Durmaz et al., 2020), learning a framework for the connection between chemical compounds and their effects on gene expression (Finlayson et al., 2020), and using varied levels of structure to learn both local and global patterns from histological images (Levy et al., 2020). Considering graph structures is a common trend across each of these latter works.

Durmaz et al. propose a framework to use subgraph mining to identify functional dysregulation patterns in cancer. They perform unsupervised learning by probabilistically mining graph structures of protein-protein interactions. To this end, they utilized subgraph frequency and random walk approaches. Their approach recovers pathways included in expert-knowledge graphs, and, through clustering, points towards the biological significance of functionally dysregulated pathways.

Understanding the ways in which chemical structure can lead to different molecular activities could greatly enhance therapeutic development and mechanistic understanding of existing therapeutics. Despite these immense advantages, accurately understanding the relationship between chemical structure and molecular activity has proven to be a challenging problem in the general sense. Finlayson et al. employ an approach to train a set of neural networks to learn how to associate the structure of a given small molecule with its effect on changes in gene expression. This method attempts to jointly optimize representations of chemical structure and the transcriptional changes resulting from exposure to these chemicals. Despite observing mixed performance when attempting to generalize to new tissues, this method shows great potential to make progress on a longstanding, challenging problem and may lead to the ability to more effectively perform *in silico* prioritization of molecules to elicit specific transcriptional responses.

Digital pathology has seen an immense amount of activity where deep learning and convolutional neural networks have been applied to analyze pathology images. One unique challenge in digital pathology has been that whole slide images are too large for many of these neural network approaches to process. Levy et al. propose methods that use a combination of topological domain analysis and graph neural networks to reduce the need to break whole slide images into smaller patches of images that are computationally tractable; this latter approach is lossy yet common. Importantly, their topological analysis allows Levy et al. to quantify a graph neural network's quality of fit and help determine regions of interest. The combination of topological domain analysis and graph neural networks showed significant improvement over traditional convolutional neural networks applied to the task.

3 Data Integration with Applications to Cancer

One of the most genetically, functionally, and medically heterogeneous diseases afflicting humans in modern times is cancer. This disease, typified by one's own cells growing and dividing uncontrollably while evading the immune system to form tumors, is still challenging to detect, diagnose, and treat due to the various molecular mechanisms involved and diverse medical presentations. The papers we highlight here have employed biomedical data integration specifically to address cancer-specific challenges. In general, multiple distinct data modalities can be integrated via novel techniques to more effectively use poorly labeled or unlabeled data. Data sources that can be examined together include: molecular 'omics data (genomics, transcriptomics, proteomics, metabolomics etc.), medical imaging data, free text, and longitudinal outcomes data. The inclusion and integration of new and novel data sources can help examine and understand various biological processes, many of which have been implicated in cancer progression.

Scott et al. highlight the lack of heterogeneity in discovery populations and subsequent inability to accurately translate biomarkers for general use in the clinic. To address this challenge, the authors attempted to leverage heterogeneity, both biological and technical, across independent cohorts to find biomarkers more likely to generalize. By utilizing a primary dataset that includes 23 different cancers and combining it with 57 independent microarray datasets, they found the gene KRT8 to be significantly hypomethylated in the 57 independent datasets and overexpressed in 22 out of 23 cancers. Scott et al. then performed additional validation steps, including single-cell analyses, immunohistochemistry of tumor biopsies, and finally, detecting levels of KRT8 in the serum of patients with pancreatic cancer vs. healthy controls. While they have not yet shown its ability as a predictive marker for cases who have not yet been identified by other means, these validation steps show great potential for translational applicability.

Durmaz et al.'s approach to subgraph analyses allowed for the examination of single cancers using The Cancer Genome Atlas (TCGA) pan-cancer data. Their approach enabled the data-driven identification of patient clusters across different TCGA disease codes based on related dysregulation patterns and led to elucidating significant differences between survival for various disease codes, including lower grade gliomas and uterine cancer. The survival differences illustrate the potential of applying pathway-based functional networks to stratify cancer as compared to traditional gene-centric models. Additionally, considering cancer-relevant dysregulation at the pathway level versus the gene level provides additional insight into disease etiology.

Similarly, Levy et al.'s combination of topological data analysis with a graph neural network allows for identification of regions of interest in whole slide pathology images. The authors were subsequently able to measure the degree of overlap between regions of interest in a tumor and in the adjacent normal tissue and then associate these regions of interest with clinical outcomes by means of cancer staging. Their approach allows for human-readable highlighted regions of interest as well as a prediction of cancer stage where they found they were able to predict advanced colon cancer staging and positive lymph nodes at >0.9 AUC.

4 Discussion

Pattern recognition has already had and will continue to have a large role in understanding both biology and medicine. Technological developments are leading to larger and more varied biomedical datasets. Both novel and repurposed methodologies must be developed and applied to these data in order to derive insights that can drive more precise patient care, yield novel therapeutics, guide earlier interventions and in general provide greater understanding of biomedicine.

The work highlighted here targets these developments. Finlayson et al. aim to make the identification of therapeutically-relevant small molecules possible at faster speeds, and Durmaz et al. aspire to characterize the molecular mechanisms of cancer development and progress through computation that considers graph structure in protein interaction networks. Levy et al. propose novel methods to precisely extract regions of interest from histopathology images and to identify prognostic predictors to enable more precise patient care. Finally, Scott et al. identified a biomarker that may help lead to earlier and more accurate diagnoses of cancer. Each of these works is guided by the common theme of using pattern recognition to go beyond computational performance and to drive biomedical discovery and understanding.

References

Ching, T., Himmelstein, D. S., Beaulieu-Jones, B. K., Kalinin, A. A., Do, B. T., Way, G. P., Ferrero, E., Agapow, P.-M., Zietz, M., Hoffman, M. M., Xie, W., Rosen, G. L., Lengerich, B. J., Israeli, J., Lanchantin, J., Woloszynek, S., Carpenter, A. E., Shrikumar, A., Xu, J., … Greene, C. S. (2018). Opportunities and obstacles for deep learning in biology and medicine. *Journal of the Royal Society, Interface / the Royal Society*, *15*(141). https://doi.org/10.1098/rsif.2017.0387

Durmaz, A., Henderson, T. A. D., Bebek, G. (2020). "Frequent Subgraph Mining of Functional Interaction Patterns Across Multiple Cancers." *Pac Symp Biocomput.* To appear.

Eagle, N., Macy, M., & Claxton, R. (2010). Network diversity and economic development. *Science*, *328*(5981), 1029–1031.

Finlayson, S. G., McDermott, M. B. A., Pickering, A. V., Lipnick, S. L., Kohane, I. S. (2020). "Cross-modal representation alignment of molecular structure and perturbation-induced transcriptional profiles." *Pac Symp Biocomput.* To appear.

Lakhani, C. M., Tierney, B. T., Manrai, A. K., Yang, J., Visscher, P. M., & Patel, C. J. (2019). Repurposing large health insurance claims data to estimate genetic and environmental contributions in 560 phenotypes. *Nature Genetics*, *51*(2), 327–334.

Lee, H., Huang, A. Y., Wang, L.-K., Yoon, A. J., Renteria, G., Eskin, A., Signer, R. H., Dorrani, N., Nieves-Rodriguez, S., Wan, J., Douine, E. D., Woods, J. D., Dell'Angelica, E. C., Fogel, B. L., Martin, M. G., Butte, M. J., Parker, N. H., Wang, R. T., Shieh, P. B., … Nelson, S. F. (2020). Diagnostic utility of transcriptome sequencing for rare Mendelian diseases. *Genetics in Medicine: Official Journal of the American College of Medical Genetics*, *22*(3), 490–499.

Levy, J., Haudenschild, C., Barwick, C., Christensen, B., Vaickus, L. (2020). "Topological Feature Extraction and Visualization of Whole Slide Images using Graph Neural Networks." *Pac Symp Biocomput.* To appear.

Le Goallec, A., Tierney, B. T., Luber, J. M., Cofer, E. M., Kostic, A. D., & Patel, C. J. (2020). A systematic machine learning and data type comparison yields metagenomic predictors of infant age, sex, breastfeeding, antibiotic usage, country of origin, and delivery type. *PLoS Computational Biology*, *16*(5), e1007895.

Mandel, J. C., Kreda, D. A., Mandl, K. D., Kohane, I. S., & Ramoni, R. B. (2016). SMART on FHIR: a standards-based, interoperable apps platform for electronic health records. *Journal of the American Medical Informatics Association: JAMIA*, *23*(5), 899–908.

Pasco, P. M. D., Ison, C. V., Muñoz, E. L., Magpusao, N. S., Cheng, A. E., Tan, K. T., Lo, R. W., Teleg, R. A., Dantes, M. B., Borres, R., Maranon, E., Demaisip, C., Reyes, M. V. T., & Lee, L. V. (2011). Understanding XDP through imaging, pathology, and genetics. *The International Journal of Neuroscience*, *121 Suppl 1*, 12–17.

Scott, M. K. D., Ozawa, M. G., Chu, P., Limaye, M., Nair, V. S., Schaffert, S., Koong, A. C., West, R., Khatri, P. (2020). "A multi-scale integrated analysis identifies KRT8 as a pan-cancer early biomarker." *Pac Symp Biocomput.* To appear.

Turro, E., Astle, W. J., Megy, K., Gräf, S., Greene, D., Shamardina, O., Allen, H. L., Sanchis-Juan, A., Frontini, M., Thys, C., Stephens, J., Mapeta, R., Burren, O. S., Downes, K., Haimel, M., Tuna, S., Deevi, S. V. V., Aitman, T. J., Bennett, D. L., … Ouwehand, W. H. (2020). Whole-genome sequencing of patients with rare diseases in a national health system. *Nature*, *583*(7814), 96–102.

Yu, K.-H., Zhang, C., Berry, G. J., Altman, R. B., Ré, C., Rubin, D. L., & Snyder, M. (2016). Predicting non-small cell lung cancer prognosis by fully automated microscopic pathology image features. *Nature Communications*, *7*, 12474.

Frequent Subgraph Mining of Functional Interaction Patterns Across Multiple Cancers

Arda Durmaz [1,5], Tim A. D. Henderson [2], and Gurkan Bebek [1–4,*]

[1] Systems Biology and Bioinformatics Graduate Program,
[2] Computer and Data Sciences Department,
[3] Center for Proteomics and Bioinformatics,
[4] Nutrition Department,
Case Western Reserve University, 10900 Euclid Ave., Cleveland OH 44106, USA
[5] The Department of Translational Hematology and Oncology Research, Taussig Cancer Institute,
Cleveland Clinic, 9500 Euclid Ave., Cleveland, OH 44195, USA
[*] Correspondence should be addressed to gurkan.bebek@case.edu.
{axd497,tadh,gurkan.bebek}@case.edu

Abstract. Molecular mechanisms characterizing cancer development and progression are complex and process through thousands of interacting elements in the cell. Understanding the underlying structure of interactions requires the integration of cellular networks with extensive combinations of dysregulation patterns. Recent pan-cancer studies focused on identifying common dysregulation patterns in a confined set of pathways or targeting a manually curated set of genes. However, the complex nature of the disease presents a challenge for finding pathways that would constitute a basis for tumor progression and requires evaluation of subnetworks with functional interactions. Uncovering these relationships is critical for translational medicine and the identification of future therapeutics. We present a frequent subgraph mining algorithm to find functional dysregulation patterns across the cancer spectrum. We mined frequent subgraphs coupled with biased random walks utilizing genomic alterations, gene expression profiles, and protein-protein interaction networks. In this unsupervised approach, we have recovered expert-curated pathways previously reported for explaining the underlying biology of cancer progression in multiple cancer types. Furthermore, we have clustered the genes identified in the frequent subgraphs into highly connected networks using a greedy approach and evaluated biological significance through pathway enrichment analysis. Gene clusters further elaborated on the inherent heterogeneity of cancer samples by both suggesting specific mechanisms for cancer type and common dysregulation patterns across different cancer types. Survival analysis of sample level clusters also revealed significant differences among cancer types ($p < 0.001$). These results could extend the current understanding of disease etiology by identifying biologically relevant interactions.

Supplementary Information: Supplementary methods, figures, tables and code are available at https://github.com/bebeklab/FSM_Pancancer.

Keywords: Frequent Subgraph Mining, Pan-Cancer, Transcriptomics; Proteomics

1 Introduction

Cancer is an inherently complex and heterogeneous disease. New technologies provided a comprehensive list of genomic and epigenetic aberrations for tumor growth and proliferation (1–4). This knowledge base could provide a more comprehensive view of how signaling events alter homeostasis within cells, between cells, or the microenvironment. The multiple omics measurements collected could be integrated to identify mechanisms or specific functions relevant to cancer (5) where shared genomic features across cancers have been identified (1, 6, 7), some of which were through integrative methods to analyze multiple -omics datasets (8–11), While these gene-centric approaches report valuable insights, the biology behind their prognostics or stratification might be more complicated, leading to poor treatment options or reproducibility. For example, gliomas with mutated *IDH1* and *IDH2* have improved prognosis compared to gliomas with wild-type *IDH* (12). As a result, mutant-selective *IDH1* inhibitors were developed, but this drug strategy could make tumor progression worse (13–16). Other arguments are made over the validity of geneset-based biomarkers (17–19). Random genesets were shown to stratify patients into subgroups, contradicting the use of these geneset based methods (20, 21). Pathway-based approaches, on the other hand, could uncover functionally relevant mechanisms of oncogenic alterations to improve treatment options (4).

The availability of pan-cancer data allowed the simultaneous analysis of multiple cancer types. However, the multifaceted view of cancer hinders these efforts to uncover comprehensive maps of cancer for each cancer type. Sanchez-Vega et al. (4) were able to map 57% of tumors to at least one expert-curated signaling pathway targetable by currently available drugs. The ten expert-curated pathways in this study are a great resource but do not cover the alterations across all cancers. Leiserson et al. (22) focused on gene-level perturbations to find subnetworks common across cancer types but the identified subnetworks are not restricted to cover the same set of samples, which can mask subpopulations of samples with different genes mutated in the given subnetwork. An unsupervised approach that mines networks for a dynamic group of patients could bring a more comprehensive map and would provide improved insight into our understanding of tumor growth and treatment opportunities.

One of the commonly used methods in graph data mining is frequent subgraph mining (FSM). FSM provides a means to extract frequently occurring patterns in a graph database. For instance in the setting for protein-protein interactions (PPIs), one can define a graph for each cancer patient based on expressed proteins and mine for commonly occurring interactions across patients (23). FSM has been widely used in a variety of applications, including the identification of common metabolic pathways and clusters (24–26). Multiple algorithms have been developed to overcome challenges inherently present in subgraph mining regarding both memory and subgraph isomorphism issues (27–30). The general approach for mining frequently occurring patterns in a graph database is to grow candidate patterns either in depth-first search or breadth-first search manner and check whether the required support is achievable. One drawback of using FSM-based methods is the computational requirement since the subgraph isomorphism problem is NP-Complete (31).

One other methodology for utilizing global network topology is the random walks with restarts (RWR) on finite graphs. RWR algorithm is the simulation of a random walker jumping from node to node in the interaction network with the given parameters similar to the PageRank algorithm (32). A modified version of this approach is used to prioritize the local neighborhood by allowing the random walk to restart from specified seed nodes. This approach has been widely used for candidate gene prediction or disease-disease similarity measurements (33–35).

In this paper, we describe an integrative -omics approach to pan-cancer analysis using FSM coupled with biased random walks utilizing genomic alterations, gene expression, and PPI networks. We use FSM to identify frequently occurring interaction patterns to provide a better understanding of functional alterations across multiple cancer types while accounting for the complex interaction topology of cancer. Our goal is to integrate PPI networks with somatic alterations and gene expression profiles to infer molecular networks representing dysregulation in cancers. More specifically, we extract subnetworks that are frequent in the population and in close proximity to the mutated genes. In our analysis, we investigate TCGA samples for 32 cancer types. We present patient clusters across all cancer types as well as patient classifications of individual cancers based on these networks. We identify mechanisms that are shared across tumor types and unique to individual cancers.

2 Methods

2.1 Pan-cancer Dataset and Omic Databases

We have downloaded TCGA single nucleotide variation (SNV) data from UCSC Xena (36). Additionally, we have filtered out samples with mutations of more than 800 to reduce the possible effects of hypermutators. PPI network was downloaded from StringDB version 10.5 and filtered to include edges with confidence scores > 0.4 with the remaining number of nodes being 17473 (37). Pathways were downloaded from the Reactome database. We excluded pathways with genes of less than 8 (38).

2.2 Biased Random Walks with Restarts

Biased random walks are applied to each sample separately by considering the mutated genes as seeds hence prioritizing local neighborhood of genomic alterations (See Supplementary Method Section S1.1). In this process, nodes with high degrees will intrinsically have increased probability values/traversed more often, to capture nodes with a statistically significant association with the seed set of nodes, we compared these results to a null distribution generated by applying the biased random walks to thousand randomly generated seed sets keeping the number of seeds equal to the original seed set. p-values for each node are obtained by comparing the steady-state probability vector to the null distribution per gene. Multiple hypothesis testing corrections are done using the Bonferroni method and genes with p-values < 0.1 are kept. Restart probability for the biased random walk is chosen as 0.6 to not restrict the networks towards the neighbors of seed sets. However, note that restart probability can be fine-tuned specifically to each network but 0.6 generally performs well across biological networks. Furthermore, since biased random walks can also identify spurious significant nodes solely due to the topology of the network, we have extracted connected components with the number of genes > 3.

2.3 Frequent Subgraph Mining

We developed an efficient method to sample for frequently occurring subgraphs across pan-cancer samples (Algorithm 1). The goal of frequent subgraph mining is to discover all subnetworks of graphs in the database which recur at least k times (39, 40). The database is a collection of undirected gene networks assembled as described in Section 2.1. The parameter k in Algorithm 1 is called the *minimum support*. A subgraph is considered "frequent" (and *supported*) if it recurs at least k times.

In this analysis pipeline, we have applied biased random walks over the PPI network for each sample separately using the somatic alterations as seed sets. Following the RWR, FSM can be applied with two approaches: Mining a single graph database generated by merging the RWR results over all the samples or mining graph databases generated separately for each sample. A Sample-specific network can be generated by filtering the combined network to include nodes found significant for the current sample. Simply this approach will result in subnetworks with the specified

```
Param: Mutation Matrix V of size g × p          // Each column is a set of mutations from one sample
Param: Interaction Network W of size g × g      // Interaction matrix stored as an adjacency matrix
Param: Minimum Support k                        // minimum network size to be discovered
Result: Set of k-frequent subgraphs S
// biased random walk profiles
for i    1 to p do
    P    RWR(V[, i], W)                  // perform RWR per sample using each sample's mutation set.
    CC    ConnComp(P)                    // Find the connected component
    RES    Append(CC)                    // Save this network for this patient
end
// sample-specific graph databases
for i    1 to p do
    PD    []
    PD[i]    RES[i]                      // For all RWR networks identified with mutations of the patient
    for j    1 to p do
        for e ∈ RES[j] do
            if e ∈ RES[i] then
                PD[j]    Append(e)       // Merge to create a sample-specific network database
            end
        end
    end
    D    Append(PD)
end
// frequent subgraph mining of sample-specific graph databases
for i    1 to p do
    PD    D[i]                           // for each patient's network
    for h ∈ GetCand(PD)                  // collect frequent subgraphs
     do
        if GetSupp(h) ≥ k                // if support for subgraphs is larger than k
         then
            S    Append(h)               // include this subnetwork in the results
        end
    end
end
return S
```

Algorithm 1: High-level algorithm for the proposed framework. The input matrices V and W have sizes $g \times p$ and $g \times g$ respectively. g is the number of genes and p is the number of samples in the pan-cancer data. k is the minimum number of samples a frequent subnetwork recurs. The algorithm returns the set of k-frequent subgraphs.

support that are also present in the current sample. FSM results for sample-specific networks are then merged and duplicate networks are filtered. We have chosen to run FSM in sample-specific approach since applying FSM over an all-sample database (a single graph database including all the edges from all the samples) will lead to bias in the identified subgraphs due to the subset of the samples having a high number of dysregulated patterns.

Applying the algorithm above to our problem naively is not practical. It involves solving several difficult sub-problems, including candidate subgraph generation and subgraph isomorphism. Furthermore, many frequent subgraphs would overlap with each other (41) returning exponentially large similar subgraphs (42). Our FSM approach resolves these problems in two ways. First, it uses a highly optimized method for candidate generation which prunes unsupported supergraphs (39). Second, instead of collecting all frequent subgraphs, a sample of graphs is collected using the

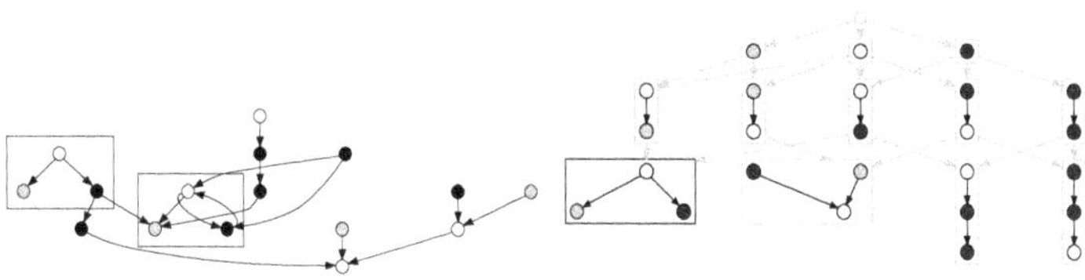

(a) A colored (labeled) digraph. *(b) A 2-frequent connected subgraph lattice.*

Fig. 1: *Figure (b) is a connected subgraph lattice of the graph in Figure (a) including only the subgraphs with 2 or more embeddings in Figure (a). The boxed nodes in the graph show the embeddings of the boxed subgraph in the lattice. In the figure, the colors (black, gray, and white) are standing in for labels on the vertices (Adapted from (39)).*

GRAPLE algorithm (42). GRAPLE models the set of frequent subgraphs as a *lattice* where the graphs in the lattice are connected by their subgraph and supergraph relationships (see Figure 1). Frequent subgraphs are sampled from the lattice by taking random walks on the lattice. For full details see (39, 42), a related approach can be found in (43).

We have extensively tested the FSM algorithm to validate our approach and also compared it to previous methods (43–45). We tested parameter k on various benchmark datasets (Supplementary Table S3) and validated k's effect on run time and subnetwork discovery. We ran these simulations in a non-heuristic mode to recover all subnetworks. We comprehensively compared our performance with GRAMI (Supplementary Table S4). GRAMI finds a slightly different number of patterns because it uses undirected graph search (otherwise GRAMI's run time suffers). Our tool outperformed GRAMI.

2.4 Integrating Gene Expression Measurements to FSM Framework

We integrated the somatic mutations with gene expressions using the same -omics dataset and interaction network from (46). The integration of gene expression is done in two steps. First, in the biased random walk step, the transition probabilities are assigned based on the euclidean norm of z-scores of interacting genes. This scheme prioritized genes with high dysregulation compared to the population in addition to seed sets. Furthermore, for functional relevance, we have applied dimension reduction followed by clustering with PAM and pathway enrichment (PAM-Clusters, Figure S10) (47). To apply the dimension reduction, each identified subgraph is assigned an average dysregulation score (matrix of frequent subgraphs vs samples) (23).

2.5 Functional Analysis

To associate biological mechanisms with frequent subgraphs, we utilize clustering, non-linear dimension reduction, pathway enrichment, and survival analysis. Since FSM is done in a sample-specific manner, identified subgraphs contain redundant interactions (repeated interactions across multiple subnetworks). We apply greedy clustering to remove these redundant interactions by grouping highly connected nodes (48). In this process, we find high modularity partitions of our networks.

For survival analysis, we utilized unsupervised clustering using the frequent subgraphs as features. Frequent subgraphs mined using gene expression integration are assigned dysregulation scores using the average euclidean norms of standardized gene expressions for a single fsg $\sqrt{\sum_i^n g_i^2}$ and samples are clustered using PAM on dimension reduced space (47, 49). For FSGs identified using only SNVs, we assigned the frequency of matching genes in the FSG and the sample as a score and employed PAM. However note that for clustering samples with matching gene frequencies as scores, we did not use non-linear dimension reduction.

3 Results

3.1 Pan-cancer Subgraphs

FSM has identified 43k unique subgraphs with sizes between 6-60 edges across the 90% of the pan-cancer dataset with support 20 corresponding to 0.3% of samples (Figure 2). Identified subgraphs covered more than 40% of the genes in the protein-protein interaction networks.

(a) Frequent subgraph with highest frequency *(b) Frequent subgraph with largest size*

Fig. 2: *Sample frequent subgraphs mined from the pan-cancer dataset. Each edge in the given subgraph is represented in at least 20 common set of samples.*

3.2 Pathway Enrichment

To elaborate on the functional relevance of the identified subgraphs, we have clustered the merged subgraph network using a greedy approach (FSG-Clusters) (48). More specifically, frequent subgraphs are merged into a single network, and clustering is applied. This method can be seen as filtering the initial protein-protein interaction network to include edges that show frequent interaction patterns. However, note that this scheme is not similar to simply filtering the edges that have minimum support level but in subgraph space. A total of 106 clusters was identified (See Suppl. tables). To filter clusters without functional relevance, we have removed clusters with node size smaller than 10 and larger than 400. Pathway enrichment analysis using the Reactome Pathways has identified a total of 620 significant subnetworks using a p-value threshold of 0.01, including previously identified mechanisms: PI3K Cascade, Cytokine Signaling, DNA Repair, Signaling by NOTCH (Figure 3).

3.3 Disease Enrichment

To evaluate the representation of cancer types in identified clusters we have done enrichment analysis for each patient as well (Figure S3). Multiple clusters showed few over-representation in terms

Fig. 3: *Top three enrichment results of identified clusters sorted by root pathways (FSG-Clusters).*

of predefined disease types. These clusters also showed few or no pathway enrichment which might suggest small subnetworks stratifying patients in combination with broad dysregulation patterns.

Samples with lower-grade glioma (LGG) are represented across different clusters similar to breast cancer samples. However, increased representation for LGG samples in clusters 85 and 63 is visible. Cluster 85 is mostly associated with CSF2RA-B metabolism, which are cytokines related to macrophage, granulocyte differentiation, and production. An earlier study showed how intercellular microglia polarization signaling through CSF2 (GM-CSF) and IFNG are the molecules that drive microglia towards the M1 phenotype (50). Cluster 63, on the other hand, is related mostly to NOTCH signaling, p75NTR degradation through NRIF interactions (Figure S11). In contrast, breast cancer patients show increased representation in clusters 23, 24, 35, and 44. Given clusters correspond to lipid metabolism (known risk factor for developing cancer (51)), membrane trafficking, cytoskeletal related processes, *SEMA3A*, *SEMA4D* signaling, which might related to increased Metastasis in breast cancer (52). Patients with skin disorders are mainly represented in clusters 47 and 102. Pathway enrichment for the clusters identifies degradation of the extracellular matrix, O-linked glycosylation, and collagen biosynthesis. On the other hand, uveal melanoma patients are enriched for cluster 89, which shows dysregulation in GPCR signaling, the main biological processes impacted by the recurrent mutations in uveal melonama (53). Thyroid cancer patients show the most specific enrichment for cluster 80, showing functional relevance in the regulation of RAS by GAPs, and MAPK pathways, key signaling pathways in both initiation and progression of medullary thyroid carcinoma (54). Prostate cancer patients are mainly enriched for clusters 35, 44, and 23, showing enrichment for Rho GTPase activation of PAK, cleavage of cell adhesion proteins through apoptosis, *SEMA3A*, and *SEMA4D* related signaling. Head and neck cancer patients also show dysregulation across a large number of clusters discovered but show the highest enrichment for cluster 106, similar to breast cancer and LGG patients.

3.4 Comparison of Pan-cancer FSM Networks

Oncogenic Signaling Pathways of Pan-cancer To further elaborate on the utility of the proposed method we have compared the genes in identified frequent subgraphs to previously established expert-curated pathways (4). We have recovered 65% of genes covering 90% of pathways reported in the curated list including *EGFR, TP53, PIK3CA, PTEN* matching various mechanisms. To further compare against previously curated pathways, we have utilized cancer hallmark genesets (55, 56). Frequent subgraphs cover 100% of the hallmark gene sets with at least 1 overlapping gene. Interestingly FSG clusters cover multiple pathways and pathways are covered by multiple FSG clusters as well both for oncogenic signaling pathways and hallmark gene sets. This further elaborates on the complexity of cancer and the interaction topology (Figure S1). We identified additional genes, novel to the curated pathway database as well suggesting the importance of system-level identification of functional mechanisms and the complexity of cancer progression (See Suppl. Figure S2).

HotNet2 Pan-cancer Subnetworks We also compared our method to HotNet2 (22), which aims to find subnetworks significantly enriched in given alterations across the pan-cancer dataset. However, the main difference is that HotNet2 focuses on gene-level perturbations and looks for subnetworks covering a wide range of samples in the dataset. More specifically in the subnetworks identified by HotNet2, different subsets of samples can show alterations in different nodes of the subnetwork. In contrast, our methodology aims to identify subnetworks for all samples meaning that in the identified subnetworks a common set of samples show dysregulation for all the nodes in the subnetwork. When we compare to HotNet2 subnetworks, we observe that clusters 63, 70, 80, and 90 correspond to 5 subnetworks out of 15 relating to $BRAF$, RAS, $PIK3CA$ subnetwork, $KDM6A$, $MLL2$, $MLL3$ subnetwork, SWI/SNF complex, $BAP1$ complex and cell adhesion networks respectively using overrepresentation analysis (See Suppl. Tables.). However, a comparison of FSGs prior to clustering results in 12 subnetworks to be significantly enriched. This suggests that different groups of patients show dysregulation in separate parts of a larger network that are combined into a single cluster based on intermediary interactions.

3.5 Functional Classification of Pancancer Samples

We calculated dysregulation scores for each subnetwork to stratify the cancer samples. We set the support level (k) to 8 for this purpose to increase the number of samples identified during the FSM run since with larger support of 20, many of the samples drop out. As expected, the number of unique frequent subgraphs increased dramatically to 135k, increasing the noise inherent in the frequent subgraph space (14k unique genes). However, dimension reduction shows clear separation of cancer types (See Figures 4, 5 and Suppl. Figures S4, S5 and S6).

While some cancers are spread across multiple clusters (e.g. BRCA), some cancers were separated based on tissue, which reflects implicit biological processed and their alterations (e.g. Uveal melanoma, brain tumors, LIHC, PCPG, THCA etc.) (See also Suppl. Figure S12). Most importantly, survival differences (Figure 4) clearly exist across cancers and cancer subtypes. LGG is split into clusters 11 and 14, where 14 represents GBM-like LGG samples with significant survival differences (57). BRCA clusters 3, 4, 5, 6, 8, and 26 show significant survival differences in these groupings, which reflect previous findings (17, 58).

Significant features between clusters are obtained by comparison of subnetwork dysregulation scores using 1 vs all approach with p-value threshold after Bonferroni correction set as 0.01. Pathway enrichment is done on genes in the significant subnetworks. Pathway enrichment also shows functionally relevant mechanisms. For instance, clusters 9, 21, and 22 representing Stomach Adenocarcinoma, Rectum Adenocarcinoma, and Colon Adenocarcinoma are significantly enriched for genes related to O-linked glycosylation **(Fig.S6)**. However separately from Rectum and Colon Adenocarcinomas, Stomach Adenocarcinoma is highly enriched for *Defective CSF2RA/CSF2RB causes pulmonary surfactant metabolism dysfunction* pathway, which has been previously associated with Stomach Adenocarcinomas. Interestingly, there is a clear separation of functional mechanisms between clusters: 1, 5, 6, 8, 9, 10, 20, 21, 22, 26 (Group 1), and the rest. More specifically, the second group of cancers is all associated with mechanisms related to signaling events such as RAF/MAP kinase cascades, FGFR signaling, and PI3K Cascade, but the first group is not. These results provide a strong validation for the FSM approach presented.

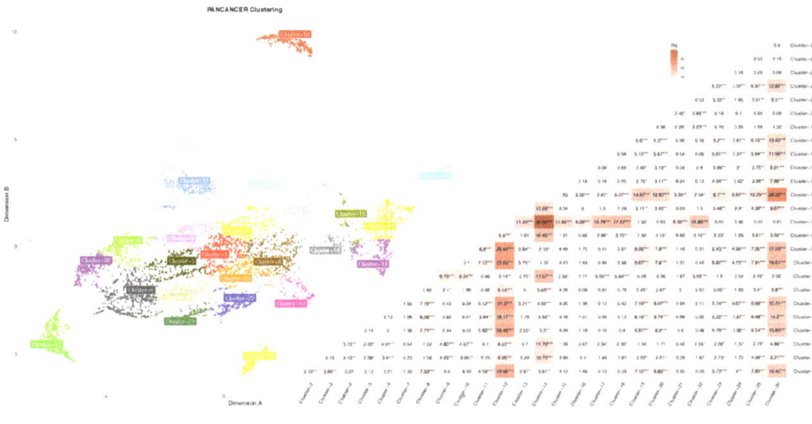

Fig. 4: ***Left:*** *UMAP dimensionality reduction on scored frequent subgraph matrix. Samples clusters are labeled and colored based on labels.* ***Right:*** *Pairwise survival differences using log-rank test are shown for FSM patient clusters shown on the left.*

3.6 Analysis of Single Cancers using Pan-cancer Frequent Subgraphs

We have shown further utility of FSGs mined using the pan-cancer dataset to stratify patients into subtypes. We have applied FSG level clustering using PAM and identified significant survival differences **(Fig. S7)**. The significant results were seen for two separate cancers, Lower Grade Gliomas (LGG) and Uterine Cancer suggest that subnetworks mined using the pan-cancer dataset is able to capture subtype-specific functional networks. This further shows the comprehensive nature of our networks identified in this framework.

3.7 Single Cancer Analysis with the FSM Framework

While individual cancer analysis using the pan-cancer FSGs are possible (as shown above Section 3.6), the FSM framework we present can be applied to a single cancer type as well. For this

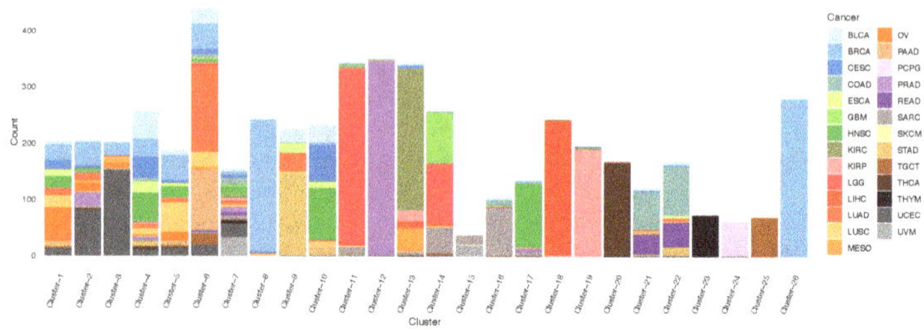

Fig. 5: Disease profile for each UMAP cluster is shown. The number of patients for each cancer type is stacked on each bar. TCGA disease codes are listed in Supplementary Table S1.

purpose we analyzed glioblastoma multiforme (GBM) samples only. In a recent study, we used a more simplified FSM framework to cluster individual cancer types and successfully found subtypes for breast cancer and GBM (23). Using our new approach, we have identified 1.2k frequent subgraphs with a total of 5 clusters representing the frequent subgraph network and covering 60% of the GBM samples. The spectrum of the pathway dysregulation in the clusters corresponded to Cytokine Signaling, TRAF6 mediated IRF7 activation, PI3K/AKT signaling, and PIP3 signaling. Interestingly, cluster-2 covered a large fraction of dysregulated pathways and, cluster-4 was enriched specifically for the AKT related pathways. However clusters 3 and 5 showed no pathway enrichment which requires further analysis.

3.8 Survival Differences of Patients Represented in PAM-Clusters

We have investigated the patient groups that correspond to each cluster identified using gene expression and SNP datasets. Pairwise comparison of survival curves show high significance between clusters **(Fig.5)**. For example, cluster-8, which is represented mostly by BRCA patients, shows a significant difference when compared against clusters 1, 4, 5, 6 that are composed of mixed disease types of OV, UCEC, HNSC, LUSC, LUAD, BRCA. Furthermore, the difference between clusters 8 and 26 for BRCA patients only might represent subtype differences as well. Similarly clusters 11 and 14 represent 2 distinct LGG patient clusters with significant survival differences. Interestingly however BRCA, LUAD, LUSC, HNSC, UCEC, and OV cancer types are heterogeneously divided into different clusters suggesting common molecular mechanisms driving the diseases and requires further investigation.

4 Discussion

We have applied frequent subgraph mining coupled with random walk with restarts to the pan-cancer dataset. The application of the FSM with patient-level constraints allowed us to extract interaction patterns functionally relevant to cancer progression. Identified patterns might prove useful for novel targeting strategies especially patient-specific targets due to increased sensitivity in regulatory pattern identification.

The approach proposed in the context of mining functionally important subgraphs is more efficient compared to our initial methodology published (23) both in terms of runtime and coverage.

Biased random walks significantly decrease the search space by reducing the number of edges per patient and applying the FSM separately for each patient as given above ensures that each sample is represented. Furthermore, the use of biased random walks allowed us to increase the sensitivity of our approach by considering the mutational signatures as a network. More specifically, each graph database is obtained based on the mutated genes but frequent subgraphs do not necessarily contain mutated genes but are associated with mutated genes. Additionally, as given above the proposed approach is more comprehensive in comparison with other methods available since gene-level enrichment-based methods or prior knowledge do not take into account the complex interaction patterns relevant to cancer progression.

In comparison to previous methods and established biomarkers, the proposed method underlines the complex interaction patterns present in defining different cancer groups. For example, SEMA3A has been previously associated with breast cancer metastases through the promotion of osteoblast differentiation in MCF-7 cell lines (59). Colony-stimulating factor has also been associated with glioma progression previously and identification of CSF2RA is an important observation (60). p75 neurotrophin receptor also is a crucial regulator of glioma progression leading to cytoskeletal modifications (61). Analysis of GBM patients only increased the sensitivity of frequent subgraphs. PI3K/AKT is responsible for drug resistance for malignant glioma patients, suggesting a critical biomarker in targeted therapies (62).

Furthermore, we have shown that the proposed approach is able to elucidate increased functional relevance by strictly enforcing frequency requirements hence decreasing false positives in contrast with previously established methods that either focus on gene-level approaches or do not consider the underlying topology of the patient data.

Finally, our approach is able to stratify patients of individual cancers based on pancancer frequent subgraphs. In this unsupervised approach, we were able to find significant survival differences in patient groups of LGG and Uterine Cancer. This further validates our approach and shows utility for future cancer studies.

Bibliography

[1] Ciriello, G. et al. *Nature genetics*, 45(10):1127, 2013.
[2] Lawrence, M.S. et al. *Nature*, 499(7457):214, 2013.
[3] Kandoth, C. et al. *Nature*, 502(7471):333, 2013.
[4] Sanchez-Vega, F. et al. *Cell*, 173(2):321–337, 2018.
[5] Werner, H.M.J., Mills, G.B. and Ram, P.T. *Nat Rev Clin Oncol*, 11(3):167–176, 2014.
[6] Hoadley, K.A. et al. *Cell*, 158(4):929–944, 2014.
[7] Hoadley, K.A. et al. *Cell*, 173(2):291–304, 2018.
[8] Bailey, M.H. et al. *Cell*, 173(2):371–385, 2018.
[9] Tamborero, D. et al. *Scientific reports*, 3:2650, 2013.
[10] Hofree, M. et al. *Nature methods*, 10(11):1108, 2013.
[11] Shen, R., Olshen, A.B. and Ladanyi, M. *Bioinformatics*, 25(22):2906–2912, 2009.
[12] Cohen, A.L., Holmen, S.L. and Colman, H. *Curr Neurol Neurosci Rep*, 13(5):345, May 2013.
[13] Dolgin, E. *Cancer Discov*, 9(8):992, Aug 2019.
[14] Sulkowski, P.L. et al. *Sci Transl Med*, 9(375), 02 2017.
[15] Johannessen, T.C.A. et al. *Mol Cancer Res*, 14(10):976–983, 10 2016.
[16] Tateishi, K. et al. *Cancer Cell*, 28(6):773–784, Dec 2015.
[17] van 't Veer, L.J. et al. *Nature*, 415(6871):530–6, Jan 2002.

[18] Paik, S. et al. *N Engl J Med*, 351(27):2817–26, Dec 2004.
[19] Parker, J.S. et al. *J Clin Oncol*, 27(8):1160–7, Mar 2009.
[20] Venet, D. et al. *PLoS Comput Biol*, 7(10):e1002240, 2011.
[21] Dhawan, A. et al. *bioRxiv*, page 203729, 2017.
[22] Leiserson, M.D. et al. *Nature genetics*, 47(2):106, 2015.
[23] Durmaz, A. et al. In *PSB 2017*, pages 402–413. World Scientific, 2017.
[24] Koyutürk, M., Grama, A. and Szpankowski, W. *Bioinformatics*, 20(suppl_1):i200–i207, 2004.
[25] Huan, J. et al. In *CSB*, pages 227–238. World Scientific, 2006.
[26] Zhang, X. and Wang, W. In *null*, page 32. IEEE, 2007.
[27] Kuramochi, M. and Karypis, G. In *Data Mining, 2001. ICDM 2001*, pages 313–320, 2001.
[28] Yan, X. and Han, J. In *Data Mining, 2002. ICDM 2003.*, pages 721–724. IEEE, 2002.
[29] Nijssen, S. and Kok, J.N. In *Proceedings of the tenth ACM SIGKDD*, pages 647–652, 2004.
[30] Ranu, S. and Singh, A.K. In *Data Engineering, 2009. ICDE'09.*, pages 844–855. IEEE, 2009.
[31] Garey, M.R. *Computers and intractability*, 1979.
[32] Can, T., Çamoglu, O. and Singh, A.K. In *Proceedings of the 5th international workshop on Bioinformatics*, pages 61–68. ACM, 2005.
[33] Köhler, S. et al. *AJHG*, 82(4):949–958, 2008.
[34] Erten, S. et al. *BioData mining*, 4(1):19, 2011.
[35] Guo, H. et al. *Scientific reports*, 5:10857, 2015.
[36] Goldman, M. et al. *bioRxiv*, page 326470, 2018.
[37] Szklarczyk, D. et al. *NAR*, 43(D1):D447–D452, 2014.
[38] Fabregat, A. et al. *NAR*, 46(D1):D649–D655, 2017.
[39] Henderson, T.A.D. *Frequent subgraph analysis and its software engineering applications*. Doctoral dissertation, Case Western Reserve University, 2017.
[40] Cheng, H. et al. In *Frequent Pattern Mining*, pages 307–338. Springer Publ., 2014.
[41] Yan, X. and Han, J. In *Proceedings of the Ninth ACM SIGKDD*, pages 286–295, 2003.
[42] Henderson, T.A.D. and Podgurski, A. In *International Workshop on Software Analytics*. ACM, 2016.
[43] Chaoji, V. et al. *Stat. Anal. Data Min.*, 1(2):67–84, jun 2008.
[44] Saha, T.K. and Hasan, M.A. In *2014 IEEE BigData*, pages 72–79, Oct 2014.
[45] Al Hasan, M. and Zaki, M. In *SIAM 2009*, volume 2, pages 646–657, 12 2009.
[46] Li, T. et al. *Nature methods*, 14(1):61, 2017.
[47] McInnes, L., Healy, J. and Melville, J. *arXiv preprint arXiv:1802.03426*, 2018.
[48] Blondel, V.D. et al. *J. Stat. Mech. Theory Exp.*, 2008(10):P10008, 2008.
[49] Rdusseeun, L. and Kaufman, P.J. 1987.
[50] Li, W. et al. *CNS Neurol Disord Drug Targets*, 12(6):750–62, Sep 2013.
[51] Long, J. et al. *Am J Cancer Res*, 8(5):778–791, 2018.
[52] Yang, Y.H. et al. *PLoS One*, 11(2):e0150151, 2016.
[53] Vivet-Noguer, R. et al. *Cancers (Basel)*, 11(7), Jul 2019.
[54] Cote, G.J., Grubbs, E.G. and Hofmann, M.C. *Recent Results Cancer Res*, 204:1–39, 2015.
[55] Subramanian, A. et al. *PNAS*, 102(43):15545–15550, 2005.
[56] Liberzon, A. et al. *Cell systems*, 1(6):417–425, 2015.
[57] Chen, R. et al. *Neurotherapeutics*, 14(2):284–297, 04 2017.
[58] Shimoni, Y. *PLoS Comput Biol*, 14(2):e1006026, 02 2018.
[59] Shen, W.W. et al. *International journal of clinical and experimental pathology*, 8(2):1584, 2015.
[60] Mueller, M.M., Herold-Mende, C.C. et al. *Am. J. Pathol.*, 155(5):1557–1567, 1999.
[61] Johnston, A.L. et al. *PLoS biology*, 5(8):e212, 2007.
[62] Stupp, R. et al. *NEJM*, 352(10):987–996, 2005.
[63] Elseidy, M. et al. *Proc. VLDB Endow.*, 7(7):517–528, March 2014.

Pacific Symposium on Biocomputing 2021

Cross-modal representation alignment of molecular structure and perturbation-induced transcriptional profiles

Samuel G. Finlayson[1,2,*], Matthew B.A. McDermott[2,*], Alex V. Pickering[3], Scott L. Lipnick[3], Isaac S. Kohane[3,†]

[1]*Department of Systems, Synthetic, and Quantitative Biology, Harvard Medical School, Boston, MA*
[2]*Department of EECS, Massachusetts Institute of Technology, Cambridge, MA*
[3]*Department of Biomedical Informatics, Harvard Medical School, Boston, MA*
[*]*Co-first author*
[†]*E-mail: isaac_kohane@harvard.edu*

Modeling the relationship between chemical structure and molecular activity is a key goal in drug development. Many benchmark tasks have been proposed for molecular property prediction, but these tasks are generally aimed at specific, isolated biomedical properties. In this work, we propose a new cross-modal small molecule retrieval task, designed to force a model to learn to associate the structure of a small molecule with the transcriptional change it induces. We develop this task formally as multi-view alignment problem, and present a coordinated deep learning approach that jointly optimizes representations of both chemical structure and perturbational gene expression profiles. We benchmark our results against oracle models and principled baselines, and find that cell line variability markedly influences performance in this domain. Our work establishes the feasibility of this new task, elucidates the limitations of current data and systems, and may serve to catalyze future research in small molecule representation learning.

Keywords: Representation Learning, Therapeutics, Gene Expression, Deep Learning, Information Retrieval

1. Introduction

Identifying molecules that are likely to have a specific biological effect is a cornerstone of drug discovery and a key component of efforts to achieve precision medicine. Classically, computational profiling of small molecules has centered on predicting affinities for specific biological targets, using tools ranging from biophysics-driven techniques such as molecular docking[1] to literature-mined annotations.[2] Small molecule modeling has also recently become a major area of interest in deep learning, a trend catalyzed by graph neural networks[3] and benchmarking datasets.[4] Graph neural networks allow for end-to-end modeling of molecular graphs,[5–8] and have yielded state-of-the-art performance on certain tasks.[9,10] In addition, deep learning approaches have been used at a more global scale modeling cross-molecule relationships.[11] To date, deep learning efforts in this space have generally focused on two extremes: highly local, biochemical prediction problems, which test the model's ability to predict specific chemical properties, and more global, clinical modeling tasks, such as indication or side effect prediction. Missing from the field, however, are benchmark tasks between these two extremes, that test the ability of deep models to encode rich, general representations of a molecule's broad-spectrum effect on cellular biology.

In parallel to these developments, connectivity mapping has emerged as a alternative approach for drug development.[12] In connectivity mapping, compounds are foremost characterized not by individual chemical properties or downstream targets, but by the broad transcriptional effects they induce in cells. Connectivity mapping begins by first developing a large dataset of *perturbational signatures* of molecules by physically treating cell lines with these molecules, then measuring the resultant changes in gene expression. These datasets are then compared to one or more *query signatures*, which are typically differential gene expression (GE) signatures representing disease states that investigators hope to reverse. Various public datasets have been curated to enable these efforts,[13,14] and researchers have sought to use these for drug repurposing, precision medicine, and analysis of gene expression data in general.[15–22]

Connectivity mapping is promising because it can be used to search for new indications of drugs without making any specific *a priori* assumptions about their mechanism of action. However, the typical framework for connectivity mapping is limited by the fact that it can only query against drugs that have already been profiled using the transcriptional assay. In other words, connectivity mapping is – in principle – very flexible with respect to the disease signatures they accept as a query, but is transductive rather than inductive with respect to the target small molecule signatures. This is the perfect complement to structure-based computational chemistry, which is typically inductive to new drug structures but can only make predictions for diseases with known targets.

In this work, we combine these two fields, by using deep chemical embedders to learn the transcriptional space encoded by CMAP profiling. More specifically, we train coordinated networks to jointly embed chemical structures and perturbational gene expression profiles such that learned chemical representations are most similar to the encodings of the transcriptional patterns they induce.[a]. Note that this task naturally fills the gap in inductive molecular modelling identified previously; by tasking the model to produce highly similar embeddings for chemical structures and the perturbational profiles they induce, we force the model to learn a *transcriptome-wide* reflection of the drug's action on the cell. We then evaluate these chemical representations by using gene expression signatures as queries into the embedding space and recovering their corresponding compounds. (See Figure 1). Crucially, the evaluation is set up such that the validation and test set compounds and cell lines are not used in training, which allows us to test the ability of the model to generalize to new drugs and cell lines.

In the rest of this work, we first offer some background on joint embedding alignment, then detail the methods used in this work. Finally, we walk through the results and discussion of these experiments, then close with concluding thoughts. A version of this work, with supplementary material present, can be found here: `https://github.com/sgfin/molecule_ge_coordinated_embeddings/blob/master/paper_08_2020.pdf`.

[a]Code available here: `https://github.com/sgfin/molecule_ge_coordinated_embeddings`

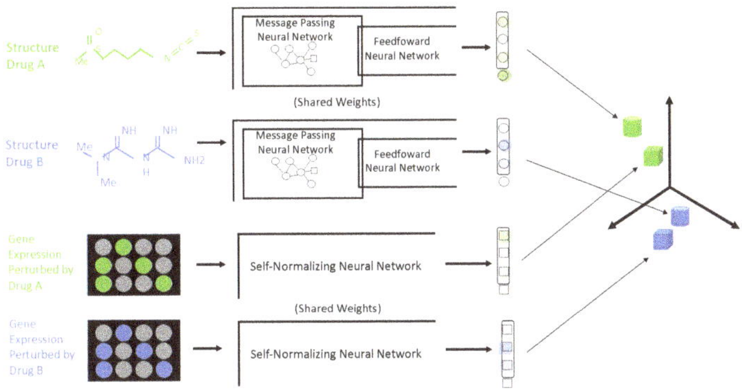

Fig. 1. Our representation learning method. Neural networks are trained to embed gene expression profiles close to the small molecule structures that induce them. Given a cross-modal alignment, gene expression signatures can be used as queries to rank chemical structures by their likelihood to induce such a signature.

2. Background

In *multi-view representation alignment*, embeddings of two associated data modalities are learned separately but in a coordinated manner, such that the resulting embeddings are similar. These methods have been used in comparing images to text but also in other domains.[23,24] In this work, we learn aligned representations such that small molecules are embedded in close proximity to the differential gene expressions they induce. Multi-view representation alignment can be achieved through a variety of methods, including classical methods, such as *canonical correlation analysis* (CCA)[25] and methods using distance, similarity, correlation, or ranking based penalties during training.[26] Ranking-based methods for multi-view representation alignment, such as that described by Deng at al,[27] allow the incorporation of ranking information into the training procedure, which may be important in tasks such as gene expression where perturbation signals may be small relative to baseline state. In addition, the field of rank-based embedding learning is intertwined with a broader literature of uni-modal embedding learning, which pioneered such architectures as Twin[28,29] and Triplet networks,[30] which optimize embeddings to bring similar data together while driving dissimilar data apart. An analysis of best practices of these architecture can be found in Wu et al.[31]

3. Methods

3.1. *Dataset & Tasks*

Data Acquisition and Subsetting All data in this study comes from the LINCS Consortium/NIH Next-Generation Connectivity Map Level 3 L1000 data.[14] This dataset features

978-dimensional gene expression profiles from a variety of human cell lines treated with chemical and genetic perturbations. To ensure support over possible drugs, our data cut uses the most frequent 8 cell lines split into train, validation, and test sets such that no cell line or drug in the train set appears in the validation or test sets. To mitigate non-random missingness, we included only drugs assayed in all cell lines, and limited experiments to those incubated with small molecules for 24 hours at a dose of $10\mu m$. Final statistics of these data are shown in Table A1. For drug structures, we used the SMILES[32] structures provided by LINCS, canonicalized using RDKIT.[33]

Preprocessing and Feature Engineering Gene expression intensity values from the training, validation, and test sets were centered and scaled at the gene-level based on the mean and standard deviation of each gene intensity across the training set. We augmented each gene expression profile with three additional sets of features: the corresponding gene expression intensities from a control signature on the same plate, the $\log_2$ fold-change between the perturbation and control signatures, and the difference between these gene expression signatures. For use in our baseline and oracle models, we also computed numerical representations of each small molecule: Morgan extended-connectivity fingerprints[34] and the output of the ChemProp network.[6]

Detailed Task Description Our goal is to learn embedders which map molecular structures and gene expression profiles into a vector space such molecular structure embeddings are close to the gene expression profile they induce while being far from other gene expression profiles (Figure 1). Formally, given a collection of gene expression signatures $\mathcal{G}$, chemical structures $\mathcal{M}$, and similarity function $\mathrm{Sim} : \mathbb{R}^d \times \mathbb{R}^d \to \mathbb{R}$, we seek to learn a gene expression embedder $E_g : \mathcal{G} \to \mathbb{R}^d$ and chemical embedder $E_m : \mathcal{M} \to \mathbb{R}^d$ to maximize $\mathrm{Sim}(E_g(g_i), E_m(m_j))$ while simultaneously minimizing $\mathrm{Sim}(E_g(g_i), E_m(m_{\neg j}))$, where gene expression g_i was induced by molecule m_j. Unless otherwise specified, the similarity function can be assumed to be Pearson Correlation in our experiments. Across our baseline and oracle methods, we realize many variants of E_g and E_m.

3.2. *Baseline and Oracle Methods*

Nearest Neighbor Baseline Nearest-neighbor (NN) methods have been previously shown to establish strong baselines for machine learning tasks on the L1000 data.[35,36] In our cross-modal, information retrieval (IR) context, traditional NN methods are not applicable, so we employ the following "double NN" baseline: given a gene expression profile as a query, we first identify the nearest gene expression profile in the train set and look up its corresponding small molecule. We then take this small molecule (from the train set) as a query, and return the most structurally similar drug from the *test* set as our final prediction.

In particular, given a mapping $G2M : \mathcal{G} \to \mathcal{M}$ from gene expression profiles to the small molecule that induced them, and a molecular embedding E_m (which may include molecular fingerprints, Chemprop embeddings, or embeddings learned from other models), we define embedder $E_g : g_{\text{query}} \mapsto E_m(G2M(\arg\max_{g_{\text{tr}} \in \mathcal{G}_{\text{train}}} \mathrm{Sim}(g_{\text{query}}, g_{\text{tr}})))$. Then, we perform information retrieval (IR) analyses with such embedders as usual.

Canonical Correlation Analysis Baseline Given training matrices of transcriptional $\mathcal{G}_{\text{train}}$

and molecular $\mathcal{M}_{\text{train}}$ encodings, we can learn a set of linear mappings $E_g : \mathcal{G}_{\text{train}} \to \mathbb{R}^d$ and $E_m : \mathcal{M}_{\text{train}} \to \mathbb{R}^d$ via d-dimensional CCA such that these mappings optimize the correlation between elements of $\mathcal{G}_{\text{train}}$ and $\mathcal{M}_{\text{train}}$.

Note that this procedure requires a default numerical representation for molecules, which, as with other methods, can be either fingerprints, ChemProp embeddings, or learned embeddings by our learning model (described below). CCA can also be performed atop other embedding systems to further optimize embedding results. CCA was performed using SciKit Learn,[37] using 50 components, chosen to optimize validation set performance via a grid-search over a range of 5-125 components, run for 1000 iterations to ensure convergence.

Oracle Models The central objective of our task is to learn small molecule embeddings that can stand in as surrogates for their corresponding gene expression signatures. To provide a rough upper-bound for expected performance on this task, we also implemented two "oracle" models, each of which queries test set GE signatures against pseudo-"chemical embeddings" that are in reality the average GE signatures from each test set drug when it was measured on either (1) the *train set cell lines*,[b] to simulate an embedder that perfectly associates all structures to perturbational profiles, but cannot generalize beyond the train set cell lines, or (2) the *test set cell lines*, which simulates a model of the same capabilities but able to generalize perfectly to the test set as well. These oracle models are still dependent on the underlying gene expression signature representation, so further innovation could offer improved upper bounds for this task. Formally, given $G2M$ mapping gene expression profiles to their corresponding perturbing molecule, we define oracle embeddings $E_g^{\text{train}} : g_{\text{query}} \mapsto \text{Avg}(\{g_i \in \mathcal{G}_{\text{train}} | G2M(g_i) = G2M(g_{\text{query}})\})$, and $E_g^{\text{test}} : g_{\text{query}} \mapsto \text{Avg}(\{g_i \in \mathcal{G}_{\text{test}} | G2M(g_i) = G2M(g_{\text{query}})\})$.

3.3. *Deep Coordinated Metric Learning Approach*

For our learned model, we realize E_g as a self-normalizing neural network (of size dictated by hyperparameter search), and E_m as a directed message-passing neural network (D-MPNN), initialized by the Chemprop system, followed by a feed-forward output layer whose shape was dictated via hyperparameter search.[6] To train these architectures, we use a margin-based quadruplet loss, building on Wu et al's adaptive margin loss.[31] The base of the adapted margin loss is defined over two data points i and j as $\text{mar}_{\alpha,\beta} := (\alpha + y_{i,j}(D_{ij} - \beta))_+$, where D is distance function (here euclidean distance), α defines a permissible margin of separation, β controls the boundary between positive and negative pairs, and $y_{i,j}$ is an indicator variable equal to 1 if i and j are of the same class and 0 otherwise (α and β were tuned as hyperparameters).

Given two pairs of matching gene expression and molecular structure embeddings, $(g_A, m_A), (g_B, m_B)$, our quadruplet loss is defined as the sum of the margin losses between all cross-modality pairs of embeddings: $\ell_{\text{quad}} = \text{mar}_{\alpha,\beta}(g_A, m_A) + \text{mar}_{\alpha,\beta}(g_A, m_B) + \text{mar}_{\alpha,\beta}(g_B, m_A) + \text{mar}_{\alpha,\beta}(g_B, m_B)$. The network is thus optimized to bring the positive embedding within the margin of the anchor and negative embedding outside the margin. For sampling these two

[b]Note that we can do this as we limited our choice of drugs to those that *were* measured in all 8 cell lines, even though our actual data split prohibits training on any drug that appears in the validation or test sets.

pairs (an analog of negative sampling for a more traditional triplet network), we first sample one matching pair, choose the molecular structure for the other pair based on the distance-weighted negative sampling scheme described in Wu et al, which was successful with their margin-based approach,[31] then fill in the other gene expression profile to match the sampled molecular structure. To make this process computationally efficient we pre-computed the average distance in *average post-perturbational gene expression space* between every pair of small molecule structures in the dataset. We additionally tried other losses, including two varieties of traditional triplet losses, and a quintuplet loss, but ultimately found the quadruplet loss to be most performant via our hyperparameter search.

Training and Hyperparameter Selection Each model was trained on a Nvidia GeForce GTX 1080 GPU. Early stopping was used to select the model with the best mean reciprocal rank on the validation set. Hyperparameter tuning via the Bayesian Hyperopt library[38] was performed over a wide range of possible hyperparameters, including network depth and width (parametrized by the size of the first hidden layer and a growth rate), learning rate, number of epochs, batch size, margin and β parameters, triplet model orientation (i.e., gene first or compound first), activation function/network type (e.g., SNNN vs. unconstrained fully connected network), and dropout, with no early stopping for hyperparameter search runs. The optimal hyperparameters from this search are shown in Appendix Section 7.2.

3.4. *Experiments*

We designed a range of experiments with two purposes: First, we sought to evaluate if and how our deep coordinated representation learning method offers improvements over principled baselines. This entails a quantitative performance comparison against baseline methods and ablated versions of our model. Second, we introspect into the representations learned by training on this new task, to better understand the challenges and utility of the general framework. This entails a quantitative performance comparison against oracle models, a statistical analysis probing the ability of our models to generalize to new structures, and a qualitative exploration of the changes in chemical representations that are induced by our training scheme.

Quantitative performance analyses began by computing the embeddings of gene expression signatures and chemical structures in the test set, using the baseline, oracle, and deep coordinated methods defined in Sections 3.2 and 3.3. Using each embedded gene expression signature as a query, we ranked all chemical structures in the test set based on their proximity to that gene expression signature in the embedding space. These rankings were then used to compute standard information-retrieval metrics (precision-recall curves, MR, MRR, and Hits at/H@ 10 or 100). For our ablation analyses, we repeated the above experiments using various combinations of raw and learned GE and chemical representations.

In addition to the information retrieval analyses, we probed the generalizability of our representations by analysing the statistical relationship between the average retrieval performance for each chemical structure and the structural similarity to the most similar chemical in the training set. Similarity was measured via the Tanimoto distance between all pairs of molecular fingerprints in the train and test set. We further examined performance vs. chemical specificity, using the number of genes that a molecule, on average, affected as a measure of specificity. Finally, we visualized the latent space of our chemical embedder (versus their pre-trained representations learned by ChemProp), and noted the relative position of drugs with the same mechanism of action (MOA) in each latent space.

4. Results & Discussion

4.1. *Quantitative IR Experiments*

Baseline and Proposed Method Information retrieval results from the baseline and proposed methods are reported in Figure 2. This figure shows that our model variants offer significant performance improvements over either of the CCA or NN baselines, and even approach the performance of the train-set only oracle model. All models fall dramatically short of the test-set generalizability oracle, which indicates that while our tasks offer significant improvement over baseline models here, there are still major gains possible, primarily by focusing on improving the generalizability to the novel cell-types of the test set.

In addition, we show various ablation studies over the baseline models in Table 1 to probe what gene or molecular representations would make them better or worse. Strikingly, we can note that uniformly using aligned representations (meaning representations based on our multi-view alignment neural network architecture) offers significant improvements over other representations, indicating that even with a baseline approach such as the double nearest neighbor (D-NN) model, improvements to the embedding quality translate to notable improvements to IR performance. Notably, this is true both for GE and Chemical embedders, with Aligned-Aligned representations yielding optimal performance for both D-NN and CCA query mechanisms. Additionally, it is also clear that CCA is the preferred query metric, over either raw correlation (Corr) based lookups or D-NN based lookups.

Oracle Model Analysis Results from the oracle models are reported in Figure A1 and Table 2. As expected, oracle models using GE signatures from the test cell line greatly outperformed those using signatures from the train cell lines. This stark difference suggests that one of the largest barriers to performance here is the generalization gap between different cell lines. This further motivates for the curation of larger, cell-line heterogeneous datasets in the future.

In addition, aligned embeddings modestly improved the performance of test set oracles, and greatly improved the performance of train set oracles, consistent with the GE embedders learning slightly more generalizable representations. Of note, our proposed model achieved comparable results as the oracle model that leveraged raw GE signatures from the train cell lines. More specifically, our approach yielded slightly worse than the oracle on metrics (MRR, H@10) that emphasize early rankings, and slightly better on metrics (MR, H@100) that focus on more aggregate results. This is also apparent in the precision-recall plot, which shows the aligned embeddings curve starting out slightly below that of the raw/train oracle curve but then moving rapidly above it as further results are considered.

4.2. *Introspection Analyses*

Figure 3A contains the results of our experiments comparing performance to distance from the training set. Regardless of the measure of chemical similarity, compound retrieval performance

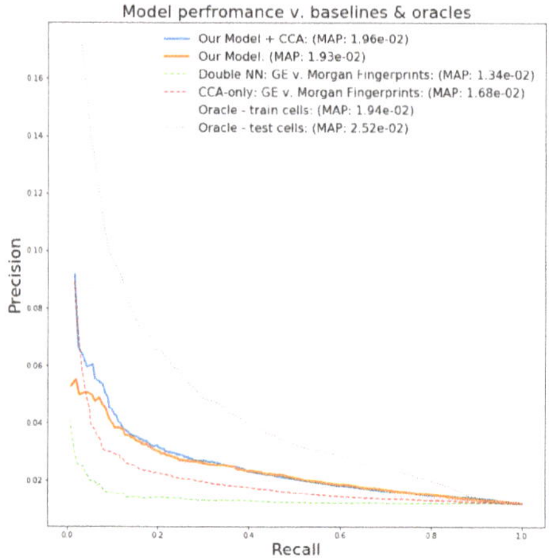

Fig. 2. Precision Recall curves for drug identification given gene expression signatures, across various baselines (dashed lines), oracles (dotted lines) and our model (solid lines).

Table 1. IR metrics across various configurations of the model/baselines. 'Chemprop' refers to pretrained model from Yang et al.[6] 'Aligned' indicates representations learned from our method (see Section 3.3). MR=median rank, MRR=mean reciprocal rank, H@K=Hit/Recall at K.

GE	Chemical	Method	MR	MRR	H@10	H@100
Raw	Morgan FP	D-NN	206	0.025	0.037	0.240
Raw	Chemprop	D-NN	211	0.025	0.035	0.254
Raw	Aligned	D-NN	189	0.033	0.047	0.290
Aligned	Morgan FP	D-NN	214	0.025	0.041	0.256
Aligned	Chemprop	D-NN	196	0.022	0.037	0.278
Aligned	Aligned	D-NN	137	0.039	0.072	0.402
Raw	Morgan FP	CCA	180	0.027	0.045	0.303
Raw	Chemprop	CCA	184	0.024	0.040	0.294
Raw	Aligned	CCA	134	0.039	0.076	0.412
Aligned	Morgan FP	CCA	177	0.027	0.050	0.319
Aligned	Chemprop	CCA	163	0.028	0.051	0.334
Aligned	Aligned	CCA	130	**0.048**	**0.093**	0.425
Aligned	Aligned	Corr	**126**	0.042	0.085	**0.432**

was inversely correlated with distance from the training set. As can be seen in Supplementary Figure A2, the same trend held with learned gene expression embeddings, and was present but much weaker using raw gene expression profiles.

Table 2. IR metrics for the various oracle methods.

Oracle Model	MR	MRR	H@10	H@100
Oracle - Train Cell Lines (Raw GE)	138	0.057	0.101	0.400
Oracle - Train Cell Lines (Embed)	110	0.064	0.128	0.466
Oracle - Test Cell Line (Raw GE)	82	0.076	0.147	0.565
Oracle - Test Cell Line (Embed)	79	0.093	0.160	0.568
Our Approach	126	0.042	0.085	0.432

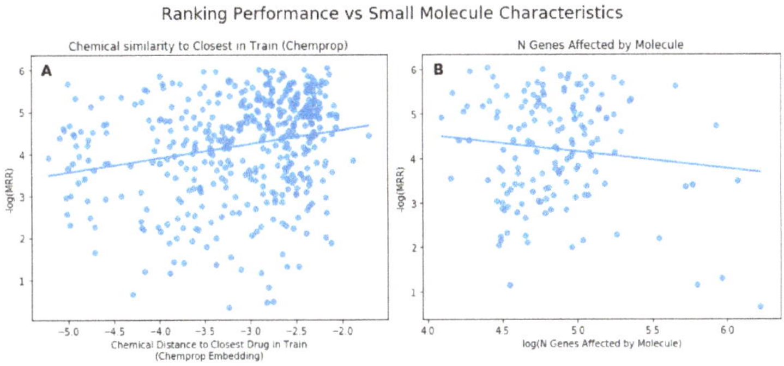

Fig. 3. Left: Performance (lower is better) vs. structural distance to the nearest compound in the training set. This plot demonstrates that compounds more structurally dissimilar to the train set show mildly worse performance than those that are more similar. See Appendix Figure A2 for analogous plots for four additional measures of distance from the training set. Right: Average Performance vs. # of Genes deferentially expression following treatment with the molecule of interest, showing that compounds that have broader transcriptomic effects are better retrieved by this method.

Figure 3B depicts the relationship between the transcriptional specificity of a compound and its ability to be retrieved using our analysis. As can be seen, there is a mild negative correlation, implying that molecules that affect the expression of many genes are easier to retrieve using this approach. Note that this observation is concordant with our findings on the difficulty of generalizing to new cell lines – drugs that affect a small, targeted set of genes are more likely to be cell line specific, and as our model is forced to surmount a significant generalization gap in evaluation, such cell-line specific signals are largely wiped out.

In addition, our analysis of the changes induced in the embedding space, shown in Supplementary Figure A4, reveal that our model's embeddings of molecules appear to better cluster shared MOAs than do the raw ChemProp embeddings, from which our model is initialized. This suggests that, as hypothesized regarding the nature of this task, our model is learning rich representations of the underlying molecules, though additional work remains to investigate this effect more thoroughly.

4.3. *Future Work*

We see several opportunities for further work on this task. First, expanding our data coverage, across molecules, cell lines, dosages, and treatment durations will allow us to measure and improve generalizability here. Second, exploring additional strategies to use the over-sampled nature of these data (e.g., ensembling together control and perturbational signatures to reduce variance) could be beneficial. Third, a more robust exploration of model architectures, losses, and deep metric learning/negative sampling methods, could offer improvements here.[26] Additionally, other styles of multimodal embedding could be explored, such as the use of cycle generative adversarial network, which in particular would enable us to adopt a semi-supervised approach.[39,40] The use of interpretability methods, particularly those used for graph analyses,[41] as well as additional studies interrogating how our model's performance changes with the amount of available training data could also be insightful here. Fourth, we recommend exploring methods to improve cell-line generalizability, e.g. incorporating information across many cell lines when forming predictions. Finally, we also note that while our analyses only examine small-molecule therapeutics, similar methods could also be applied to other modalities, such as RNA-based therapies.

5. Conclusion

We present a new task: cross-modal multi-view alignment between drug structures and perturbational gene expression profiles, which links molecular structure to an objective, functional readout of drugs with very broad biomedical relevance. We profile state-of-the-art representation learning methods on this new task, and inspect the learned chemical embeddings. We find that this modeling task induces an embedding space reflective of drug mechanism of action – which is not explicitly included in the training regime – and see modest generalization to both new structures and a new biological environment. Our oracle experiments demonstrate major performance gaps when trying to generalize to new tissues. We hope that this new benchmark task will catalyze future research and ultimately help enable a rapid, in silico compound prioritization methods.

6. Acknowledgements

The authors thank Connor Coley and Kyle Swanson for providing the pretrained chemical embedder from Yang et al.[6] S.G.F. was supported by training grant T32GM007753 from the National Institute of General Medical Science. M.B.A.M. was funded in part by National Institutes of Health: National Institutes of Mental Health grant P50-MH106933 as well as a Mitacs Globalink Research Award. The content is solely the responsibility of the authors.

References

1. T. Hansson, C. Oostenbrink and W. van Gunsteren, Molecular dynamics simulations, *Current opinion in structural biology* **12**, 190 (2002).
2. M. Krallinger, O. Rabal, A. Lourenco, J. Oyarzabal and A. Valencia, Information retrieval and text mining technologies for chemistry, *Chemical reviews* **117**, 7673 (2017).

3. P. W. Battaglia, J. B. Hamrick, V. Bapst, A. Sanchez-Gonzalez, V. Zambaldi, M. Malinowski, A. Tacchetti, D. Raposo, A. Santoro, R. Faulkner *et al.*, Relational inductive biases, deep learning, and graph networks, *arXiv preprint arXiv:1806.01261* (2018).
4. Z. Wu, B. Ramsundar, E. N. Feinberg, J. Gomes, C. Geniesse, A. S. Pappu, K. Leswing and V. Pande, Moleculenet: a benchmark for molecular machine learning, *Chemical science* **9**, 513 (2018).
5. S. Kearnes, K. McCloskey, M. Berndl, V. Pande and P. Riley, Molecular graph convolutions: moving beyond fingerprints, *Journal of computer-aided molecular design* **30**, 595 (2016).
6. K. Yang, K. Swanson, W. Jin, C. W. Coley, P. Eiden, H. Gao, A. Guzman-Perez, T. Hopper, B. Kelley, M. Mathea *et al.*, Analyzing learned molecular representations for property prediction, *Journal of chemical information and modeling* (2019).
7. Y.-C. Lo, S. E. Rensi, W. Torng and R. B. Altman, Machine learning in chemoinformatics and drug discovery, *Drug discovery today* **23**, 1538 (2018).
8. B. Ramsundar, S. Kearnes, P. Riley, D. Webster, D. Konerding and V. Pande, Massively multitask networks for drug discovery, *arXiv preprint arXiv:1502.02072* (2015).
9. F. A. Faber, L. Hutchison, B. Huang, J. Gilmer, S. S. Schoenholz, G. E. Dahl, O. Vinyals, S. Kearnes, P. F. Riley and O. A. Von Lilienfeld, Prediction errors of molecular machine learning models lower than hybrid dft error, *Journal of chemical theory and computation* **13**, 5255 (2017).
10. J. Gilmer, S. S. Schoenholz, P. F. Riley, O. Vinyals and G. E. Dahl, Neural message passing for quantum chemistry, *Proceedings of the 34th International Conference on Machine Learning-Volume 70* , 1263 (2017).
11. M. Zitnik, M. Agrawal and J. Leskovec, Modeling polypharmacy side effects with graph convolutional networks, *Bioinformatics* **34**, i457 (2018).
12. A. Musa, L. S. Ghoraie, S.-D. Zhang, G. Glazko, O. Yli-Harja, M. Dehmer, B. Haibe-Kains and F. Emmert-Streib, A review of connectivity map and computational approaches in pharmacogenomics, *Briefings in bioinformatics* **19**, 506 (2017).
13. J. Lamb, E. D. Crawford, D. Peck, J. W. Modell, I. C. Blat, M. J. Wrobel, J. Lerner, J.-P. Brunet, A. Subramanian, K. N. Ross *et al.*, The connectivity map: using gene-expression signatures to connect small molecules, genes, and disease, *science* **313**, 1929 (2006).
14. A. Subramanian, R. Narayan, S. M. Corsello, D. D. Peck, T. E. Natoli, X. Lu, J. Gould, J. F. Davis, A. A. Tubelli, J. K. Asiedu *et al.*, A next generation connectivity map: L1000 platform and the first 1,000,000 profiles, *Cell* **171**, 1437 (2017).
15. A. Musa, S. Tripathi, M. Kandhavelu, M. Dehmer and F. Emmert-Streib, Harnessing the biological complexity of big data from lincs gene expression signatures, *PloS one* **13**, p. e0201937 (2018).
16. T.-P. Liu, Y.-Y. Hsieh, C.-J. Chou and P.-M. Yang, Systematic polypharmacology and drug repurposing via an integrated l1000-based connectivity map database mining, *Royal Society open science* **5**, p. 181321 (2018).
17. N. R. Clark, K. S. Hu, A. S. Feldmann, Y. Kou, E. Y. Chen, Q. Duan and A. Maayan, The characteristic direction: a geometrical approach to identify differentially expressed genes, *BMC bioinformatics* **15**, p. 79 (2014).
18. Y. Donner, S. Kazmierczak and K. Fortney, Drug repurposing using deep embeddings of gene expression profiles, *Molecular pharmaceutics* **15**, 4314 (2018).
19. A. B. Dincer, S. Celik, N. Hiranuma and S.-I. Lee, Deepprofile: Deep learning of cancer molecular profiles for precision medicine, *bioRxiv* , p. 278739 (2018).
20. L. Rampášek, D. Hidru, P. Smirnov, B. Haibe-Kains and A. Goldenberg, Dr. vae: improving drug response prediction via modeling of drug perturbation effects, *Bioinformatics* **35**, 3743 (2019).
21. J. Cheng, Q. Xie, V. Kumar, M. Hurle, J. M. Freudenberg, L. Yang and P. Agarwal, Evaluation of analytical methods for connectivity map data (World Scientific, 2013) pp. 5–16.

22. R. Lopez, J. Regier, M. B. Cole, M. I. Jordan and N. Yosef, Deep generative modeling for single-cell transcriptomics, *Nature methods* **15**, 1053 (2018).
23. T.-M. H. Hsu, W.-H. Weng, W. Boag, M. McDermott and P. Szolovits, Unsupervised multimodal representation learning across medical images and reports, *arXiv preprint arXiv:1811.08615* (2018).
24. K. Hassani and A. H. Khasahmadi, Contrastive multi-view representation learning on graphs, *arXiv preprint arXiv:2006.05582* (2020).
25. H. Hotelling, Relations between two sets of variates (Springer, 1992) pp. 162–190.
26. Y. Li, M. Yang and Z. M. Zhang, A survey of multi-view representation learning, *IEEE Transactions on Knowledge and Data Engineering* (2018).
27. C. Deng, Z. Chen, X. Liu, X. Gao and D. Tao, Triplet-based deep hashing network for cross-modal retrieval, *IEEE Transactions on Image Processing* **27**, 3893 (2018).
28. G. Koch, R. Zemel and R. Salakhutdinov, Siamese neural networks for one-shot image recognition, *ICML deep learning workshop* **2** (2015).
29. T. M. Filzen, P. S. Kutchukian, J. D. Hermes, J. Li and M. Tudor, Representing high throughput expression profiles via perturbation barcodes reveals compound targets, *PLoS computational biology* **13**, p. e1005335 (2017).
30. E. Hoffer and N. Ailon, Deep metric learning using triplet network, *International Workshop on Similarity-Based Pattern Recognition* , 84 (2015).
31. C.-Y. Wu, R. Manmatha, A. J. Smola and P. Krahenbuhl, Sampling matters in deep embedding learning, *Proceedings of the IEEE International Conference on Computer Vision* , 2840 (2017).
32. D. Weininger, A. Weininger and J. L. Weininger, Smiles. 2. algorithm for generation of unique smiles notation, *Journal of chemical information and computer sciences* **29**, 97 (1989).
33. G. Landrum *et al.*, Rdkit: Open-source cheminformatics (2006).
34. D. Rogers and M. Hahn, Extended-connectivity fingerprints, *Journal of chemical information and modeling* **50**, 742 (2010).
35. R. Hodos, P. Zhang, H.-C. Lee, Q. Duan, Z. Wang, N. R. Clark, A. Maayan, F. Wang, B. Kidd, J. Hu *et al.*, Cell-specific prediction and application of drug-induced gene expression profiles, *Pacific Symposium on Biocomputing* **23** (2017).
36. M. McDermott, J. Wang, W. N. Zhao, S. D. Sheridan, P. Szolovits, I. Kohane, S. J. Haggarty and R. H. Perlis, Deep learning benchmarks on l1000 gene expression data, *IEEE/ACM transactions on computational biology and bioinformatics* (2019).
37. F. Pedregosa, G. Varoquaux, A. Gramfort, V. Michel, B. Thirion, O. Grisel, M. Blondel, P. Prettenhofer, R. Weiss, V. Dubourg *et al.*, Scikit-learn: Machine learning in python, *Journal of machine learning research* **12**, 2825 (2011).
38. J. Bergstra, D. Yamins and D. D. Cox, Making a science of model search: Hyperparameter optimization in hundreds of dimensions for vision architectures, *Proceedings of the 30th International Conference on International Conference on Machine Learning - Volume 28* , p. I115I123 (2013).
39. R. Felix, V. B. Kumar, I. Reid and G. Carneiro, Multi-modal cycle-consistent generalized zero-shot learning, *Proceedings of the European Conference on Computer Vision (ECCV)* , 21 (2018).
40. M. B. A. McDermott, T. Yan, T. Naumann, N. Hunt, H. Suresh, P. Szolovits and M. Ghassemi, Semi-Supervised Biomedical Translation with Cycle Wasserstein Regression GANs, p. 8 (2018).
41. Z. Ying, D. Bourgeois, J. You, M. Zitnik and J. Leskovec, Gnnexplainer: Generating explanations for graph neural networks, *Advances in neural information processing systems* , 9244 (2019).

Topological Feature Extraction and Visualization of Whole Slide Images using Graph Neural Networks

Joshua Levy*, Christian Haudenschild
Quantitative Biomedical Sciences, Geisel School of Medicine at Dartmouth
Lebanon, NH 03756, USA
Email: joshua.j.levy.gr@dartmouth.edu, christian.c.haudenschild.jr.gr@dartmouth.edu

Clark Barwick
School of Mathematics, University of Edinburgh
Edinburgh, EH9 3FD, United Kingdom
Email: clarkbar@gmail.com

Brock Christensen
Department of Epidemiology, Department of Molecular and Systems Biology,
Geisel School of Medicine at Dartmouth
Lebanon, NH 03756, USA
Email: brock.c.christensen@dartmouth.edu

Louis Vaickus
EDIT, Department of Pathology, Dartmouth Hitchcock Medical Center
Lebanon, NH 03756, USA
Email: louis.j.vaickus@hitchcock.org

Whole-slide images (WSI) are digitized representations of thin sections of stained tissue from various patient sources (biopsy, resection, exfoliation, fluid) and often exceed 100,000 pixels in any given spatial dimension. Deep learning approaches to digital pathology typically extract information from sub-images (patches) and treat the sub-images as independent entities, ignoring contributing information from vital large-scale architectural relationships. Modeling approaches that can capture higher-order dependencies between neighborhoods of tissue patches have demonstrated the potential to improve predictive accuracy while capturing the most essential slide-level information for prognosis, diagnosis and integration with other omics modalities. Here, we review two promising methods for capturing macro and micro architecture of histology images, Graph Neural Networks, which contextualize patch level information from their neighbors through message passing, and Topological Data Analysis, which distills contextual information into its essential components. We introduce a modeling framework, *WSI-GTFE* that integrates these two approaches in order to identify and quantify key pathogenic information pathways. To demonstrate a simple use case, we utilize these topological methods to develop a tumor invasion score to stage colon cancer.

Keywords: Topological Data Analysis; Graph Neural Networks; Whole Slide Images; Tumor Invasion; Uncertainty

---

* To whom correspondence should be addressed.

1. Introduction

Large-scale architectural motifs and repetitive patterns of functional tissue sub-units (eg. cells, connective tissue, extracellular matrix) form the basis of histopathology. While normal tissue is relatively homogenous, cancer contains disordered structures / phenotypes that reflect driving genetic alterations. As neoplastic transformation progresses, the extent of infiltration and destruction of normal tissue is used to grade and stage cancers. Practitioners of histopathology are thus highly sensitive to disruptions in normal structure. A wide variety of computational methods have been developed to augment traditional histological inspection[1] by reducing time and personnel costs associated with manual slide screening. These emerging techniques have also demonstrated the potential for identifying novel disease pathways and previously unrecognized morphologies.

Deep learning has been particularly successful in digital pathology [2]. In comparison to prior modeling techniques that use handcrafted features, deep learning applies parameterized filters and pooling mechanisms via convolutional neural networks (CNN) to capture and integrate lower level image features into successively higher levels of complexity [3]. These approaches have been used to automatically stage liver fibrosis [4], identify morphological features correspondent with somatic alterations[5], assess urine slides for bladder cancer[6], and circumvent costly chemical staining procedures[7,8], amongst many others[9]. Many research groups are developing high-throughput clinical pipelines to take advantage of these healthcare technologies. Validating and scaling these technologies is essential for successful deployment[10].

As a result of the gigapixel resolution of Whole Slide Images (WSI), which contain a diverse range of tissue and morphological features, researchers typically must partition the WSI into smaller sub-images. These sub-images are then evaluated separately via the deep learning model for classification or segmentation tasks, from which their results may be aggregated for slide-level inferences. Aggregation via a CNN incorporates excessive whitespace and places unnecessary dependence on the orientation and positioning of the tissue section[11]. Alternatively, a 'bag of images' approach can be taken, in which patch representations are aggregated using autoregressive or attention-based mechanisms to generate a whole slide representation, ignoring non-tissue regions[12–14]. These integrative approaches may be highly stochastic and insufficiently reproducible / reliable to be properly integrated into the clinic or with other omics-based modalities. These methods may additionally undervalue the higher order context between a patch and its immediate neighbors which may be vitally important to the targeted prediction.

Graphs are mathematical constructs that model pairwise relationships between entities. Accordingly, graphs are well suited to model dyadic relationships between single patches (nodes) in a WSI as defined by their spatial distance/correlation (edges). Graph Neural Networks (GNN) have been developed to encapsulate information from adjacent tissue regions/cells in order to inform the representation of the current patch of interest. GNN naturally capture the intermingling of various tissue sub-compartments while remaining permutationally invariant (the ordering/rotation of patches on slides does not impact prediction). While square-grid convolutions over WSI sub-images propagate information within a fixed neighborhood of patches and require consistent ordering of patches [11], GNN relax the convolutional operator to aggregate information across an unfixed number of neighbors to update the patch-level embedding[15].

Prior GNN research on WSIs center graph nodes on cells under the assumption that cell-cell interactions are the most salient points of information[16]. However, this approach underappreciates the diagnostic/prognostic information conveyed by tissue macro-architectural structures. Constructing cell-centered graphs are limited by cell detection accuracy (a surprisingly difficult problem) and more importantly, incorporating all cells in a graph model is subject to complexity constraints. Despite these potential limitations, there remain numerous techniques to study WSI using GNN at various scales [17]. Here, we seek methods to explain graph convolution results post-hoc to elucidate mechanisms by which tissue regions interact.

Topological Data Analysis (TDA) quantifies the underlying shape and structure of data by collapsing persistent topological structures [18]. TDA is well-suited for summarizing Whole Slide Graphs (WSI fitted by a GNN; e.g. WSG) to identify and relate key tissue architectures, regions of interest, and their intermingling. However, the sheer quantity, complexity, and dimensionality of histology data makes interpretation challenging. A recently developed TDA-tool, Mapper[18], alleviates this issue by providing a succinct summary of high-dimensional data to elucidate obscured relationships. Mapper projects the data to a lower dimensionality, packs the data into overlapping sets, which are then clustered to form a simplified, easily interpretable graph. Unlike pooling approaches that are built into the deep learning model and must be pre-specified, Mapper is generalizable and can be configured to study WSI information at multiple resolutions after fitting a GNN model. These models have the capacity to provide higher order descriptors of information flow for any GNN model, greatly simplifying analysis. Abstractions can then be analyzed to learn new disease biology through interrogation of patch-level embeddings. While TDA methods have previously been applied to high dimensional omics data[19–21] and histopathology images[22,23], to our knowledge, there have been no applications of TDA methods to GNN models fit on histological data, where these methods may be of great benefit.

Colorectal Cancer (CRC) is a common cancer with approximately 150,000 new cases annually in the United States and an estimated 63% 5-year survival rate. CRC most commonly arises from dysplastic adenomatous polyps with somatic alterations in the APC pathway or the mismatch repair (MMR) pathway[24]. The Colon is divided into distinct layers including epithelium, lamina propria, submucosa, muscularis propria, pericolic fat, and serosa (in certain anatomical locations). Tumor staging comprises tissue and nodal stages with higher numerals indicating a greater depth of invasion and greater number of lymph nodes (LN) involved by the tumor, respectively.

We present a Whole Slide Image GNN Topological Feature Extraction workflow (*WSI-GTFE)*[*], for applying topological methods to interrogate a WSI GNN fit and demonstrate its utility in determining colon cancer stage. As a simple use case for these methods, we apply Mapper to quantitate the degree of tumor invasion into deeper sub-compartments of the colon and corroborate these tumor invasion scores (TIS) with disease staging to form an interpretable predictive score. The demonstration featured in this paper outlines some of the many potential applications of topological methods in the analysis of WSI GNN models.

---

[*] Software available on GitHub at the following URL: https://github.com/jlevy44/WSI-GTFE

2. Materials and Methods

2.1. *Data Acquisition and Processing*

We selected Colon (n=172) and accompanying Lymph Node (n=84) resection slides from 36 patients at Dartmouth Hitchcock Medical Center. Briefly, samples were grossed, embedded in paraffin blocks, sliced into five-micron sections and scanned using the Leica Aperio-AT2 scanner at 20x, stored in SVS image format, from which our in-house pipeline *PathFlowAI*[10] was utilized to extract and preprocess the slides into NPY format. A board-certified pathologist provided coarse segmentation maps in the following categories: 1) epithelium, 2) submucosa, 3) muscularis propria, 4) fat, 5) serosa, 6) debris, 7) inflammation, 8) lymph node, and 9) cancer. We extracted 2.69 million 256x256 pixel patches correspondent to these slides.

2.2. *Overview of Framework*

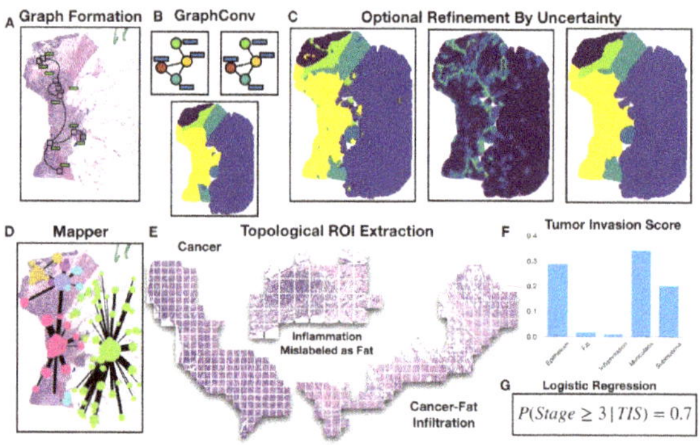

Fig. 1. WSI-GTFE Framework: a) patch-level CNN embeddings extracted using *PathFlowAI* form graph via their spatial adjacency; b) targets (eg. colon sub-compartments) predicted using successive applications of graph convolutions; c) highly uncertain regions (middle) from noisy prediction map (left) may be reassigned (right); d) Mapper summarizes GNN embeddings over WSI as a graph; e) Meaningful histology (ROI) captured as Mapper graph nodes; f) Functional relationships between Cancer and other ROI, weighted edges Mapper graph, mined to form *TIS* vector; g) *TIS* used in prediction model to form interpretable staging score (odds ratios and log-odds probability), demonstrates type of relationships that may be extracted using TDA

The *WSI-GTFE* framework (**Figure 1**), provides methods to summarize the intermingling of tissue sub-compartments via a two-stage CNN-GNN model, followed by utilization of TDA methods:

1. Learning patch-level CNN embeddings and constructing spatial adjacency graph (**Figure 1A**)
2. Contextualizing patch level embeddings via an unsupervised or supervised GNN (**Figure 1B**)
3. Optionally refining the patch level embeddings through estimation of uncertainty in patch-level classification tasks (**Figure 1C**)
4. Applying Mapper to pool patches into overlapping Regions of Interest (ROI) (**Figure 1D-E**)
5. Estimating the degree of information flow and intermingling between the regions (**Figure 1F**)
6. Optionally using measures of information flow as additional markers for clinical or molecular associations (**Figure 1G**)

2.2.1. *Estimation of Patch-Level Embeddings*

A WSI (an RGB array on the order of 100,000 pixels in any spatial dimension) $\vec{X}$, is comprised of a collection of sub-images, $\{\vec{x_i}\}$. A neural network maps each sub-image to a low dimensional embedding or representation, $\vec{z_i}$, via the following mapping $f: X \to Z$, $\vec{z_i} = f(\vec{x_i})$. Patch-level features may be extracted using pretrained CNNs such as ImageNet, which has learned a huge collection of convolutional filters and features correspondent to 1000 common objects such as dogs, cats and birds [25]. Features may also be acquired using unsupervised approaches such as variational autoencoders (VAE)[26] or self-supervised techniques such as contrastive predictive coding (CPC)[14] or SimCLR[27]. Finally, patch-level features may be learned after pretraining on histology targets of interest, such as classified objects or ROI. We utilized both an ImageNet-pretrained CNN as well as a CNN we pretrained for tissue sub-compartment classification task, generating two separate sets of patch-level embeddings for comparison.

2.2.2. *Contextualizing Patch-Level Embeddings via GNN*

Graphs are represented via following expression $G = (V, E, A, X)$. The set of nodes/patches or vertices V are related to each other via edgelist E. Alternatively, the edgelist may be represented by a sparse adjacency matrix A, of which binary indicators A_{ij} depict a relationship between node i and node j. Node/patch-level embeddings or features are represented by attribute matrix X. A WSI may be encoded as a graph by storing patch level embeddings ($\{\vec{z_i}\}$, index i for select patch) in the attribute matrix X (m patches by n embedding dimensions) and recording spatial adjacency (via a k or radius nearest neighbors) of all patch coordinates as A. GNN utilize message passing operations to update the embeddings of nodes by their neighbors via the following convolution operation [28]:

$$\vec{z_i^*} = \vec{z_i} * g = \gamma\left(\vec{z_i}, \text{SCATTER}_{j \in N}\,\phi(\vec{z_i}, \vec{z_j})\right) \quad (1)$$

The embeddings of the neighbors of patch i, in neighborhood N, are themselves updated via some parameterized functional ϕ, which is scattered in parallel across GPUs, and then aggregated to update the embedding of patch i via the parameterized operation γ. Information from neighboring patches are passed as such. Multiple applications of these convolutions expand the neighborhood from which information is propagated. Additional pooling mechanisms, *AGG*, such as DiffPool or MinCutPool[29,30], serve to aggregate the patch level representations into cluster or slide-level representations:

$$\vec{z} = AGG\left(\{\vec{z_i^*}\}\right) \quad (2)$$

There exist multiple modeling objectives for updating these embeddings, which include: 1) node-level classification, where $\vec{y} = f\left(\vec{z_i^*}\right)$, trained via the cross-entropy loss, 2) unsupervised node-level measures such as Deep Graph Infomax[31] and spectral clustering objectives, and 3) graph-level supervised, eg. $\vec{y} = f(\vec{z})$, or 4) graph-level self-supervised objectives. For demonstration purposes, we learn patch-level classification of colon sub-compartments and predict these sub-compartments on held-out slides after initialization of an adjacency matrix of patches, which could be used to pretrain whole-slide level objectives. From the fitted GNN model, intermediate patch-level or cluster-level (when applying pooling operations) embeddings may be extracted for further analysis. While we constructed WSG from the spatial adjacency of patches in this work, this *WSI-GTFE* method is agnostic of WSG creation approach. These graphs also may be built using cell / nucleus detection methods, though such methods are beyond the scope of this work.

2.2.3. *Optional Refinement of Patch-Level Predictions*

Graph convolutions aim to contextualize patches with their neighbors and as such are able to smooth the map of predictions across a slide. However, small deformities in an otherwise homogenous decision map, (for instance, pockets of inflammation that were not captured by the pathologist's relatively coarse annotations), may be a source of signal noise. To further smooth the classification map of patches across a slide, assigned patch-level labels may be refined using label propagation techniques. Dropout[32] methods randomly set predictors at a particular neural network layer to 0 with a certain probability, while DropEdge[33] randomly prunes edges of a graph, which in this case corresponds to the adjacency matrix of the WSI. While both of these techniques have been utilized to improve the generalization of graph neural networks through perturbations to the input data and intermediate outputs, applications of these techniques during prediction may be used to make multiple posterior draws of a patch-level categorial distribution for class label assignment. Both the variance of the predictive posterior distribution after numerous posterior draws and the entropy in the class labels after averaging the results for a sample after application of SoftMax layer may be used as estimates of uncertainty in prediction[34]. Nodes that exhibit high uncertainty may be pruned and the remaining class labels may be propagated to the unlabeled patches.

2.2.4. *Application of Mapper to Extract Regions of Interest*

Once a GNN model has been fit, post-hoc model explanation techniques such as visualization of the attention weights or the use of GNNExplainer[35] to identify important subgraphs for classification can be performed. However, these may be difficult to interpret because they attempt to summarize complex interactions between high-dimensional data at the scale of thousands of patches per WSI. The complexity of such visualizations makes them difficult to understand and highlights the need for a simplified visualization.

Because similarity-based distances between patch-level GNN embeddings reflect higher-order connectivity and perceptually similar histological information, topological methods (such as Mapper) can compress this data to its essential structures while revealing the most salient aspects. For a given WSI, Mapper operates on the resulting point cloud of the patch-level GNN embeddings to first project the points to a lower dimensional space via techniques such as PCA, UMAP or NCVis [36] (referred to as *Projecting, f*). Once the data is projected, it is separated into overlapping sets (*Covering, U*), the number of which determines the resolution of the data summary. In each set, a *Clustering* algorithm (e.g. hierarchical clustering) is applied to the datapoints. The output of applying Mapper to this structure is a graph, where a node represents a cluster of WSI patches and an edge represent the degree of shared patches between the clusters[37]. This Mapper graph summarizes higher-order architectural relationships between patches and their shared histological information. In our framework, we refer to the nodes (collection of patches) as ROI, and the topological connectivity between the ROIs as their functional connectedness or "intermingling". For instance, if a tumor ROI was connected to an ROI of the submucosa, we would say that the tumor has invaded (intermingled with) the submucosa. The degree of intermingling is quantified by the amount of overlap as defined by covering U and weighted by the incidence of cancer in each ROI. The expressiveness of this summary graph may be modified by selection of different *Filter*, *Cover*, and *Cluster* parameters which allows the user to interrogate ROIs in the WSI at different scales (a degree of flexibility beyond that of currently existing GNN pooling operations). We implemented

Mapper using the *Deep Graph Mapper* implementation; however, python-based *Kepler-Mapper* and *giotto-tda* also present software solutions that may be readily employed [37–39].

2.2.5. *Associating ROI Connectivity with Clinical Outcomes*

Once ROIs have been extracted using Mapper, measures of functional relatedness between the regions may be correlated with slide-level clinical outcomes. In our simple use case, we developed a Tumor Invasion Score (TIS) that measures the degree of overlap between the tumor and an adjacent tissue region. To construct this score, we first decompose each ROI into a vector encoding the frequency of each predicted tissue sub-compartment, $\vec{c_i}$ (counts of patch class assignment). The amount of overlap, as learned by Mapper's *Cover* operation, between two ROI (ROI_i and ROI_j, frequency vectors $\vec{c_i}$ and $\vec{c_j}$ respectively) is w_{ij}. The intermingling between different tumor sub-compartments for a pair of ROI may be expressed as $A_{ij} = w_{ij}\vec{c_i} \otimes \vec{c_j}$. Given Mapper graph G, with edge-list E, ($e_{ij} \in E$), the final pairwise associations between the regions are given by: $I = \sum_{e\in E} \frac{A_{ij}+A_{ij}^T}{2}$. To measure tumor invasion/infiltration of the surrounding sub-compartments, we select the row of this matrix corresponding to the tumor: $TIS = \overrightarrow{I_{tumor}}$. These vectors may be stacked across patients to form a design matrix, $\overleftrightarrow{X}$, which may be associated with binary or continuous outcomes, $\vec{y}$. Here, we utilize Logistic Regression to associate TIS with cancer staging greater than 2 for the Colon samples, and positive Lymph Node status for Lymph Nodes.

2.3. *Experimental Details*

As proof of concept, we first pre-trained colon (comprised of epithelium, submucosa, muscularis propria, fat, serosa, inflammation, debris and cancer) and lymph node (fat, lymph node, cancer) classification networks with 10-fold cross validation (partitioning 10 separate training (82%), validation (8%) and test (10%) sets), evaluated using the area under the receiver operating curve (AUC/AUROC) and a weighted F1-Score. We extracted features from the penultimate layer of a ResNet50 neural network for about 2.7 million images per fold (26.9 million embedding extractions across 10-folds), using the pretrained network and separately from an ImageNet-pretrained model. After extraction of image features, we constructed graph datasets through calculation of the spatial adjacency (k-nearest neighbors) between the patches and storing node level embeddings into the attribute matrix. We created and trained a GNN that featured four graph attention layers, interspersed with ReLU activation functions [40] and DropEdge layers, followed by one layer of DropOut and finally a linear layer with SoftMax activation for node level prediction (**Figure 2A**). Models were generated using the pytorch-geometric [28] framework and trained using Nvidia v100 GPUs. For each cross-validation fold, we updated the parameters of our GNN through backpropagation of Cross-Entropy loss for node level classification on the training slides, while evaluating the potential to generalize on the validation set of *Whole Slide Graphs* (WSG) through evaluation of the F1-score. We saved the model parameters correspondent to the training epoch with the highest validation F1-score and extracted graph-node level embeddings and predictions on the validation and test sets of slides for each cross-validation fold. We refined patch-level predictions for all validation and test slides for the models with ImageNet-pretrained image features. To evaluate node-level tissue sub-compartment classification, we then calculated AUROC and F1-Score fit statistics across test slides. Finally, we applied Mapper to extract ROIs and TIS for all test slides

(**Figure 2B**), of which 10-fold cross validation was applied over a non-penalized Logistic Regression model to estimate concordance with tumor stage and lymph node status. We also fit similar models to the frequencies of assignment of tissue sub-compartments for each case and combined relative frequencies of tissue sub-compartments with their *TIS* values to yield a final model. To evaluate the parsimony of the logistic regression model for alignment with our expectation that tumor infiltrating the fat corresponds to a high stage, we fit a generalized linear mixed effects model to the *TIS* scores, clustered by patient, and inspected the regression coefficients for quantitating the nature of these functional relationships.

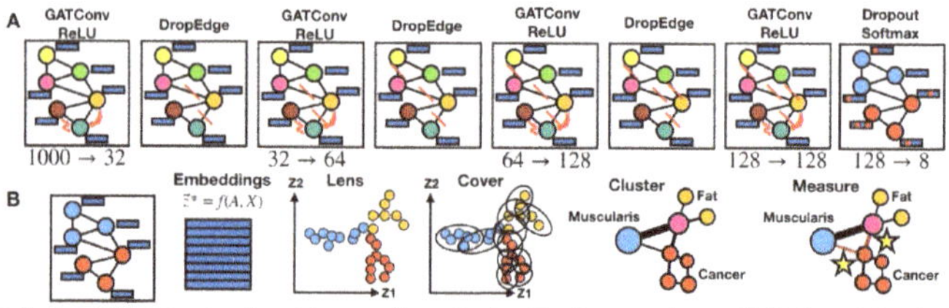

Fig. 2. Methods: a) Neural network architecture for node classification experiment; 1000-d patch-level embeddings pass through graph attention convolutions, ReLU and DropEdge layers which alter dimensionality of patch embeddings while routing information from neighbors; attention between blue node and neighborhood is characterized using red curves; pruned edges are portrayed using red lines; b) once GNN classification model has been fit, GNN embeddings are extracted; lens function projects them to lower-dimensionality; patches are *covered* and *clustered* to reveal high-level measurable relationships between muscularis propria, fat and cancer

3. Results

3.1. *Patch-Level Classification and Embeddings*

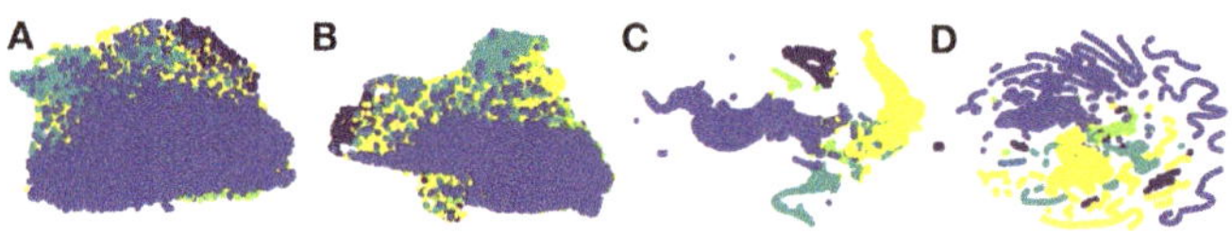

Fig. 3. UMAP projection of penultimate layer of neural network for one select colon slide; nodes colored by true sub-compartment; a) ImageNet-pretrained CNN embeddings of patches; b) colon-pretrained CNN embeddings of patches; c) updated GNN patch embeddings after ImageNet extraction; d) updated GNN patch embeddings after colon-pretrained CNN extraction

An acceptable patch-level classification performance indicates room to further interrogate the slides for functional relationships between patches. In **Table 1**, we present 10-fold CV AUROC and F1-Score statistics on held-out test slides for patch level classification. The pretrained CNN for colon segmentation yielded moderately low performance metrics, while pretraining for lymph node yielded much higher scores. After feature extraction and training of the GNN taking into account the information from neighboring patches, scores increased substantially. Pretraining the CNN on Colon-specific targets had little impact on the classification model after fitting the GNN, suggesting that information contained from the patch surroundings is sufficient to contextualize that particular patch, or that the pretrained ImageNet is generalizable enough to histology images. Inspection of the patch-level embeddings (**Figure 3**), further corroborate that the original CNN does little to delineate the different classes of colon tissue, while the GNN embeddings demonstrate clear separation between these sub-compartments.

Table 1. GNN node classification results for colon (n=172 slides; 2,116,396 images) and LN (n=84; 570,326 images); averaged across slides; confidence assessed via 1000-sample non-parametric bootstrap

10-Fold CV (n=256) Node Classification	AUC ± SE	F1-Score ± SE
Colon CNN-Only	0.75±0.0054	0.43±0.0079
Colon GNN ImageNet	0.95±0.0026	0.81±0.006
Colon Prediction Refinement	n/a	0.83±0.0063
Colon GNN Pretrained	0.96±0.0031	0.82±0.0074
LN CNN-Only	0.91±0.0069	0.8±0.013
LN GNN ImageNet	0.96±0.0067	0.89±0.014
LN GNN Pretrained	0.97±0.0049	0.9±0.014

3.2. *Tumor Staging via Mapper Derived Invasion Scores*

Figure 4 illustrates the extracted Mapper graph of representative low stage and higher stage slides. As compared to the lower stage slide, the *TIS* score derived for the higher stage indicates higher intermingling of tumor with regions of fat and is confirmed by pathologist annotations. We extracted an average of 32 ROIs from each WSI (range 5 – 155 ROI). Inspections of ROIs indicated some with clusters of tissue finely localized to histological tissue layer, and a few ROI that went undetected by the initial pathologist inspection (e.g. pockets of inflammation and tumor seeding in the fat).

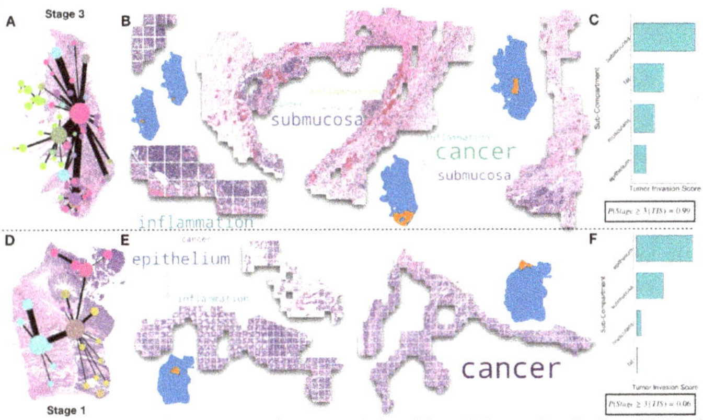

Figure. 4. Example Topological Feature Extraction on two Colon slides; a) Mapper visualization of *WSG* of a stage 3 tumor; each vertex corresponds to an ROI, placement of the vertex reflects center of mass and thickness of edge connecting two points reflects topological overlap; b) example of four ROIs from the stage 3 slide; image patches are stitched back together; location is depicted in slide; composition of ROI is denoted with word clouds, where size of word is proportional to percentage makeup of ROI; c) actual *TIS* score for slide, prevalent invasion in the submucosa, fat and muscularis to reveal deep invasion; actual score from classifier gives 99% probability of advanced stage; d) Mapper visualization for stage 1 slide; e) left-most ROI demonstrates epithelial crypts with inflammation in lower right pocket; f) actual reported *TIS* score denotes invasion of epithelium with 6% probability of advanced stage

Table 2. Ten-Fold AUROC statistics for unpenalized logistic regression prediction model on held out test data across all slides for colon (n=172) and LN (n=84); three head columns indicate whether advanced staging was predicted using aggregates of colon sub-compartment assignments (Region Counts); invasion (Tumor Invasion Scores); or Both to lend complementary information; models with main effects and interactions were considered; confidence assessed via 1000-sample non-parametric bootstrap

AUC (10-Fold CV)	**Region Counts Only**		**Tumor Invasion Scores Only**		**Both**	
Stage > 2; LN Positive	**Main Effects**	**Second Order Terms**	**Main Effects**	**Second Order Terms**	**Main Effects**	**Second Order Terms**
Colon GNN Pretrained	0.89±0.028	0.85±0.031	0.84±0.032	0.89±0.028	0.91±0.022	0.85±0.031
Colon GNN Not Pretrained	0.89±0.027	0.88±0.028	0.86±0.029	0.89±0.027	0.91±0.023	0.88±0.028
LN GNN Pretrained	0.76±0.077	0.88±0.046	0.88±0.046	0.76±0.077	0.88±0.046	0.88±0.046
LN GNN Not Pretrained	0.89±0.051	0.92±0.039	0.92±0.034	0.89±0.051	0.92±0.037	0.92±0.039

Table 3. Taking into account clustering on the patient level, odds-ratios derived from GLMM (ICC=0.21, n=172) fit on *TIS* scores derived from GNN that utilized colon-pretrained CNN embeddings; odds ratios indicate risk of advanced progression given tumor invasion of region

	Stage $\geq$ **3**		
Predictors	*Odds Ratios*	*CI*	*p*
(Intercept)	0.52	0.29 – 0.93	**0.027**
***TIS:* Epithelium**	0.82	0.43 – 1.56	0.539
***TIS:* Fat**	7.54	2.93 – 19.38	**<0.001**
***TIS:* Muscularis**	1.68	1.02 – 2.77	**0.043**
***TIS:* Serosa**	1.43	0.33 – 6.28	0.632
***TIS:* Submucosa**	1.23	0.57 – 2.63	0.597

TIS scores correlated very well with Tumor staging. Ten-fold CV AUC was 0.91 for advanced Colon cancer staging and 0.92 for positive Lymph Nodes (**Table 2**). The frequency of sub-compartment instance and tumor invasion was also able to predict cancer stage when considered in isolation. Taken together, *TIS* and compartment localization achieved a higher AUC score, which speaks to the complementary information that each approach was able to provide to form a more complete picture of tumor progression. From the *TIS* scores, we were able to derive odds ratios (*OR*; measure of association between exposure and outcome, greater than one indicates adverse risk) as to their relation to tumor staging using linear mixed effects models (clustered on individual). As expected, fat interaction was highly associated with progression to a stage 3 or higher (**Table 3**). Importantly, invasion of the muscularis propria, an adjacent and superficial region to the fat, had a statistically significant odds ratio commensurate with its depth in the colon.

4. Discussion

Graph Neural Networks are increasingly promising approaches for studying WSI (and other gigapixel scale images) at multiple scales of inference through propagation of patch-wise information. However, when employing GNNs, the route of propagation often becomes obfuscated by the sheer quantity of patches being studied. This, in turn may make it difficult for researchers, clinicians or biologists to accept or understand these graph neural network technologies and their predictions. However, the compartmentalized and repetitive nature of tissue means that histology images can be greatly simplified via grouping of spatially adjacent subimages with perceptually similar and complementary input features. We have introduced methods from TDA to capture and reduce these motifs. In colon histology, we distilled information across WSI to better quantitate how the intermingling of different tissue sub-compartments inform disease stage. These results warrants investigation of other spatially driven processes, such as identifying ROIs correspondent to spatial transcriptomics[41] and integration with high-dimensional omics data types.

Mapper has proven to be a useful TDA tool for elucidating high-level topology of the WSI. However, Mapper is highly dependent on the *Filter* function, *Cover* and *Cluster* parameters and algorithms to generate a topological map. While these features offer flexibility to study the WSI at multiple resolutions, which includes expanding the large range of ROI extracted, full exploration of the parameter space to identify an ideal range of parameters for Mapper graphs for the slide in study are beyond the scope of this work.

We also assessed the impact of domain-specific pre-training of a CNN on the resulting GNN predictions. Our preliminary results showed negligible impact on GNN accuracy. Integration of signal from the surrounding tissue context via GNNs may therefore be sufficient to overcome domain differences between histology images and real-world images (ImageNet). Further experimentation is needed on more nuanced examples to test this hypothesis.

There are a few limitations to our study. We assumed that GNNs are able to adequately capture patch-level information and their surrounding tissue architecture. The accuracy of our model was constrained by relatively coarse physician annotations that tended to ignore small structures like veins in the fat region of the lymph nodes, or small pockets of inflammation bounded by other tissue compartments, thereby reducing the accuracy of the model. However, inspection of regions with high uncertainty and label propagation allowed for correction of some of these issues. We also acknowledge the possibility of bias in given cross-validation folds. While we stratified the slides by whether they were representative of high or low stage, slides may contain different macroarchitectural features, and may, for instance, be completely devoid of serosa (which is only present in certain regions of the abdomen), which made it difficult to predict its presence. The colon WSI sections were analyzed from 36 patients (representing 256 slides). We acknowledge that there were repeated measurements taken across different slides from the same patient, the results of these sections may be correlated. While in our final inference on the *TIS* scores we account for this using mixed effects modeling, extracting samples from different patients would have been preferred to reduce cluster-level effects (data and pathologist time allowing). Due to the number of free parameters, we did not perform robust hyperparameter scans over the GNN.

In the future, we intend to utilize extracted GNN features contained within our ROIs to better identify the core topological structures that form a pathologist's understanding of a slide[18]. Simplicial complexes represent series of points, lines, triangles and higher-dimensional tetrahedra. Persistence diagrams discover topological features in the form of simplicial complexes that persist over wide changes in proximity between points. These approaches can be readily applied to GNN embeddings to establish "barcodes" of various ROIs contained within the slide[42,43], which may be used to supplement existing efforts to hash WSI to further assess the composition of other slides by the presence of characteristic topologies[44]. In addition to utilizing persistence based TDA methods, we aim to apply the aforementioned methods to GNN embeddings after applying graph pooling layers to identify topology and ROIs which may be related to molecular targets of interest, dense omics profiles and unlabeled clusters of tissue.

5. Conclusion

As multimodal deep learning approaches become increasingly important, GNNs are emerging as an attractive modeling tool for WSI representation where proper integration and association with slide-level outcomes is required. Conveniently, these approaches learn to identify key information pathways which may be simplified and visualized using TDA tools such as Mapper. Our method, *WSI-GTFE*, presents a framework from which to flexibly summarize the key insights acquired from fitting any GNN model to histological data. We hope that topological methods continue to see usage and integration with their deep learning graph counterparts for WSI level histological analyses given the benefits they provide in terms of model interpretability, quantitation of tissue compartment interaction, and potential for new biological discovery and disease prognostication.

References

1. Komura, D. & Ishikawa, S. Machine Learning Methods for Histopathological Image Analysis. *Computational and Structural Biotechnology Journal* **16**, 34–42 (2018).
2. LeCun, Y., Bengio, Y. & Hinton, G. Deep learning. *Nature* **521**, 436–444 (2015).
3. Krizhevsky, A., Sutskever, I. & Hinton, G. E. ImageNet Classification with Deep Convolutional Neural Networks. in *Advances in Neural Information Processing Systems 25* (eds. Pereira, F., Burges, C. J. C., Bottou, L. & Weinberger, K. Q.) 1097–1105 (Curran Associates, Inc., 2012).
4. Yu, Y. *et al.* Deep learning enables automated scoring of liver fibrosis stages. *Scientific Reports* **8**, 16016 (2018).
5. Coudray, N. *et al.* Classification and mutation prediction from non–small cell lung cancer histopathology images using deep learning. *Nat Med* **24**, 1559–1567 (2018).
6. Vaickus, L. J., Suriawinata, A. A., Wei, J. W. & Liu, X. Automating the Paris System for urine cytopathology—A hybrid deep-learning and morphometric approach. *Cancer Cytopathology* **127**, 98–115 (2019).
7. Levy, J. J. *et al.* A Large-Scale Internal Validation Study of Unsupervised Virtual Trichrome Staining Technologies on Non-alcoholic Steatohepatitis Liver Biopsies. *bioRxiv* 2020.07.03.187237 (2020) doi:10.1101/2020.07.03.187237.
8. Levy, J., Jackson, C., Sriharan, A., Christensen, B. & Vaickus, L. Preliminary Evaluation of the Utility of Deep Generative Histopathology Image Translation at a Mid-sized NCI Cancer Center. in 302–311 (2020).
9. Levy, J. J., Jackson, C. R., Haudenschild, C. C., Christensen, B. C. & Vaickus, L. J. PathFlow-MixMatch for Whole Slide Image Registration: An Investigation of a Segment-Based Scalable Image Registration Method. *bioRxiv* 2020.03.22.002402 (2020) doi:10.1101/2020.03.22.002402.
10. Levy, J., Salas, L. A., Christensen, B. C., Sriharan, A. & Vaickus, L. J. PathFlowAI: A High-Throughput Workflow for Preprocessing, Deep Learning and Interpretation in Digital Pathology. *Pacific Symposium on Biocomputing* **25**, 403–414 (2020).
11. Tomita, N. *et al.* Attention-Based Deep Neural Networks for Detection of Cancerous and Precancerous Esophagus Tissue on Histopathological Slides. *JAMA Netw Open* **2**, e1914645–e1914645 (2019).
12. van der Laak, J., Ciompi, F. & Litjens, G. No pixel-level annotations needed. *Nature Biomedical Engineering* **3**, 855–856 (2019).
13. Hao, J., Kosaraju, S., Tsaku, N., Song, D. & Kang, M. PAGE-Net: Interpretable and Integrative Deep Learning for Survival Analysis Using Histopathological Images and Genomic Data. *Pacific Symposium on Biocomputing* **25**, 355–366 (2020).
14. Lu, M., Chen, R., Wang, J., Dillon, D. & Mahmood, F. *Semi-Supervised Histology Classification using Deep Multiple Instance Learning and Contrastive Predictive Coding.* (2019).
15. Wu, Z. *et al.* A Comprehensive Survey on Graph Neural Networks. (2019).
16. Chen, R. J. *et al.* Pathomic Fusion: An Integrated Framework for Fusing Histopathology and Genomic Features for Cancer Diagnosis and Prognosis. (2019).
17. Adnan, M., Kalra, S. & Tizhoosh, H. R. Representation Learning of Histopathology Images Using Graph Neural Networks. in 988–989 (2020).
18. Chazal, F. & Michel, B. An introduction to Topological Data Analysis: fundamental and practical aspects for data scientists. *arXiv:1710.04019 [cs, math, stat]* (2017).
19. Wang, T., Johnson, T., Jie, Z. & Huang, K. Topological Methods for Visualization and Analysis of High Dimensional Single-Cell RNA Sequencing Data. *Pacific Symposium on Biocomputing* **24**, 350–361 (2019).
20. Nicolau, M., Levine, A. J. & Carlsson, G. Topology based data analysis identifies a subgroup of breast cancers with a unique mutational profile and excellent survival. *PNAS* **108**, 7265–7270 (2011).
21. Lum, P. Y. *et al.* Extracting insights from the shape of complex data using topology. *Scientific Reports* **3**, 1236 (2013).
22. Lawson, P., Schupbach, J., Fasy, B. T. & Sheppard, J. W. Persistent homology for the automatic classification of prostate cancer aggressiveness in histopathology images. in *Medical Imaging 2019: Digital Pathology* vol. 10956 109560G (International Society for Optics and Photonics, 2019).
23. Lawson, P., Sholl, A. B., Brown, J. Q., Fasy, B. T. & Wenk, C. Persistent Homology for the Quantitative Evaluation of Architectural Features in Prostate Cancer Histology. *Scientific Reports* **9**, 1139 (2019).
24. Peltomäki, P. Deficient DNA mismatch repair: a common etiologic factor for colon cancer. *Hum Mol Genet* **10**, 735–740 (2001).
25. Deng, J. *et al.* ImageNet: a Large-Scale Hierarchical Image Database. in 248–255 (2009). doi:10.1109/CVPR.2009.5206848.
26. Karim, M. R. *et al.* Deep learning-based clustering approaches for bioinformatics. *Brief Bioinform* doi:10.1093/bib/bbz170.
27. Chen, T., Kornblith, S., Norouzi, M. & Hinton, G. A Simple Framework for Contrastive Learning of Visual Representations. (2020).
28. Fey, M. & Lenssen, J. E. Fast Graph Representation Learning with PyTorch Geometric. *arXiv:1903.02428 [cs, stat]* (2019).
29. Bianchi, F. M., Grattarola, D. & Alippi, C. Spectral Clustering with Graph Neural Networks for Graph Pooling. *arXiv:1907.00481 [cs, stat]* (2020).
30. Ying, R. *et al.* Hierarchical Graph Representation Learning with Differentiable Pooling. *arXiv:1806.08804 [cs, stat]* (2019).
31. Veličković, P. *et al.* Deep Graph Infomax. *arXiv:1809.10341 [cs, math, stat]* (2018).
32. Gal, Y. & Ghahramani, Z. Dropout as a Bayesian Approximation: Representing Model Uncertainty in Deep Learning. *arXiv:1506.02142 [cs, stat]* (2016).
33. Rong, Y., Huang, W., Xu, T. & Huang, J. DropEdge: Towards Deep Graph Convolutional Networks on Node Classification. *arXiv:1907.10903 [cs, stat]* (2020).
34. Houlsby, N., Huszár, F., Ghahramani, Z. & Lengyel, M. Bayesian Active Learning for Classification and Preference Learning. *arXiv:1112.5745 [cs, stat]* (2011).
35. Ying, R., Bourgeois, D., You, J., Zitnik, M. & Leskovec, J. GNNExplainer: Generating Explanations for Graph Neural Networks. (2019).
36. Artemenkov, A. & Panov, M. NCVis: Noise Contrastive Approach for Scalable Visualization. (2020).
37. Bodnar, C., Cangea, C. & Liò, P. Deep Graph Mapper: Seeing Graphs through the Neural Lens. *arXiv:2002.03864 [cs, stat]* (2020).
38. Tauzin, G. *et al.* giotto-tda: A Topological Data Analysis Toolkit for Machine Learning and Data Exploration. *arXiv:2004.02551 [cs, math, stat]* (2020).
39. Veen, H. J. van, Saul, N., Eargle, D. & Mangham, S. W. Kepler Mapper: A flexible Python implementation of the Mapper algorithm. *Journal of Open Source Software* **4**, 1315 (2019).
40. Agarap, A. F. Deep Learning using Rectified Linear Units (ReLU). *arXiv:1803.08375 [cs, stat]* (2019).
41. He, B. *et al.* Integrating spatial gene expression and breast tumour morphology via deep learning. *Nature Biomedical Engineering* 1–8 (2020) doi:10.1038/s41551-020-0578-x.
42. Zhao, Q., Ye, Z., Chen, C. & Wang, Y. Persistence Enhanced Graph Neural Network. in *International Conference on Artificial Intelligence and Statistics* 2896–2906 (2020).
43. Adams, H. *et al.* Persistence Images: A Stable Vector Representation of Persistent Homology. *Journal of Machine Learning Research* **18**, 1–35 (2017).
44. Aygüneş, B. *et al.* Graph convolutional networks for region of interest classification in breast histopathology. in *Medical Imaging 2020: Digital Pathology* vol. 11320 113200K (International Society for Optics and Photonics, 2020).

A multi-scale integrated analysis identifies KRT8 as a pan-cancer early biomarker

Madeleine K. D. Scott
Biophysics Program, Department of Medicine
Stanford University, Stanford, CA, USA
Email: scottmk@stanford.edu

Michael G. Ozawa, Pauline Chu
Department of Pathology
Stanford University, Stanford, CA, USA
Email: mgozawa@stanford.edu; pchu@stanford.edu

Maneesha Limaye
Department of Pediatrics, Children's Hospital of OC
University of Irvine, California, Orange, CA, USA
Email: maneesha.r.limaye@gmail.com

Viswam S. Nair
Clinical Research Division,
Fred Hutch Cancer Center, Seattle, WA, USA
Email: v846531n@stanford.edu

Steven Schaffert
Institute for Immunity, Transplantation and Infection
Stanford University, Stanford, CA, USA
Email: schaffert@stanford.edu

Albert C. Koong
Department of Radiation Oncology
Stanford University, Stanford, CA, USA
Email: akoong@mdanderson.org

Robert West
Department of Pathology
Stanford University, Stanford, CA, USA
Email: rbwest@stanford.edu

Purvesh Khatri
Institute for Immunity, Transplantation and Infection
Stanford University, Stanford, CA, USA
Email: pkhatri@stanford.edu

An early biomarker would transform our ability to screen and treat patients with cancer. The large amount of multi-scale molecular data in public repositories from various cancers provide unprecedented opportunities to find such a biomarker. However, despite identification of numerous molecular biomarkers using these public data, fewer than 1% have proven robust enough to translate into clinical practice. One of the most important factors affecting the successful translation to clinical practice is lack of real-world patient population heterogeneity in the discovery process. Almost all biomarker studies analyze only a single cohort of patients with the same cancer using a single modality. Recent studies in other diseases have demonstrated the advantage of leveraging biological and technical heterogeneity across multiple independent cohorts to identify robust disease biomarkers. Here we analyzed 17149 samples from patients with one of 23 cancers that were profiled using either DNA methylation, bulk and single-cell gene expression, or protein expression in tumor and serum. First, we analyzed DNA methylation profiles of 9855 samples across 23 cancers from The Cancer Genome Atlas (TCGA). We then examined the gene expression profile of the most significantly hypomethylated gene, *KRT8*, in 6781 samples from 57 independent microarray datasets from NCBI GEO. *KRT8* was significantly over-expressed across cancers except colon cancer (summary effect size=1.05; $p < 0.0001$). Further, single-cell RNAseq analysis of 7447 single cells from lung tumors showed that genes that significantly correlated with *KRT8* ($p < 0.05$) were involved in p53-related pathways. Immunohistochemistry in tumor biopsies from 294 patients with lung cancer showed that high protein expression of KRT8 is a prognostic marker of poor survival (HR = 1.73, p = 0.01). Finally, detectable KRT8 in serum as measured by ELISA distinguished patients with pancreatic cancer from healthy controls with an AUROC=0.94. In summary, our analysis demonstrates that KRT8 is (1) differentially expressed in several cancers across all molecular modalities and (2) may be useful as a biomarker to identify patients that should be further tested for cancer.

Keywords: Meta-analysis; Cancer; Diagnostic; Methylation

1. Introduction

Most of the public health burden of cancer results from our inability to detect tumors before they become untreatable[1]. For instance, non-small cell lung cancer (NSCLC), the leading cause of cancer deaths worldwide, progresses from early to advanced stages over a year[2]. Early detection of NSCLC is shown to substantially improve survival through surgical resection of the tumor[3]; however, after the cancer has metastasized, surgical intervention does not improve patient outcomes[4]. This critical need for early cancer biomarkers motivated the creation of consortiums like the TCGA[5]. Since the first TCGA data was released in 2006, there have been hundreds of putative molecular biomarkers proposed across all cancer types, with most focusing on gene expression biomarkers[6,7]. However, most gene signature biomarkers were identified in only one cancer type or subtype, and very few ever proved to be viable for clinical use[7,8]. Many proposed signatures failed to translate into clinical practice because they could not be replicated in outside cohorts or performed poorly when clinical data was considered[9]. DNA methylation profiles have been shown to carry additional information to either genomic or expression data[10,11]. Yang *et al.* demonstrated that TCGA methylation data could identify clinically relevant subsets of patients with breast cancer that could not be classified by gene expression[12]. Others have documented the prognostic ability of other epigenetic signatures in colon, lung, and pancreatic cancer[13-15]. However, the bulk of putative methylation biomarkers are limited to a single disease and face the same clinical translation issues as gene expression biomarkers[16]. To increase the probability that a methylation biomarker is useful in clinical practice, it is critical to demonstrate a robust functional and translational relevance of the differentially methylated genes in multiple cohorts[17]. Additionally, the focus on single-cancer biomarkers has raised concerns about the potential to overlook common epigenetic drivers of cancer[18].

In this study, we performed a pan-cancer analysis of TCGA DNA methylation data from 9855 tissue samples across 23 cancers to inform subsequent gene expression, proteomic, and clinical outcome analyses. The methylation samples were divided into discovery (2019 samples across 10 cancers) and validation (7836 samples across 21 cancers). *KRT8* was the most significant differentially methylated gene across cancers. We next examined the gene expression profile of *KRT8* in 6781 samples from 57 independent microarray datasets in five solid tumor cancers (breast, colon, pancreatic, ovarian and lung) from NCBI GEO[19], and found *KRT8* to be universally overexpressed. Our analysis of intra-cellular gene-*KRT8* expression correlations in 7447 single cells derived from lung tumor biopsies found *KRT8* is correlated with genes involved in p53-related pathways. We validated these correlations in gene expression microarrays of 1276 tissue biopsies from patients with lung cancer. We examined the prognostic relevance of tumor *KRT8* protein in 294 tissue microarrays (TMAs) from patients with lung cancer. We then calculated the prognostic value of tumor *KRT8* gene expression in pancreatic cancer

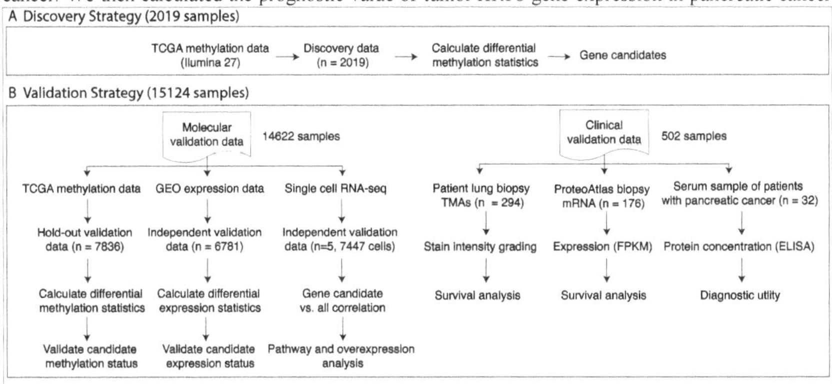

Figure 1. Analysis overview. (A) DNA methylation data from TCGA for 10 cancers comprising 2019 samples profiled using the Illumina 27 platform were used for discovery. **(B)** Validation data comprised of 15,124 samples profiled using either DNA methylation, bulk and single cell gene expression, or tumor and serum protein.

with data from Protein Atlas. Finally, we validated the potential of *KRT8* as non-invasive biomarker with serum KRT8 in 32 pancreatic patients and 6 healthy controls from Stanford Hospital. An overview of this analysis is displayed in **Figure 1**.

2 Methods

2.1. Data Collection from Public Repositories – TCGA and GEO

All methylation and transcriptome data used in our analyses are publicly available. We downloaded all available DNA methylation data the TCGA data portal (https://tcga-data.nci.nih.gov/tcga/tcgaHome2.jsp) irrespective of cancer on May 19, 2018. We excluded data for cancers where less than two non-cancerous samples were profiled, which resulted in DNA methylation data for 9855 samples across 23 cancers. For DNA methylation profiling, samples from these 23 cancers were profiled using either the Infinium HM27 array (27,578 CpG site targeting probes) or Infinium HM450 array (485,577 CpG site targeting probes). All data was generated and processed by The Cancer Genome Atlas research network as described previously[5,18]. We used data profiled on the HM27 array as our discovery cohort (10 cancers, 2019 samples) and data profiled on the HM450 array (21 cancers, 7836 samples) as validation.

For gene expression, we downloaded whole transcriptome data for 6,781 tumor biopsies across 57 independent datasets profiled using microarrays from the NCBI GEO. All datasets were required to measure gene expression in a minimum of two non-cancerous tissue samples. These tumor biopsies came from a patient with breast, lung, pancreatic, ovarian or colon cancer.

2.2. Data Processing and Effect Size Estimation

We ensured all downloaded gene expression data was log2-transformed. For each gene, we calculated change in expression in a tumor biopsy as Hedges' g with adjustment for small sample size because it captures both the fold change and variance. We have previously used Hedges' g to generate robust gene signatures with diagnostic and prognostic value[20,21]. We used the random-effects inverse variance meta-analysis using Dersimonian-Laird method to calculate a summary effect size (ES) across datasets for each gene[22]. We chose Dersimonian-Laird as our previous work has shown it to be a good compromise between more conservative meta-analysis methods (Sidik–Jonkman, Hedges–Olkin, empiric Bayes, restricted maximum likelihood) and lenient methods (Hunter–Schmidt)[22]. If multiple probes mapped to a gene, the effect size for each gene was summarized via the fixed effect inverse-variance model. We corrected p-values for summary effect-sizes for multiple hypotheses testing using Benjamini-Hochberg false discovery rate (FDR) correction.[23] We removed one cancer at a time and applied both meta-analysis methods at each iteration to avoid influence of a specific cancer with a large sample size on the results.

2.3. Survival Analysis and Modeling

We used a right-censored model to fit survival data with the survival package in the R statistical computing environment (Version 3.5.1). We fit univariate and multivariate Cox proportional hazards models onto survival data using the coxph function. We confirmed the proportional hazard assumption with the cox.zph function.

2.4. Human Plasma Samples

Our study includes 32 human EDTA blood plasma samples collected between January 2007 and October 2011 from identically staged patients with advanced pancreatic ductal adenocarcinoma treated at Stanford University Medical Center under an institutional review board-approved protocol. All plasma samples were collected from untreated (*de novo*) patients with biopsy- proven pancreatic adenocarcinomas. Median age at blood collection was 68 years (range 37-84 years). As a control group, 6 additional plasma samples were collected from age- matched, healthy volunteers under an IRB-approved protocol. Immediately after acquisition, blood samples were centrifuged and aliquots of plasma stored at -80°C.

2.4. Enzyme-linked immunosorbent assay (ELISA)

The serum biomarker concentration was measured with a commercially available human protein sandwich enzyme immunoassay kit with two mouse monoclonal antihuman antibodies (R&D Systems, Inc., Minneapolis, MN, USA). All serum samples from patients and standards were incubated in microplate wells coated with the first mouse monoclonal anti-human biomarker antibody. After washing, a second antihuman biomarker antibody labeled with peroxidase (HRP) was added for subsequent incubation. The reaction between HRP and substrate (hydrogen peroxide and tetramethylbenzidine) resulted in color development and the intensities were measured with a microplate reader at an absorbance of 450 nm. Concentrations of serum biomarkers were determined against a standard curve.

2.5. Single cell data collection and processing

We downloaded count matrices of 52,698 single cells from the tumor microenvironment of five lung cancer patient samples from Array Express (E-MTAB-6149)[24]. Of the total 52,698 cells, 7,447 originated from the tumor. We calculated the Pearson correlation between expression of *KRT8* and all other measured genes within each tumor cell. For each KRT8-gene correlation, we required non-zero expression of both genes in a minimum of 25 cells. We removed correlations with a p-value ≥ 0.05.

2.6. KRT8 Expression in Patients with Pancreatic Cancer from The Protein Atlas

We downloaded prognostic information for 176 pancreatic cancer patients stratified by tumor KRT8 expression from The Protein Atlas[25] (https://www.proteinatlas.org/ENSG00000170421-KRT8/pathology/tissue/pancreatic+cancer). We stratified patients based on median KRT8 expression of the cohort. Patient samples originated from the TCGA data repository. All counts are reported as Fragments Per Kilobase of exon per Million reads (FPKM).

2.7. TMA cohort, and immunohistochemistry

Patient samples were retrieved from the surgical pathology archives at the Stanford Department of Pathology and linked to a clinical database using the Cancer Center Database and STRIDE Database tools from Stanford. Patients who had surgically treated disease and paraffin embedded samples from 1995 through June, 2010 were included. Surgical specimens that contained viable tumor from slides were reviewed by a board-certified pathologist (RBW) to build the Stanford Lung Cancer TMA as described previously. The area of highest tumor content was marked for coring blocks corresponding to the slides using 0.6 mm cores in duplicate arrays as previously described[26]. These cores were aligned by histology and stage and negative controls included a variety of benign and malignant tissues that included normal non-lung tissue, abnormal non-lung tissue, placental markers, and normal lung[26]. Normal lung consisted of a specimen adjacent, but distinct, from tumor over the years 1995 through 2010 to assess the variability of staining by year. OligoDT analysis was performed on the finished array to assess the architecture of selected cores and adequacy of tissue content prior to target immunohistochemistry (IHC) analysis. Serial 4 μm sections were cut from FFPE specimens and processed for IHC using the Ventana BenchMark XT automated immunostaining platform (Ventana Medical Systems/Roche, Tucson, AZ). Rabbit monoclonal anti-Cytokeratin 8 (phospho S431) antibody was obtained from Abcam (ab109452, Burlingame, CA). Mouse monocolonal Anti-Cytokeratin 8 antibody was also obtained from Abcam (ab9023, Burlingame, CA). The intensity of KRT8 immunostaining was graded from 1-4 as determined by an independent pathologist who was blinded to patient outcome.

3. Results

3.1. Integrated analysis of TCGA data identifies KRT8 as hypomethylated across cancers

We identified 23 cancers that had methylation data and at least two healthy controls per cancer from TCGA. We split the resulting 9855 samples into discovery cohorts (2019 samples from 10 cancers profiled using the Illumina 27 platform) and the validation cohorts (7836 samples from 21 cancers profiled using the Illumina 450 platform) for validation. In order to avoid the potential influence of a single cancer on the results due to unequal sample sizes or other unknown confounding factors among cohorts, we performed a ''leave-one-cancer-out'' analysis. We hypothesized that the resulting set of methylation sites, irrespective of the set of cancers analyzed, would constitute a robust methylation signature across cancers. We identified 1,801 differentially methylated genes (1,081 hyper- and 720 hypomethylated, FDR < 5%) across all cancers (**Figure 1A and Supplementary Figure 1A**). We did not remove differentially methylated sites with significant heterogeneity for two reasons. First, heterogeneity is expected due to known heterogeneity within and between cancers. Second, we have previously shown that when combining across multiple datasets, filtering by heterogeneity removes higher proportion of true positives than false positives[22]. In the validation cohorts, which used Illumina 450 platform, we found 1083 out of 1,801 sites were differentially methylated across all cancers (FDR < 5%; **Figure 1B** and **Supplementary Figure 1B**).

Our discovery analysis found several previously reported differentially methylated genes. The hypomethylated genes across all cancers in the discovery cohort included *CLDN4*[27] (discovery ES = -1.86, p = 8.0e-7; validation ES = -0.56, p = 1.55e-06) and *SFN*[28] (discovery ES = -0.96, p = 2.01e-7; validation ES = -0.94, p = 9.4e-10) that have been previously shown to promote cancer cell proliferation (Ehrlich 2009), whereas the hypermethylated genes included known tumor suppressors such as *SOX1*[29] (discovery ES = 1.05, p = 3.4e-08; validation ES = 1.08, p = 4.4e-22), *TWIST*[30] (discovery ES = 0.89, p = 1.5e-5; validation ES = 0.59, p = 4.3e-16), and *GATA4*[31] (discovery ES = 0.92, p = 1.7e-6; validation ES = 0.38, p = 3.77e-12). *KRT8* was the most statistically significant hypomethylated gene after multiple hypothesis correction (discovery ES = -1.71, p = 3.2e-7, FDR=9.15e-6), but was unchanged in renal clear cell carcinoma. *KRT8* was also hypomethylated in the validation cohorts across all cancers except pheochromatoma/paraganglioma and melanoma (validation ES=-0.69, p = 3.3e-15, FDR = 4.0e-14).

3.2. Multi-cohort gene expression analysis demonstrates KRT8 is over-expressed in five cancers

Hypomethylation and hypermethylation typically lead to over- and under-expression of the corresponding gene, respectively.[32] Therefore, we hypothesized that hypo- or hyper-methylated genes across multiple cancers will be over- or under-expressed across multiple cancer compared to control samples. Arguably, we could use gene expression data for the same samples from TCGA. However, we decided to use gene expression data from completely independent cohorts from a different source to increase stringency of our analysis. Therefore, to test this hypothesis, we downloaded 57 microarray gene expression datasets from the NCBI GEO[19] comprising of 6781 samples (4870 cases, 1911 controls) obtained from human tissue biopsies of five cancers: breast, colon, lung adenocarcinoma, ovarian, or pancreatic. These 57 datasets included broad biological and technical heterogeneity, such as treatment protocols, demographics, collection year, and microarray platforms to further increase the stringency of our analysis and identify robust signals that persist despite these potential sources of noise. Differential gene expression meta-analysis across all 6781 samples identified overexpression of known oncogenes such as *ERBB2* (ES =0.51, p = 6.22e-13), *KRAS* (ES =0.43, p = 2.90e-9), *CCND1* (ES = 0.25, p = 7.34e-3), and *VEGFA* (ES = 0.42, p = 2.19e-06). Housekeeping genes did not show a change xin expression between control and cancer, such as *B2M* (ES = 0.12, p = .25), *HBS1L* (ES = -0.08, p = 0.15), or *EMC7* (ES = 0.18, p = 0.09)[33,34].

Next, we calculated the Spearman correlation between the discovery methylation ES and gene expression ES in the 1,801 differentially methylated genes as -0.21 (p=1.27e-19), which in line with previous studies[35] that examined intra-sample methylation-expression correlation (**Supplementary Figure 2**). Finally, we found that hypomethylation of *KRT8* led to overexpression in multiple cancers compared to healthy samples (ES=1.05, p=2.8e-27, FDR=2.0e-24). *KRT8* was over-expressed in pancreatic cancer (ES=0.69, p=4.02e-08), ovarian cancer (ES=1.61, p=1.93e-03), lung cancer (ES=1.55, p=1.95e-13), and breast cancer (ES=0.88, p=7.82e-10), but not in colon cancer (ES = 0.14, p = 0.38).

3.3. *KRT8 overexpression is associated with a chemotherapy-resistant phenotype in vitro*

Chemotherapy resistance is responsible for more than 80% of cancer-related mortality. We investigated whether increased *KRT8* expression is associated with chemotherapy resistance. We downloaded 100 samples in seven datasets from NCBI GEO across six cancers that contained both chemotherapy-resistant and chemotherapy-sensitive cell lines. *KRT8* was consistently overexpressed across all chemo-resistant cancer cell lines (summary effect size=0.76, p=0.035; **Supplementary Figure 3**). This result demonstrates a consistent association between *KRT8* expression and chemotherapy resistance *in vitro*.

3.4. *Single cell analysis of KRT8 expression*

Single cell gene expression data has allowed researchers to probe intra-cellular gene-gene correlations, which in turn suggest gene interactions or a common regulator. We analyzed intra-cellular correlations between every gene and *KRT8* with single cell RNA sequencing data of 7447 cells from tumor biopsies of five lung cancer patients. To calculate intra-cell gene-gene correlations, we correlated the expression of each gene to *KRT8* expression in every cell. Several other keratin genes were positively correlated with KRT8. For example, *KRT18* and *KRT7* had Pearson correlation of 0.59 and 0.55, respectively,

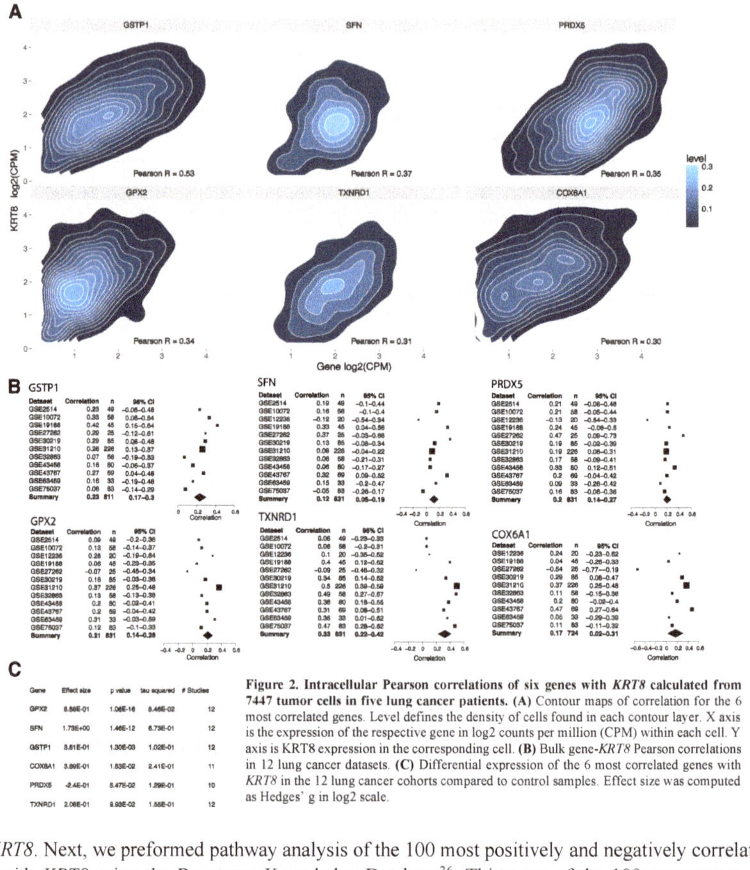

C

Gene	Effect size	p value	tau squared	# Studies
GPX2	8.58E-01	1.06E-16	8.46E-02	12
SFN	1.73E+00	1.46E-12	6.73E-01	12
GSTP1	3.61E-01	1.30E-03	1.02E-01	12
COX6A1	3.89E-01	1.53E-02	2.41E-01	11
PRDX5	-2.4E-01	5.47E-02	1.29E-01	10
TXNRD1	2.06E-01	9.93E-02	1.56E-01	12

Figure 2. Intracellular Pearson correlations of six genes with *KRT8* calculated from 7447 tumor cells in five lung cancer patients. (A) Contour maps of correlation for the 6 most correlated genes. Level defines the density of cells found in each contour layer. X axis is the expression of the respective gene in log2 counts per million (CPM) within each cell. Y axis is KRT8 expression in the corresponding cell. **(B)** Bulk gene-*KRT8* Pearson correlations in 12 lung cancer datasets. **(C)** Differential expression of the 6 most correlated genes with *KRT8* in the 12 lung cancer cohorts compared to control samples. Effect size was computed as Hedges' g in log2 scale.

with *KRT8*. Next, we preformed pathway analysis of the 100 most positively and negatively correlated genes with *KRT8* using the Reactome Knowledge Database[36]. Thirty out of the 100 genes were not annotated in the Reactome Knowledge Database. We identified six significantly enriched pathways, each of which has been previously implicated in cancer progression (**Figure 2A**). The top three significantly enriched pathways were comprised of six unique genes: *GSTP1, PRDX5, GPX2, TXNRD1, SFN, COX6A1* (**Supplementary Table 1**). Each of these six genes had an intra-cellular correlation with

KRT8 expression ≥0.30 (**Figure 2A**). All genes except *GSTP1* are annotated in Reactome as involved in p53 signal transduction (**Supplementary Table 1**). However, *GSTP1* is known to be a direct transcriptional target of p53[37], further supporting the association between *KRT8* and genes involved in

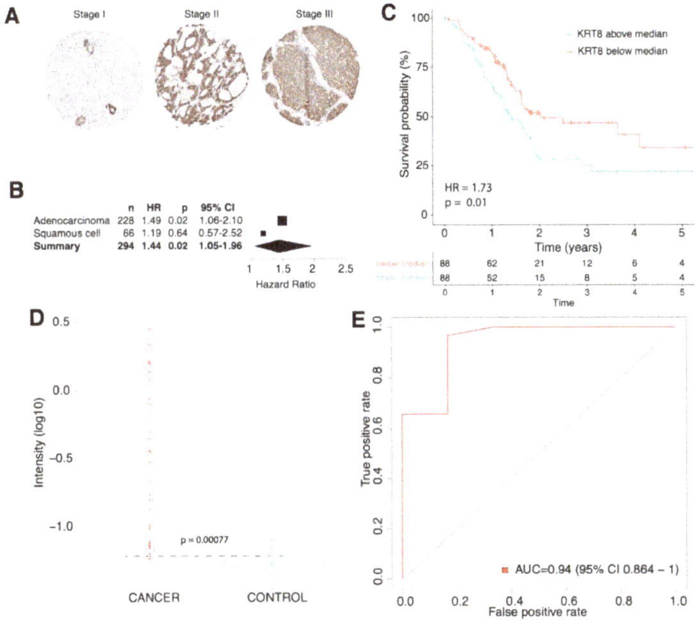

Figure 3. Protein measurement of KRT8 in cancer. A. IHC of lung adenocarcinoma TMAs for KRT8. **B.** Cox proportional hazard of 294 lung cancer samples stratified by KRT8 concentration. **C.** Survival of 176 patients with pancreatic cancer stratified by KRT8 expression relative to the median of the cohort. **D-E.** Violin (**D**) and ROC (**E**) plots of serum KRT8 as measured by ELISA in patients with pancreatic cancer and healthy controls. Dashed line represents the ELISA detection threshold. Width of a violin plot indicates density of samples, where each dot represents a sample.

the p53 pathway.

We next examined the correlation between the six genes and *KRT8* in bulk lung adenocarcinoma gene expression data from microarrays of 1276 lung biopsy samples from 12 datasets. All genes were significantly correlated with *KRT8* at sample level (**Figure 2B**). All genes except *PRDX5* were overexpressed in lung adenocarcinoma compared to healthy patients (**Figure 2C**). The majority of these six genes were additionally overexpressed across 5505 microarray samples from four cancers (breast, colon, ovarian, and pancreatic; **Supplementary Table 2**).

3.5. Protein expression of KRT8 is associated with mortality in patients with lung adenocarcinoma

Given robust hypomethylation of *KRT8* across 9,855 samples from 23 cancers, over-expression across 6,781 biopsies from 5 cancers, strong association with chemo-resistance, and sustained correlation with p53-regulated genes both at single-cell and sample levels, we investigated whether *KRT8* is also expressed at protein-level in tumor biopsies, and whether it is associated with survival in patients with either lung adenocarcinoma or lung squamous cell carcinoma. We stained tissue microarrays (TMAs) containing 294 lung tumors (228 lung adenocarcinoma, 66 lung squamous cell carcinoma) resected from patients at Stanford Hospital for *KRT8* protein (**Supplementary Table 3**). An expert pathologist (MO) rated the maximum intensity of cancerous cell *KRT8* staining in each TMA (**Figure 3A**). Out of the 294 samples, 5 (1.7 %) scored as 1+, 35 (11.9%) as 1-2+, 55 (18.7%) as 1-3+, 8 (2.72%) as 2+, 85 (28.9%) as 2-3+ and 106 (36.1%) as 3+. In a multivariable cox regression model, *KRT8* intensity was a significant predictor of mortality after adjusting for sex and age at diagnosis in lung adenocarcinoma (Hazard Ratio = 1.49, 95% CI = 1.06 – 2.10, p=0.02), but not in squamous cell (Hazard Ratio = 1.19, 95% CI = 0.57 – 2.52, p=0.65; **Figure 3B**).

3.6. Elevated RNA expression of KRT8 is associated with mortality in patients with pancreatic cancer

Next, we investigated whether *KRT8* tumor gene expression is a prognostic marker of survival. We downloaded *KRT8* expression and corresponding survival data for 176 patients with stage I-IV pancreatic cancer from Human Protein Atlas (**Supplementary Table 4**). We classified patients as either "High *KRT8*" or "Low *KRT8*" if their *KRT8* expression was above or below the median *KRT8* expression of the cohort (363.5 FPKM), respectively. Patients in the "High *KRT8*" group had an increased risk of mortality (cox proportional hazard ratio = 1.73 p = 0.01 **Figure 3C**).

3.7. Serum KRT8 discriminates between healthy and pancreatic patients

Finally, we explored the potential of KRT8 as a minimally invasive biomarker. We measured *KRT8* concentration in serum of 32 biopsy-confirmed patients with pancreatic ductal adenocarcinoma and six healthy controls by enzyme-linked immunosorbent assays (ELISA). Samples were collected from Stanford Hospital (**Supplementary Table 5**). The mean KRT8 concentration was significantly higher in the pancreatic cancer patients compared to that of healthy controls (p = 7.7e-4; **Figure 3D**). Samples were considered KRT8+ if they had a measured KRT8 value about the detectability limit of the ELISA (0.06 RLU). KRT8+ status distinguished patients with pancreatic cancer from healthy controls with an area under the curve (AUC) of 0.94 (**Figure 3E**) and an area under the precision recall curve (AUPRC) of 0.99 (**Supplementary Figure 4**).

4. Discussion

Only a fraction of molecular cancer biomarkers published in academic literature are reproducible in follow-up studies. The first step to identifying a robust biomarker is to ensure that the discovery phase has included a heterogeneous set of samples, platforms, and measurement technologies. Here, we identified KRT8 as such a biomarker by integrating DNA methylation profiling of 2019 samples across 10 cancers from the TCGA. We then validated that KRT8 is a robust biomarker on 7836 samples in 21 cancers measured with a different DNA methylation platform within the TCGA. We next analyzed the diagnostic and prognostic value of tumor *KRT8* gene and protein expression as well as serum KRT8 using ELISA in over 7000 samples spanning 10 years, multiple platforms, and data repositories.

Pan-cancer methylation findings have been hindered by questions about batch effects and platform bias[38]. In this work, we used samples run on Illumina 27 platform as our discovery data and Illumina 450 as validation. KRT8 was significantly hypomethlyated in both platforms, suggesting it is robust to platform bias. While TCGA has gene expression data, we chose to use microarray samples from the NCBI GEO to ensure that our findings would be robust to data type, batch effect, and platform.

Single cell analysis has broadened our understanding of tumor heterogeneity, but it can be difficult to interpret the immediate translational value of a single time point scRNA-seq analysis. Here, we show that intra-cellular gene-gene correlations can suggest overlooked gene functions.

Our study has several limitations. First, it does not include the entirety of all cancer data available in the public sphere, and thus presents an incomplete picture of *KRT8* across all data. However, this study used 17149 samples across 23 cancers, which still includes significant amount of biological, clinical, and technical heterogeneity in the real-world patient population. Further, we have previously shown that 4-5 independent datasets with a total of approximately 200-250 samples substantially increases the probability of validation in independent cohorts[22]. Second, we only required two control samples in the methylation discovery analysis, which could have led to false positive or patient-specific effects within a dataset. However, the integration of all the discovery cohorts and independent validation using Illumina 450 methylation platform substantially mitigated the effect of a single cancer outlier. In addition, our rigorous downstream analysis of gene expression from 6781 samples in 57 datasets from 5 cancers provide strong evidence of the robustness of our analyses. Third, we chose only the top gene and validated it here. It is possible that other genes may provide equal or greater prognostic value than *KRT8*. However, our aim is to demonstrate the value of the framework we propose here and thus we explored only the most promising gene, *KRT8*. Forth, we do not provide any indication of the mechanism underlying the prognostic value of *KRT8*. It may be as straightforward as increasing epithelial cancer cell numbers results in more *KRT8* released into the bloodstream, or perhaps there is a more complex biological phenomenon at work. These questions can only be answered with follow-up hypothesis-driven research.

Previous studies have highlighted the contribution of *KRT8* in the progression of gastric and kidney cancer. *KRT8* has also been proposed as a biomarker in lung cancer. However, *KRT8* has never been shown to be overrepresented across cancers in a multi-omic analysis. One GEO dataset (GSE15932) contained expression from peripheral blood samples. In this dataset, *KRT8* expression was able to differentiate cancerous from healthy patients, suggesting that circulating *KRT8* RNA may be a candidate for a diagnostic blood biomarker. Biomarkers not only have diagnostic and prognostic implications, but are also helpful for measurement of treatment responses, surveillance for tumor recurrence and guiding clinical decisions. For many cancers, there is not a single blood biomarker; others like pancreatic cancer have one or two unreliable screening biomarkers. Here we show the potential use of serum KRT8 protein as a blood biomarker in pancreatic cancer. Given that we identified *KRT8* as overexpressed across cancers, it stands to reason that KRT8 may be useful as a peripheral biomarker in other cancers as well.

Most importantly, this work demonstrates a strategy to translate large molecular analyses into specific, clinically relevant hypotheses. Omics sciences enable complex biological systems to be visualized in a holistic and integrative manner. Application of systems biology to interpret large multidimensional omics data across cancer types will enable the robust identification of biomarkers that share common pathophysiology, which can potentially be further explored for pan-cancer interventions

Ethics approval and consent to participate

All aspects of this study were approved by the Stanford Institutional Review Board in accordance with the Declaration of Helsinki guidelines for the ethical conduct of research. The reference number for the approval is IRB-20170. A waiver of informed consent was obtained for the subjects in this study according to Stanford's Institutional Review Board policy since this was a retrospective study of both alive and deceased patients, many of whom were lost to follow-up.

Availability of data and materials

All supplemental figures and tables are available as part of the medRxiv preprint (doi: 10.1101/2020.10.01.20205450). Microarray data are available from the NCBI GEO at: https://www.ncbi.nlm.nih.gov/geo/. The accession numbers and corresponding links for the individual studies are listed in Supplemental Table 6.

Funding

This work was supported in part by grants RO1 AI125197-01, U19AI109662, and U19AI057229 from the National Institute for Allergy and Infectious Diseases to P.K; and training grant F30 HL149252-01A1 from the National Heart, Lung, and Blood Institute to M.K.D.S.

5. References – Complete list available on medRxiv preprint

1 Hiom, S. C. Diagnosing cancer earlier: reviewing the evidence for improving cancer survival. *Br J Cancer* **112 Suppl 1**, S1-5, doi:10.1038/bjc.2015.23 (2015).

2 de Groot, P. M., Wu, C. C., Carter, B. W. & Munden, R. F. The epidemiology of lung cancer. *Transl Lung Cancer Res* 7, 220-233, doi:10.21037/tlcr.2018.05.06 (2018).

What about the environment? Leveraging multi-omic datasets to characterize the environment's role in human health

Kristin Passero
Huck Institutes of the Life Sciences, The Pennsylvania State University
University Park, PA 16802
Email: kxp642@psu.edu

Shefali Setia-Verma
Department of Genetics and Institute for Biomedical Informatics, University of Pennsylvania
Philadelphia, PA 19104
Email: shefali.setiaverma@pennmedicine.upenn.edu

Kimberly McAllister
Program Administrator
Genes, Environment, and Health Branch
Division of Extramural Research and Training
National Institute of Environmental Health Sciences
P.O. Box 12233 (MD EC-21)
Research Triangle Park, NC 27709
Email: mcallis2@niehs.nih.gov

Arjun Manrai
Computational Health Informatics Program, Boston Children's Hospital
Department of Biomedical Informatics, Harvard Medical School
Boston, MA 02115
Email: manrai@post.harvard.edu

Chirag Patel
Department of Biomedical Informatics, Harvard Medical School
Boston, MA 02115
Email: chirag@hms.harvard.edu

Molly Hall
Department of Veterinary and Biomedical Sciences, The Pennsylvania State University
University Park, PA 16802
Email: mah546@psu.edu

The environment plays an important role in mediating human health. In this session we consider research addressing ways to overcome the challenges associated with studying the multifaceted and ever-changing environment. Environmental health research has a need for technological and methodological advances which will further our knowledge of how exposures precipitate complex phenotypes and exacerbate disease.

Keywords: Environment; Health Outcomes; Multi-omics.

1. The complexities of environmental health research

The environment is increasingly seen as a casual or moderating factor that governs aspects of complex disease etiology (Hall, Moore, & Ritchie, 2016; Manrai et al., 2017). Since there is a great breadth of environmental risk factors, researchers classify exposures into three categories: (Stingone, Buck Louis, et al., 2017; Wild, 2012) internal exposures arising from endogenous processes (e.g. metabolism, inflammation), intrinsic qualities (e.g. body morphology), or microorganisms living in or on an individual (e.g. microbes colonizing the gut) that affect the body's cellular environment (Wild, 2012). Specific external exposures are extrinsic and "target" the body directly. Examples include infectious agents, diet and substance use, pollutants, and occupational exposures (Martin Sanchez, Gray, Bellazzi, & Lopez-Campos, 2014; Wild, 2012). Lastly, general external exposures are broad characteristics, such as which geography and climate a person resides in, socioeconomic indicators, or psychosocial exposures, that affect both the individual and, to a degree, the experience of internal and specific external exposures (Wild, 2012). Household income, work-life balance, healthcare access, or home rurality are general external exposures.

A comprehensive assessment of environmental risk factors remains challenging as the environment is dynamic. Exposure presence and intensity change over time. Environmental risk is a cumulative measure acquired throughout the lifespan and beginning from conception (Manrai et al., 2017; Stingone, Buck Louis, et al., 2017). Longitudinal investigation of exposures is crucial for research investigating vulnerability periods, such as the prenatal period, where exposures impart their most salient effects on health. The within-person heterogeneity of exposures is a major limitation in the field of human exposure research, as timing and intensity may be difficult to capture without consistent monitoring (Manrai et al., 2017; van Tongeren & Cherrie, 2012). Sources of environmental data are diverse. Environmental data may be obtained from surveys or can rely on a collection of 'omics level data, such as the metabolome and the microbiome, when quantifying measures such as exogenous chemical exposure, internal metabolism, or gut microbial diversity. Other sources of information about environmental circumstances may come from purchasing history, food expenditures, mobile phones, social media, or home sensors (Martin Sanchez et al., 2014; van Tongeren & Cherrie, 2012).

Another limitation in environmental health research is the relative dearth of data analytic tools, databases and ontologies, and standardized practices which would aid in the assessment of high-dimensional exposure data (Bocato, Bianchi Ximenez, Hoffmann, & Barbosa, 2019; Manrai et al., 2017; Martin Sanchez et al., 2014; Stingone, Buck Louis, et al., 2017). Researchers seeking to utilize big environmental data would benefit from the development of methods and infrastructure to investigate environmental underpinnings of disease. This includes the curation of high-information environmental datasets (e.g. the HELIX study (Vrijheid et al., 2014)), analytical techniques to assess multivariate, longitudinal data or environmental mixtures (Manrai et al., 2017; Patel, 2017), and curation of database/development of ontologies for known environmental risk factors and their associations (Manrai et al., 2017; Martin Sanchez et al., 2014).

2. Progress made in environmental health research

Environmental health research is a multidisciplinary field and its past successes have utilized various approaches and data types. A study of gene-by-environment interaction found that subjects sharing regional ancestry but living in different regions, showed many differentially expressed genes, whose expression was correlated with fine-scale air pollution (Favé et al., 2018). In a closer look, they identified four quantitative trait loci where transcription was moderated by pollution level (Favé et al., 2018). Other approaches have leveraged environment-wide datasets and found associations between exposures and phenotypes. For example, an environment-wide association study (EWAS) found that blood serum antioxidants, vitamin D, and intense physical activity were associated with abdominal obesity in both sexes (Wulaningsih et al., 2017), and a meta-analysis of EWAS performed on the National Health and Nutrition Examination Surveys from 1999-2012 identified alcohol consumption and urinary cesium as associated with systolic and diastolic blood pressure respectively (McGinnis, Brownstein, & Patel, 2016). The microbiome is increasingly seen as a player in human health (Young, 2017). An investigation of Type I Diabetes onset in infants found that prior to diagnosis, gut microbial diversity decreased and microbe metabolite production reflected a shift towards nutrient transport rather than biosynthesis (Kostic et al., 2015). Machine learning (ML) methods have been applied to probe how pollutant exposures within urban areas affect academic performance (Stingone, Pandey, Claudio, & Pandey, 2017). Another study used ML to create environmental risk scores for oxidative stress which were associated with cardiovascular phenotypes (Park, Zhao, & Mukherjee, 2017).

Metabolomics is useful when assessing environmental risk factors as it can detect both internal exposures (e.g. proinflammatory molecules) and chemicals or toxins (Bloszies & Fiehn, 2018). Computational tools to enable untargeted metabolomics studies, which will aid researchers seeking to agnostically profile the environment, are emerging (Domingo-Almenara et al., 2019; Pirhaji et al., 2016). Other open-source software developed for the quality-control, analysis, and visualization of general environment-wide data (Hernandez-Ferrer et al., 2019; Lucas et al., 2019) are also becoming available to researchers. Future projects will benefit from the curation of environment-wide databases for blood (Barupal & Fiehn, 2019), urine (Jia et al., 2019), and the indoor built environment (Dong et al., 2019) as guides for future, larger-scale metabolomics projects. Finally, the most comprehensive assessment of environment may be achieved through rigorous biomonitoring. Jiang and colleagues (2018) conducted an impressive study by fitting participants with wearable devices which collected longitudinal data on climate, biotic, and abiotic factors. They found the human environment of microbial and chemical exposure varied widely across geographical location and season, even within the same individual (Jiang et al., 2018).

There is much evidence that the environment impacts human health, with disease risk arising from many sources: pollutants, industrial chemicals, lifestyle habits, social climate, etc. Yet the challenges of collecting and analyzing environmental data remain. Different sources of environmental data may need different methodological standards and techniques for effective research. Thus, researchers need user-friendly tools to handle pre-processing, quality assessments, and analysis of various data types. There also remain the questions of which environmental data are most informative when predicting health outcomes, and how we can integrate these various sources of data to define environment-wide risk. There are many opportunities for researchers to develop or improve existing methodologies and advance environmental health research.

3. In this session

Demonstrating the breadth found within environmental health research, our selected publications address key areas of environmental health research: (1) metabolomic profiling and pipeline development and (2) the role of sociodemographic in the prediction of complex health outcomes.

Aguilar, McGuigan, and Hall have developed a semi-automated pipeline for processing and analyzing NMR data. Their method uses open-source software, making it accessible to researchers and easy to document, thereby improving reproducibility and replication capabilities. After applying their pipeline to assess how smoking perturbs human metabolism, they identified associations between various metabolites which past research suggests are implicated in cardiac, pulmonary, and neural diseases. Furthermore, metabolites showing ostensibly differential concentrations between smokers and non-smokers were used as input for a random forest model. This technique found metabolic heterogeneity between and within smoking classes, identifying several unique metabolic profiles which distinguished subsets of smokers and non-smokers. Their study emphasizes how a single exposure, such as smoking, may precipitate complex phenotypic outcomes. Furthermore, it leveraged the metabolome in a joint assessment of the internal and external environment. Smoking was linked to changes in the internal environment, which may in turn affect physiology. Additionally, profiling the metabolome identified within smokers an exogenous pollutant absorbed by tobacco plants. Aguilar et al. highlight how multiple sources of environmental risk may act in concert to develop complex phenotypes.

While the former study evaluates how an acute environmental risk factor is associated with multiple metabolic phenotypes, the environment also exerts influence at a societal and geographical level. Makridis, Strebel, and Alerovitz assessed how different geographic granularities of sociodemographic data affect prediction of mortality in veterans hospitalized due to COVID-19. Their social variables included ZIP-code-level, county-level, or state-level population density, healthcare access, and distributions of age, race/ethnicity, occupation, and education. They noted that in linear models using comparable demographic variables measured county-level or state-level, demographics differed in the effect sizes and significance in association with COVID-19 cumulative cases and deaths. When predicting veteran mortality attributed to COVID-19 using a linear XGBoost algorithm, county-level and ZIP-code level data had negligible differences in prediction accuracy, yet outperformed state-level prediction. Yet interestingly, the features most important in the county-level model differed from that of the ZIP code-level model. The granularity of the environmental data is important when predicting outcomes in a region. Social environmental data may be collected at multiple hierarchies – e.g. state, county, ZIP code – and the demographics at each level may carry different information pertaining to health outcomes, which may be important when trying to design and implement public health policies.

Together, these papers highlight the nuanced relationship the environment has with human disease. The environment has an unavoidable influence on life yet remains difficult to characterize and quantify. It has many dimensions (e.g. internal, specific external, general external), a hierarchical organization (e.g. environment at the individual, home, neighborhood, county, etc. levels), and is dynamic which makes parsing the relevant components which contribute to disease risk challenging. Answering *what*, *when*, and *how* environmental factors affect health requires collecting data that reflects environmental diversity. This may be achieved by collecting environment-wide data covering multiple domains, capturing exposures longitudinally, or, as Makridis et al. imply, considering environmental data at different organizational hierarchies. Simultaneously, researchers must develop and evaluate ways to handle data heterogeneity, model environmental mixtures and interactions, and assess risk at various levels.

References

Barupal, D. K., & Fiehn, O. (2019). Generating the blood exposome database using a comprehensive text mining and database fusion approach. *Environmental Health Perspectives*, *127*(9). https://doi.org/10.1289/EHP4713

Bloszies, C. S., & Fiehn, O. (2018, April 1). Using untargeted metabolomics for detecting exposome compounds. *Current Opinion in Toxicology*. Elsevier B.V. https://doi.org/10.1016/j.cotox.2018.03.002

Bocato, M. Z., Bianchi Ximenez, J. P., Hoffmann, C., & Barbosa, F. (2019). An overview of the current progress, challenges, and prospects of human biomonitoring and exposome studies. *Journal of Toxicology and Environmental Health - Part B: Critical Reviews*, *22*(5–6), 131–156. https://doi.org/10.1080/10937404.2019.1661588

Domingo-Almenara, X., Montenegro-Burke, J. R., Guijas, C., Majumder, E. L.-W., Benton, H. P., & Siuzdak, G. (2019). Autonomous METLIN-Guided In-source Fragment Annotation for Untargeted Metabolomics. *Analytical Chemistry*, *91*(5), 3246–3253. https://doi.org/10.1021/acs.analchem.8b03126

Dong, T., Zhang, Y., Jia, S., Shang, H., Fang, W., Chen, D., & Fang, M. (2019). Human Indoor Exposome of Chemicals in Dust and Risk Prioritization Using EPA's ToxCast Database. *Environmental Science & Technology*, *53*(12), 7045–7054. https://doi.org/10.1021/acs.est.9b00280

Favé, M. J., Lamaze, F. C., Soave, D., Hodgkinson, A., Gauvin, H., Bruat, V., ... Awadalla, P. (2018). Gene-by-environment interactions in urban populations modulate risk phenotypes. *Nature Communications*, *9*(1), 827. https://doi.org/10.1038/s41467-018-03202-2

Hall, M. A., Moore, J. H., & Ritchie, M. D. (2016). Embracing Complex Associations in Common Traits: Critical Considerations for Precision Medicine. *Trends in Genetics : TIG*, *32*(8), 470–484. https://doi.org/10.1016/j.tig.2016.06.001

Hernandez-Ferrer, C., Wellenius, G. A., Tamayo, I., Basagaña, X., Sunyer, J., Vrijheid, M., & Gonzalez, J. R. (2019). Comprehensive study of the exposome and omic data using rexposome Bioconductor Packages. *Bioinformatics*, *35*(24), 5344–5345. https://doi.org/10.1093/bioinformatics/btz526

Jia, S., Xu, T., Huan, T., Chong, M., Liu, M., Fang, W., & Fang, M. (2019). Chemical Isotope Labeling Exposome (CIL-EXPOSOME): One High-Throughput Platform for Human Urinary Global Exposome Characterization. *Environmental Science & Technology*, *53*(9), 5445–5453. https://doi.org/10.1021/acs.est.9b00285

Jiang, C., Wang, X., Li, X., Inlora, J., Wang, T., Liu, Q., & Snyder, M. (2018). Dynamic Human Environmental Exposome Revealed by Longitudinal Personal Monitoring. *Cell*, *175*(1), 277–291. https://doi.org/10.1016/j.cell.2018.08.060

Kostic, A. D., Gevers, D., Siljander, H., Vatanen, T., Hyötyläinen, T., Hämäläinen, A. M., … Xavier, R. J. (2015). The dynamics of the human infant gut microbiome in development and in progression toward type 1 diabetes. *Cell Host and Microbe*, *17*(2), 260–273. https://doi.org/10.1016/j.chom.2015.01.001

Lucas, A. M., Palmiero, N. E., McGuigan, J., Passero, K., Zhou, J., Orie, D., … Hall, M. A. (2019). CLARITE Facilitates the Quality Control and Analysis Process for EWAS of Metabolic-Related Traits. *Frontiers in Genetics*, *10*. https://doi.org/10.3389/fgene.2019.01240

Manrai, A. K., Cui, Y., Bushel, P. R., Hall, M., Karakitsios, S., Mattingly, C. J., … Patel, C. J. (2017). Informatics and Data Analytics to Support Exposome-Based Discovery for Public Health. *Annual Review of Public Health*, *38*(1), 279–294. https://doi.org/10.1146/annurev-publhealth-082516-012737

Martin Sanchez, F., Gray, K., Bellazzi, R., & Lopez-Campos, G. (2014). Exposome informatics: considerations for the design of future biomedical research information systems. *Journal of the American Medical Informatics Association*, *21*(3), 386–390. https://doi.org/10.1136/amiajnl-2013-001772

McGinnis, D. P., Brownstein, J. S., & Patel, C. J. (2016). Environment-Wide Association Study of Blood Pressure in the National Health and Nutrition Examination Survey (1999–2012). *Scientific Reports*, *6*(1), 30373. https://doi.org/10.1038/srep30373

Park, S. K., Zhao, Z., & Mukherjee, B. (2017). Construction of environmental risk score beyond standard linear models using machine learning methods: application to metal mixtures, oxidative stress and cardiovascular disease in NHANES. *Environmental Health*, *16*(1), 102. https://doi.org/10.1186/s12940-017-0310-9

Patel, C. J. (2017). Analytic Complexity and Challenges in Identifying Mixtures of Exposures Associated with Phenotypes in the Exposome Era. *Current Epidemiology Reports*, *4*(1), 22–30. https://doi.org/10.1007/s40471-017-0100-5

Pirhaji, L., Milani, P., Leidl, M., Curran, T., Avila-Pacheco, J., Clish, C. B., … Fraenkel, E. (2016). Revealing disease-associated pathways by network integration of untargeted metabolomics. *Nature Methods*, *13*(9), 770–776. https://doi.org/10.1038/nmeth.3940

Schiffman, C., Petrick, L., Perttula, K., Yano, Y., Carlsson, H., Whitehead, T., … Dudoit, S. (2019). Filtering procedures for untargeted LC-MS metabolomics data. *BMC Bioinformatics*, *20*(1), 334. https://doi.org/10.1186/s12859-019-2871-9

Stingone, J. A., Buck Louis, G. M., Nakayama, S. F., Vermeulen, R. C. H., Kwok, R. K., Cui, Y., … Teitelbaum, S. L. (2017). Toward Greater Implementation of the Exposome Research Paradigm within Environmental Epidemiology. *Annual Review of Public Health*, *38*(1), 315–327. https://doi.org/10.1146/annurev-publhealth-082516-012750

Stingone, J. A., Pandey, O. P., Claudio, L., & Pandey, G. (2017). Using machine learning to identify air pollution exposure profiles associated with early cognitive skills among U.S. children. *Environmental Pollution*, *230*, 730–740. https://doi.org/10.1016/j.envpol.2017.07.023

van Tongeren, M., & Cherrie, J. W. (2012, March). An integrated approach to the exposome. *Environmental Health Perspectives*. https://doi.org/10.1289/ehp.1104719

Vrijheid, M., Slama, R., Robinson, O., Chatzi, L., Coen, M., van den Hazel, P., … Nieuwenhuijsen, M. J. (2014). The human early-life exposome (HELIX): Project rationale and design. *Environmental Health Perspectives*. Public Health Services, US Dept of Health and Human Services. https://doi.org/10.1289/ehp.1307204

Wild, C. P. (2012, February). The exposome: From concept to utility. *International Journal of Epidemiology*. Int J Epidemiol. https://doi.org/10.1093/ije/dyr236

Wulaningsih, W., Van Hemelrijck, M., Tsilidis, K. K., Tzoulaki, I., Patel, C., & Rohrmann, S.

(2017). Investigating nutrition and lifestyle factors as determinants of abdominal obesity: an environment-wide study. *International Journal of Obesity*, *41*(2), 340–347. https://doi.org/10.1038/ijo.2016.203

Young, V. B. (2017, March 15). The role of the microbiome in human health and disease: An introduction for clinicians. *BMJ (Online)*. BMJ Publishing Group. https://doi.org/10.1136/bmj.j831

Semi-automated NMR Pipeline for Environmental Exposures: New Insights on the Metabolomics of Smokers versus Non-smokers

Morris A. Aguilar†, John McGuigan

Huck Institutes of the Life Sciences, The Pennsylvania State University,
512 Wartik, University Park, PA 16802, USA
Email: m0rris@psu.edu

Molly A. Hall, Ph.D., M.S.
512A Wartik Laboratory
University Park, PA 16801, USA
Email: mah546@psu.edu

Environmental exposure pathophysiology related to smoking can yield metabolic changes that are difficult to describe in a biologically informative fashion with manual proprietary software. Nuclear magnetic resonance (NMR) spectroscopy detects compounds found in biofluids yielding a metabolic snapshot. We applied our semi-automated NMR pipeline for a secondary analysis of a smoking study (MTBLS374 from the MetaboLights repository) (n = 112). This involved quality control (in the form of data preprocessing), automated metabolite quantification, and analysis. With our approach we putatively identified 79 metabolites that were previously unreported in the dataset. Quantified metabolites were used for metabolic pathway enrichment analysis that replicated 1 enriched pathway with the original study as well as 3 previously unreported pathways. Our pipeline generated a new random forest (RF) classifier between smoking classes that revealed several combinations of compounds. This study broadens our metabolomic understanding of smoking exposure by 1) notably increasing the number of quantified metabolites with our analytic pipeline, 2) suggesting smoking exposure may lead to heterogenous metabolic responses according to random forest modeling, and 3) modeling how newly quantified individual metabolites can determine smoking status. Our approach can be applied to other NMR studies to characterize environmental risk factors, allowing for the discovery of new biomarkers of disease and exposure status.

Keywords: Environmental Exposure; Metabolomics; Cigarette Smoke; Bioinformatics

1. Introduction

Cigarette smoke (CS) is made of harmful constituents that cause many diseases.[1] Additionally, there are many indicators that CS exposure has led to increased medical costs and loss of productivity

† Morris A. Aguilar was supported on the PSU/NIDDK funded Integrative Analysis of Metabolic Phenotypes (IAMP) Predoctoral Training Program (T32DK120509). Research reported in this publication was supported by the National Cancer Institute of the National Institutes of Health under Award Number R01CA239256. This work was supported by the USDA National Institute of Food and Agriculture and Hatch Appropriations under Project #PEN04275 and Accession #1018544, Huck Institutes for the Life Sciences, Penn State Cancer Institute, and the Dr. Frances Keesler Graham Early Career Professorship.

over a lifespan.[2] The thousands of reactive oxidative species (ROS) generated from burning cigarettes are found in the gaseous state and are responsible for CS related pathogenesis.[3] The ROS damage epithelial cell linings by disrupting oxidative-sensitive metabolism and triggering DNA damage.[4] The effects of CS on immunity can be both pro-inflammatory and suppressive.[5] CS derived ROS can lead to neuronal damage[6], atherosclerosis, increases predisposition of cardiovascular events[7] and inhibit tumor suppressive mechanisms.[1] Metabolomic interrogations of CS exposure may help investigators further understand the pathogenesis of several diseases strongly associated with CS exposure. Metabolomics studies the small molecules from biological samples that can reveal metabolic changes following environmental exposures.[8,9] With respect to the genome, transcriptome, and proteome, metabolomics generally involves the small molecule compounds that are metabolized by enzymes; the metabolome can act synergistically with other "-omic" layers as well.[10] Unlike other "-omics," metabolomics reveals biochemical states and best represents the molecular phenotype.[8] Additionally, metabolomic studies of disease can reveal new biomarkers, understudied pathways, and prognosis measures to improve risk stratification.[11,12]

A previous metabolomic study of CS that incorporated NMR and MS data derived from human blood serum found metabolites associated with chronic obstructive pulmonary disease, cardiovascular disease and cancer.[6,7,13] This study by Kaluarachchi et al. is unique because it is the only study to date that used 1D 1H NMR on human blood serum for CS exposure from which 3 metabolites were reported.[13] The raw NMR data for this human blood serum CS exposure study (n = 112) is publicly available on the MetaboLights repository as MTBLS374.[8,13] The raw MTBLS374 data was originally analyzed with proprietary software to identify and quantify metabolites.

Although commercial software are popular, they often lack advanced editing, require iterative steps, and involve arbitrary adjustments based on subjective user judgement.[14] Previous studies indicate that this manual method is prone to false positive metabolite identification that increases as more metabolites are quantified.[15,16] The NMR analysis described here incorporates several R and Python packages to aid in the detection of additional metabolites that were previously unreported. We created novel random forest classification (RF) models from the quantitative metabolite data and the unprofiled spectra to classify smoking status. Furthermore, our RF classification decision trees reveal the statistical importance of the detected biomarkers, and findings were supported by pathway enrichment analysis.

Here we demonstrate how an environmental exposure like smoking and its metabolic effects can be quantified and modeled with NMR data via open source packages. With our pipeline, we quantified 79 previously undetected metabolites in this dataset. With the metabolite quantification data generated from our pipeline, we developed 2 high fidelity models that classified between the smoking classes. Our pipeline increases transparency of user set analysis parameters and unifies existing open source packages for spectral processing and multivariate analyses.

2. Methods

2.1. *Data Set*

The MTBLS374 dataset that was used for this study was acquired from the MetaboLights repository and contains 1D 1H NMR spectra of human blood serum from 112 participants.[8,13] The original study also incorporated mass spectroscopy and lipoprotein fraction data in addition to NMR data to identify biochemical differences in smoking classes.[13] They found that the metabolites they detected indicated that smoking exposure impacted glutathione, bilirubin and lipids. The authors suggested that their metabolic enrichment pathways were related to chronic obstructive pulmonary disease, cardiovascular diseases, and cancer.[13] There were 55 (27 females, 28 males) smoker class samples and 57 (28 female, 29 male) never smoker class samples. The participants were from Hamburg, Germany who had a body mass index (BMI) within a healthy range and no clinical history of heart, lung diseases and chronic diseases. The MTBLS374 data set sample labels were limited to gender and smoking status (smoker/never smoker) due to adherence of participant privacy policies; however, the original study included BMI, age, and drug intake in their confounding analysis. The 1H 1D NMR spectroscopy data was generated with the Carr-Purcell-Meiboom-Gill pulse sequence with the following parameters: relaxation delay of 4 s, a mixing time of 0.01 s, a spin–echo delay of 0.3 ms, 128 loops and a free induction 3.067 s of decay acquisition time, total of 32 scans recorded into 96 thousand data points with a spectral width of 20 ppm.[13]

2.2. *Pipeline*

The innovation of the pipeline lies in its capability of extracting metabolomic data from raw data NMR data in a semi-automated fashion (i.e., the arduous task of metabolite identification/quantification has been made automated, yet some parameter choices are still needed by the user). Open-source packages are unified to promote analysis reproducibility for the complex multistep analytical process of quantifying metabolic effects of environmental exposures. Typically, proprietary graphical user interface (GUI) software requires one set of software to edit the raw data to remove instrumental artifacts, a separate GUI application for metabolite quantification, and a separate statistical analysis software. These software do not record the repetitive and arbitrary user decisions to manipulate the data which is not conducive to analysis reproducibly. The proprietary software offers limited automation tools thereby constraining the user to iterative processes. The pipeline we describe here addresses the multiple steps data processing (Figure 1) and analysis challenges in environmental exposure metabolomics. We uploaded scripts to this pipeline to GitHub (github.com/HallLab/MTBLS374_smoking_study_secondary_analysis). We will describe the application of our pipeline to cigarette smoke exposure below.

2.2.1 *Preprocessing and Spectral Analysis*

Before metabolites are identified and quantified, the first step in our pipeline is to preprocess the NMR data (i.e., data editing to enhance signal-to-noise ratio and minimize instrumental artifacts). This preprocessing is accomplished with the PepsNMR[14] R package. A user may set parameters before bulk preprocessing of NMR data. The raw NMR data was first pre-processed (Figure 1a) so that the NMR data can be interpreted by subsequent analysis packages. The NMR spectra were zero-filled, Fourier transformed, zero phase corrected, first phase corrected, warping, binning, and normalized semi-autonomously by using the PepsNMR presets.[14] We corrected for pH-induced chemical shifts with the warping and binning functions provided by PepsNMR. The NMR spectra were normalized with constant sum normalization which is recommended for sera.[14] The regions corresponding to the water peak at 4.5 - 5.1 ppm were removed. The resulting output was pre-processed NMR data (Figure 1b) that can be utilized as input for subsequent analyses. Data clustering was observed with multivariate principal components analysis (PCA) analysis including samples who were categorized as smoking classes, and quality control class. The pre-processed binned spectral data was also used to generate random forest classification models with k-fold (k=10) validation with the Scikit-learn (0.22.1) python package.[17]

2.2.2 *Identification and Quantification*

The binned spectral data were tested for significant spectral differences between the smoking classes. Between classes, each corresponding bin had a non-normal distribution thus warranting the Wilcoxon Rank Sum Test and Bonferroni adjustment (α =0.05) (Figure 1c).[18] The spectral positions of the significant bins (Figure 1c) were cross referenced from a pure metabolite standard from HMDB to build a list of compounds that rDolphin[19] (a profiling tool for 1H-NMR-based studies) automatically detects and quantifies within the preprocessed NMR data according to metabolite multiplicity and chemical shifts (Figure 1d).

2.2.3 *Analysis*

The metabolite identification and quantification data output from rDolphin (Figure 1e) were used for t-tests and as features to train a second random forest classifier with k-fold (k=10) validation. The metabolite data was piped to the MetaboAnalyst R package (3.0.3) for data transformation such as normalization by sum, log transform and pareto scaling for t-tests (Figure 1f).[20] Finally, the transformed metabolite data were used for metabolic pathway enrichment based on ontologies from the Kyoto Encyclopedia of Genes and Genomes (KEGG) Pathway Data Base and conducted via MetaboAnalyst. The enrichment analysis had 2-fold filter criteria.

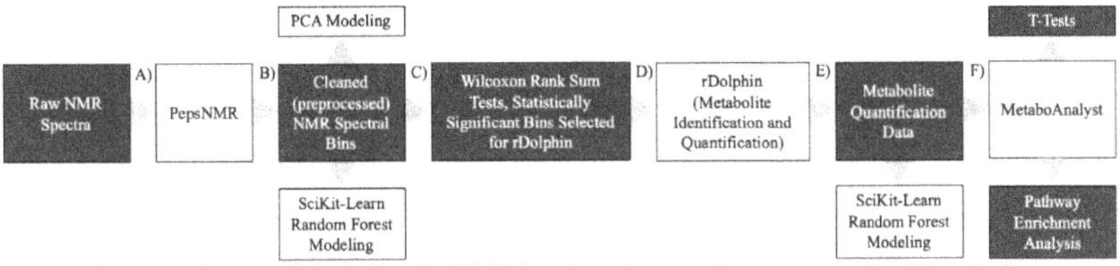

Fig 1. Semi-automatied pipeline for NMR based environmental exposure studies. The pipeline connected open source packages (white boxes). Outputs are represented in gray boxes.

3. Results

A PCA was conducted on the pre-processed NMR spectral data to reveal clustering patterns based on smoking status and gender (Figure 2). Results from the PCA with the smoking classes indicate that the clusters overlap more so than the gender-based classes. PC1 and PC2 explain 77.0% and 13.3% of the variance for the gender and smoking status groups. The PCA results suggest that the gender classes may be a confounding factor. Logistic regression to test if gender was a significant predictor of smoking status in our data set yielded a non-significant (p-value: 0.40) predictor of smoking status.

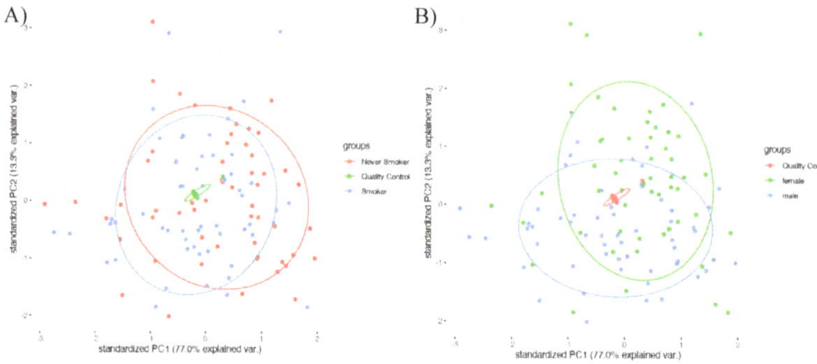

Fig. 2. PCA Clustering of Smoking Status (A) and Gender Classes (B). PC 1 and PC2 are represented on the x-axis and y-axis, respectively. A) The PCA plot clustered the data points according to the female (green) and male (blue) classes according to PC 1 and PC 2. B) The PCA plot clusters the data points according to the smoker class (blue) and never smoker class (red). Both plots have a quality control class (red in subplot A and green in subplot B) to gauge technical variance.

To assess which NMR peaks warrant metabolite identification and quantification, the NMR spectral bins from 0.0 ppm to 10.0 ppm between the smoking classes were tested for significant differences in 467 spectral bins. For the Wilcoxon Rank Sum test, each bin was compared to its corresponding position in the NMR spectra between classes, i.e., the bin at position 1 ppm from the smoking class was only compared to the bin at position 1 ppm for the never smoker class. Each of the 467 non-normal spectral bins were tested for significance with the Wilcoxon Rank Sum test and 32 bins were significant when Bonferroni-adjusted (α: 0.05) (Figure 3). Spectral bins passing this threshold were investigated for metabolite identification and quantification via the rDolphin peak aligner.

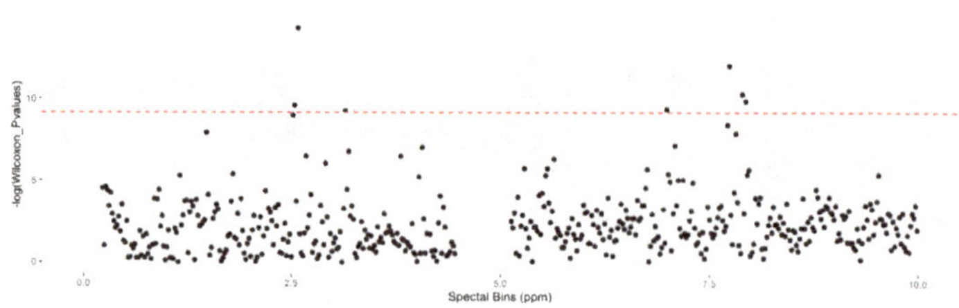

Fig. 3. Manhattan plot of spectral bin associations with smoking status. The NMR spectrum for each sample was represented on the x-axis from 0 – 10 ppm and divided into bins with widths of 0.02 ppm and the y-axis represents the -log (10) of the p-value. The red line represents the Bonferroni significance threshold (alpha: 0.05, 467 tests). The absence of data points between at 4.5 - 5.1 ppm was expected due to the removal of the water signal.

After metabolite quantification, the 79 putatively identified metabolites and their relative concentrations were sum normalized, log transformed and pareto scaled for univariate two tailed t-tests. When the smoking classes were compared, 6 compounds were significant after Bonferroni adjustment (Figure 4). The significant metabolites include: Indole-3-propionicacid (p-value: 5.24×10^{-6}), Indoxyl sulfate (p-value: 6.57×10^{-6}), N-Acetyl-L-aspartic (p-value: 1.27×10^{-5}), xanthine (p-value: 3.36×10^{-5}), L-tryptophan (p-value: 7.36×10^{-5}) and L-histidine (p-value: 0.00010336).

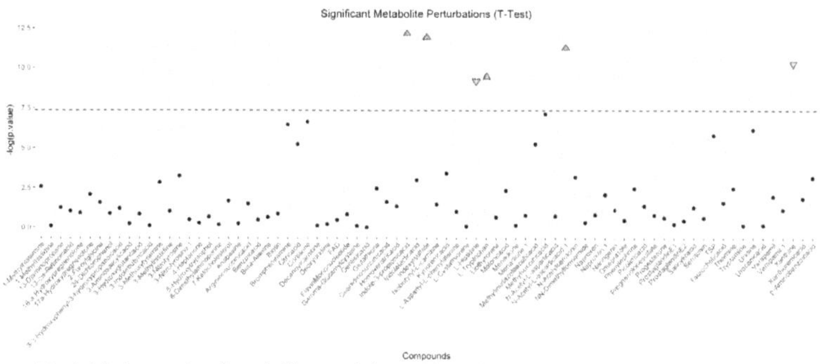

Fig. 4. Manhattan plot of metabolite associations with smoking status. The Manhattan plot displays the metabolites on the x-axis and their -log(10) p-values on the y-axis. The red line represents the Bonferroni corrected significance threshold. The blue and yellow triangles represent increased and decreased metabolites.

Two types of RF models were generated and were trained with either spectral data or quantitative metabolic data (Figure 5). For smoking status, the models demonstrated an AUC of 0.76 (SD: 0.15) for spectral bins (Figure 5a) and an AUC of 0.86 (SD: 0.14) for quantified metabolites (Figure 5c). For gender, the models demonstrated an AUC 0.70 (SD: 0.15) for spectral bins (Figure 5b) and AUC of 0.41 (SD: 0.13) for quantified metabolites (Figure 5d).

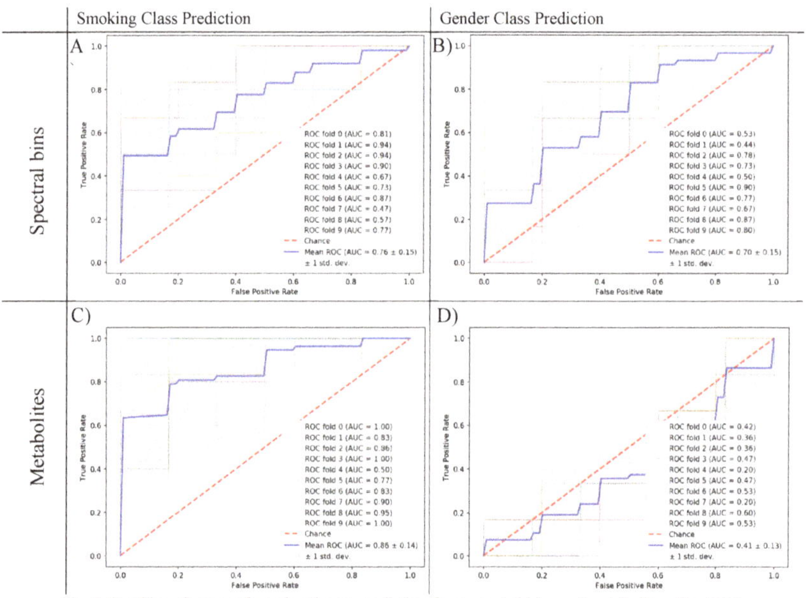

Fig. 5. Smoking classes and gender classes prediction from spectral bins and metabolites. The ROC curves represent the RF models' ability to discriminate between case and control and characterizes the model's true positive and false positive rates. The plots also depict the model for every k-fold cross validation and the thick blue line represents the mean ROC curve derived from the cross validated models. A set of RF models were created by using the NMR spectral bins (467 per sample) as features. Another RF model set was created using the quantified metabolite data generated from the compound detection.

We created the decision tree from the RF model trained on the quantitative metabolic data that predicted smoking classes (Figure 6). When the RF model was trained it iteratively split the smoking classes into two branches but not all splits are perfect. Gini impurity represents the quality of the split between smoking classes at a node and a perfect split between classes at a node has a value of 0 like the terminal nodes in (Figure 6). 2,4-dichlorophenol (Figure 6a), 3-nitrotyrosine (Figure 6b), and xanthurenic acid (Figure 6c) have a gini impurity of 0.1, 0.36, and 0.23, respectively. The gini impurity at the 3-nitrotyrosine node indicates that the metabolite is not always perturbed for the smoking class which reveals smoking exposure metabolic heterogeneity. Also, at each node the percent of samples in the dataset that fulfill the quantitative threshold is given for each metabolite in the tree. The multivariate RF model indicates how combinations of metabolic perturbations occur depending on CS exposure which is more representative the highly interconnected metabolic biology of humans.

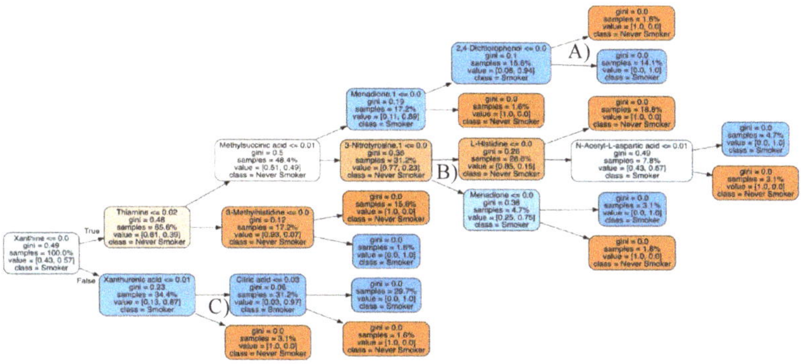

Fig. 6. Metabolite random forest model for smoking classes prediction. This metabolite-based RF model has a decision tree that places each metabolite at a node and branches according to a Boolean quantitative threshold; when a condition was true the node branches upwards and if the condition was false the node branches downwards. Notable metabolites in the tree include A) 2,4-dichlorophenol, B) 3-nitrotyrosine, and c) xanthurenic acid. The decision tree emphasizes that several unique combinations of biomarkers differentiate smoking classes.

To determine which metabolic pathways were significantly perturbed, we performed enrichment tests on the 79 metabolites we quantified (that were found in the statistically significant spectral bins) and were mapped to known metabolic pathways from the KEGG database. The top 15 metabolic pathways that were perturbed between smoking classes are listed (Figure 7). The Bonferroni-adjusted significant pathways were aminoacyl-tRNA biosynthesis, histidine metabolism, purine metabolism, and beta-alanine metabolism. At most, the significantly enriched pathways have two metabolite hits which means that 2 of the metabolites we newly quantified are known to participate in that metabolic pathway.

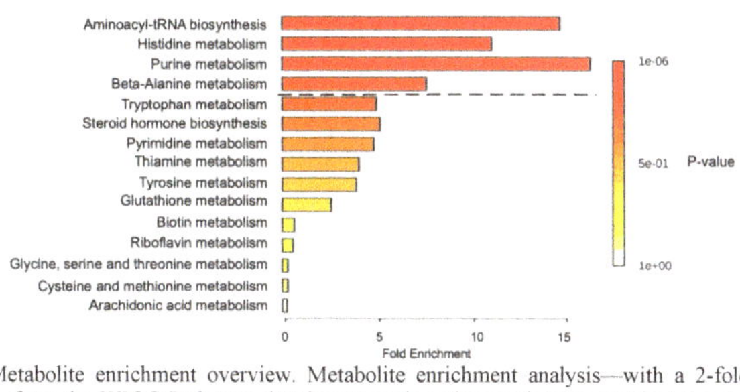

Fig. 7. Metabolite enrichment overview. Metabolite enrichment analysis—with a 2-fold change criterion—from the KEGG Pathways data base reveals pathways that are enriched due to smoking

status. The metabolic pathways above the black dashed line represents statistical significance after Bonferroni adjusted ($\alpha = 0.05$) multiple test correction.

4. Discussion

Environmental exposures can perturb the complex human metabolome, and it is difficult to quantify the numerous metabolic pathways with NMR data using proprietary software with limited automation features and no record of data transformation. We demonstrated the technical feasibly of describing the metabolome when affected by an environmental exposure like CS by unifying open source NMR packages. The MTBLS374 NMR data set was originally used to quantify 3 specific metabolites; however, the NMR spectra of each human blood serum sample was representative of thousands of metabolites that are expected to be found.[21] We demonstrated our pipeline's potential to increase the number of quantified metabolites.

To understand the global metabolomic differences between the smoking status classes and the gender classes, PCA was performed. The PCA cluster based on spectral data indicated more distinct separation between the gender-based classes than smoking exposure classes. The female and male groups have clusters that overlap with one another (Figure 1a), which suggests there may be more spectral differences related the metabolic sexual dimorphism which has been demonstrated previously.[22] The pooled quality control classes clustered more tightly relative to the gender and smoking based classes, and we expected the quality control samples to display very little variance between one another and the variance that that we do detect likely came from variance from the NMR instrumentation.

We found 6 significant metabolites, all of which were not previously identified in the data set, however, we did only detect 1 out of the 3 metabolites the original authors found in the NMR data. We used a new computational approach involving semi-automated pre-processing and automated metabolite quantification open source packages as opposed to proprietary software like the original authors. Therefore, we did not necessarily expect to detect the same metabolites from the NMR data. Of the significantly perturbed metabolites from Figure 3, Indole-3-propionicacid is known to be neuroprotective antioxidant[23] and more likely to be affected in smokers with atherosclerosis.[24] Indoxyl sulfate is a known cardiotoxin and uremic toxin.[25] A previous study found that indoxyl sulfate is lower in smokers' blood serum, while here we found it was elevated.[26] N-Acetyl-L-aspartic acid is one of the most concentrated compounds in the brain for myelin[27] and a previous study found that this metabolite is decreased in the left hippocampus tissue in smokers.[28] In our analysis we found that N-Acetyl-L-aspartic acid was elevated in blood serum. Xanthine is involved in the purine degradation pathway.[29] The xanthine oxidase enzyme is elevated in smokers and it produces uric acid by consuming xanthine as a precursor molecule.[30] We found that xanthine was significantly decreased in blood serum which might be due to its consumption of elevated xanthine oxidase. L-Tryptophan is an amino acid that is a precursor to hormones and neurotransmitters[31] and has been found to be downregulated in those attempting to quit smoking.[32] In our study we found that L-Tryptophan was significantly elevated which might play a role in cigarette smoking related behavior. L-Histidine is an essential amino acid and is a precursor to an inflammatory agent, histamine.[33] L-Histidine is depressed in smokers without chronic obstructive pulmonary disease (COPD) versus those with COPD suggesting its consumption for histamine production thereby increasing inflammatory response.[34] In our study, L-Histidine is significantly decreased suggesting that we might detect markers of inflammation in blood serum due to CS exposure. The significant perturbations of these 6 metabolites reinforces how CS exposure contributes to pathologies relating to ROS metabolism, cardiac damage, neural toxicity, and inflammatory response. Given that CS exposure perturbs individual metabolites it follows that it was possible to classify smoking exposure classes based on these perturbations.

The metabolite-based RF model that predicted smoking status has a decision tree that found novel relationships between metabolites. 2,4-Dichlorophenol (Figure 6a) is a known hazardous air pollutant and is a soil pollutant that tobacco plants can absorb.[35,36] Within the context of other metabolites, 2,4-dichlorophenol is a necessary smoking class decision node. Smoking is associated with a decrease in 3-nitrotyrosine levels of plasma proteins and vascular endothelial dysfuction.[37] 3-Nitrotyrosine (Figure 6b) was not significant within our univariate t-tests but in a multivariate context 3-nitrotyrosine was a necessary decision node for smoking classes. Although there is an inverse metabolic relationship between xanthine and neuronal uptake of xanthurenic acid[38] on the path towards the terminal node (Figure 5c), there is no documented relation of these two metabolites with respect to smoking exposure. The root nodes in the decision tree (Figure 6) begin with a high gini impurity and terminate with 0 impurity. This means that each terminal node is dependent on the node path leading back to the root metabolite in the tree. In other words, these metabolite changes were dependent on one another to yield a metabolic profile indicative of the smoking classes. The combinations of these metabolites have not been previously documented and suggests a heterogenous response to a smoking exposure. These metabolite combinations used to classify smoking exposure status may be indicative of interconnected perturbations of metabolic pathways. Nevertheless, the decision tree found a statistical relationship and did not relate metabolites to mapped metabolic pathways.

We conducted a pathway enrichment analysis to relate how the metabolic perturbations we quantified relate to previously empirically derived metabolic pathways. In the enrichment analysis we included the 79 putatively identified compounds we quantified from NMR data. The original study found that the aminoacyl-tRNA biosynthesis was one of the top significantly enriched pathways which we replicated in this automated analysis. Another smoking exposure blood serum based mass spectroscopy study also corroborated the enrichment of aminoacyl-tRNA biosynthesis.[39] Nonetheless we found purine, histidine, and biotin pathways to be enriched which was not previously described for human samples with CS exposure. These three pathways that we newly derived from NMR data is supported by a previous mass spectrometry blood serum based smoking study in a mouse model.[40] A smoking exposure NMR study on mouse lung tissue extracts also found purine and histidine pathway perturbations likely due to cell injury.[41] In particular the purine pathway perturbation might be due to CS related DNA damage and cell injury.[4] The original study's enrichment analysis was supplemented by mass spectroscopy data, which may contribute to divergence in enrichment results.

Although this study demonstrates that our pipeline can reveal more NMR generated metabolomic information about environmental exposures, we did not uncover all of the possible metabolic perturbations. The significant results from the univariate analysis described here provided a limited viewing window into the CS exposure metabolome because it does not describe the interconnected reality of human metabolism. The RF decision tree begins to describe interconnected metabolism and suggests that multiple combinations of metabolites are associated with the smoking classes. However, these combinations are not to be interpreted as being the only metabolites that are perturbed. Given that the public repository did not include the BMI, age, and drug intake data from the original study, we were not able to do additional confounder tests. Scalability of the pipeline becomes limited with data sets larger than MTBLS374 given that the preprocessing package (PepsNMR) and peak alignment package (rDolphin) where not coded with multicore support. Next steps include testing this pipeline on other NMR based environmental exposure studies to classify disease status, replicating major findings, and describing novel findings. Nonetheless, our unified pipeline overcame the limitations of manual NMR pre-processing and quantification and has enabled us to extract valuable metabolomic findings regarding smoking exposure.

5. Conclusion

Here we demonstrate how an environmental exposure like smoking and its metabolic effects can be quantified and modeled with NMR data. Our approach of filtering spectral bins via multiple tests informed which metabolites were automatically quantified. The RF modeling reveals how several unique combinations of metabolites are associated with smoking classes. This suggests there are more than one combination of metabolite perturbations associated with smoking and a heterogenous response to smoking exposure. Several of the metabolites that belong to these combinations have a known relationship to smoking and/or cellular damage. The novelty of our analysis approach lies in breaking from the conventional manual analysis methods and promoting study reproducibility.

References

1. Das, S. K. Harmful health effects of cigarette smoking. *Mol. Cell. Biochem.* **253**, 159–165 (2003).
2. Max, W. The Financial Impact of Smoking on Health-Related Costs: A Review of the Literature. *Am. J. Health Promot.* **15**, 321–331 (2001).
3. Huang, M.-F., Lin, W.-L. & Ma, Y.-C. A study of reactive oxygen species in mainstream of cigarette. *Indoor Air* **15**, 135–140 (2005).
4. Valavanidis, A., Vlachogianni, T. & Fiotakis, K. Tobacco Smoke: Involvement of Reactive Oxygen Species and Stable Free Radicals in Mechanisms of Oxidative Damage, Carcinogenesis and Synergistic Effects with Other Respirable Particles. *IJERPH* (2009).
5. Lee, J., Taneja, V. & Vassallo, R. Cigarette Smoking and Inflammation: Cellular and Molecular Mechanisms. *J. Dent. Res.* **91**, 142–149 (2012).
6. Swan, G. E. & Lessov-Schlaggar, C. N. The Effects of Tobacco Smoke and Nicotine on Cognition and the Brain. *Neuropsychol. Rev.* **17**, 259–273 (2007).
7. Ambrose, J. A. & Barua, R. S. The pathophysiology of cigarette smoking and cardiovascular disease: An update. *J. Am. Coll. Cardiol.* **43**, 1731–1737 (2004).
8. Haug, K. *et al.* MetaboLights: a resource evolving in response to the needs of its scientific community. *Nucleic Acids Res.* doi:10.1093/nar/gkz1019.
9. Dona, A. C. *et al.* Precision High-Throughput Proton NMR Spectroscopy of Human Urine, Serum, and Plasma for Large-Scale Metabolic Phenotyping. *Anal. Chem.* **86**, 9887–9894 (2014).
10. Sun, Y. V. & Hu, Y.-J. Integrative Analysis of Multi-omics Data for Discovery and Functional Studies of Complex Human Diseases. *Adv. Genet.* **93**, 147–190 (2016).
11. Vignoli, A. *et al.* NMR-based metabolomics identifies patients at high risk of death within two years after acute myocardial infarction in the AMI-Florence II cohort. *BMC Med.* **17**, 3 (2019).
12. Hendriks, M. M. W. B. *et al.* Data-processing strategies for metabolomics studies. *TrAC Trends Anal. Chem.* **30**, 1685–1698 (2011).
13. Kaluarachchi, M. R., Boulangé, C. L., Garcia-Perez, I., Lindon, J. C. & Minet, E. F. Multiplatform serum metabolic phenotyping combined with pathway mapping to identify biochemical differences in smokers. *Bioanalysis* **8**, 2023–2043 (2016).

14. Martin, M. *et al.* PepsNMR for 1H NMR metabolomic data pre-processing. *Anal. Chim. Acta* **1019**, 1–13 (2018).
15. Walker, L. R. *et al.* Unambiguous metabolite identification in high-throughput metabolomics by hybrid 1D 1H NMR/ESI MS1 approach. *Magn. Reson. Chem.* **54**, 998–1003 (2016).
16. Tredwell, G. D., Behrends, V., Geier, F. M., Liebeke, M. & Bundy, J. G. Between-Person Comparison of Metabolite Fitting for NMR-Based Quantitative Metabolomics. *Anal. Chem.* **83**, 8683–8687 (2011).
17. Pedregosa, F. *et al.* Scikit-learn: Machine Learning in Python. *J. Mach. Learn. Res.* **12**, 2825–2830 (2011).
18. Wilcoxon, F. Individual Comparisons by Ranking Methods. in *Breakthroughs in Statistics: Methodology and Distribution* (eds. Kotz, S. & Johnson, N. L.) 196–202 (Springer, 1992). doi:10.1007/978-1-4612-4380-9_16.
19. Cañueto, D., Gómez, J., Salek, R. M., Correig, X. & Cañellas, N. rDolphin: a GUI R package for proficient automatic profiling of 1D 1H-NMR spectra of study datasets. *Metabolomics* **14**, 24 (2018).
20. Xia, J., Sinelnikov, I. V., Han, B. & Wishart, D. S. MetaboAnalyst 3.0—making metabolomics more meaningful. *Nucleic Acids Res.* **43**, W251–W257 (2015).
21. Wishart, D. S. *et al.* HMDB 4.0: the human metabolome database for 2018. *Nucleic Acids Res.* **46**, D608–D617 (2018).
22. Krumsiek, J. *et al.* Gender-specific pathway differences in the human serum metabolome. *Metabolomics* **11**, 1815–1833 (2015).
23. Chyan, Y. J. *et al.* Potent neuroprotective properties against the Alzheimer beta-amyloid by an endogenous melatonin-related indole structure, indole-3-propionic acid. *J. Biol. Chem.* **274**, 21937–21942 (1999).
24. Cason, C. A. *et al.* Plasma microbiome-modulated indole- and phenyl-derived metabolites associate with advanced atherosclerosis and postoperative outcomes. *J. Vasc. Surg.* **68**, 1552-1562.e7 (2018).
25. PubChem. Indoxyl sulfate. https://pubchem.ncbi.nlm.nih.gov/compound/10258.
26. Viaene, L. *et al.* Heritability and Clinical Determinants of Serum Indoxyl Sulfate and p-Cresyl Sulfate, Candidate Biomarkers of the Human Microbiome Enterotype. *PLOS ONE* **9**, e79682 (2014).
27. Nordengen, K., Heuser, C., Rinholm, J. E., Matalon, R. & Gundersen, V. Localisation of N-acetylaspartate in oligodendrocytes/myelin. *Brain Struct. Funct.* **220**, 899–917 (2015).
28. Gallinat, J. *et al.* Abnormal Hippocampal Neurochemistry in Smokers: Evidence From Proton Magnetic Resonance Spectroscopy at 3 T. *J. Clin. Psychopharmacol.* **27**, 80–84 (2007).
29. PubChem. Xanthine. https://pubchem.ncbi.nlm.nih.gov/compound/1188.
30. Shah, A. A., Khand, F. & Khand, T. U. Effect of smoking on serum xanthine oxidase, malondialdehyde, ascorbic acid and α-tocopherol levels in healthy male subjects. *Pak. J. Med. Sci.* **31**, 146–149 (2015).
31. Slominski, A. *et al.* Conversion of L-tryptophan to serotonin and melatonin in human melanoma cells. *FEBS Lett.* **511**, 102–106 (2002).
32. Bowen, D. J., Spring, B. & Fox, E. Tryptophan and high-carbohydrate diets as adjuncts to smoking cessation therapy. *J. Behav. Med.* **14**, 97–110 (1991).
33. PubChem. Histidine. https://pubchem.ncbi.nlm.nih.gov/compound/6274.
34. Diao, W. *et al.* Disruption of histidine and energy homeostasis in chronic obstructive pulmonary disease. *Int. J. Chron. Obstruct. Pulmon. Dis.* **14**, 2015–2025 (2019).
35. Talano, M. A. *et al.* Phytoremediation of 2,4-dichlorophenol using wild type and transgenic tobacco plants. *Environ. Sci. Pollut. Res.* **19**, 2202–2211 (2012).
36. Laurent, F., Canlet, C., Debrauwer, L. & Pascal-Lorber, S. Metabolic fate of [14C]-2,4-dichlorophenol in tobacco cell suspension cultures. *Environ. Toxicol. Chem.* **26**, 2299–2307 (2007).
37. Jin Hongjun *et al.* Smoking, COPD, and 3-Nitrotyrosine Levels of Plasma Proteins. *Environ. Health Perspect.* **119**, 1314–1320 (2011).
38. Gobaille, S. *et al.* Xanthurenic acid distribution, transport, accumulation and release in the rat brain. *J. Neurochem.* **105**, 982–993 (2008).
39. Liu, G., Lee, D. P., Schmidt, E. & Prasad, G. Pathway Analysis of Global Metabolomic Profiles Identified Enrichment of Caffeine, Energy, and Arginine Metabolism in Smokers but Not Moist Snuff Consumers. *Bioinforma. Biol. Insights* **13**, 1177932219882961 (2019).
40. Cruickshank-Quinn, C. I. *et al.* Transient and Persistent Metabolomic Changes in Plasma following Chronic Cigarette Smoke Exposure in a Mouse Model. *PLoS ONE* **9**, (2014).
41. JZ, H. *et al.* Metabolite Signatures in Hydrophilic Extracts of Mouse Lungs Exposed to Cigarette Smoke Revealed by 1H NMR Metabolomics Investigation. *Metabolomics Open Access* **5**, (2015).

How Much Does the (Social) Environment Matter? Using Artificial Intelligence to Predict COVID-19 Outcomes with Socio-demographic Data*

Christos A. Makridis
Arizona State University, MIT Sloan School of Management, Department of Veterans Affairs
Washington, DC, 20005
Email: christos.a.makridis@gmail.com

Anish Mudide
Phillips Exeter Academy
Exeter, NH 03833
Email: amudide@gmail.com

Gil Alterovitz
Brigham and Women's Hospital/Harvard Medical School, Boston, MA 02115 and
Department of Veterans Affairs, Washington, DC, 20005
Email: ga@alum.mit.edu

While the coronavirus pandemic has affected all demographic brackets and geographies, certain areas have been more adversely affected than others. This paper focuses on Veterans as a potentially vulnerable group that might be systematically more exposed to infection than others because of their co-morbidities, i.e., greater incidence of physical and mental health challenges. Using data on 122 Veteran Healthcare Systems (HCS), this paper tests three machine learning models for predictive analysis. The combined LASSO and ridge regression with five-fold cross validation performs the best. We find that socio-demographic features are highly predictive of both cases and deaths—even more important than any hospital-specific characteristics. These results suggest that socio-demographic and social capital characteristics are important determinants of public health outcomes, especially for vulnerable groups, like Veterans, and they should be investigated further.

Keywords: Artificial Intelligence, Coronavirus, COVID-19, Machine Learning, Veterans.

1. Introduction

Following months of the coronavirus ("COVID-19") pandemic, a large body of research has emerged quantifying the contribution of individual characteristics towards exposure of the virus (Britton et al., 2020; Martin et al., 2020). Moreover, there is also increasing evidence that certain vulnerable groups have been affected more adversely than others, especially minorities (Pan et al., 2020). However, researchers have struggled to obtain bias-free, reliable, and externally-valid predictions on representative datasets (Wynants et al., 2020).

---

* All replication files are available here: https://github.com/amudide/COVID-Sociodemographics-AI

The majority of studies have focused on the role of individual-level factors, but a separate vein of research in computational social science has found that socio-economic factors also play an important role in mediating the spread of the virus (Makridis and Wu, 2020; Ding et al., 2020; Barrios et al., 2020). For example, the Joint Economic Committee (JEC) in the United States Congress has focused on quantifying social capital and its important implications for economic outcomes and well-being (JEC, 2018). Moreover, social capital has also been associated with community-level health outcomes (Gordeev and Egan, 2015; Kolak et al., 2020).

We use machine learning (ML) and artificial intelligence (AI) methods on a combination of socio-demographic and social capital data to investigate the importance of local factors in explaining coronavirus health outcomes among Veterans. Given that Veterans are a vulnerable group and exhibit more mental and physical health challenges than non-Veterans, even within the same organization (Schult et al., 2019), community characteristics may play an important role in mediating the effects of the pandemic. For example, Veterans in communities with greater social capital may engage in more preventative health investments, which would bolster their immunity and recovery to viruses.

Using three different estimators—naïve ordinary least squares (OLS), ridge with cross validation (CV), and least absolute shrinkage and selection operator (LASSO) with ridge and CV—we predict coronavirus case and mortality outcomes using data on 122 Veteran Healthcare Systems (HCS). While our OLS specification performs well in-sample, it exhibits weak out-of-sample behavior. Instead, the ridge regression with LASSO for feature selection performs the best with an R-squared of 0.617 (0.471) when we predict coronavirus cases (deaths). We show that socio-demographic features matter more than any standard hospital features, such as its patient satisfaction or the number of services provided. These results are important for at least two reasons. First, we can obtain reasonable model accuracy on such a small sample. Second, we show that socio-demographic features matter even more than hospital features. This suggests that further research and predictive modeling on infectious diseases must incorporate socio-demographic and social capital characteristics if these models are going to be useful for policymakers and clinicians.

This paper contributes to a growing literature about the importance of socio-economic factors for understanding health outcomes for the spread of infectious diseases (Amarasingham et al., 2010; Navathe et al., 2018; Bejan et al., 2018; Makridis et al., 2020). The inclusion of socio-demographic variables at a geographic-level can improve the performance of otherwise standard ML models of the virus, but disaggregating between county and ZIP code does not make much of a difference. However, disaggregating between state and county does make a significant difference. The result is evidence that the community and the resulting healthcare infrastructure is determined at more of a county-level, rather than a ZIP code-level. Even if residential decisions take place at a more granular level, public health interventions may reside at a more aggregate level.

Our paper also contributes directly to an existing and timely research agenda on the effects of COVID-19 and the identification of individuals who are more exposed to it than others. For example, age has emerged as one of the most important comorbidities (Zhou et al., 2020; Richardson et al., 2020). Similarly, Makridis and Wu (2020) show that social capital---that is, measures that describe the quality and strength of ties and relationships within a community---plays an important mediating role over the duration of the pandemic: counties with higher social capital have systematically lower infections and a slower spread of infection even after controlling for demographic characteristics and population density.

2. Data and Measurement

2.1. *Location-specific demographic characteristics*

Our socio-demographic data comes from the Census Bureau's 2014-2018 American Community Survey. The Census provides demographic characteristics, e.g.: the race distribution, the population density, the share male, the age distribution (the share under age 18, age 25-44, age 45-64, and 65+), the share married, the education distribution (the share with less than a high school degree, some college, and college or more), the income distribution (the share with less than $15,000, $15-29,000, $30-39,000, $40-49,000, $50-59,000, $60-99,000, $100-149,000, over $150,000), and the poverty rate (the share of people living in poverty under age 18, age 18-64, and 65+).[a]

2.2. *Hospital coronavirus cases, mortality, and features*

We use the Department of Veterans Affairs (VA) Facilities API.[b] We observe the number of services that a hospital provides for Veterans and the average satisfaction. We also observe the logged number of coronavirus cases and deaths within an HCS.[c] While we observe 1,297 VA health facilities, our coronavirus cases and deaths are available at only 122 HCS, which consist of multiple VA medical facilities. We map VA health facilities into an HCS by taking the weighted average of features in each health facility in an HCS using the number of Veterans in the area as our weight.

3. Methods

We use three standard statistical estimators: naïve ordinary least squares (OLS), ridge with cross validation (CV), and least absolute shrinkage and selection operator (LASSO) with ridge CV. A ridge estimator is given by the following:

$$\hat{\beta}^{RIDGE} = \arg\min\{(y - X\beta)^2 + \lambda||\beta||^2\} \quad (1)$$

This, unlike a standard OLS estimator, which only minimizes the sum of square error, inserts an additional term, λ, that biases certain features over others in the regression. While this will lead to "biased" parameter estimates, the fit is better because more important features are given additional weight. Moreover, the regularization term, λ, prevents overfitting since the model cannot adjust too

[a] We also include county data from the Joint Economic Committee (JEC) social capital index (JEC, 2018).
[b] See: https://developer.va.gov/explore/facilities/docs/facilities.
[c] See: https://www.accesstocare.va.gov/Healthcare/COVID19NationalSummary.

many feature weights without the $||\beta||^2$ term getting too big. We also experiment with a LASSO estimator for feature selection followed by ridge.

4. Results

We begin by reporting the performance of our ridge regression with CV and ridge regression with LASSO feature selection and CV in Table 1. We omit the performance of our OLS regression: since we did not do CV on it, the in-sample fit is artificially high because of the common problem of overfitting. Note that our measure of model performance is R-squared, rather than the more common Regression Receiver Operating Characteristic (RROC) curve plot that is more common for predicting continuous variables (Hernandez-Orallo, 2013). We find that the models for predicting mortality perform worse than for infections. One reason for this stems from the fact that deaths are relatively infrequent, so there is less variation available for prediction.

We also observe that the combined LASSO and ridge CV regression performs better than a standard ridge CV regression regardless of the performance metric that we focus on. For example, Panel A shows that the R-squared for LASSO and ridge is 0.617 and it is 0.564 for ridge both when the outcome variable is logged coronavirus cases. The R-squared values for logged deaths for our two respective models are 0.471 and 0.401. Turning towards Panel B, the RMSE for the LASSO and ridge CV regression is 6.6% (7.1%) lower when predicting cases (deaths). Note that since we are predicting $\log_2(cases)$, being 0.9 off, for example, translates to being a factor of $2^{0.9} = 1.86$ off. We find similar patterns in Panel C, which shows the MAE by model.

These differences between the two estimators emerge because of at least two reasons: (i) our sample of HCS is small ($N = 122$), (ii) and we have a large and multi-dimensional feature set, especially for demographic characteristics. Although ridge regressions allow features to contribute differently to the RMSE, it may still not perform optimally when there are nearly as many variables as there are observations. By applying LASSO, we can select only the most predictive variables and include them in a subsequent ridge regression. This performs the best.

Table 1: COVID-19 Model Performances

Outcome Variable =	log(Coronavirus Cases)	log(Coronavirus Deaths)
Panel A: Coefficient of Determination (R-squared)		
Ridge CV	0.564	0.401
LASSO + Ridge CV	0.617	0.471
Panel B: Root Mean Square Deviation (RMSE)		
Ridge CV	0.958	1.115
LASSO + Ridge CV	0.895	1.035
Panel C: Mean Absolute Error (MAE)		
Ridge CV	0.770	0.903
LASSO + Ridge CV	0.707	0.838

Given that the LASSO and ridge CV regression performs the best, we now treat this model as the baseline and examine the most important features for the coronavirus cases and deaths prediction problems in Figures 1 and 2, respectively. Note that all variable importance coefficients are measured in absolute value so that we can focus on the relative magnitudes.

We find that the share of the population between ages 55 and 64 is the most predictive, followed by the share of the population working in professional services and in sales occupations. The poverty rate for those over the age of 65, the male unemployment rate, the employment share in construction, agriculture / mining, and the employment share in education / health all enter as important predictors too. We find similar results when our outcome variable is coronavirus deaths, but several notable differences emerge. While certain variables, like dropping out of high school, were highly associated with deaths, they were not with cases (see Figure 1 and 2). There may be hidden variables associated with this, and other social capital variables, that lead to such worse health outcomes for such individuals.

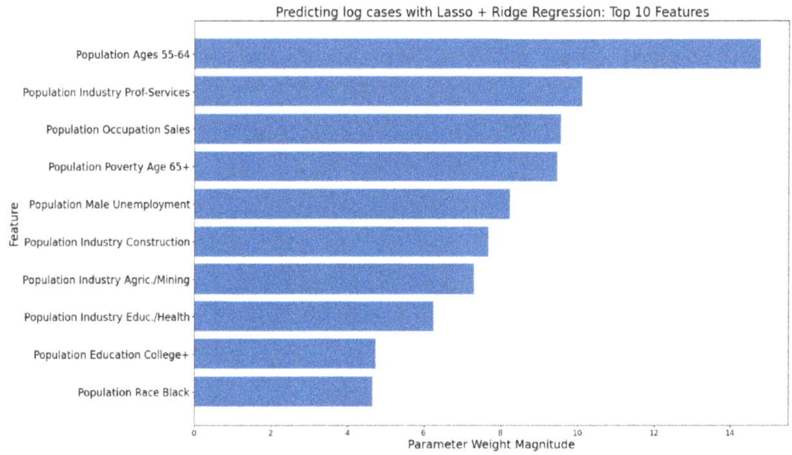

Fig. 1. Predicting log cases with Lasso + Ridge Regression: Top 10 Features.

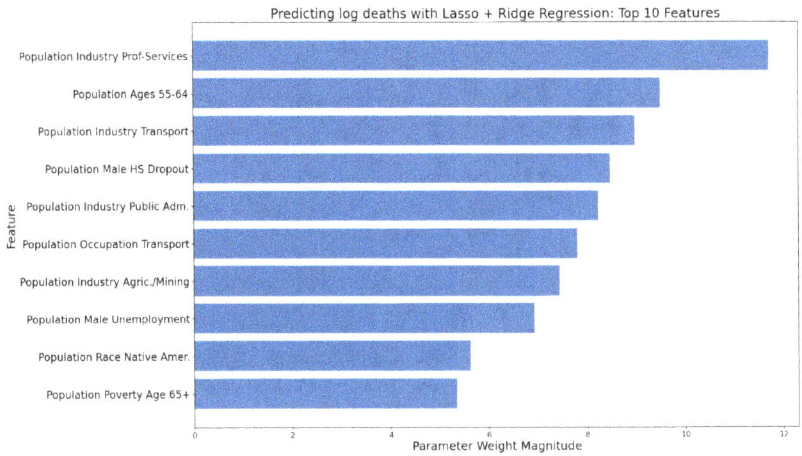

Fig. 2. Predicting log deaths with Lasso + Ridge Regression: Top 10 Features.

5. Conclusion

While there are now many predictive models aimed at understanding the role that co-morbidities play in explaining coronavirus outcomes, there is little research that explores the outcomes among Veterans and the specific role that social capital and other socio-economic local factors play as mediating forces. We estimate three predictive models for coronavirus cases and deaths at a healthcare system (HCS) level, aggregating across 1,297 VA medical facilities. We find that the combined LASSO and ridge with CV regression performs the best. Importantly, while HCS characteristics matter, socio-demographic characteristics also matter greatly and are more important than any of the hospital features. This suggests that public health interventions, especially towards vulnerable groups, must account for the role of an individual's environment and surrounding.

References

1. Amarasingham, R., Moore, B. J., Tabak, Y. P., Drazner, M. H., Clark, C. A., Zhang, S., Reed, W. G., Swanson, T. S., Ma, Y., and Halm, E. A. (2010). An Automated Model to Identify Heart Failure Patients at Risk for 30-day Readmission or Death Using Electronic Medical Record Data. Medical Care, 48(11):981–988.
2. Barrios, J. M., Benmelech, E., Hochberg, Y. V., and Zingales, L. (2020). Civic capital and social distancing during the covid-19 pandemic. NBER working paper.
3. Bejan, C. A., Angiolillo, J., Conway, D., Nash, R., Shirey-Rice, J. K., Lipworth, L., Cronin, R. M., Pulley, J., Kripalani, S., Barkin, S., Johnson, K. B., and Denny, J. C. (2018). Mining 100 million notes to find homelessness and adverse childhood experiences: 2 case studies of rare and severe social determinants of health in electronic health records. Journal of the American Medical Informatics Association, 25(1):61–71.
4. Britton, T., Ball, F., and Trapman, P. (2020). A mathematical model reveals the influence of population heterogeneity on herd immunity to SARS-CoV-2. Science, 23.
5. Ding, W., Levine, R., Lin, C., and Xie, W. (2020). Social Distancing and Social Capital: Why U.S. Counties Respond Differently to COVID-19. NBER working paper.
6. Dingel, J. I. and Neiman, B. (2020). How many jobs can be done at home. Journal of Public Economics.
7. Gordeev, V. S., and Egan, M. (2015). Social cohesion, neighbourhood resilience, and health: evidence from New Deal for Communities programme. Lanclet, 386(2): S39.
8. Hernandez-Orallo, J. (2013). ROC curves for regression. Pattern Recognition, 46.
9. Joint Economic Committee (JEC). (2018). "The geography of social capital in America." SCP Report No. 1-18.
10. Kolak, M., Bhatt, J., Park, Y. H., Padron, N. A., and Molefe, A. (2020). Quantification of Neighborhood-Level Social Determinants of Health in the Continental United States. Journal of the American Medical Association Network Open, 3(1).
11. Makridis, C. A. and Wu, C. (2020). Ties that Bind (and Social Distance): How Social Capital Helps Communities Weather the COVID-19 Pandemic. SSRN working paper.
12. Makridis, C. A., Zhao, D., Bejan, A. C., and Alterovitz, G. (2020b). Leveraging Machine Learning to Characterize the Role of Socio-economic Determinants of Physical Health and Well-being Among Veterans. SSRN working paper.
13. Martin, C. A., Jenkins, D. R., Minhas, J. S., Gray, L. J., Tang, J., Williams, C., Sze, S., Pan, D., Jones, W., Verma, R., Knapp, S., Major, R., Davies, M., Brunskill, N., Wiselka, M., Brightling, C., Khunti, K., Haldar, P., and Pareek, M. (2020). Socio-demographic heterogeneity in the prevalence of COVID-19 during lockdown is associated with ethnicity and household size: Results from an observational cohort study. The Lanclet.
14. Navathe, A. S., Zhong, F., Lei, V. J., Chang, F. Y., Sordo, M., Topaz, M., Navathe, S. B., Rocha, R. A., and Zhou, L. (2018). Hospital readmission and social risk factors identified from physician notes. Health Services Research, 53(2):1110–1136.

15. Pan, D., Sze, S., Minhas, J. S., Bangash, M. N., Pareek, N., Divall, P., Williams, C. M., Oggioni, M. R., Squire, I. B., Nellums, L. B., Hanif, W., Khunti, K., and Pareek, M. (2020). The impact of ethnicity on clinical outcomes in COVID-19: A systematic review. Lanclet, 23(100404).
16. Richardson, S., Hirsch, J. S., Narasimhan, M., Crawford, J. M., McGinn, T., Davidson, K. W., and Northwell, C.-. R. C. (2020). Presenting Characteristics, Comorbidities, and Outcomes Among 5700 Patients Hospitalized With COVID-19 in the New York City Area. Journal of the American Medical Association, 323(20):2052–2059.
17. Schult, T. M., Schmunk, S. K., Marzolf, J. R., and Mohr, D. R. (2019). The Health Status of Veteran Employees Compared to Civilian Employees in Veterans Health Administration. Military Medicine, 184(7-8): e218–e224.
18. Wynants, L., Van Calster, B., Collins, G. S., Riley, R. D., Heinze, G., Schuit, E., Bonten, M. M. J., Damen, J. A. A., Debray, T. P. A., De Vos, M., Dhiman, P., Haller, M. C., Harhay, M. O., Henckaerts, L., Kreuzberger, N., Lohmann, A., Luijken, K., Ma, J., Navarro, C. L., Reitsma, J. B., Sergeant, J. C., Shi, C., Skoetz, N., Smits, L. J. M., Snell, K. I. E., Sperrin, M., Spijker, R., Steyerberg, E. W., Takada, T., van Kuijk, S. M. J., van Royen, F. S., Wallisch, C., Hooft, L., Moons, K. G. M., and van Smeden, M. (2020). Prediction models for diagnosis and prognosis of covid-19: Systematic review and critical appraisal. British Medical Journal, 369.
19. Zhou, F., Du, R., Fan, G., Liu, Y., Liu, Z., Xiang, J., Wang, Y., Song, B., Gu, X., Guan, L., Wei, Y., Li, H., Wu, X., Xu, J., Tu, S., Zhang, Y., Chen, H., and Cao, B. (2020). Clinical course and risk factors for mortality of adult inpatients with COVID-19 in Wuhan, China: a retrospective cohort study. The Lanclet, 395(10299):1054–1062.

Bioinformatics of corals: Investigating heterogeneous omics data from coral holobionts for insight into reef health and resillience

Lenore J. Cowen
Department of Computer Science, Tufts University,
Medford, MA 02155, USA
E-mail: cowen@cs,tufts,edu

Judith Klein-Seetharaman
College of Applied Science and Engineering, Colorado School of Mines,
Golden, CO 80401
E-mail: judithklein@mines.edu

Hollie Putnam
Department of Biological Sciences, University of Rhode Island,
Kingston, RI 02881, USA
E-mail: hputnam@uri.edu

Coral reefs are home to over 2 million species and provide habitat for roughly 25% of all marine animals, but they are being severely threatened by pollution and climate change. A large amount of genomic, transcriptomic and other -omics data from different species of reef building corals, the uni-cellular dinoflagellates, plus the coral microbiome (where corals have possibly the most complex microbiome yet discovered, consisting of over 20,000 different species), is becoming increasingly available for corals. This new data present an opportunity for bioinformatics researchers and computational biologists to contribute to a timely, compelling, and urgent investigation of critical factors that influence reef health and resilience. This paper summarizes the content of the Bioinformatics of Corals workshop, that is being held as part of PSB 2021. It is particularly relevant for this workshop to occur at PSB, given the abundance of and reliance on coral reefs in Hawaiʻi and the conference's traditional association with the region.

Keywords: coral reefs, coral holobiont, non-model organisms, functional genomics, genotype to phenotype, genome and environment, workshop.

1. Introduction, Background and Motivation

Corals are important natural resources that are key to the oceans' vast biodiversity and provide economic, cultural, and scientific benefits. Coral colonies are comprised of clonal cnidarian polyps that depend on a symbiotic relationship with algae in the family Symbiodiniaceae.[1] The dinoflagellate algae harvest light and synthesize nutrients in exchange for shelter and nitrogen sources.[2] Coral reefs cover only 0.1% of the ocean floor, but are home to the largest density of animals on earth, rivaling rain forest habitats in species diversity.[3] The symbiosis, which was originally thought to primarily include endosymbiotic algae, is now known to extend to a much more complex community than anticipated with thousands of bacteria, bacteriophages, viruses and fungi, in addition to Symbiodiniaceae.[4,5] Thus, corals are more like cities than individual animals, as they provide factories, housing, restaurants, nurseries, and more for an entire ecosystem, both at the micro and macro levels. The entirety of the organism community in a coral is referred to as a *holobiont*.

The environmental sensitivity and symbiotic biological complexity of corals makes understanding the genomic variability that influences vulnerability and resilience of local coral reef systems very challenging.[2] However, improving this understanding has taken on increasing urgency, as coral reefs are declining rapidly due to the consequences of climate change. For example, mass coral bleaching, or the expulsion of the symbiotic algae due primarily to thermal stress driven by marine heatwaves, is resulting in substantial coral mortality.[6] Fortunately, a large amount of genomic, transcriptomic and other omics data from different species of reefbuilding corals (e.g.,[7]), the uni-cellular dinoflagellates,[8] and the highly diverse coral microbiome,[9] is becoming increasingly available for corals.[10–14] This is a terrific opportunity for bioinformatics researchers and computational biologists to contribute to a timely, compelling and urgent investigation of critical factors that influence reef health and resilience.[15]

We have recruited some of the premier experts who are working on bioinformatics of coral reefs to participate in our workshop already. We will introduce this exciting topic to the PSB community, with the goal of energizing collaborations and approaches to address the compelling problems in this captivating and complex system. It is particularly relevant for this session to occur in Hawai'i given the abundance of and reliance on coral reefs in the region. Coral genomes from this location show some of the highest complexity to date,[16] exemplifying the bioinformatic challenges faced by the field in the study of the coral metaorganism. This convergence of complex multi-organism data and critical need to address this globally declining ecosystem provides a timely and impactful topic for a Workshop at PSB 2021.

2. Workshop Presenters

The workshop consists of four invited presentations, and then short contributed talks. The invited speakers are:

- Christian Voolstra, Ph.D. (Professor of Genetics of Adaptation in Aquatic Systems, Department of Biology, University of Konstanz, Germany)
- Ross Cunning, Ph.D. (Research Scientist, John G. Shedd Aquarium, Chicago, USA)
- Zachary Fuller, Ph.D. (Postdoctoral Fellow, Dept. of Biological Sciences, Columbia University, USA)
- Cheong Xin (CX) Chan, Ph.D. (Senior Research Fellow, Australian Centre for Ecogenomics, School of Chemistry and Molecular Biosciences, The University of Queensland, Australia)

3. Invited Presenters' Abstracts

The abstracts of the invited presentations appear below.

The metaorganism frontier - we are not alone

Christian Voolstra (University of Konstanz, Germany)

Recent years have brought a changing imperative in life sciences sparked by the revolution of genomic tools to study the molecular composition and functional organization of organisms. The development of next-generation sequencing changed our understanding of microbial diversity associated with organisms and environments. There are now a multitude of studies that support the notion that a host-specific microbiome associates with multicellular organisms and provides functions related to metabolism, immunity, and environmental adaptation, among others. Consequently, interactions and communication mechanisms of members in this metaorganism presumably play a major role in maintaining host health, organismal homeostasis, and resilience to environmental disturbance. The seminar will highlight and discuss recent efforts to investigate coral metaorganism function and evolution using a suite of ecological, physiological, and molecular approaches.

Genotype by genotype by environment interactions in the conservation of reef corals

Ross Cunning (Shedd Aquarium, Chicago, USA)

Current conservation goals for reef-building corals under climate change involve boosting desirable traits like heat tolerance and fast growth in natural and restored coral populations. This may be accomplished through a number of interventions including symbiotic manipulation, selective propagation and breeding, and assisted gene flow. However, the success of these interventions depends on understanding how the desired traits are controlled by the genes of the coral host, its algal symbionts, and the environment (i.e., genotype by genotype by environment interactions). Here I will describe research aimed at characterizing these interactions in the growth and thermal tolerance phenotypes of several Caribbean coral species through both laboratory and field approaches. In experimental manipulations of algal symbionts in the coral *Montastraea cavernosa*, different symbiont taxa modulated host gene expression, contributing to differences in thermal tolerance. In the endangered staghorn coral *Acropora cervicornis*, variability in thermal tolerance was linked to specific alleles for coral genes associated with the heat stress response. In the field, large-scale reciprocal transplant experiments in partnership with reef restoration practitioners are also revealing genotype by environment interactions, which, along with new technologies to quantify thermal tolerance, are being used to identify high-performing and resilient individuals across whole managed coral populations. This phenotypic catalog, combined with whole genome sequencing and analysis, will help determine the genomic basis of key performance traits, and guide effective intervention strategies for coral conservation under climate change.

Genome-wide association study (GWAS) of bleaching tolerance in a Great Barrier Reef coral

Zachary Fuller (Columbia University)

Although reef-building corals are rapidly declining worldwide, there is considerable variation in bleaching response and heat tolerance within populations, which is in part heritable. To map the genetic basis of this variation and develop individual predictors of bleaching in the wild, we conducted a genome-wide association study (GWAS) of bleaching in *Acropora millepora* from the Great Barrier Reef. We first generated a chromosome-scale genome assembly and obtained whole genome sequences for over 200 phenotyped samples collected at 12 reefs, across which we found little population structure. We show that we can reliably impute genotypes in low-coverage sequencing data with a modestly sized reference haplotype panel to obtain millions of high confidence single nucleotide polymorphism (SNP) calls. Testing 6.8 million SNPs for association with bleaching, we show that no single variant reaches genome-wide significance. However, we show a polygenic score constructed from the GWAS estimates is a significant predictor of bleaching. We then demonstrate the feasibility of such an approach by scaling up our GWAS to an increased sample size of more than 1000 whole-genome sequenced and phenotyped individuals. These results thus set the stage for the use of genomic-based prediction in coral conservation strategies.

Understanding genome evolution of coral symbionts

Cheong Xin Chan (University of Queensland, Australia)

The ecological success of corals in nutrient-poor waters relies on photosynthetic algal symbionts (Symbiodiniaceae) for supply of fixed carbon as energy, and nutrients. The evolution of these algae and its implications on coral evolution remains little known. Genomes of Symbiodiniaceae present a bioinformatics challenge, because of their large sizes (1-5 Gbp) and highly idiosyncratic features. In this talk, I will present our recent effort to generate *de novo* genome assemblies from diverse Symbiodiniaceae species and their free-living relative, and to develop a customised computational workflow for predicting genes from these genomes. Comparative analysis reveals high sequence and structural divergence, and conserved lineage-specific gene families of unknown function. I will also present the use of an alignment-free approach to capture comprehensive phylogenetic signal from these whole-genome sequences. Our results highlight the rapid evolution of coral symbionts that comprise an extensive phylogenetic diversity, and elucidate how selection acts within the context of a complex genome structure to facilitate local adaptation. These outcomes provide an important reference for research of coral holobionts and their resilience in changing environments.

Acknowledgments

The organizers thank the US National Science Foundation, HDR grants 1939249, 1939263, 1939699, 1939795, and 1940169 for suppporting us in the organization of this workshop.

References

1. T. C. LaJeunesse, J. E. Parkinson, P. W. Gabrielson, H. J. Jeong, J. D. Reimer, C. R. Voolstra and S. R. Santos, Systematic revision of Symbiodiniaceae highlights the antiquity and diversity of coral endosymbionts, *Current Biology* **28**, 2570 (2018).
2. H. M. Putnam, K. L. Barott, T. D. Ainsworth and R. D. Gates, The vulnerability and resilience of reef-building corals, *Current Biology* **27**, R528 (2017).
3. O. Hoegh-Guldberg, E. S. Poloczanska, W. Skirving and S. Dove, Coral reef ecosystems under climate change and ocean acidification, *Frontiers in Marine Science* **4**, p. 158 (2017).
4. L. L. Blackall, B. Wilson and M. J. van Oppen, Coral—the world's most diverse symbiotic ecosystem, *Molecular Ecology* **24**, 5330 (2015).

5. C. M. Dunphy, T. C. Gouhier, N. D. Chu and S. V. Vollmer, Structure and stability of the coral microbiome in space and time, *Scientific reports* **9**, p. 6785 (2019).
6. T. P. Hughes, K. D. Anderson, S. R. Connolly, S. F. Heron, J. T. Kerry, J. M. Lough, A. H. Baird, J. K. Baum, M. L. Berumen, T. C. Bridge *et al.*, Spatial and temporal patterns of mass bleaching of corals in the Anthropocene, *Science* **359**, 80 (2018).
7. C. Shinzato, K. Khalturin, J. Inoue, Y. Zayasu, M. Kanda, M. Kawamitsu, Y. Yoshioka, H. Yamashita, G. Suzuki and N. Satoh, Eighteen coral genomes reveal the evolutionary origin of Acropora strategies to accommodate environmental changes, *Molecular Biology and Evolution* (2020).
8. A. R. Mohamed, C. X. Chan, M. A. Ragan, J. Zhang, I. Cooke, E. E. Ball and D. J. Miller, Close relationship between coral-associated chromera strains despite major differences within the Symbiodiniaceae, *bioRxiv* , p. 825992 (2019).
9. A. Hernandez-Agreda, W. Leggat, P. Bongaerts, C. Herrera and T. D. Ainsworth, Rethinking the coral microbiome: simplicity exists within a diverse microbial biosphere, *MBio* **9**, e00812 (2018).
10. E. Meyer and V. M. Weis, Study of cnidarian-algal symbiosis in the "omics" age, *The Biological Bulletin* **223**, 44 (2012).
11. S. J. Robbins, C. M. Singleton, C. X. Chan, L. F. Messer, A. U. Geers, H. Ying, A. Baker, S. C. Bell, K. M. Morrow, M. A. Ragan *et al.*, A genomic view of the reef-building coral Porites lutea and its microbial symbionts, *Nature Microbiology* **4**, 2090 (2019).
12. R. Cunning, R. Bay, P. Gillette, A. C. Baker and N. Traylor-Knowles, Comparative analysis of the Pocillopora damicornis genome highlights role of immune system in coral evolution, *Scientific Reports* **8**, 1 (2018).
13. S. Planes, D. Allemand, S. Agostini, B. Banaigs, E. Boissin, E. Boss, G. Bourdin, C. Bowler, E. Douville, J. M. Flores *et al.*, The Tara Pacific expedition—A pan-ecosystemic approach of the "-omics" complexity of coral reef holobionts across the Pacific Ocean, *PLoS biology* **17**, p. e3000483 (2019).
14. Z. L. Fuller, V. J. Mocellin, L. A. Morris, N. Cantin, J. Shepherd, L. Sarre, J. Peng, Y. Liao, J. Pickrell, P. Andolfatto *et al.*, Population genetics of the coral Acropora millepora: Toward genomic prediction of bleaching, *Science* **369** (2020).
15. P. A. Cleves, A. Shumaker, J. Lee, H. M. Putnam and D. Bhattacharya, Unknown to known: Advancing knowledge of coral gene function, *Trends in Genetics* **36**, 93 (2020).
16. A. Shumaker, H. M. Putnam, H. Qiu, D. C. Price, E. Zelzion, A. Harel, N. E. Wagner, R. D. Gates, H. S. Yoon and D. Bhattacharya, Genome analysis of the rice coral Montipora capitata, *Scientific reports* **9**, 1 (2019).

Establishing the reliability of algorithms

Lara Mangravite
Sage Bionetworks
Seattle, WA, USA
Email: lara.mangravite@sagebionetworks.org

Sean D. Mooney
Department of Biomedical Informatics and Medical Education, University of Washington
Seattle, WA, USA
Email: sdmooney@uw.edu

Iddo Friedberg
Bioinformatics and Computational Biology Program, Department of Veterinary Microbiology and Preventive Medicine, Iowa State University
Ames, IA, USA

Justin Guinney
Sage Bionetworks
Seattle, WA, USA
Email: justin.guinney@sagebionetworks.org
idoerg@gmail.com

As rich biomedical data streams are accumulating across people and time, they provide a powerful opportunity to address limitations in our existing scientific knowledge and to overcome operational challenges in healthcare and life sciences. Yet the relative weighting of insights *vs.* methodologies in our current research ecosystem tends to skew the computational community away from algorithm evaluation and operationalization, resulting in a well-reported trend towards the proliferation of scientific outcomes of unknown reliability. Algorithm selection and use is hindered by several problems that persist across our field. One is the impact of the self-assessment bias, which can lead to mis-representations in the accuracy of research results. A second challenge is the impact of data context on algorithm performance. Biology and medicine are dynamic and heterogeneous. Data is collected under varying conditions. For algorithms, this means that performance is not universal -- and need to be evaluated across a range of contexts. These issues are increasingly difficult as algorithms are trained and used on data collected in the real-world, outside of the traditional clinical research lab. In these cases, data collection is neither supervised nor well controlled and data access may be limited by privacy or proprietary reasons. Therefore, there is a risk that algorithms will be applied to data that are outside of the scope of the intent of the original training data provided. This workshop will focus on approaches that are emerging across the researcher community to quantify the accuracy of algorithms and the reliability of their outputs.

Keywords: benchmarking; algorithm assessment; open science; translational research

1. Introduction

Despite intensive efforts to utilize this data to optimize healthcare, relatively few methods have been adequately validated and clinically deployed. The reasons for this are technical, scientific, social and business related. On the technical side this includes inaccessibility of gold-standard datasets for robust validation, heterogeneity in data collected from distributed sources, contextual relevance of biological observations across samples, poor algorithmic reproducibility and community-acceptance of biased approaches for assessing methods. Reproducibility and transparency are two methods which support development of reliable biomedical claims that can both generate new knowledge and apply it to advance health care. Although these approaches have become firmly established and increasingly practiced over the past decade, they do not fully address the question of transferability in biomedical research findings or algorithms. This topic builds from the types of work described in the PSB 2017 Session on Methods to Ensure Reproducibility in Biomedical Research, which was developed in reaction to both the announcement of the data sharing initiatives of the Biden Cancer program and the NEJM data parasite commentary, focused on methods that individual researchers were taking to assure reproducibility within their own work. This session will discuss general methods for open community-based methods to benchmark algorithms, including the use of crowd-sourced challenges[1-3] as a tool for the unbiased assessment of tools and algorithms.

The public health, economic, and social justice crises that have occurred in 2020 have brought an urgency to the question of rapid, reliable algorithm assessments. The global COVID-19 pandemic has provided an urgent need to rapidly optimize healthcare practices, establish public health practices for prevention and monitoring, and identify drugs and vaccines to use in prevention and treatment. The urgency of this situation is at odds with the typical pace through which scientific knowledge is developed, established and integrated into care. Further, the social justice crisis underlies the known issues with medical algorithms that initiate biases or may propagate those established in the underlying data.

2. Workshop goals and organization

This workshop, **Establishing Reliability in Algorithms**, at the 2021 Pacific Symposium on Biocomputing is designed to stimulate conversation around mechanisms that our community can use to objectively establish the reliability of algorithms. This will include community mechanisms for evaluation as well as mechanisms for use by individual researchers within the context of independent research programs. The workshop will provide three examples of existing approaches and then stimulate an open discussion that will be actively guided and moderated by the organizers. The conversation should extend to a discussion of potential mechanisms for establishing standards that enforce greater accountability across the community.

Topics to be covered in the presented materials will include:

Predictive analytics in healthcare: The COVID-19 pandemic has highlighted an urgent need for healthcare systems to learn from and with each other. Clinical analytics teams are implementing predictive analytics methods that use algorithms trained on electronic health record (EHR) and other data to improve patient care and lower costs. While these methods have the promise of being impactful in delivering on precision medicine and managing population health, their real world accuracy over time is not well understood[4-7]. It is the case in most areas of biomedicine that the evaluation of methods across multiple data sets should be transparent and used to establish their replicability and reliability. Due to differences across clinical sites in practice, population, and data capture, the question of reliability may be less evident and requires an understanding of the context - and potential impact - of deploying an algorithm within a particular system.

Regulatory Science: Another area where analytical methods are directly impacting health care is in support of regulatory filings for new drugs and devices. Data derived from both EHR and from remote monitoring devices are increasingly utilized in this capacity. Recognizing the need to objectively assess the accuracy of methodologies used in the development of regulatory filings, the FDA introduced PrecisionFDA[8], an objective benchmarking program in 2015, which has built from an original focus on genomic processing methods. The proprietary nature of this work introduces barriers to data collection or sharing that make traditional approaches to algorithm assessment unsatisfying. Approaches that can support objective evaluation of results arising from closed data sources are required. Acknowledgement of these needs are represented from the FDA by their Spring 2020 solicitation for community input towards the modernization of their data strategy[9].

Molecular Modeling and Analytics: Biomedical researchers are routinely generating genomic, proteomic, epigenomic, imaging, and other emerging molecular data types comprising billions of data-points. Community benchmarking approaches such as the DREAM Challenges or the

Critical Assessment experiments have predominantly focused in this domain[10-14], where fewer commercial interests impact the sharing of data or knowledge. An evaluation of benchmarking practices within this domain can help to identify lessons from the successes, current gaps in practice, and the development of sustained standards for community-based algorithm assessment.

A moderated discussion will follow that will cover the following topics:

- successes and lessons learned to-date from community benchmarking practices - indicating what impact these approaches to establish reliable outcomes have had on subsequent research or translation practices
- early lessons learned from healthcare method implementation
- emerging approaches in data sharing and in algorithm development and assessment that are addressing the issue of appropriate algorithm interpretation
- community needs and potential solutions for addressing algorithm reliability
- Development of better gold standards in biomedicine and approaches to overcome sub-optimal gold standards

3. References:

1. Ellrott, K. et al. Reproducible biomedical benchmarking in the cloud: lessons from crowd-sourced data challenges. Genome Biol. 20, 195 (2019).
2. Bender, E. Challenges: Crowdsourced solutions. Nature 533, S62–4 (2016).
3. Saez-Rodriguez, J. et al. Crowdsourcing biomedical research: leveraging communities as innovation engines. Nat. Rev. Genet. 17, 470–486 (2016).
4. EHR DREAM Challenge. synapse.org/ehr_dream_challenge_mortality doi:10.7303/syn18405991.
5. Kahn, M. G. et al. A Harmonized Data Quality Assessment Terminology and Framework for the Secondary Use of Electronic Health Record Data. EGEMS (Wash DC) 4, 1244 (2016).
6. Beaulieu-Jones, B. K. et al. Privacy-Preserving Generative Deep Neural Networks Support Clinical Data Sharing. Circ. Cardiovasc. Qual. Outcomes 12, e005122 (2019).
7. Chen, J., Chun, D., Patel, M., Chiang, E. & James, J. The validity of synthetic clinical data: a validation study of a leading synthetic data generator (Synthea) using clinical quality measures. BMC Med. Inform. Decis. Mak. 19, 44 (2019).
8. https://precision.fda.gov/
9. Modernizing the Food and Drug Administration's Data Strategy; Public Meeting; Request for Comments. https://www.federalregister.gov/documents/2020/01/08/2020-00071/modernizing-the-food-and-drug-administrations-data-strategy-public-meeting-request-for-comments

10. Marbach D. et. Al. Wisdom of crowds for robust gene network inference. Nat Methods. 2012 Aug; 9(8): 796–804.
11. Trister, A. D., Buist, D. S. M. & Lee, C. I. Will Machine Learning Tip the Balance in Breast Cancer Screening? JAMA Oncol (2017) doi:10.1001/jamaoncol.2017.0473.
12. Keller, A. et al. Predicting human olfactory perception from chemical features of odor molecules. Science 355, 820–826 (2017).
13. Radivojac, P., Clark, W., Oron, T. *et al.* A large-scale evaluation of computational protein function prediction. Nat Methods 10, 221–227 (2013).
14. Zhou N, Siegel ZD, Zarecor S, Lee N, Campbell DA, et al. (2018) Crowdsourcing image analysis for plant phenomics to generate ground truth data for machine learning. PLOS Computational Biology 14(7): e1006337

Making Tools that People Will Use: User-Centered Design in Computational Biology Research

Mary Goldman
UC Santa Cruz Genomics Institute,
University of California, Santa Cruz, 1156 High Street,
Santa Cruz, CA 95064, USA
Email: mary@soe.ucsc.edu

Nils Gehlenborg
Biomedical Informatics, Harvard Medical School, 10 Shattuck Street,
Boston, MA 02115, USA
Email: nils@hms.harvard.edu

User-Centered Design (UCD) focuses on deeply understanding the needs of users and ensuring these needs are met by tools and software. UCD methodology aims to make tools easier to use, reduce time spent in development and the need for user support, as well as make it easier to create and maintain documentation. The goal of UCD is to ultimately make a tool that meets user needs and is a pleasure to use. This workshop will give an overview of UCD and several examples of how UCD practices are already being used at several institutions. Attendees will leave with ideas of how to incorporate UCD into their tool development as well as general resources to get started.

Keywords: User Centered Design, User Experience, UX/UI, Usability

1. Introduction, Background, and Motivation

Effective software tools are needed in computational biology to help understand the results of computational analyses, visualize multi-scale datasets and mathematical models, and generate insights[1]. As progress in artificial intelligence in the biomedical space is manifesting itself primarily in the form of augmented intelligence, it is becoming even more important to design approaches that enable efficient interactions between powerful software and human experts. Additionally, users of computational biology tools are diverse, often with particular needs unique to their area of research. Developing tools that address the specialized needs of practitioners, either a few individuals or a larger given field, is commonly known as User-Centered Design (UCD)[2]. The goal of UCD is to create a product that satisfies users' needs, has an interface that is easy-to-use, and, in general, is a tool that people want to use.

User-Centered Design is design based upon an explicit understanding of users, tasks, and environments, and is driven and refined by iterative user-centered evaluation[2]. While UCD increases the number of users and their satisfaction with a tool, UCD also can reduce development costs/time as well as the effort required for user support and documentation[3]. UCD can be applied to tools that are simple or complex, that are made for only a few users or a few thousand users, or that are interacted with via a graphical user interface (GUI), command line interface (CLI), or application programming interface (API).

While UCD has been broadly applied across many industries, it typically has not been applied in computational biology and bioinformatics research, where the focus tends to be more on the optimization of underlying algorithms and the computational core of the software[4-6]. This approach often neglects a tool's usability, in the interface components of the tool as well as in user's workflow both inside and outside the tool.

This workshop will be an overview of how UCD has been successfully applied to computational biology tools and bioinformatics resources. Speakers will discuss how they have incorporated various aspects of this discipline into their tool development, including tips for success and lessons learned. Attendees will leave with an understanding of common UCD practices as well as models of how they might apply them to their own tools.

2. Workshop Presenters

The three-hour workshop will begin with an overview presentation of User-Centered Design, and will followed by four presentations. The workshop will conclude with a panel discussion session, which will be moderated by Nils Gehlenborg and Mary Goldman.

2.1 Workshop Speakers

Ljubomir Bradic (Sage Bionetworks)
Ljubomir is the Director of Design at Sage Bionetworks, where he leads the design of Sage Bionetwork's data sharing and collaboration platform, as well as the Digital Health platform. He specializes in complex problem spaces in environments with fluid requirements, particularly early startups. His deep understanding of the software development life cycle comes from being a startup founder and previous experience as a product manager and developer.

Jeremy Kriegel (Audible, Inc)
Jeremy was previously the UX Lead for the Broad Institute and is currently the Director of User Experience at Audible, Inc. Over the past two decades, he has worked on user experience problems as a consultant and as a part of internal teams at organizations that range from start-ups to Fortune 100 companies.

Zinaida Perova (EMBL-EBI)
Zinaida Perova is a Project Lead at the European Bioinformatics Institute. Her work is aimed to further expand the PDX Finder resource to cover other patient-derived cancer models, such as cancer cell lines and organoids. She has a PhD and postdoctoral experience in Biological Sciences.

Galabina Yordanova (EMBL-EBI)
Galabina is a User Experience Architect at the European Bioinformatics Institute. She is currently working on the Data Submission process for the Human Cell Atlas and on the COVID-19 Data Portal: covid19dataportal.org. She has worked in the field of product management and user experience design for the last 15 years, bringing her expertise to a variety of online products and services.

2.2 Panel Moderators

Nils Gehlenborg (Harvard Medical School)
Nils has over 15 years of experience in applying user-centered design approaches in biomedical data visualization tool development and has played a central role in the establishment of the VIZBI and BioVis meetings.

Mary Goldman (UC Santa Cruz Genomics Institute)
Mary has been working in genomics for ten years, both for the UCSC Genome Browser and UCSC Xena. She began her role in User-Centered Design five years ago and leads the User Centered Design Working Group at the UC Santa Cruz Genomics Institute.

3. Speakers Abstracts

Systems Design Methods for Building Bioinformatics Applications
Ljubomir Bradic
This talk will introduce Systems Design principles for creating long lasting and scalable ecosystems of bioinformatics applications. Based on methods used at Sage Bionetworks, we will demonstrate how commercial software industry design techniques and methodologies have been adapted to deliver software that supports open science initiatives. We will also cover design and organizational best practices for working in resource constrained environments.

User-Centered Design for the Broad Institute
Jeremy Kriegel
User-Centered Design recognizes that all tool development starts with an understanding of the user and a real-world problem they experience. Good design needs an in-depth understanding of users' tasks, motivations, goals, and steps they take, as well as an overall grasp of the context in which they use a tool and what other technologies they rely on. UCD can be applied at any stage of development of a project but is ideally incorporated through the entire process. I will talk about several User-Centered Design practices that can be applied to development of a computational biology tool, from inception to launch and ongoing efforts. Applying these methods helps ensure that a tool fits well into a user's workflow, addresses their needs, and is a pleasure to use.

User centric development of the PDX Finder database

Zinaida Perova

PDX Finder is an open and comprehensive global database of patient derived xenograft models and data. PDX Finder was developed using User-Experience Design methods through iterative collaboration between users and developers from the start thorough the entire life of the project. We worked as a multi-disciplinary team to understand the user's problems, needs and general scope of the project. This talk will lay out the UX methods that were used from conception to development to release, including defining user personas, identifying stakeholders, outlining the user journey, as well as how we used various workshops to refine standards, needs and database design before implementation. Finally, this talk will end with the usability sessions and metrics we developed and measured to ensure that PDX Finder met user needs.

User research - how discovery and evaluation helps guide the development of our data portals

Galabina Yordanova

In the field of bioinformatics software development, there is sometimes a tendency to quickly move on with system architecture specification of what looks like the obvious solution for a tool or a service. It is rare to dedicate time on understanding the workflows and needs of the intended users of those tools or systems. Building knowledge about user needs and getting feedback on whether suggested solutions will help teams build tools and services which help researchers do their work in a more efficient way. Sharing the insights of those findings helps to align multi-disciplinary or international teams, so that everyone is working towards a common outcome.

I will talk about the user research methods we applied for two of our projects - the Human Cell Atlas data platform and the COVID-19 Data portal. I will share our experience and how these user research activities helped align the team, improve our understanding of user needs and guide the development of those portals.

4. Conclusion

This workshop will highlight a number of User-Centered Design methods, strategies, and tools currently being used to help design and create computational bioinformatics tools and resources. By supporting scientists with better tools, that are easy-to-use and fit well within a user's workflow, we enable them to focus on their research and advancing science.

References

1. Kumar, S. and Dudley, J. Bioinformatics software for biologists in the genomics era. *Bioinformatics* **23,** 14 (2007)
2. Norman, D.A. (2013). The design of everyday things. New York: Basic Books.
3. Mangul, S., Martin, L.S., Eskin, E. *et al.* Improving the usability and archival stability of bioinformatics software. *Genome Biology* **20,** 47 (2019).
4. *usability.gov: Improving the User Experience.* 1 October 2020: https://www.usability.gov/

5. List, M., Ebert, P., and Albrecht, F. Ten Simple Rules for Developing Usable Software in Computational Biology. *PLoS Computational Biology* 13, 1 (2017)
6. Bolchini, D., Finkelstein, A., Perrone, V., and Nagl, S. Better bioinformatics through usability analysis. *Bioinformatics* **25,** 3, (2009)

Raising the stakeholders: Improving patient outcomes through interprofessional collaborations in AI for healthcare

Carly A. Bobak
Program in Quantitative Biomedical Sciences, Dartmouth College
Hanover, NH 03755, USA
Email: Carly.A.Bobak.GR@Dartmouth.edu

Marek Svoboda
Program in Quantitative Biomedical Sciences, Dartmouth College
Hanover, NH 03755, USA
Email: Marek.Svoboda.GR@Dartmouth.edu

Kristine A. Giffin
Program in Quantitative Biomedical Sciences, Dartmouth College
Hanover, NH 03755, USA
Email: Kristine.A.Giffin@Dartmouth.edu

Dennis P. Wall
Pediatrics (Systems Medicine) and Biomedical Data Science, Stanford University
Stanford, CA 94305
Email: dpwall@stanford.edu

Jason Moore
Institute for Biomedical Informatics, University of Pennsylvania
Philadelphia, PA 19104
Email: jhmoore@upenn.edu

Research into AI implementations for healthcare continues to boom. However, successfully launching these implementations into healthcare clinics requires the co-operation and collaboration of multiple stakeholders in healthcare including healthcare professionals, administrators, insurers, legislators, advocacy groups, as well as the patients themselves. The co-operation and collaboration of these interprofessional groups is necessary not just in the final stages of launching AI based solutions in healthcare, but along each stage of the research design and analysis. In this workshop, we solicited talks from researchers who have embraced the idea of interprofessional collaboration across many different stakeholder groups at multiple stages of their research. We specifically focus on projects which included heavy collaborations from healthcare professionals, embraced the research subjects' communities as critical research partners, as well as included researchers who are advocating for systemized changes to include interprofessional stakeholders as evaluators of AI research in healthcare.

Keywords: Artificial Intelligence, Socio-technical systems, Interprofessional Collaboration, Translational Science.

1. Introduction

Artificial intelligence (AI) and other computational and bioinformatic approaches have become a critical component of biomedical research. The wealth of available medical data and pertinent research questions have driven experts across many scientific fields to begin developing computational methods to drive innovation in medical research. However, AI in healthcare is often labelled as "disruptive," a word simultaneously embracing its innovative nature, while warning against its turbulent impact on a broad range of health-care related disciplines. As a result, many healthcare stakeholders continue to be reserved, and even outright resistant, to AI advances for clinical outcomes.

Healthcare stakeholders include researchers across a variety of disciplines, clinicians, patients, insurers, legislators, lawyers, economists, UN agencies, government, private and non-profit organizations, to name a few. Reservations regarding AI healthcare research from any stakeholder group creates both hard barriers (restrictive legislation) and soft barriers (aversion to data sharing) in conducting, validating, and implementing AI approaches in the clinic. Ford et al., note "(r)esearchers who work in cultural silos are unlikely to maximize the potential of patient data"[1] and recommend meaningful stakeholder involvement is necessary at every stage of research in order to remove barriers for clinical translation.

However, there is no straightforward strategy for creating meaningful involvement mechanisms across many healthcare stakeholders. In this workshop, we aim to invite talks focusing on AI approaches in biomedical research from diverse and inclusive research teams, with expertise that spans different academic and professional disciplines, or who have collaborated with or studied the perspective of various stakeholders of computational healthcare research. Specifically, talks will emphasize both lessons learned from collaborative research and how the collaboration influenced the design, interpretation and overall positioning of the results, as well as provide advice for how other researchers can engage in their stakeholder community.

2. Effective medical research requires active involvement of medical professionals

Research into AI tools aimed at improving clinical outcomes needs to evaluate not only technical performance, but socio-technical performance outcomes. It is inevitable that the introduction of AI technologies to clinics will cause breaks and necessitate changes to existing systems.[2] Medical professionals are essential to include as active participants in AI biomedical research to design tools that minimize these breaks but also to act as diplomats and repairmen to bring AI to its full medical potential.[2] This socio-technical approach to AI research is exemplified by the 'Sepsis Watch' project led by Dr. Mark Sendak and other researchers at Duke University.[3] One of the critical factors influencing the potential of Sepsis Watch to improve septic patient outcomes was the integration of the tool into existing social and professional dynamics – and active involvement from rapid response team nurses was essential for this to occur.[2] Following observations of Sepsis Watch during its first two years of implementation, these researchers posed four key values necessary for the translation of biocomputing research into medical practice: rigorously defining the problem, building relationships with key stakeholders, respecting professional discretion, and creating an ongoing feedback loop with stakeholders.[3]

3. Study subjects and their communities must also be treated as research partners

Beyond the inclusion of medical professionals in AI research, study subjects themselves are critical collaborators whose experiences and communities influence the ability of AI driven tools to improve clinical outcomes. Dr. Lisa Vizer recently published a qualitative study which investigated the friction points of tracking health indicators of chronic disease. Vizer proposed a Conceptual Model of Shared Health Informatics (CoMSHI) that specifically identifies that tracking tools need to consider the social context of the person with chronic illness, including not only health professionals but also informal carers. They recommend that tools need to be reflective of the shared work of many community members in the tracking and monitoring of chronic illness and need to be designed to easily be used by multiple members in the participants' community as well as the participants themselves.[4]

The Wall Lab at Stanford University has embraced the idea of creating tools aimed at serving various stakeholders of the autism community. They have developed 'SuperpowerGlass,' a product based off Google Glass, as a wearable device for children with Autism Spectrum Disorder (ASD) which in real-time classifies the emotions of their family and peers while also recording interactions for additional insight. As well, they have launched a therapeutic mobile-device game called 'Guess What?' which tests children's abilities to act out and identify emotions while recording their play time as a long-term data source from which behavioral improvement can be measured. A critical aspect of both of these technologies is that insights and data are visible to parents and carers so they can also review and learn from their child's interactions. Moreover, the Wall lab has developed a crowd sourced ASD screening tool using home videos of children which alleviates the long wait times for official ASD diagnoses and allows critical early intervention for behavioral improvement. The Wall lab has also used machine learning algorithms to identify the most important questions used by clinicians in diagnosing ASD so that questionnaires and time-to-diagnosis can be shortened.[5]

Dr. Dan Gillis works as part of a team that is building computational infrastructure for the Inuit community in Rigolet, Nunatsiavut, Canada. A critical aspect of this research is working in partnership with the community to develop Inuit-led monitoring systems to understand and respond to not only classic metrics of climate change, but also to intangible losses that are priorities for the Inuit people. Involving the Inuit perspectives in the design, maintenance, and use of the monitoring system allows them to understand and mitigate the impacts of climate change in their community. Furthermore, they advocate for community-specific priorities in terms of public health and how climate change influences the health of the community considering the perspectives of researchers, public health officials, and the Inuit community itself.[6]

4. Precision medicine has accelerated the need for systems that evaluate the promise of AI research from multiple stakeholder perspectives

AI has become a critical driver of biomarker discovery and precision medicine, but there are few systems in place to evaluate the efficacy and make appropriate recommendations for these discoveries.[6] Dr. John Carethers, in partnership with the National Academies of Sciences, Engineering, and Medicine, co-authored a report proposing a roadmap to address the lack of systems for evaluating precision medicine research. The team interviewed federal regulators, insurers, developers of biomarker tests, medical professionals, and advocacy groups to identify 10 goals for establishing systems for the evaluation of precision medicine research including standardizing patient and provider information, studying different demographic groups, developing evidence-based guidelines for clinical practice, and maintaining a robust database to share findings.[7]

Dr. Amar Das has proposed an interdisciplinary, phased research framework to better evaluate AI tools and applications in healthcare, similar to the multi-phase system used to approve novel drugs. They propose the following phases: discovery and invention, technical performance and safety, efficacy and side effects, therapeutic efficacy, and safety and effectiveness. Critically, at all stages of their research framework user feedback and continuous monitoring is essential in evaluating and updating AI implementations for clinical practice.[8]

5. Conclusion

Given the current state of biocomputing, it is inevitable that AI will be a critical driver of biomedical innovation. However, it is of utmost importance that researchers engage with and secure the trust of healthcare stakeholders to maximize the potential of AI in improving patient outcomes. As Obermeyer & Lee stated, "machine learning in medicine will be a team sport, like medicine itself. But the team will need some new players [...] who can contribute meaningfully to algorithm development and evaluation."[10] It is our hope that this workshop will galvanize computational researchers to engage with stakeholders in meaningful ways and move AI from being "disruptive" to "progressive."

References

1. Ford E, Boyd A, Bowles JK, Havard A, Aldridge RW, Curcin V, Greiver M, Harron K, Katikireddi V, Rodgers SE, Sperrin M. Our data, our society, our health: A vision for inclusive and transparent health data science in the United Kingdom and beyond. Learning health systems. 2019 Jul;3(3):e10191.
2. Sendak, M., Elish, M. C., Gao, M., Futoma, J., Ratliff, W., Nichols, M., ... & O'Brien, C. (2020, January). " The human body is a black box" supporting clinical decision-making with deep learning. In Proceedings of the 2020 Conference on Fairness, Accountability, and Transparency (pp. 99-109).
3. Sendak, M. P., Ratliff, W., Sarro, D., Alderton, E., Futoma, J., Gao, M., ... & Kester, K. (2019). Sepsis Watch: A Real-World Integration of Deep Learning into Routine Clinical Care. JMIR Preprints, 15182.

4. Sawatzky, A., Cunsolo, A., Jones-Bitton, A., Gillis, D., Wood, M., Flowers, C., ... & Harper, S. L. (2020). "The best scientists are the people that's out there": Inuit-led integrated environment and health monitoring to respond to climate change in the Circumpolar North. *Climatic Change*, 1-22.
5. Vizer, L. M., Eschler, J., Koo, B. M., Ralston, J., Pratt, W., & Munson, S. (2019). "It's Not Just Technology, It's People": Constructing a Conceptual Model of Shared Health Informatics for Tracking in Chronic Illness Management. *Journal of medical Internet research, 21*(4), e10830.
6. Washington, P., Park, N., Srivastava, P., Voss, C., Kline, A., Varma, M., ... & Chrisman, B. (2019). Data-driven diagnostics and the potential of mobile artificial intelligence for digital therapeutic phenotyping in computational psychiatry. *Biological Psychiatry: Cognitive Neuroscience and Neuroimaging*.
7. Sawatzky, A., Cunsolo, A., Jones-Bitton, A., Gillis, D., Wood, M., Flowers, C., ... & Harper, S. L. (2020). "The best scientists are the people that's out there": Inuit-led integrated environment and health monitoring to respond to climate change in the Circumpolar North. *Climatic Change*, 1-22.
8. National Academies of Sciences, Engineering, and Medicine. (2016). *Biomarker tests for molecularly targeted therapies: key to unlocking precision medicine*. National Academies Press.
9. Park, Y., Jackson, G. P., Foreman, M. A., Gruen, D., Hu, J., & Das, A. K. (2020). Evaluating artificial intelligence in medicine: phases of clinical research. *JAMIA Open*.
10. Obermeyer Z, Lee TH. Lost in Thought - The Limits of the Human Mind and the Future of Medicine. N Engl J Med. September 28, 2017;377(13):1209-1211. doi:10.1056/NEJMp1705348.

Translational Bioinformatics: Integrating Electronic Health Record and Omics Data

Dokyoon Kim
Department of Biostatistics, Epidemiology, & Informatics, Institute for Biomedical Informatics, The Perelman School of Medicine, University of Pennsylvania, D202 Richards Building, 3700 Hamilton Walk Philadelphia, PA 19104, USA
Email: dokyoon.kim@pennmedicine.upenn.edu

Ju Han Kim
Department of Biomedical Sciences, Seoul National University Graduate School, Biomedical Science Building 117, 103 Daehakro, Jongro-gu, Seoul 110-799, Korea
Email: juhan@snu.ac.kr

Jason H Moore
Department of Biostatistics, Epidemiology, & Informatics, Department of Genetics, and Institute for Biomedical Informatics, The Perelman School of Medicine, University of Pennsylvania, D202 Richards Building, 3700 Hamilton Walk Philadelphia, PA 19104, USA
Email: jhmoore@upenn.edu

Translational bioinformatics (TBI) is focused on the integration of biomedical data science and informatics. This combination is extremely powerful for scientific discovery as well as translation into clinical practice. Several topics where TBI research is at the leading edge are 1) the clinical utility of polygenic risk scores, 2) data integration, and 3) artificial intelligence and machine learning. This perspective discusses these three topics and points to the important elements for driving precision medicine into the future.

Keywords: translational bioinformatics, precision medicine, data integration, artificial intelligence, machine learning, electronic health records, biobank, polygenic risk scores

1. Introduction

Translational bioinformatics (TBI) is a multi-disciplinary and rapidly emerging field of biomedical data sciences and informatics that includes the development of technologies that efficiently translate basic molecular, genetic, cellular, and clinical data into clinical products or health implications. TBI involves applying novel methods to the storage, analysis, and interpretation of a massive volume of genetics, genomics, multi-omics, and clinical data; this includes diagnoses, medications, laboratory measurements, imaging, and clinical notes. TBI bridges the gap between bench research and real-world applications to human health. Many health-related topics are increasingly falling within the scope of TBI, including rare and complex human disease, cancer, biomarkers, pharmacogenomics, drug repositioning, genomic medicine, and clinical decision support systems.

TBI in precision medicine attempts to determine individual solutions based on the genomic, environmental, and clinical profiles of each individual, providing an opportunity to incorporate individual genomic data into patient care. While a plethora of genomic signatures have successfully demonstrated their predictive power, they are merely statistically significant differences between dichotomized phenotypes (for example cases and controls of a specific disease) that are in fact severely heterogeneous phenotypes. Despite many translational barriers, connecting the molecular world to the clinical world and vice versa will undoubtedly benefit human health in the near future.

Due to the rapid pace of TBI, we assembled diverse perspectives to review the state of the art in translation bioinformatics including the clinical utility of polygenic risk scores, data integration, and artificial intelligence in medicine. We provide perspective on where the current efforts are focused and where the future is headed for biobanks in different disciplines, especially about the utility of polygenic risk scores. Additionally, special attention will be given to data integration. In particular, radiogenomics or imaging genomics is one of the primary areas that focus on the relationship between imaging phenotypes and genomics. We also discuss artificial intelligence and machine learning and how these are being used now for integrating electronic health record (EHR) and omics data as well as how we anticipate they will be used in the future. Translational bioinformatics is a fast-moving field and we believe that integrating the basic science community from genomics, bioinformatics, computer science, and statistics together with the translational community including clinical/medical informatics, pharmacogenomics, and genomic medicine will be mutually beneficial to accelerate the translational of biomedical research into precision medicine.

2. The clinical utility of polygenic risk scores

Many research programs have capitalized on these population-based registries with complementary biobanks for research linkage to the health registry including UK Biobank [1], FinnGEN [2], and deCODE [3]. EHRs and national health registries have both been adopted as clinical data sources for genetic and genomic analyses for a wide variety of diseases/conditions. The utility of these clinical data linked with genetic and genomic data has enormous potential for disease gene discovery. Much research is ongoing to identify risk factors for complex disease, evaluate the potential repurposing medications for multiple phenotypes, and the identification of novel therapeutic targets. In particular, the development of polygenic risk scores (PRS) as well as genomic risk assessments, which integrate PRS with known clinical risk factors, are an emerging area of research in large scale biobanks linked with clinical data sources. PRS is a value accumulated based on the effect sizes of multiple genetic variants across the genome and has shown great promise in the prediction of risk for many diseases [4]. Furthermore, recent studies for many diseases suggest that our knowledge of the common variants underlying diseases or phenotypes has improved to a point where polygenic risk profiling provides personal and clinical utility by identifying groups of individuals who could benefit from the knowledge of their probabilistic susceptibility to disease [5]. As more health systems and academic medical centers continue to build large scale biobanks, the opportunities for discovery in biobanks linked to clinical data sources will continue to explode.

3. Data integration

While individual analysis of omics datasets is valuable for identifying omic-phenotype associations, analyses using only one data type are not sufficient to fully elucidate complex diseases because such diseases are the end point of events cumulating with multiple variations through multi-omics biology. To better understand the genetic architecture of complex diseases, relevant strategies for integrating multi-omics data are required. Many studies have shown that an integrative systems genomics approach and addressed the idea that integration of multi-omics data can be substantially more informative than separate analyses of each single dimension of genomic data [6]. Data integration methods can be broadly categorized into two types of approaches, as follows. In multi-staged analysis, models are constructed using only two different scales at a time, in a stepwise, linear, or hierarchical manner. A multi-staged analysis would be applicable when the relationship between genotype and phenotype can be modelled in a linear manner (e.g. association of SNPs with DNA methylation) and subsequently associated with phenotypes. However, this approach is difficult to apply simultaneously to more than two types of -omics data. An alternative approach is meta-dimensional analysis (i.e. fusion of scales), which simultaneously combines all scales of data to produce complex, meta-dimensional models with multiple variables from different data types. The scale and richness of these ever-increasing data sets hold great promise, yet the complexity presents an urgent need to find effective ways to integrate diverse data from different levels of technologies to fully exploit the potential informativeness of big data. One particularly rich source of information contained in medical records are imaging data, such as MRI, CT scan, fundoscopic images, or histopathology slides. Radiogenomics or imaging genomics is one of the primary areas that focus on the relationship between imaging phenotypes and genomics. With state-of-the-art deep learning approaches, radiogenomics might offer a practical way to leverage limited and incomplete data to generate knowledge that could lead to improved decision making, and as a result, improved patient outcomes [7].

4. Artificial intelligence in medicine

The integration of genomics data with EHR data opens the door to numerous research question about the role of genomic variation in human health. Artificial intelligence and machine learning have an important role to play in answering these questions. An important challenge that computational methods are well-suited to is the definition of phenotypes that are more accurate than those provided by disease diagnoses captured in billing codes. The challenge here to find a mathematical function of laboratory measures, medication, and other information that can be used to make a more accurate diagnosis. Machine learning is ideally suited to building models of disease phenotypes. Once accurate phenotypes are derived, the next step is to perform association analysis. Genome-wide association studies in epidemiologic studies have focused almost exclusively on statistical tests of each genetic variant independent of their genomic or environmental context. This has benefits such as speed and interpretation. However, genetic variants are likely to have effects that are context-dependent and thus not captured by univariate models. Machine learning can complement statistical methods by modeling non-additive effects among multiple factors. Further, machine learning can capture heterogeneity of genetic effects that can also be quite common. The development and application of machine learning methods in biobanks is an active area of research and very much in its infancy. Issues such as choosing the right machine learning methods for the data, interpreting the results, and developing actionable validation and implementation strategies are complex and in need of future work. An emerging area addresses the first issue is automated machine learning (AutoML) that focuses on optimization algorithms for choosing the right methods for a given data set. Automated machine learning is a step towards artificial intelligence with the goal of developing algorithms that solve problems the way human analysts do. It is important to remember that the goal of machine learning is to identify those unexpected results that would be missed by parametric statistical methods.

5. Discussion

Translational bioinformatics (TBI) lives at the intersection of informatics and biomedical data science. Due to the explosion of data in molecular and cellular technologies in the 'omics era paired with the rapid increase in the access and availability to clinical information and imaging data from EHRs, the possibilities for discovery and rapid translational into clinically and biologically meaningful outcomes are tremendous. To all of these rich data, add the powerful technologies being developed in artificial intelligence and machine learning, this leads to a unique opportunity for biomedical data science to elevate in ways that are unprecedented. The future of precision medicine will be led by translational bioinformatics.

References

[1] Bycroft C, Freeman C, Petkova D, Band G, Elliott LT, Sharp K, Motyer A, Vukcevic D, Delaneau O, O'Connell J, Cortes A, Welsh S, Young A, Effingham M, McVean G, Leslie S, Allen N, Donnelly P, Marchini J: The UK Biobank resource with deep phenotyping and genomic data. Nature 2018, 562:203-9.
[2] Mars N, Koskela JT, Ripatti P, Kiiskinen TTJ, Havulinna AS, Lindbohm JV, Ahola-Olli A, Kurki M, Karjalainen J, Palta P, FinnGen, Neale BM, Daly M, Salomaa V, Palotie A, Widen E, Ripatti S: Polygenic and clinical risk scores and their impact on age at onset and prediction of cardiometabolic diseases and common cancers. Nature medicine 2020, 26:549-57.
[3] Swede H, Stone CL, Norwood AR: National population-based biobanks for genetic research. Genetics in medicine : official journal of the American College of Medical Genetics 2007, 9:141-9.
[4] Khera AV, Chaffin M, Aragam KG, Haas ME, Roselli C, Choi SH, Natarajan P, Lander ES, Lubitz SA, Ellinor PT, Kathiresan S: Genome-wide polygenic scores for common diseases identify individuals with risk equivalent to monogenic mutations. Nat Genet 2018, 50:1219-24.
[5] Torkamani A, Wineinger NE, Topol EJ: The personal and clinical utility of polygenic risk scores. Nature reviews Genetics 2018, 19:581-90.
[6] Ritchie MD, Holzinger ER, Li R, Pendergrass SA, Kim D: Methods of integrating data to uncover genotype-phenotype interactions. Nature reviews Genetics 2015, 16:85-97.
[7] Mazurowski MA: Radiogenomics: what it is and why it is important. J Am Coll Radiol 2015, 12:862-6.

www.ingramcontent.com/pod-product-compliance
Lightning Source LLC
LaVergne TN
LVHW010535300926
849162LV00002B/153
* 9 7 8 9 8 1 1 2 3 2 6 9 5 *